Lipid Analysis and Lipidomics

New Techniques and Applications

Lipid Analysis and Lipidomics

New Techniques and Applications

Editors
Magdi M. Mossoba
John K.G. Kramer
J. Thomas Brenna
Richard E. McDonald

AOCS PRESS

Champaign, Illinois

The paper used in this book is acid-free and falls within the guidelines established to ensure permanence and durability.

Library of Congress Cataloging-in-Publication Data

Lipid analysis and lipidomics : new techniques and applications / editors,
Magdi M. Mossoba ... [et al.].
p. cm.
Includes bibliographical references.
ISBN-13: 978-1-893997-85-1 (alk. paper)
1. Lipids—Analysis. 2. Chromatographic analysis. I. Mossoba, Magdi M.
[DNLM: 1. Lipids—analysis. 2. Chromatography—methods. 3. Spectrum Analysis—methods. QU 85 L76115 2006]
QP751.L542 2006
612'.01577—dc22

2005035465

Printed in the United States of America.
10 09 08 07 06 5 4 3 2 1

Contents

Preface

Lipid Analysis and Lipidomics: New Techniques and Applications is a book that will give to both experienced researches in chemistry, chemical engineering, and food science in the oils and fats industry as well as advanced students a resource that will provide an overview of the latest developments in the rapidly changing world of lipid analysis.

The Chapters were written by an impressive group of internationally recognized experts. Most authors provide a basic or theoretical background as well as the latest developments in their areas of expertise. State of the art instrumentation and novel developments in lipid applications and lipidomics are also discussed in depth. A comprehensive list of the latest applicable references is provided for each chapter.

Hyphenated techniques as well as the latest developments in several areas including fast GC, HPLC, LC-MS, SFC, chiral separation, size exclusion chromatography, TLC, multidimensional mass spectrometry, mid- and near-infrared and Raman spectroscopy as well as chemometrics are presented. Techniques to analyze a wide conglomerate of matrixes are outlined; these include global cellular lipidomes, *trans* and conjugated fatty acid isomers, fat replacers, oxidized lipids, phosholipids, waxes and sterols. Several chromatographic, spectroscopic, and mass spectral techniques are presented that are applicable to structural lipids, cellular lipids and/or bacterial lipids.

We would like to thank each author who contributed to this book as well as the staff of AOCS and members of the Books and Special Publications Committee.

Magdi M. Mossoba
John K.G. Kramer
J. Thomas Brenna
Richard E. McDonald

Part I. Overview

1 Techniques and Applications in Lipid Analysis

Nils Hinrichsen and Hans Steinhart

Institute of Biochemistry and Food Chemistry, University of Hamburg, 20146 Hamburg, Germany

Introduction

The term "lipids" describes a variety of different substances. All of these substances have in common a distinct hydrophobicity and hence good solubility in nonpolar solvents. The term includes, for example, triacylglycerols, di- and monoacylglycerols, sterols, and their esters, tocopherols, free fatty acids (FFA), phospholipids, proteolipids, CoA esters, and more. Therefore, there are distinct differences in the molecular structures of lipids. According to the diversity of the substances, there are numerous methods and applications in lipid analysis. In addition, the choice of the right method for the analysis depends not only on the substances to be determined, but also on the information that is required. The following chapter describes the most frequently used techniques and applications for the analysis of those compounds.

Extraction Methods

For nearly all lipid analyses, a purified lipid extract is needed. Because lipids normally do not appear in their free form, but embedded in a matrix, an extraction step is necessary before further analysis. In fact, this step frequently generates mistakes that lead to a false analytical conclusion. Therefore, a well-chosen method of extraction for the lipid to be determined is required.

The components obtained in the lipid extract depend on the method of extraction used, especially on the solvent. Nonpolar solvents (e.g., hexane, ether, or supercritical carbon dioxide) can be adopted for the extraction of simple neutral lipids, for example, esters of fatty acids (FAs) and acylglycerols. More complex and more polar lipids (e.g., phospholipids, lipoproteins, glycolipids, FFAs) require more polar solvents such as methanol or acetonitrile. Generally solid phase extraction (SPE) methods are advisable for complex polar and nonpolar lipid components (1).

For the extraction of fats from food and different biological matrices, digestion of the surrounding material is necessary, especially if a quantitative isolation is required. There are different "classical" digestion methods that have been used for many years. The choice of the technique must depend on the surrounding matrix. The Weibull-

Stoldt method is used for the extraction of fat from meat, fish, or oilseeds. After the surrounding proteinogenic material has been decomposed by hot hydrochloric acid, the fat is extracted in a Soxhlet-apparatus with either ether or hexane. For dairy products, an alkaline decomposition is normally used. Although these methods have proved to provide exact results regarding the total fat content (2,3), the extreme conditions (high temperature, very high or very low pH value) often lead to unintentional modifications of the molecular structure of the analytes. Therefore, if further analysis of certain lipid compounds is required, an extraction method that does not alter the structure of the analytes must be used (see scheme in Fig. 1).

In 1959 Bligh and Dyer introduced a method for the isolation of total lipid from fish muscle using chloroform and methanol as solvents (4); the method became very popular and has often been modified and improved. Disadvantages of this method include its toxicity and the cost of the solvents. The toxicity can be avoided by using other, preferentially nonhalogenated solvents (5,6).

Although the operating expense factor remains with these applications, it can be reduced by using supercritical fluid extraction (SFE) (7). The marginal alteration of lipid compounds during extraction, the ease of solvent removal from the extract, and

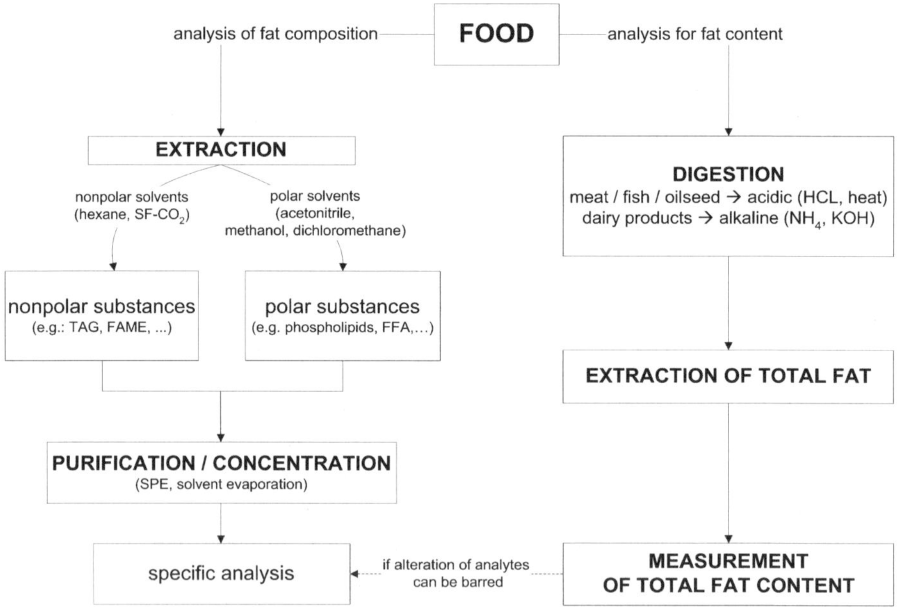

Fig. 1. Approach diagram for the analysis of fat. For the determination of the total fat content, another way of extraction has to be chosen than for the analysis of specific compounds (e.g., fatty acids, phospholipids) because digestion often leads to alterations in individual lipids.

the lack of toxicity are other advantages of SFE (8). Furthermore, the SFE-equipment can be coupled with chromatographic systems, thus providing the opportunity to automate nearly the entire lipid analysis. Although many efforts have successfully been made to shorten the length of the "classical" lipid extraction, for example, by using an ultrasonic- (9) or microwave-assisted (10–12) Soxhlet extraction, SFE remains one of the fastest methods. Modern extraction methods provide reliable results within 30–50 min. However, SFE requires expensive equipment and the savings achieved by avoiding solvents are profitable only after a multitude of extractions. Finally, it should be emphasized that no single extraction method is applicable to all matrices or all analytes. The right choice of method applied depends on various factors and is an important element of a successful analysis.

Classical Applications

A multitude of classical chemical applications can be used to evaluate the quality and chemical characteristics of lipids in food.

Chromatographic techniques have reduced the need for identification, but saponification value (SV), iodine value (IV), acidity, and peroxide value (PV) are most commonly used to characterize the quality of fats. They are used to determine the purity (SV), the number of unsaturated compounds (IV), and the FFA content. The PV is normally obtained by iodometric titration and describes the content of unsaturated FA hydroperoxides, which result from lipid oxidation. Accordingly, this measurement displays an indicator for the deterioration of fats in food. Although these and other classical methods are still in use, they are being replaced more and more by faster and more modern methods, especially in laboratories controlling foods. Free FAs, formerly evaluated by acidimetric titration, can be measured rapidly by Fourier transform infrared spectroscopy (FTIR) (13) or by gas chromatography (GC) and high-performance liquid chromatography (HPLC) (14). In the same amount of time, the last-mentioned methods not only provide information about the quantity of FFAs, but additionally characterize the exact FAs contained in the sample.

For the determination of the IV, different methods are in use. For the Hanus and Kaufmann methods, the dissolved fat is spiked with a surplus of brome, which leads to an addition reaction with the ethylenic bonds. The remaining brome, which did not react with any ethylenic bond, is used to oxidize iodide ions to iodine. This is measured by titration with a solution of sodium thiosulfate. Both the AOAC and AOCS suggest the Wijs method for common fats and oils, in which brome is substituted with iodine, which has a much lower toxicity. In addition to being time consuming, a substantial disadvantage of the IV is that not only ethylenic bonds from FAs, but also those from impurities or other lipid compounds, e.g., sterols, are measured. A reasonable substitute for the IV is the determination of fatty acid methyl esters (FAME) by GC. This method shows the complete FA composition, so that the content of unsaturated FA can easily be obtained.

For the measurement of PV, there are also various alternative methods available. PV can be determined more reliably using GC (15), chemiluminescent assays (16), or the so-called "xylenol orange method" (17) than by iodometric titration. The xylenol orange method is based on the following principle: Fe II is oxidized to Fe III by lipid hydroperoxides followed by the complexation of Fe III with xylenol orange. The resulting complex is quantified by photometric measurement. The method is very sensitive and requires only small sample quantities (18). Another method that provides very fast results is FTIR. Samples can be measured at a very high frequency, especially when FTIR is coupled with flow analysis (19).

Chromatographic Methods

Thin-Layer Chromatography (TLC)

Thin-layer chromatography is a simple analytical method that can be accomplished without expensive equipment. Although the results provided are less accurate than those achieved using GC or HPLC, TLC delivers a great deal of useful information with very little effort. The technique allows the separation of a mixture of lipids with different polarities in a single run. It is carried out on a glass or aluminium plate that is coated with an adsorbent. The most common adsorbent is silica gel G.

With the addition of calcium sulfate, silica gel is suitable for the separation of cholesterol and its esters, FFAs, phospholipids, or diacylglycerols. As an example, a TLC-chromatogram from the separation of phospholipids is shown in Figure 2.

Silica gel worked up with silver nitrate is used for the separation of FAs according to the configuration of their ethylenic bond and their degree of saturation. During the production and storage of argentated TLC-plates, it is important to take care that light is avoided; otherwise, they would blacken as a consequence of silver oxidation, thereby inhibiting proper chromatography.

Normally, the samples are applied manually to the plate using a glass syringe. In addition, there are modern sampling tools commercially available that apply more precise spots to the plates. This leads to a better reproducibility and a more accurate analysis. Regrettably those tools are expensive, so that the economic simplicity and efficiency normally associated with TLC are lacking. The plates are developed in a chamber containing an appropriate solvent mixture, depending on the analytes. After the development, the plates are removed from the chamber and the solvents are evaporated. It is possible to develop the plate again in another solvent or in another direction.

There are different methods for the visualization of the spots. The simplest way is to char the compounds by spraying the plate with 50–60% methanolic sulfuric acid or, for the detection of phospholipids, for example, with a solution of copper sulfate in aqueous phosphoric acid and charring in a drying chamber at 160°C (20). The disadvantage of this technique is the disintegration of the analytes. By using another nondestructive visualization, it is possible to scrape off the spots, dissolve the separated

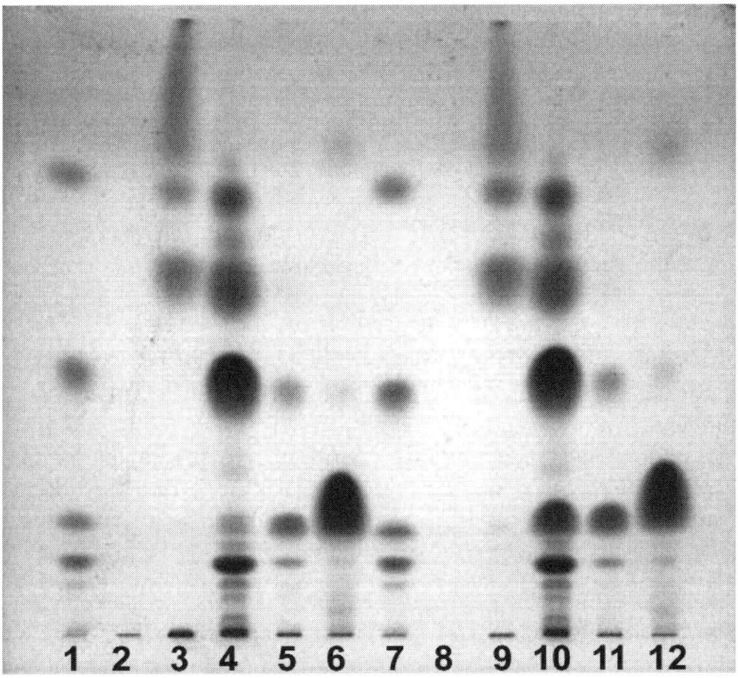

Fig. 2. Thin layer chromatography of phospholipids from pork tissues. On lanes 1 and 7, different concentrations of a standard solution containing sphingomyelin (R_f = 0.39), phosphatidylcholine (R_f = 0.44), phosphatidylserine (R_f = 0.50) and phosphatidylethanolamine (R_f = 0.63) are applied. Lipid extracts from different pork tissues were analyzed in the rest. Lanes 3–6 are duplicated in lanes 9–12. Most phospholipids can be found in liver tissue, which is applied on lanes 4 and 10. Separation was accomplished on silica gel with methyl acetate/chloroform/1-propanol/0.25% aqueous potassium chloride (25/25/25/10/9, by vol) as the mobile phase.

compounds again, and use them for further analysis [e.g., GC or mass spectrometry (MS)]. Here, a fluorescent dye, such as 2′-7′-dichlorofluorescein or Rhodamine 6G, is frequently used.

The identification of the analytes is carried out by the comparison of the R_f-values. For a correct analysis, it is necessary to apply standard substances beside the samples. In some cases, it is also necessary to supplement the samples with standards.

The quantification of the separated analytes may be carried out by densitometric or fluorimetric measurement. Recent computer developments allow a much more economical way of quantification, without the use of expensive equipment. The densitometer is replaced by a personal computer with a scanner. The TLC-plate is scanned in b/w-mode and converted to external chromatograms by software (e.g., "Image J," which is available without cost on the Internet). The emerging peaks are integrated and the following evaluation resembles that from other chromatographic methods (GC, HPLC). Disadvantages of this quantification method are a relatively short linear area for the regression line and a slow sampling rate.

A wider linear area for regression can be obtained by using microrod (Chromarod) TLC technology with TLC-flame ionization detection (FID) (the Iatroscan). It was first introduced in the late 1970s (21), and improvements continue to be made (22). It is suitable for the analysis of various lipids from biological matrices (23,24). The coated rods can be treated with reagents before chromatography like common plates. Banerjee (25) successfully transferred common plate TLC methods using silver nitrate-, oxalic acid- and iodine vapor-treatment.

However, the common TLC on plates can be coupled with various analytical equipment by scraping off the spots and dissolving the analytes again. A very useful tool for further analysis is MS, as used by Lee et al. (26), for example. If matrix-assisted laser desorption/ionization (MALDI) technology is used, scraping off and dissolving the spots is unnecessary because the TLC plates can be attached directly to the MALDI target, where the analytes are desorbed and ionized (27). To obtain larger quantities of lipids or other materials, streaking the plates may be preferred.

High Performance Liquid Chromatography (HPLC)

Nearly all methods that work with TLC can be assigned to HPLC methods because both follow the same chromatographic principles. However, HPLC offers a great number of detectors and a more accurate separation of the analytes. The technique is used for the analysis of nonvolatile, high-molecular-weight lipids in several modes, employing either partition or adsorption chromatography. The latter is obtained with silica gel or diols as the stationary phase (normal phase HPLC; NP-HPLC) for the separation of lipid classes according to the number and character of polar functional groups. The most widely used diol phase is the 1,2-dihydroxypropyl propyl ether phase. Schaefer et al. (28) introduced a method for the identification and quantitative estimation of 12 lipid classes, including paraffins, wax esters, cholesterol esters, FAME, triacylglycerols, fatty alcohols, FFA, cholesterol, 1,3-diacylglycerols, 1,2-diacylglycerols, monoacylglycerols, and FA amides. The separation was tested on various stationary phases with the result that a diol phase provided the best selectivity. Due to their sensitivity to water and the resulting decrease of selectivity with preceding numbers of analyses, NP columns have to be rinsed mainly with water-free solvents. Similar methods also provide good results for the analysis of lipid classes in human tissues (29). NP-HPLC is also utilized for the separation of phospholipids (20,30,31).

Another very common technique is reversed-phase HPLC (RP-HPLC), which is based on the principle of partition chromatography. Normally, it is applied for the separation of individual lipid components that belong to one single class. The phase consists of bonded nonpolar silanes, usually octadecylsilane (C_{18}) or octylsilane (C_8). The mobile phase is composed of numerous polar solvents, normally acetonitrile, methanol, water, and sometimes tetrahydrofuran.

RP-HPLC is a suitable method for the determination of the positional isomers of triacylglycerols (32). Amaral et al. (33) used this method for the characterization of the triacylglycerol composition of nine different cultivars of walnut. This phase can also be used for the fast analysis of FFAs (34,35). Svensson et al. (36) applied RP-HPLC for the separation of geometrical and positional isomers of octadecenoic acids. Recent inquiries showed that the technique is also suitable for the measurement of conjugated linoleic acids (CLAs) as free acids and as FAME (37). It is possible to identify geometrical isomers of CLA and their metabolites from a large variety of biological samples with a relatively low detection limit, even though the separation of the CLA geometrical isomers *Z,E* from *Z,Z* can be performed only partially and is unsatisfactory.

A much better separation of the geometrical isomers of CLA and other FAs can be obtained by silver-ion chromatography (38). Analyses have been made with up to six silver-ion columns in series to improve the selectivity of the method. However, three Ag^+-HPLC columns in series appeared to be the best compromise to obtain satisfactory resolution of most CLA isomers found in natural products (39).

Silver-ion HPLC is also used for the analysis of positional isomers of di- and triacylglycerols (40). It is possible to couple silver-ion chromatography systems with various detectors, including MS. Although this has been a well-discussed technique for lipid analysis, there are still some open questions, e.g., the influence of temperature on the retention time of the analytes (41).

The most commonly used HPLC detectors are photometric detectors that absorb in the ultraviolet (UV)-visible range. In lipid analysis, they have only limited application because most lipids absorb between 200 and 210 nm, a range in which most solvents are not transparent. Only a few eluents, such as hexane, acetonitrile, methanol, or water can be used. Analytes with conjugated bond systems offer absorption maxima with longer wavelengths and can easily be detected. CLAs, for example, have an absorption maximum at ~233 nm (42). However, use of a photodiode array detector (PDA) rather than a common UV detector is recommended. This device not only monitors a single wavelength, but also provides complete UV-spectra for every moment of the run. It is possible to verify the peak purity and the identity of the analytes by comparison of the spectra from standard substances and the sample. Nevertheless, the use of a PDA does not solve the problem of nontransparent solvents at certain wavelengths. Some analytes can be derivatized to UV-active substances to improve UV detectability. For example, Bravo et al. (43) detected different carboxylated FAs in water-microemulsion-oil systems after microwave-assisted derivatization with 2,4-dinitrophenylhydrazine, benzoyl chloride, and phenylhydrazine. Miwa (44) reported an application for the determination of esterified and unesterified FAs with UV-VIS detection after derivatization with 2-nitrophenylhydrazide.

In addition to the techniques described above, there are various methods for the detection of the analytes after separation via HPLC, including refractive index, flame ionization, and electrochemical detection devices. Kotani et al. (14) reviewed several applications for the determination of FAs with electrochemical detection. The evaporative light scattering detector (ELSD) offers a high reproducibility and is insensitive to solvent changes and polarity. The detector response depends on the analyte mass. ELS detection is convenient for several lipid components, especially when they are not satisfactorily detectable with UV detectors or PDA. The device can be coupled with all kinds of HPLC columns. It was used by Schaefer et al. (28) for the detection of different lipid classes that migrate from wrapping materials into food, after separation on a diol phase. Similar applications were made by Perona and Ruiz-Gutierrez (29) and Beermann et al. (45), who utilized ELS detection for the characterization of seed lipid compositions in plants. It is also often used for the detection of phospholipids (30) because other proper detection devices for this lipid group are very rare. Balazs et al. (46) demonstrated that ELSD has a superior precision for the determination of phospholipids from soybeans. A method using ELS detection was compared with the AOCS Official Method (Ja 7b-91) and a mixed phase method for the analysis of phospholipids, both using UV detection. Regrettably, the calibration curve in analyses using ELSD is often not linear and has to be evaluated with second-order equations, which complicates the interpretation of the measured values.

Liquid chromatography (LC)-MS has great potential in lipid analysis. In addition to the information delivered by separating and comparing retention times with those of standards, this technique provides data about the identity of the analytes through MS-spectra, although a certain amount of knowledge and analytical skills is required for a successful application. Perret et al. (35) used tandem electrospray ionization (ESI-MS-MS) for the determination of FFAs in chocolate after separation on a C_{18}-stationary phase. A similar system was developed by Zink and Mangelsdorf (47) for the elucidation of the molecular structure of phospholipids from sediments. ESI also delivers very good results in the analysis of FA acyl-CoA compounds (48). Byrdwell and Neff (49) analyzed high-molecular-weight oligomers formed from heated triolein, a triacylglycerol used as a model for dietary oils.

In the ESI-MS-MS technique, a first mass spectrometer that employs a quadrupole mass filter is tuned to allow only the analyte ion of interest through. This is taken into a collision cell where a further dissociation is accomplished. The so-called daughter-ions are then swept into another mass filter where they are separated and detected. For ESI, polar, acidic, or basic groups are required for a proper ionization. Figure 3 shows the schematic configuration of a triple quadrupole ESI-MS-MS detector that can be utilized for the analysis of ceramides and phospholipids (50). The apparatus also allows the direct injection of lipid extracts into the mass spectrometer. Thus, the analysis of lipid classes can be performed within 120 s. As an example, Figure 4 displays the parent scan of the mass number *m/z* 184, which occurs in the analysis of phospholipids (e.g., sphingomyelin and phosphatidylcholine).

In addition to ESI, the analytes can also be ionized by atmospheric pressure chemical ionization (APCI), which was used by Kemmo et al. (51) for the determination of stigmasterol peroxides.

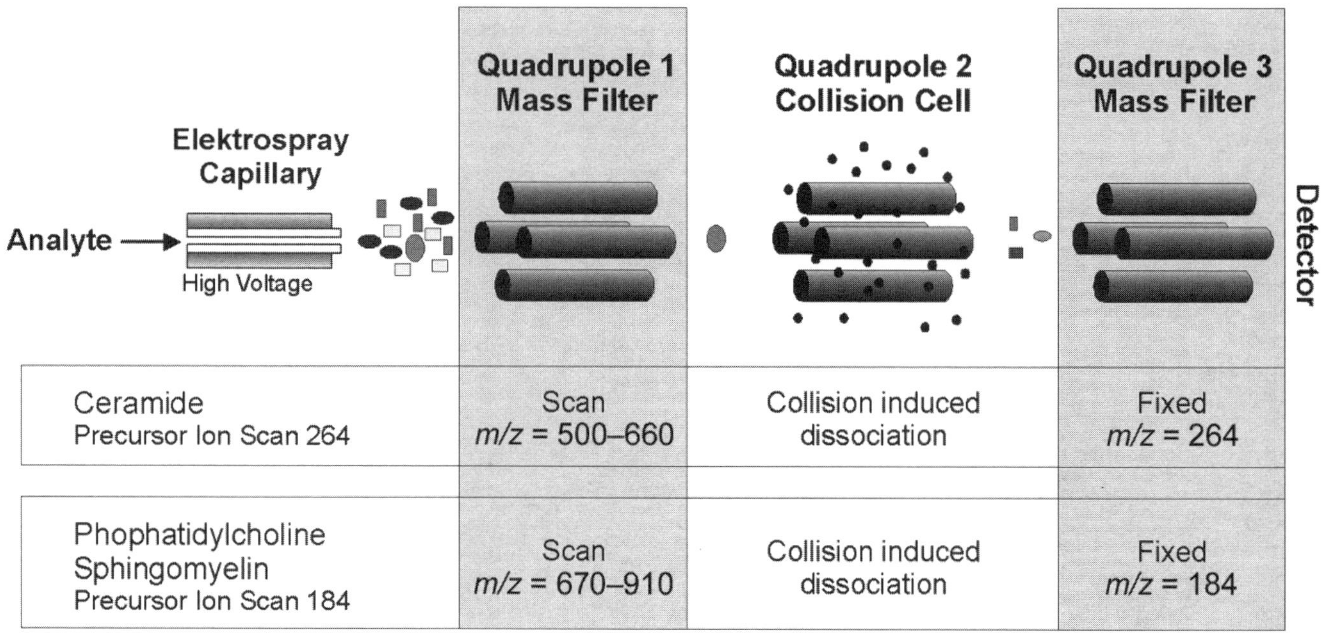

Fig. 3. Schematic configuration of a triple quadrupole electrospray ionization (ESI)-mass spectrometry (MS)/MS detector. In addition, the settings for the analysis of ceramides and phospholipids are mentioned below.

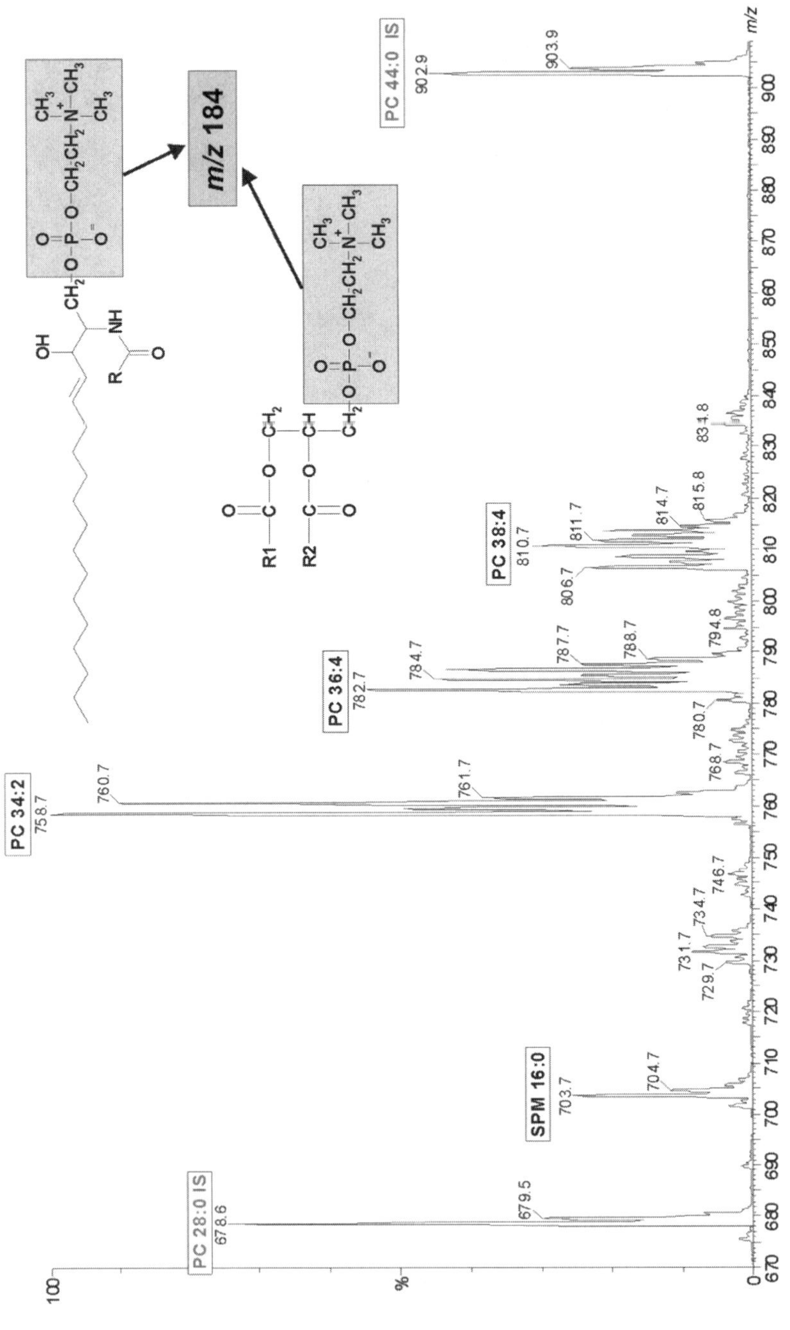

Fig. 4. Parent scan of the mass number *m/z* 184, which is utilized for the analysis of sphingomyelin and phosphatidylcholine.

Because analysis via HPLC does not require high temperatures, the technique is also suitable for the determination of thermolabile analytes, for example, hydroperoxides emerging from lipid oxidation (52) or in biological matrices.

Gas Chromatography (GC)

Unlike HPLC, the mobile phase in GC has no interactions with the analytes because preferably inert gases are used, normally hydrogen, nitrogen, or helium.

The FA composition of fats and oils is normally resolved by GC analysis of their methyl ester (FAME) derivates after transesterification. Sometimes other derivates are formed because these may improve the peak resolution. The separation of FAs is usually achieved on highly polar liquid phases, for example, cyanopropylsiloxanes or polyethyleneglycols. The analytes are separated according to their chain length and degree of saturation. Positional and geometric isomers can seldom be separated satisfactorily (e.g. $C_{16:1}$, $C_{18:2}$, $C_{18:1}$), but *trans* monoethylenic FAs are currently of much interest.

Nonpolar phases can also be employed for the FA analysis, but they usually achieve only an imperfect separation of the analytes, although the information obtained is sufficient to provide useful data about the chain lengths present in a FAME mixture. For the definite identification of FAME, it is indispensable to compare the retention times of standards with those of the samples. This should be performed on two or more columns with different polarities. Admittedly, this procedure assumes the availability of appropriate standards. Fortunately, many FAs are commercially available, although still not all. For the identification of such FAs, GC-MS and infrared techniques are advisable. The most common detection device for GC in lipid analysis is FID.

Normally the analysis of the FA composition of fats and oils by GC takes between 30 and 60 min. Narrow-bore columns ($\leq$0.15 mm in diameter), with a thin stationary phase film and a relatively short length ($\leq$15 m) combined with high-performance furnaces provide a much faster separation. Mondello et al. (53) examined 40 FAs via fast GC and fast GC-MS methods. Bondia-Pons et al. (54) used similar techniques to separate 35 species of FAME from human plasma in 3.2 min, which is less than one-tenth of the time normally required for a comparable separation.

GC can also be used for the separation of triacylglycerols according to their number of carbon atoms (55). The analysis is normally performed at high temperatures on non- or medium polar columns. For instance, Guyon et al. (56) quantified the cocoa butter equivalent added to chocolate bars using this method.

For the separation of *cis* and *trans* FAs with the same chain length, it is necessary to use a long column with a length of 100 or 200 m and a high polar stationary phase. Best results are obtained with a CP-Sil 88 or a SP-2560 column. Ratnayake et al. (57) reported that the SP-2560 capillary column has a slight advantage over the CP-Sil 88 column for the simultaneous resolution of all the FAs generally found in partially hydrogenated vegetable oils, including various *trans* FAs. However, clear results for the analysis of geometrical isomers are normally not obtained by using GC exclusively; a combination with other methods such as silver-ion TLC or HPLC is necessary. A practical overview for the identification of *trans* FAs is given by Ratnayake (58).

For the identification and quantification of complex samples, the comprehensive, two-dimensional (2D) GC (GC × GC) is also an adequate method. It is based on the coupling of two capillary columns that each contribute in a different way to the unprecedented resolving power of this technique. The 2D space chromatograms that derive from GC × GC analysis have great potential for the identification of the analytes. This is due to the fact that the plot positions, pinpointed by two retention time coordinates, give characteristic patterns for specific families of compounds that can be translated mathematically (59). Hyoetylaeinen et al. (60) described an application for the analysis of dietary FAs using this technique.

Although these techniques demonstrate good approaches to the solution of various analytical problems, they are not able to provide the information delivered by a MS detector. In addition to the retention-time data, GC-MS provides structural as well as molecular weight information about the analytes. In addition, a clean chromatography is indeed suggested, but is not compulsory for the analysis because measurements can be performed in selected ion monitoring (SIM)-mode, in which only peaks that contain fragments with a specific weight are detected. Thus, if no satisfactory separation of two peaks is possible, one of them can often be unmasked by this technique.

The lipid compounds can be examined as they appear in their matrix, but frequently they are derivatized to obtain specific information about the structure. The location of ethylenic bonds in FAME can be determined by the formation of addition compounds, such as dimethyldisulfide adducts. For example, the structural characterization of some FAs from the brains of different domestic animals was performed by Biedermann et al. (61) using such a method. Similar adducts can be formed by derivatization with trimethylsilyl ether. Unsaturated sites can also be found with derivatives that can localize the charge of the molecular ion. Most commonly, 4,4-dimethyloxazoline (DMOX) derivatives (62,63) are formed. Such applications were used for the identification of various octadienoic acid isomers, e.g., 7-*trans*,9-*cis*-CLA in cow's milk, cheese, beef, human milk, and adipose tissue (64) or various isomers of CLA in hydrogenated soybean oil (65). Similar derivatives can be formed as pyrrolidine, picolinyl esters, or nicotinate derivatives. However, DMOX derivates have the benefit that they are separable on the same polar stationary phase (Fig. 5) as FAME and sometimes even assist the separation (66).

For many analytes, derivatization before GC-MS analysis is not necessary. The identification and determination of sterols, for example, can be performed either after silylation or without any prior treatment. Bodzek et al. (67) determined a number of sterols without derivatization in consumable fats using a combination of SPE, TLC, and GC-MS. Keller and Jahreis (68) measured underivatized sterols and bile acid trimethylsilyl ether and methyl esters in feces. Even though a proper chromatographic separation was not achieved, the method provided high accuracy, because measurements were performed in SIM-mode.

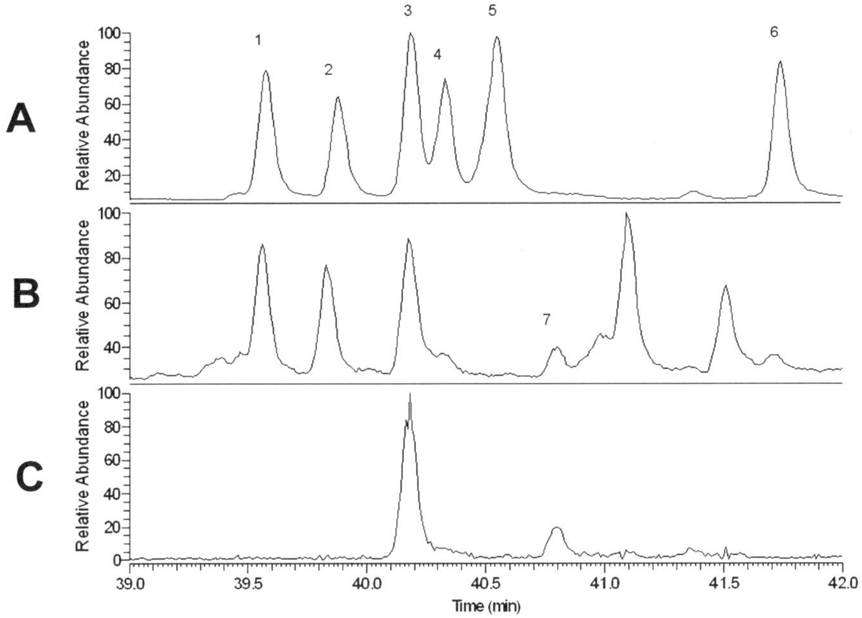

Fig. 5. (A) Partial chromatogram (total ion current; TIC) of 4,4-dimethyloxazoline (DMOX)-derivates of fatty acid standard compounds: 1: $C_{20:1n\text{-}9}$; 2: $C_{18:3n\text{-}2}$; 3: *c*9*t*11-conjugated linoleic acid (CLA); 4: $C_{21:0}$; 5: *t*10*c*12-CLA; 6: $C_{20:2n\text{-}6}$; 7: *t*9*t*11-CLA. (B) Partial chromatogram (TIC) of DMOX-derivates of a *c*9*t*11-CLA–incubated Caco-2-cell lipid extract. (C) Partial chromatogram of 333 mu trace of a *c*9*t*11-CLA–incubated Caco-2-cell lipid extract.

GC can also be coupled with IR-technologies, e.g., GC-FTIR. The FTIR-detector delivers a spectrum of every analyte included in the measured sample (a complete chromatographic separation is assumed). Thus, it delivers information about functional groups and the geographic configuration of the substances. At present there are three types of GC-FTIR interfaces commercially available: The matrix isolation (MI), direct deposition (DD), both operating at very low temperatures, and the light pipe (LP) interface have all been used. The last-mentioned is the most commonly used and consists of a glass tube with alkali halide windows. The spectra are recorded in real time when the analytes leave the GC column. DD and MI provide a better sensitivity; regrettably, they are also much more expensive. With these systems, the substances leaving the GC are trapped under vacuum on a cryogenic surface for a proximate off-line signal calculation by FTIR. The DD interface also allows on-the-fly measurement of FTIR spectra.

The FTIR method can be used for the identification of various kinds of FAME (69). For example, it is a useful tool for the identification of *trans*-FAs (70). The major drawback of this technique is the lack of sensitivity (71). As a consequence, GC-FTIR is applied mainly as an additional tool to GC-MS for isomer discrimination and structure elucidation of compounds with closely related structures. However, because the coupling of these technologies is very complex, GC-MS-FTIR equipment is expensive and requires certain analytical skills for a proper employment. Waehl et al. (72) used this technique to identify FAMEs and differentiate between the *cis* and *trans* isomers.

Counter Current Chromatography (CCC)

CCC displays a rarely employed, but simple liquid-liquid-chromatographic method for the separation of a large number of substances. The apparatus consists of a rotating coiled tubing, filled with two, nonmixable liquids that both interact with the analytes flowing through them. The chromatographic system can be coupled with the same detectors that are used for size exclusion chromatography (SEC) or HPLC.

Bousquet and le Goffic (73) used the method for the separation of PUFA. Matsuda et al. (74) applied an improved version of the method (toroidal-CCC) for the separation of phospholipids and glycolipids. Because the technique has not yet been sufficiently developed, the accuracy of the separation cannot be compared with those delivered by HPLC, for example. Because it is possible to analyze larger sample amounts, CCC may be an alternative to preparative-scale HPLC.

Size Exclusion Chromatography (SEC)

SEC is a separation technique based on the relative size of the analytes. Molecules of a specific size diffuse mechanically, partially or completely, into appropriate pores of the stationary phase. Smaller molecules that fit into the pores will eluate later than larger ones. The stationary phase consists of crosslinked macromolecules; for fat analysis, these are most frequently copolymers of styrene-divinyl benzene.

High-performance SEC is frequently used for certain analytes, such as triacylglycerol dimers, oxidized triacylglycerol monomers, mono- and diacylglycerols, free sterols, and free, partly cyclic FAs emerging from frying fat through hydrolysis and oxidation (75). The optimal equipment for the detection and quantification of these analytes is an evaporative light scattering or a refractive index detector. With the techniques mentioned above, products can be separated within 30 min in various matrices, e.g., the lipid fractions of margarines (76) and bouillon cubes (77) or microencapsulated fish oils (78). Hopia et al. (79) used SEC techniques to monitor the autoxidation of unsaturated triacylglycerols.

Supercritical Fluid Chromatography (SFC)

SFC is a technique that combines elements of HPLC (injection valve, pump, detectors, packed columns) and capillary GC (oven, capillary columns). SFC applications utilize supercritical CO_2 as the mobile phase. Because most applications are accomplished at temperatures slightly above room temperature, the technique is suitable for the analysis of thermolabile analytes. SFC can be coupled with a large number of detection devices including MS (80,81), FTIR (82), and FID. It can be used for the analysis of various lipids, e.g., FAs and their methyl esters (83), triacylglycerols (80), or cholesteryl esters of FAs (84).

Capillary columns can be used as well as packed ones. In addition, 2D and comprehensive techniques have been employed for lipid analysis, utilizing at least two columns in one analytical apparatus. Hirata and Sogabe (85) used an octadecyl silica gel (ODS), and a silica column, a UV-detector, and a FID to perform comprehensive 2D separation of FAMEs. Thus, they could improve the resolution and lower the detection limits of minor components of FAME analysis via SFC. Sandra et al. (86) used an ODS-column in the first dimension and a silver-loaded stationary phase in the second dimension for the characterization of triacylglycerols in vegetable oils. However, it should be noted that these techniques require very sophisticated instrumentation and sufficient experience for a successful employment.

Spectroscopic Methods

Fourier Transform Infrared Spectroscopy (FTIR)

IR and FTIR techniques are important tools for the resolution of the configuration and structure analysis of lipids. Functional groups, such as ethylenic bonds (conjugated as well as isolated), hydroxyl-, epoxy- and ester functions produce unique absorption bands in the mid-infrared spectral region (wave numbers 600–4000 cm^{-1}). The solvent commonly used is carbon disulfide.

FTIR is frequently used to measure the total *trans* FAs in oils (87). The determination is based on the measurement of the 966 cm^{-1} out-of-plane deformation vibration. This band is characteristic for isolated *trans* FA bonds. Regrettably, it overlaps with other features in the IR spectrum, leading to an inconsistent background. Actually, the band turns into a shoulder in the spectrum if the sample measured contains amounts of *trans* FAs $< 2\%$. The resulting low accuracy can be improved by using attenuated total reflection (ATR) cells (88). The method requires neither weighing nor the quantitative dilution in any solvents and delivers a radical improvement of the sensitivity at low *trans* FA concentrations (89). The McGill IR group also developed FTIR applications for the measurements of FFAs (90), IV (91), and saponification number (92). With the employment of a FTIR method for the measurement of the PV (93), an accurate and fast tool for monitoring the deterioration of fats was obtained.

Another interesting application area of FTIR spectroscopy is the determination of the content of solids in fats. This parameter influences the characteristics of margarines, shortenings, and other fat blends. It is normally obtained using dilatometry or nuclear magnetic resonance (NMR). Both techniques involve measurements at a series of set temperatures, making them relatively time consuming. In contrast, van de Voort et al. (94) presented a FTIR method that requires only a single measurement of the cleaned, purified, and melted sample. The method delivered a reproducibility and an accuracy comparable to those of conventional methods. Thus, the FTIR method could be used as a substitute for either the dilatometric or the NMR method to determine the content of solids. In addition, it has the advantage of a shortened analysis time by the elimination of the tempering steps required for the traditional methods.

Near Infrared Spectroscopy (NIR)

NIR technology is related to FTIR. It is frequently used in food analysis and provides fast results with only small sample amounts. NIR applications normally require little or even no sample preparation. The spectra are recorded in the area between 700 and 2500 nm, where combination and overtone bands of carbon-hydrogen, oxygen-hydrogen, and nitrogen-hydrogen are displayed. The technique is applied for the determination of moisture, fat, and protein content in various matrices, e.g., oil seeds, grains, meat, or flour. NIR techniques are employed primarily in food production where raw materials have to be checked within seconds before and during the production process to prevent faulty products and to save costs resulting from an out-of-specification batch (95).

The spectral region examined in this technique is complex. Thus, sophisticated calibrations are necessary. Although most NIR instruments have associated software to assist with the development of calibrations, adjustments have to be performed before every measurement.

There are a large number of NIR applications for food analysis available. Moisture, protein, and fat content can be measured simultaneously in different matrices (96,97). Christy et al. (98) developed a rapid method using NIR technology for the detection and quantification of adulterations in olive oil. Although it was possible to detect an adulteration with nearly 100% certainty, it was not possible to identify the adulterant exclusively with NIR. Other analytical techniques such as GC of FAME are necessary. Nevertheless, various works show that the estimation of the FA composition of fats and oils can also be performed by NIR (99–102). The method provides a simple, rapid and nondestructive means of estimating the FA composition. However, not all FAs could be measured with the accuracy yielded with other analytical methods such as GC of FAME.

Nuclear Magnetic Resonance Spectroscopy (NMR)

NMR is a frequently used method in the fats and oils industry for monitoring and quality control applications. It is employed for the simple and rapid measurement of the fat concentration in food and oil seeds. Guthausen et al. (103) described a portable NMR analyzer to measure the fat content in a packaged product without destruction of the material. The processed NMR signal was comparable to the fat content obtained by reference methods.

^{1}H NMR is used for the characterization of different lipids. For example, the technique enables food controllers to distinguish milk samples from different species of animals (104). Aerts et al. (105) detected the yields of mono- and diepoxidized products emerging from diunsaturated substrates during epoxidation reactions of FAME.

In fundamental research, ^{1}H and ^{13}C NMR are used for structure determination and confirmation of identity. In ^{13}C spectra, characteristic bands are generated by certain carbon atoms, e.g., olefinic, allylic or ω1–3 carbons. The technique requires large amounts of high-purity samples (1–50 mg). Sometimes, the plotting of different spectra against each other is essential for the clear identification of the structure. Thus, 2D spectra are generated. This technique identifies the atoms that are adjacent and clarifies which atoms couple among each other. Both NMR methods (^{1}H and ^{13}C) are very useful for structure determination of individual compounds, but provide only medium accuracy in analyzing complex mixtures.

In contrast, the ^{31}P NMR can be used for the exact determination of different phospholipids in parallel (20). Because these contain only 1 phosphor atom/molecule, which generates characteristic signals for the substances, the intensity of the signals correlates with the analyte content in the measured sample. ^{31}P NMR techniques have been used for the determination of phospholipids in different matrices, e.g., human and boar spermatozoa (106), milk (107), or lecithins and flour (20). Cremoni et al. (108) reported that not all solvents employed provide exact results. The so-called CUBO solvent (a ternary mixture of *N,N*-dimethylformamide, triethylamine, and guanidinium hydrochloride) impairs the results obtained for the phospholipid content. In general, ^{31}P NMR represents a fast and precise method for the measurement of phospholipids in different matrices. Regrettably, high sample amounts are required in all NMR techniques. When only small amounts are available, NP-HPLC with ELS detection is the more appropriate choice for analysis.

Acknowledgments

The authors acknowledge Gerhard Liebisch (Institute for Clinical Chemistry and Laboratory Medicine, University of Regensburg) for the graphical support and also André Müller and Alexandra Fliegel (Institute of Biochemistry and Food Chemistry, University of Hamburg) for critical comments and corrections.

References

1. Ruiz, J., T. Antequera, A.I. Andres, M.J. Petron, and E. Muriel, Improvement of a Solid Phase Extraction Method for Analysis of Lipid Fractions in Muscle Foods, *Anal. Chim. Acta 520*:201–205 (2004).
2. Eckoldt, J., R. Buschmann, and C. Wauschkuhn, Methods for Determination of the Total Fat Content in Meat and Sausage in Comparison, *Lebensmittelchemie 53*:120 (1999).
3. Kolar, K., U. Faure, I. Torelm, and P. Finglas, An Intercomparison of Methods for the Determination of Total Fat in a Meat Reference Material, *Fresenius J. Anal. Chem. 347*:393–395 (1993).
4. Bligh, E.G., and W.J. Dyer, A Rapid Method of Total Lipid Extraction and Purification, *Can. J. Biochem. Phys. 37*:911–917 (1959).

5. Hara, A., and N.S. Radin, Lipid Extraction of Tissues with a Low-Toxicity Solvent, *Anal. Biochem. 90*:420–426 (1978).
6. Jensen, S., L. Häggberg, H. Jörundsdóttir, and G. Odham, A Quantitative Lipid Extraction Method for Residue Analysis of Fish Involving Nonhalogenated Solvents, *J. Agric. Food Chem 51*:5607–5611 (2003).
7. Neff, W.E., F. Eller, and K. Warner, Composition of Oils Extracted from Potato Chips by Supercritical Fluid Extraction, *Eur. J. Lipid Sci. Technol. 104*:785–791 (2002).
8. Johnson, R.B., and H.J. Barnett, Determination of Fat Content in Fish Feed by Supercritical Fluid Extraction and Subsequent Lipid Classification of Extract by Thin Layer Chromatography-Flame Ionization Detection, *Aquaculture 216*:263–282 (2003).
9. Luque-Garcia, J.L., and M.D. Luque de Castro, Ultrasound-Assisted Soxhlet Extraction: An Expeditive Approach for Solid Sample Treatment. Application to the Extraction of Total Fat from Oleaginous Seeds, *J. Chromatogr. A 1034*:237–242 (2004).
10. Garcia-Ayuso, L.E., J. Velasco, M.C. Dobarganes, and M.D. Luque de Castro, Accelerated Extraction of the Fat Content in Cheese Using a Focused Microwave-Assisted Soxhlet Device, *J. Agric. Food Chem 47*:2308–2315 (1999)
11. Priego-Capote, F., and M.D. Luque de Castro, Focused Microwave-Assisted Soxhlet Extraction: A Convincing Alternative for Total Fat Isolation from Bakery Products, *Talanta 65*:98–103 (2005).
12. Priego-Capote, F., J. Ruiz-Jimenez, J. Garcia-Olmo, and M.D. Luque de Castro, Fast Method for the Determination of Total Fat and *trans* Fatty-Acids Content in Bakery Products Based on Microwave-Assisted Soxhlet Extraction and Medium Infrared Spectroscopy Detection, *Anal. Chim. Acta 517*:13–20 (2004).
13. Verleyen, T., R. Verhe, A. Cano, A. Huyghebaert, and W. De Greyt, Influence of Triacylglycerol Characteristics on the Determination of Free Fatty Acids in Vegetable Oils by Fourier Transform Infrared Spectroscopy, *J. Am. Oil Chem. Soc. 78*:981–984 (2001).
14. Kotani, A., F. Kusu, and K. Takamura, New Electrochemical Detection Method in High-Performance Liquid Chromatography for Determining Free Fatty Acids, *Anal. Chim. Acta 465*:199–206 (2002).
15. Lagarda, M.J., J.G. Manez, P. Manglano, and R. Farre, Lipid Hydroperoxides Determination in Milk-Based Infant Formulae by Gas Chromatography, *Eur. J. Lipid Sci. Technol. 105*:339–345 (2003).
16. Bunting, J.P., and D.A. Gray, Development of a Flow Injection Chemiluminescent Assay for the Quantification of Lipid Hydroperoxides, *J. Am. Oil Chem. Soc. 80*:951–955 (2003).
17. Navas, J.A., A. Tres, R. Codony, J. Boatella, R. Bou, and F. Guardiola, Modified Ferrous Oxidation-Xylenol Orange Method to Determine Lipid Hydroperoxides in Fried Snacks, *Eur. J. Lipid Sci. Technol. 106*:688–969 (2004).
18. Arab, K., and J.-P. Steghens, Plasma Lipid Hydroperoxides Measurement by an Automated Xylenol Orange Method, *Anal. Biochem. 325*:158–163 (2004).
19. Ruiz, A., M.J.A. Canada, and B. Lendl, A Rapid Method for Peroxide Value Determination in Edible Oils Based on Flow Analysis with Fourier Transform Infrared Spectroscopic Detection, *Analyst 126*:242–246 (2001).
20. Helmerich, G., and P. Koehler, Comparison of Methods for the Quantitative Determination of Phospholipids in Lecithins and Flour Improvers, *J. Agric. Food Chem. 51*:6645–6651 (2003).

21. Sipos, J.C., and R.G. Ackman, Automated and Rapid Quantitative Analysis Lipids with Chromarods, *J. Chromatogr. Sci. 16*:443–447 (1978).
22. Ogasawara, M., K. Tsuruta, and S. Arao, Flame Photometric Detector for Thin-Layer Chromatography, *J. Chromatogr. A 973*:151–158 (2002).
23. Copeman, L.A., and C.C. Parrish, Lipids Classes, Fatty Acids, and Sterols in Seafood from Gilbert Bay, Southern Labrador, *J. Agric. Food Chem 52*:4872–4881 (2004).
24. Striby, L., R. Lafont, and M. Goutx, Improvement in the Iatroscan Thin-Layer Chromatographic-Flame Ionization Detection Analysis of Marine Lipids. Separation and Quantitation of Monoacylglycerols and Diacylglycerols in Standards and Natural Samples, *J. Chromatogr. A 849*:371–380 (1999).
25. Banerjee, A.K., Chromarod-Iatroscan TLC FID as an Analytical Technique Applications in Lipid Analysis, *Chem. Environ. Res. 5*:289–294 (1996).
26. Lee, C.-S., Y.-G. Kim, H.-S Joo, and B.-G. Kim, Structural Analysis of Lipid A from *Escherichia coli* O157:H7:K Using Thin-Layer Chromatography and Ion-Trap Mass Spectrometry, *J. Mass Spectrom. 39*:514–525 (2004).
27. Ivleva, V.B., Y.N. Elkin, B.A. Budnik, S.C. Moyer, P.B. O'Connor, and C.E. Costello, Coupling Thin-Layer Chromatography with Vibrational Cooling Matrix-Assisted Laser Desorption/Ionization Fourier Transform Mass Spectrometry for the Analysis of Ganglioside Mixtures, *Anal. Chem. 76*:6484–6491 (2004).
28. Schaefer, A., T. Küchler, T.J. Simat, and H. Steinhart, Migration of Lubricants from Food Packagings. Screening for Lipid Classes and Quantitative Estimation Using Normal-Phase Liquid Chromatographic Separation with Evaporative Light Scattering Detection, *J. Chromatogr. A 1017*:107–116 (2003).
29. Perona, J.S., and V. Ruiz-Gutierrez, Quantification of Major Lipid Classes in Human Triacylglycerol-Rich Lipoproteins by High-Performance Liquid Chromatography with Evaporative Light-Scattering Detection, *J. Sep. Sci. 27*:653–659 (2004).
30. Descalzo, A.M., E.M. Insani, and N.A. Pensel, Light-Scattering Detection of Phospholipids Resolved by HPLC, *Lipids 38*:999–1003 (2003).
31. Neron, S., F. El Amrani, J. Potus, and J. Nicolas, Separation and Quantification by High-Performance Liquid Chromatography with Light Scattering Detection of the Main Wheat Flour Phospholipids During Dough Mixing in the Presence of Phospholipase, *J. Chromatogr. A 1047*:77–83 (2004).
32. Momchilova, S., K. Tsuji, Y. Itabashi, B. Nikolova-Damyanova, and Kuksis, A., Resolution of Triacylglycerol Positional Isomers by Reversed-Phase High-Performance Liquid Chromatography, *J. Sep. Sci. 27*:1033–1036 (2004).
33. Amaral, J.S., S.C. Cunha, M.R. Alves, J.A. Pereira, R.M. Seabra, and B.P.P. Oliveira, Triacylglycerol Composition of Walnut (*Juglans regia* L.) Cultivars: Characterization by HPLC-ELSD and Chemometrics, *J. Agric. Food Chem. 52*: 7964–7969 (2004).
34. Kotani, A., T. Fuse, and F. Kusu, Determination of Plasma Free Fatty Acids by High-Performance Liquid Chromatography with Electrochemical Detection, *Anal. Biochem. 284*:65–69 (2000).
35. Perret, D., A. Gentili, S. Marchese, M. Sergi, and L. Caporossi, Determination of Free Fatty Acids in Chocolate by Liquid Chromatography with Tandem Mass Spectrometry, *Rapid Commun. Mass Spectrom. 18*:1989–1994 (2004).
36. Svensson, L., L. Sisfontes, G. Nyborg, and R. Blomstrand, High-Performance Liquid Chromatography and Glass Capillary Gas Chromatography of Geometric and Positional Isomers of Long Chain Monounsaturated Fatty Acids, *Lipids 17*:50–59 (1982).

37. Murru, E., E. Angioni, G. Carta, M.P. Melis, S. Spada, and S. Banni, in *Advances in Conjugated Linoleic Acid Research, Vol. 2*, edited by J.L. Sébédio, W.W. Christie, and R. Adlof, AOCS Press, Champaign, IL, 2003, pp. 94–100.
38. Delmonte, P., M.P. Yurawecz, M.M. Mossoba, C. Cruz-Hernandez, and J.K.G. Kramer, Improved Identification of Conjugated Linoleic Acid Isomers Using Silver-Ion HPLC Separations, *J. Assoc. Off. Anal. Chem. Int. 87*:563–568 (2004).
39. Sehat, N., R. Rickert, M.M. Mossoba, J.K.G. Kramer, M.P. Yurawecz, J.A.G. Roach, R.O. Adlof, K.M. Morehouse, J. Fritsche, K.D. Eulitz, H. Steinhart, and Y. Ku, Improved Separation of Conjugated Fatty Acid Methyl Esters by Silver Ion-High-Performance Liquid Chromatography, *Lipids 34*:407–413 (1999).
40. Fevrier, P., A. Binet, L. Dufosse, R. Gree, and F. Yvergnaux, Separation of Glyceride Positional Isomers by Silver Ion Chromatography, *J. Chromatogr. A 923*:53–57 (2001).
41. Adlof, R., and G. List, Analysis of Triglyceride Isomers by Silver-Ion High-Performance Liquid Chromatography. Effect of Column Temperature on Retention Times, *J. Chromatogr. A 1046*:109–113 (2004).
42. Yurawecz, M.P., and K.M. Morehouse, Silver-Ion HPLC of Conjugated Linoleic Acid Isomers, *Eur. J. Lipid Sci. Technol. 103*:609–913 (2001).
43. Bravo, B., G. Chavez, N. Pina, F. Ysambertt, N. Marquez, and A. Caceres, Developing an On-Line Derivatization of FAs by Microwave Irradiation Coupled to HPLC Separation with UV Detection, *Talanta 64*:1329–1334 (2004).
44. Miwa, H., High-Performance Liquid Chromatographic Determination of Free Fatty Acids and Esterified Fatty Acids in Biological Materials as Their 2-Nitrophenylhydrazides, *Anal. Chim. Acta 465*:237–255 (2002).
45. Beermann, C., A. Green, M. Moebius, J.J. Schmitt, and G. Boehm, Lipid Class Separation by HPLC Combined with GC FA Analysis: Comparison of Seed Lipid Compositions from Different *Brassica napus* L. Varieties, *J. Am. Oil Chem. Soc. 80*:747–753 (2003).
46. Balazs, P.E., P.L. Schmit, and B.F. Szuhaj, High-Performance Liquid Chromatographic Separations of Soy Phospholipids, *J. Am. Oil Chem. Soc. 73*:193–197 (1996).
47. Zink, K.-G., and K. Mangelsdorf, Efficient and Rapid Method for Extraction of Intact Phospholipids from Sediments Combined with Molecular Structure Elucidation Using LC-ESI-MS-MS Analysis, *Anal. Bioanal. Chem. 380*:798–812 (2004).
48. Mauriala, T., K.H. Herzig, M. Heinonen, J. Idziak, and S. Auriola, Determination of Long-Chain Fatty Acid Acyl-Coenzyme A Compounds Using Liquid Chromatography-Electrospray Ionization Tandem Mass Spectrometry, *J. Chromatogr. B 808*:263–268 (2004).
49. Byrdwell, W.C., and Neff, W.E., Electrospray Ionization MS of High M.W. TAG Oligomers, *J. Am. Oil Chem. Soc. 81*:13–26 (2004).
50. Liebisch, G., B. Lieser, J. Rathenberg, W. Drobnik, and G. Schmitz, High-Throughput Quantification of Phosphatidylcholine and Sphingomyelin by Electrospray Ionization Tandem Mass Spectrometry Coupled with Isotope Correction Algorithm, *Biochim. Biophys. Acta 1686*:108–117 (2004).
51. Kemmo, S., V. Ollilainen, A.-M. Lampi, and V. Piironen, Determination of Stigmasterol Hydroperoxides Using High-Performance Liquid Chromatography-Mass Spectrometry with Atmospheric Pressure Chemical Ionization, *Czech J. Food Sci. 22*:144–146 (2004).
52. Steenhorst-Slikkerveer, L., A. Louter, H.-G. Janssen, and C. Bauer-Plank, Analysis of

Nonvolatile Lipid Oxidation Products in Vegetable Oils by Normal-Phase High-Performance Liquid Chromatography with Mass Spectrometric Detection, *J. Am. Oil Chem. Soc. 77*: 837–845 (2000).

53. Mondello, L.C., A. Casilli, P.Q. Tranchida, R. Costa, B. Chiofalo, P. Dugo, and G. Dugo, Evaluation of Fast Gas Chromatography and Gas Chromatography-Mass Spectrometry in the Analysis of Lipids, *J. Chromatogr. A 1035*:237–247 (2004).
54. Bondia-Pons, I., A.I. Castellote, and M.C. Lopez-Sabater, Comparison of Conventional and Fast Gas Chromatography in Human Plasma Fatty Acid Determination, *J. Chromatogr. B 809*:339–344 (2004).
55. Zarubova, M.,V.S. Filip, J. Smidrkal, T. Kutek, and I. Piska, The Separation of Triacylglycerols Using Unpolar and Medium Polar Capillary Columns, *Czech J. Food Sci. 22*:123–126 (2004).
56. Guyon, F., S. Destouesse, J. Moustirats, M. Esclapez, M.-H. Salagoity, and B. Medina, Alternative Method for the Quantification by Gas Chromatography Triacylglycerol Class Analysis of Cocoa Butter Equivalent Added to Chocolate Bars, *J. Agric. Food Chem 52*:2770–2775 (2004).
57. Ratnayake, W.M.N., L.J. Plouffe, E. Pasquier, and C. Gagnon, Temperature-Sensitive Resolution of *cis*- and *trans*-Fatty Acid Isomers of Partially Hydrogenated Vegetable Oils on SP-2560 and CP-Sil 88 Capillary Columns, *J. Assoc. Off. Anal. Chem. Int. 85*:1112–1118 (2002).
58. Ratnayake, W.M.N., Overview of Methods for the Determination of *trans* Fatty Acids by Gas Chromatography, Silver-Ion Thin-Layer Chromatography, Silver-Ion Liquid Chromatography, and Gas Chromatography/Mass Spectrometry, *J. Assoc. Off. Anal. Chem. Int. 87*:523–539 (2004).
59. Mondello, L., A. Casilli, P.Q. Tranchida, P. Dugo, and G. Dugo, Detailed Analysis and Group-type Separation of Natural Fats and Oils Using Comprehensive Two-Dimensional Gas Chromatography, *J. Chromatogr. A 1019*:187–196 (2003).
60. Hyoetylaeinen, T., M. Kallio, M. Lehtonen, S. Lintonen, P. Peraejoki, M. Jussila, and M.-L. Riekkola, Comprehensive Two-Dimensional Gas Chromatography in the Analysis of Dietary Fatty Acids, *J. Sep. Sci. 27*:459–467 (2004).
61. Biedermann, W., E. Luecker, J. Poerschmann, S. Lachhab, U. Truyen, and A. Hensel, Structural Characterisation of Some Fatty Acids from the Brain as Biomarkers of BSE Risk Material, *Anal. Bioanal. Chem. 379*:1031–1038 (2004).
62. Christie, W.W., G.W. Robertson, W.C. McRoberts, and J.T.G. Hamilton, Mass Spectrometry of the 4,4-Dimethyloxazoline Derivatives of Isomeric Octadecenoates (Monoenes), *Eur. J. Lipid Sci. Technol. 102*:23–29 (2000).
63. Hamilton, J.T., and W.W. Christie, Mechanisms for Ion Formation During the Electron Impact-Mass Spectrometry of Picolinyl Ester and 4,4-Dimethyloxazoline Derivatives of Fatty Acids, *Chem. Phys. Lipids 105*:93–104 (2000).
64. Yurawecz, M.P.R., J.A.G. Roach, N. Sehat, M.M. Mossoba, J.K.G. Kramer, J. Fritsche, H. Steinhart, and Y. Ku, A New Conjugated Linoleic Acid Isomer, 7 *trans*,9 *cis*-Octadecadienoic Acid, in Cow Milk, Cheese, Beef and Human Milk and Adipose Tissue, *Lipids 33*:803–809 (1998).
65. Jung, M.Y., and M.O. Jung, Identification of Conjugated Linoleic Acids in Hydrogenated Soybean Oil by Silver Ion-Impregnated HPLC and Gas Chromatography-Ion Impacted Mass Spectrometry of Their 4,4-Dimethyloxazoline Derivatives, *J. Agric. Food Chem. 50*:6188–6193 (2002).

66. Mossoba, M.M., M.P. Yurawecz, and J.A.G. Roach, Confirmation of *trans* 18:1 Fatty Acid Isomers. Reply to Comments, *J. Am. Oil Chem. Soc. 75*:423 (1998).
67. Bodzek, D., W. Bakowski, T. Wielkoszynski, B. Janoszka, B. Jaremczuk, R. Tarnawski, and K. Tyrpien, TLC and GC-MS Determination of Cholesterol in Consumable Fats, *Acta Chromatographica 8*:122–143 (1998).
68. Keller, S., and G. Jahreis, Determination of Underivatized Sterols and Bile Acid Trimethyl Silyl Ether Methyl Esters by Gas Chromatography-Mass Spectrometry-Single Ion Monitoring in Faeces, *J. Chromatogr B 813*:199–207 (2004).
69. Mossoba, M.M., R.E. McDonald, M.P. Yurawecz, and J.K.G. Kramer, Application of On-Line Capillary GC-FTIR Spectroscopy to Lipid Analysis, *Eur. J. Lipid Sci. Technol. 103*:826–830 (2001).
70. Steinhart, H., R. Rickert, and K. Winkler, *Trans* Fatty Acids (TFA): Analysis, Occurrence, Intake and Clinical Relevance, *Eur. J. Med. Res. 8*:358–362 (2003).
71. Visser, T., FT-IR Detection in Gas Chromatography, *Trends Anal. Chem. 21*:627–636 (2002).
72. Waehl, H.G., S.Y. Habel, N. Schmieder, and H.M. Liebich, Identification of *cis/trans* Isomers of Methyl Ester and Oxazoline Derivatives of Unsaturated Fatty Acids Using GC-FTIR-MS, *J. High Res. Chromatogr. 17*:543–548 (1994).
73. Bousquet, O., and F. Le Goffic, Counter-Current Chromatographic Separation of Polyunsaturated Fatty Acids, *J. Chromatogr. A 704*:211–216 (1995).
74. Matsuda, S., K. Matsuda, and Y. Ito, Separation of Phospholipids and Glycolipids Using Analytical Toroidal-Coil Counter-Current Chromatography. II. Comparison of the Hydrophobicity Between *Mycoplasma fermentans* and Human-Brain Lipids, *J. Liq. Chromatogr. 26*:1135–1147 (2003).
75. Dobarganes, M.C., and G. Marquez-Ruiz, Analytical Evaluation of Fats and Oils by Size-Exclusion Chromatography, *Analusis 26*:M61–M66 (1998).
76. Caponio, F., T. Gomes, and M.T. Bilancia, Measurement of Degradation of the Lipid Fraction in Margarines, *Eur. Food Res. Technol. 216*:83–87 (2003).
77. Caponio, F., T. Gomes, and D. Delcuratolo, Qualitative and Quantitative Characterisation of the Lipid Fraction of Bouillon Cubes, *Eur. Food Res. Technol. 215*:200–203 (2002).
78. Marquez-Ruiz, G., J. Velasco, and C. Dobarganes, Evaluation of Oxidation in Dried Microencapsulated Fish Oils by a Combination of Adsorption and Size Exclusion Chromatography, *Eur. Food Res. Technol. 211*:13–18 (2000).
79. Hopia, A.I., A.M. Lampi, V.I. Piironen, L.E.T. Hyvonen, and P.E. Koivistoinen, Application of High-Performance Size-Exclusion Chromatography to Study the Autoxidation of Unsaturated Triacylglycerols, *J. Am. Oil Chem. Soc. 70*:779–784 (1993).
80. Laakso, P., and P. Manninen, Identification of Milk Fat Triacylglycerols by Capillary Supercritical Fluid Chromatography-Atmospheric Pressure Chemical Ionization Mass Spectrometry, *Lipids 32*:1285–1295 (1997).
81. Medvedovici, A., and F. David, Comprehensive pSFC/pSFC-MS for the Characterization of Triglycerides in Vegetable Oils, *LC GC Eur. 16*:32–345 (2003).
82. Hellgeth, J.W., J.W. Jordan, L.T. Taylor, and M.A. Khorassani, Supercritical Fluid Chromatography of Free Fatty Acids with On-Line FTIR Detection, *J. Chromatogr. Sci. 24*:183–188 (1986).
83. Senorans, F.J., and E. Ibanez, Analysis of Fatty Acids in Foods by Supercritical Fluid Chromatography, *Anal. Chim. Acta 465*:131–144 (2002).

84. Hishimoto, S.H., and Y. Hirata, A Simple Flow Switching Interface for Two-Dimensional Capillary Supercritical Fluid Chromatography, *Analusis 26*:M23–M26 (1998).
85. Hirata, Y., and I. Sogabe, Separation of Fatty Acid Methyl Esters by Comprehensive Two-Dimensional Supercritical Fluid Chromatography with Packed Columns and Programming of Sampling Duration, *Anal. Bioanal. Chem. 378*:1999–2003 (2004).
86. Sandra, P., A. Medvedovici, Y. Zhao, and F. David, Characterization of Triglycerides in Vegetable Oils by Silver-Ion Packed-Column Supercritical Fluid Chromatography Coupled to Mass Spectroscopy with Atmospheric Pressure Chemical Ionization and Coordination Ion Spray, *J. Chromatogr. A 974*:231–241 (2002).
87. Ma, K., F.R. Van de Voort, J. Sedman, and A.A. Ismail, *Trans* Fatty Acid Determination in Fats and Margarine by Fourier Transform Infrared Spectroscopy Using a Disposable Infrared Card, *J. Am. Oil Chem. Soc. 76*:1399–1404 (1999).
88. Mossoba, M.M., M.P. Yurawecz, P. Delmonte, and J.K.G. Kramer, Overview of Infrared Methodologies for *trans* Fat Determination, *J. Assoc. Off. Anal. Chem. Int. 87*:540–544 (2004).
89. Milosevic, M., V. Milosevic, J.K.G. Kramer, H. Azizian, and M.M. Mossoba, Determining Low Level of *trans* Fatty Acids in Foods Using an Improved ATR-FTIR Procedure, *Lipid Technol. 16*: 252–255 (2004).
90. Al-Alawi, A., F.R. Van De Voort, and J. Sedman, New FTIR Method for the Determination of FFA in Oils, *J. Am. Oil Chem. Soc. 81*:441–446 (2004).
91. Man, Y.B.C., G. Setiowaty, and F.R. Van de Voort, Determination of Iodine Value of Palm Oil by Fourier Transform Infrared Spectroscopy, *J. Am. Oil Chem. Soc. 76*:693–699 (1999).
92. Van de Voort, F.R., J. Sedman, G. Emo, and A.A. Ismail, Rapid and Direct Iodine Value and Saponification Number Determination of Fats and Oils by Attenuated Total Reflectance/Fourier Transform Infrared Spectroscopy, *J. Am. Oil Chem. Soc. 69*:1118–1123 (1992).
93. Ma, K., F.R. Van de Voort, A.A. Ismail, H. Zhuo, and B. Cheng, Monitoring Peroxide Value in Fat Liquor Manufacture by Fourier Transform Infrared Spectroscopy, *J. Am. Oil Chem. Soc. 77*:681–685 (2000).
94. Van de Voort, F.R., K.P. Memon, J. Sedman, and A.A. Ismail, Determination of Solid Fat Index by Fourier-Transform Infrared Spectroscopy, *J. Am. Oil Chem. Soc. 73*:411–416 (1996).
95. Kurowski, C., D. Timm, U. Grummisch, U. Meyhack, and H. Grunewald, The Benefits of Near Infrared Analysis for Food Product Quality, *J. Near Infrared Spec. 6*:A343–A348 (1998).
96. Zhang, H.-Z., W. Zeng, and T.-C. Lee, Simultaneous Determination of Moisture, Protein and Fat in Fish Meal Using Near-Infrared Spectroscopy, *Food Sci. Technol. Res. 6*:19–23 (2000).
97. Cozzolino, D., I. Murray, R. Paterson, and J.R. Scaife, Visible and Near-Infrared Reflectance Spectroscopy for the Determination of Moisture, Fat and Protein in Chicken Breast and Thigh Muscle, *J. Near Infrared Spec. 4*:213–223 (1996).
98. Christy, A.A., S. Kasemsumran, Y. Du, and Y. Ozaki, The Detection and Quantification of Adulteration in Olive Oil by Near-Infrared Spectroscopy and Chemometrics, *Anal. Sci. 20*:935–940 (2004).
99. Sato, T., A.A. Maw, and M. Katsuta, NIR Reflectance Spectroscopic Analysis of the FA Composition in Sesame (*Sesamum indicum* L.) Seeds, *J. Am. Oil Chem. Soc. 80*:1157–1161 (2003).

100. Sato, T., New Estimation Method for Fatty Acid Composition in Oil Using Near Infrared Spectroscopy, *Biosci. Biotechnol. Biochem. 66*:2543–2548 (2002).
101. Baye, T., and H.C. Becker, Analyzing Seed Weight, Fatty Acid Composition, Oil, and Protein Contents in *Vernonia galamensis* Germplasm by Near-Infrared Reflectance Spectroscopy, *J. Am. Oil Chem. Soc. 81*:641–645 (2004).
102. Gonzalez-Martin, I., C. Gonzalez-Perez, N. Alvarez-Garcia, and J.M. Gonzalez-Cabrera, On-Line Determination of Fatty Acid Composition in Intramuscular Fat of Iberian Pork Loin by NIRs with a Remote Reflectance Fibre Optic Probe, *Meat Sci. 69*:243–248 (2004).
103. Guthausen, A., G. Guthausen, A. Kamlowski, H. Todt, W. Burk, and D. Schmalbein, Measurement of Fat Content of Food with Single-Sided NMR, *J. Am. Oil Chem. Soc. 81*:727–731 (2004).
104. Brescia, M.A., V. Mazzilli, A. Sgaramella, S. Ghelli, F.P. Fanizzi, and A. Sacco, ^{1}H NMR Characterization of Milk Lipids: A Comparison Between Cow and Buffalo Milk, *J. Am. Oil Chem. Soc. 81*:431–436 (2004).
105. Aerts, H.A.J., and P.A. Jacobs, Epoxide Yield Determination of Oils and Fatty Acid Methyl Esters Using ^{1}H NMR, *J. Am. Oil Chem. Soc. 81*:841–846 (2004).
106. Lessig, J., C. Gey, R. Suss, J. Schiller, H.-J. Glander, and J. Arnhold, Analysis of the Lipid Composition of Human and Boar Spermatozoa by MALDI-TOF Mass Spectrometry, Thin Layer Chromatography and ^{31}P NMR Spectroscopy, *Comp. Biochem. Phys. B 137B*:265–277 (2004).
107. Murgia, S., S. Mele, and M. Monduzzi, Quantitative Characterization of Phospholipids in Milk Fat via ^{31}P NMR Using a Monophasic Solvent Mixture, *Lipids 28*:585–591 (2003).
108. Cremonini, M.A., L. Laghi, and G. Placucci, Investigation of Commercial Lecithin by ^{31}P NMR in a Ternary CUBO Solvent, *J. Sci. Food Agric. 84*:786–790 (2004).

Part II. Mass Spectral Techniques/Lipidomics

2 An Overview of Modern Mass Spectrometry Methods in the Toolbox of Lipid Chemists and Biochemists

Robert A. Moreau

Eastern Regional Research Center, Agricultural Research Service, United States Department of Agriculture, Wyndmoor, PA 19038

Introduction

The concept of mass spectrometry was developed ~100 years ago at the Cavendish Laboratory of the University of Cambridge, by Joseph John Thomson and colleagues (1). Thomson and his student, Francis William Aston, received Nobel Prizes in Physics in 1906 and 1920, respectively, for their pioneering studies that gave birth to the field of mass spectrometry. During the 1930s, mass spectrometry became a valuable tool for organic chemists. In the 1970s, gas chromatography-mass spectrometry (GC-MS) emerged as a powerful and popular tool for lipid structural identification. During the 1980s, searchable databases containing structural information for many fatty acids and other lipids became available (2). In the 1980s, several forms of "soft" ionization were developed, and some provided excellent interfaces for high performance liquid chromatography (HPLC) and MS. The combination is referred to as HPLC-MS or simply LC-MS. In the 1990s, these LC-MS interfaces were commercialized, miniaturized, and "married" to powerful user friendly PC-based software; these tools have been applied to many areas of lipid research. In recent years, matrix-assisted laser desorption-ionization time-of-flight mass spectrometry (MALDI-TOF MS) became a valuable new tool for "proteomics" research, and several new applications were developed for lipid research.

Most modern mass spectrometers have three basic components (Fig. 1). In the first component, uncharged organic molecules (M) are ionized to either a positively or negatively charged ion ($M^{+\bullet}$ or M^{-}). In the second step, ions are analyzed or separated on the basis of their mass (m/z, mass to charge ratio). In the final step, ions are detected, sometimes qualitatively and other times quantitatively, via a detector. Much of this chapter will compare and contrast the major strategies for ionization, with a brief description of some of the modern instrumental strategies for ion separation and detection.

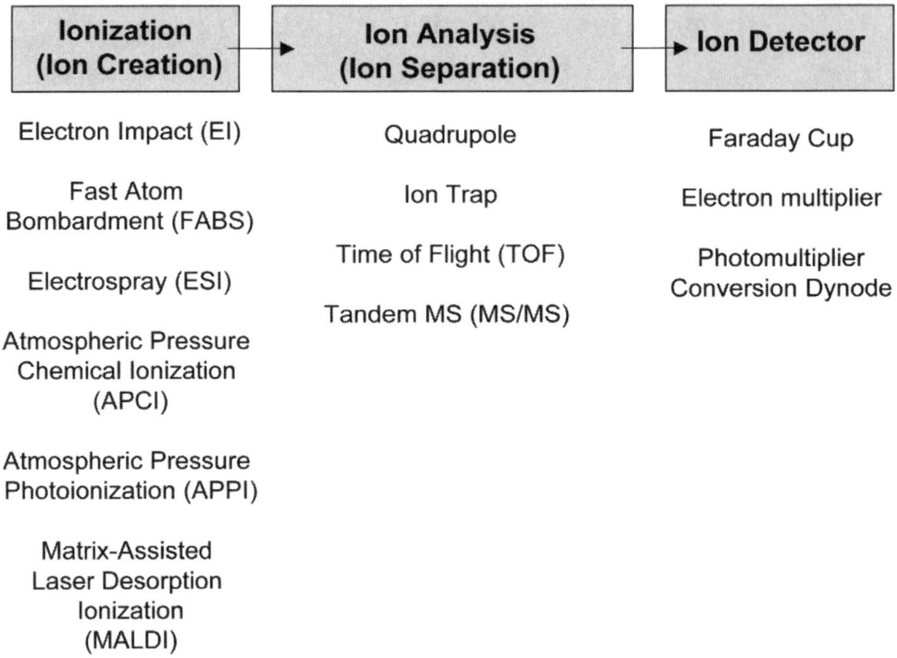

Fig. 1. The three essential components of a modern mass spectrometer system and examples of some of the most important technological strategies for each that have been applied to lipid research.

"Hard" GC-MS-EI and LC-MS-EI

In the field of lipid chemistry, the first wave of advances in mass spectrometry involved GC-MS technology and occurred in the 1970s. The Hewlett-Packard Company of Houston, TX introduced the popular HP 5992a bench top GC-MS in 1976 (2). This instrument, combined with newly perfected capillary columns, made the tool of mass spectrometry readily available to many more lipid chemists than ever before. These early GC-MS instruments employed an ionization method called "electron impact" (EI), which involves bombarding molecules with electrons, creating positively charged ions when the molecules eject their own electrons (Fig. 2). Electron impact is considered a "hard" ionization method, meaning that the energy of the electrons (usually 70 eV) is high enough to cause considerable and highly repeatable fragmentation patterns from small molecules. It is best suited for lipids with a molecular weight <400–500 because thermal degradation makes it difficult to obtain meaningful spectra from larger molecules. For the analysis of fatty acids, the free acids are most often converted to their methyl esters (FAMEs), which are sufficiently volatile to be easily separated by many types of

GC columns (3). In addition to methyl esters, fatty acids can be converted to nitrogen-containing esters, such as picolinyl esters or dimethyloxazoline (DMOX) derivatives, which provide fragments that make it easier to deduce the position of double bonds in polyunsaturated fatty acids (PUFAs) (4) (Fig. 3). For both types of esters, a small molecular ion ($M^{+\bullet}$) is generated by removing one electron from the intact molecule, thus leaving it with a positive charge and a molecular weight identical to that of the intact molecule (M), and numerous fragment ions (F^+). The fragment ion with the greatest abundance (largest peak) is sometimes designated as the "base peak" and the height of all other peaks is reported as a relative percentage (relative to the base peak which is designated as 100%).

Although EI ionization methods have been interfaced mainly with gas chromatographs, Waters Corporation (Milford, MA) developed an EI-MS, called a ThermaBeam^{TM} MS, designed to interface to an HPLC. We employed this method to identify two unusual lipids [diferuloylputrescine (DFP) and *p*-coumaroylferuloylputrescine (CFP)] in lipid extracts from corn bran and corn fiber (5). The topic of LC-MS-EI and its application to the elucidation of the structures of several lipids was reviewed recently (6).

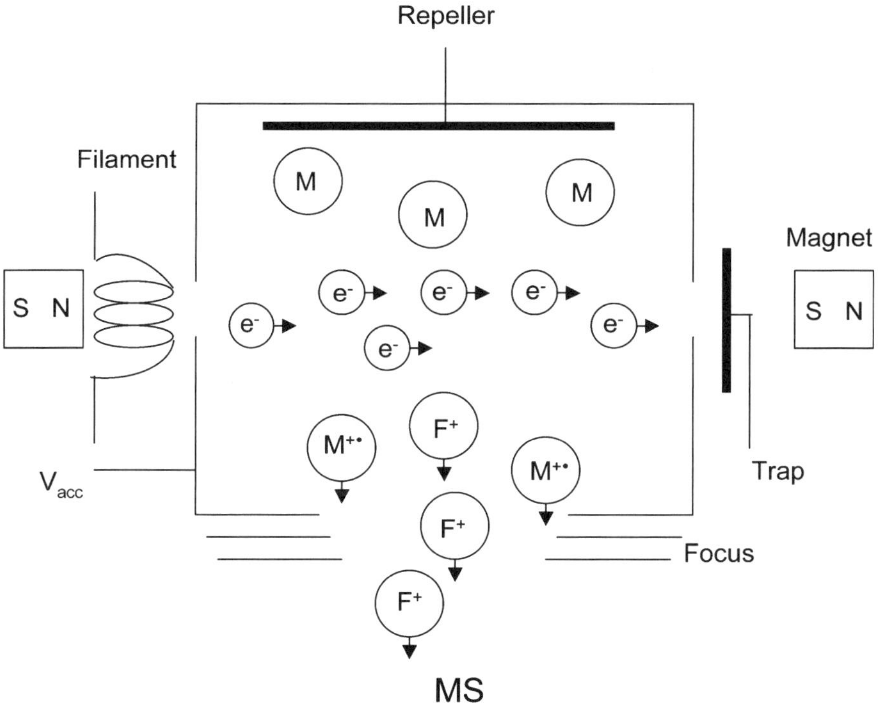

Fig. 2. The mechanism of electron impact (EI) ionization. Abbreviations: M = neutral molecules, e^- = electrons, $M^{+\bullet}$ = molecular ion, F^+ = fragment ion.

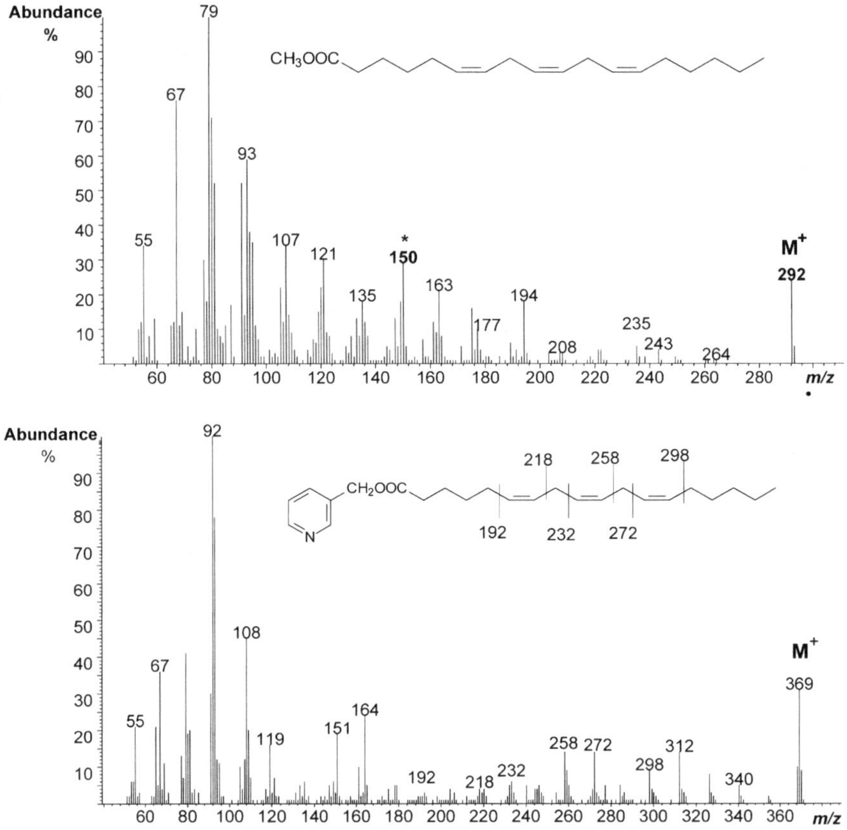

*For fatty acid methyl esters, a peak at *m/z* = 150 appears to be a characteristic for fatty acids with an n-6 terminal moiety, whereas one at *m/z* = 108 defines an n-3 terminal group

Fig. 3. Examples of electron impact (EI) fragmentation of fatty acid esters of γ-linolenic acid (6, 9, 12-octadecatrenoic acid). (A) Fatty acid methyl ester; (B) fatty acid picolinyl ester. Reproduced from (3) with permission.

In addition to being linked to GC and HPLC, EI-MS can also be utilized by introducing a purified sample via direct probe. We employed direct probe EI-MS to identify several unknown lipids in *Frankia,* a bacterium that fixes nitrogen in the roots of several trees (7,8). Our studies identified these unknowns as hopanoids, a class of sterol-like pentacyclic triterpenes that appear to function as sterol equivalents in many species of bacteria (7,8). Although EI is usually considered too "hard" for phospholipid structural determination, Klein et al. (9,10) used it to study the structure of intact phospholipids.

One of the powerful features of GC-MS-EI (and LC-MS-EI) is the fact that the spectra obtained can be compared easily with those in public databases using powerful PC-based software. This ability to compare quantitatively spectra of unknown compounds with those of well-characterized lipid standards is a valuable tool for the elucidation of unknown structures. The two most popular software packages are the NIST (National Institute of Standards and Technology) Reference Database and the Wiley Registry of Mass Spectral Data. The NIST database currently contains 147,194 chemical structures and 174,948 EI spectra (11). The Wiley Registry of Mass Spectral Data, 7th ed., which also includes the NIST Database, is a searchable database that currently contains 338,000 chemical compounds and 390,000 different EI mass spectra (12). Most modern GC-MS-EI and HPLC-MS-EI systems are equipped with one or both of these searchable databases. As EI spectra are obtained, their degree of similarity can rapidly be compared (reported in units of relative percentage of identity) to the thousands of other spectra in the database.

"Soft" (Gentle) Ionization Methods, First Used Mainly for Large Biomolecules Such as Proteins and Peptides, and Well-Suited for Intact Lipids

Although EI is often considered to be a "hard" ionization technique because it causes considerable and reproducible fragmentation patterns for small lipids (e.g., fatty acids and free sterols), its usefulness is often limited for larger lipids (triacylglycerols, glycolipids, and phospholipids) because it causes too much fragmentation and often no molecular ions. In contrast, "soft" ionization methods are more useful for larger lipids because they produce molecular ions and have limited fragmentation.

Many types of "soft" ionization techniques have been developed and employed for lipid research (Table 1). Figure 4 illustrates the type of spectra that are generated when a larger lipid (a sophorolipid, molecular weight 688) is analyzed by a "hard" ionization method (LC-MS-EI) and also by a "soft" ionization method involving atmospheric pressure chemical ionization (LC-MS-APCI). The mechanism of soft ionization by APCI will be described in detail in a later section of this chapter. Note that the "hard" method produced many small fragments and no molecular ions and the "soft" method produced larger fragments and a discernible $[M+H]^+$ ion. In addition to generating positive ions, most ionization methods can also generate, separate, and detect negative ions $[M-H]^-$. For simplicity, each of the figures in this chapter portrays only positive ions, but the reader should be aware that for some lipid applications, operating in the negative mode can be advantageous.

Chemical ionization (CI) is a soft ionization technique that has been applied to many types of lipids. In CI, ions are formed when the sample reacts with an ionized reagent gas or its products. When CI is used with GC-MS methods, electrons generated via EI ionize the reagent gas (methane, isobutene, or ammonia) and the

TABLE 1
Mass Spectrometry Ionization Methods That Have Been the Most Influential for Lipid Analysis

Method	Decade developed	Number of references of recent papers using this method for lipid research	Reference
Group 1 (A very popular hard ionization technique usable by generalists with FAMEs, often linked to GC)			
Electron impact (GC/MS)	1960s	>1000	3,4
Group 2 (Soft ionization techniques usable by specialists and now used infrequently with lipids)			
Chemical ionization	1970s	>100	3,4
Thermospray	1980s	Few	14
Field desorption	1960s	Few	14
Plasma desorption	1960s	Few	14
Laser desorption	1960s	Few	14
Group 3 (Soft ionization technique used extensively by specialists for lipids in the past but less so today)			
Fast atom bombardment	1980s	336	20
Group 4 (Soft ionization techniques usable by generalists, atmospheric pressure, often linked to liquid chromatography)			
Electrospray	1980s	387	25
Atmospheric pressure chemical ionization	1970s	96	30
Atmospheric pressure photoionization	2000s	4	34
Group 5 (Soft ionization technique usable by generalists for proteomics and beginning to be used with lipids)			
Matrix assisted laser desorption ionization	1980s	277 123	39 44

ionized gas protonates the analyte molecules. Ammonia has been the most common CI gas for lipids; in the positive mode, the main peaks observed are MH^+ and $[M+NH_4]^+$. The topic of chemical ionization of fatty acids and their isomers will be discussed in detail in Chapter 6 of this volume. Although it is most often used for small lipids such as fatty acid esters, in 1983 Crawford and Plattner (13) reported that ammonia could be used as a CI agent for intact diacylphosphatidylcholine. APCI-MS is a specialized form of CI that will be described later in this chapter.

Thermospray was the first soft ionization method that was developed as an LC-MS interface. Thermospray was developed by Vestel, and the first commercial units appeared in the early 1980s (14) (Table 1). Although several other soft ionization techniques have been developed, the five in Figure 1 are the most successful and most widely utilized in the field of lipid research. The first, fast-atom-bombardment (FAB) MS, is a "soft" ionization technique that has been valuable for lipid analysts. In general, FAB instruments are more expensive and FAB methods are less user friendly than some of the more modern soft ionization methods, and

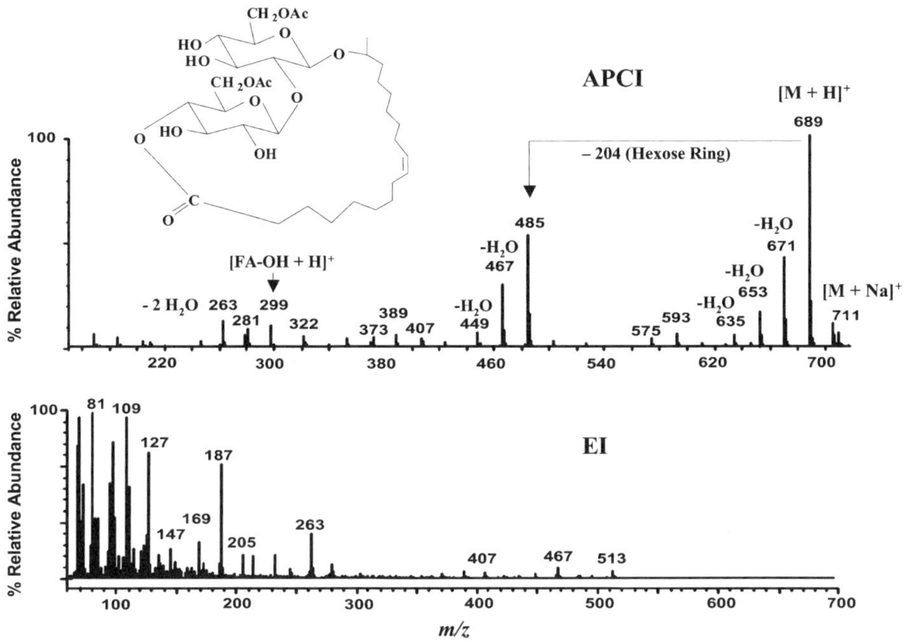

Fig. 4. A comparison of the type of fragmentation of a large intact lipid (a sophorolipid, molecular weight 688) that is caused by a "hard" ionization method such as electron impact liquid chromatography-mass spectrometry-electron impact (LC-MS-EI) vs. a "soft" ionization method, such as LC-MS-atmospheric pressure chemical ionization (APCI) (positive mode). Abbreviations: M, molecular ion; FA, fatty acid. Reproduced from (2) with permission.

their use has been declining in recent years. The next two soft ionization techniques, electrospray ionization (ESI) and APCI have been extremely valuable for lipid applications. The fourth, atmospheric pressure photoionization (APPI), is a very new technique, with very few publications for lipids; it is likely to become very important for lipid analysts. ESI, APCI, and APPI are all considered to be "atmospheric pressure ionization methods" (API) because the ionization occurs at atmospheric pressure. Several LC-MS manufacturers sell easily interchangeable ionization chambers that allow the use of ESI, APCI, and APPI (each in either the positive or negative ion mode) with the same instrument by attaching the appropriate spray chamber. The final soft ionization method that will be discussed is MALDI-MS. MALDI-MS requires a specialized instrument and because it involves mixing the analyte with a matrix and exposing the mixture to a laser, MALDI is not easily interfaced to an HPLC.

Soft Ionization via FAB-MS

FAB-MS was developed by Barber (15) in ~1981 (Fig. 5). The method involves mixing the sample with a suitable matrix. The matrix is then placed in a vacuum chamber and bombarded with abeam of atoms, often cesium, at bombardment energies of 2–30 keV. The interaction of the atoms with the matrix adds a proton $[M+H]^+$ or removes a proton $[M-H]^-$ from the analyte molecules (M). $[M+H]^+$ and $[M-H]^-$ are sometimes called pseudomolecular ions. In addition to protonated and deprotonated ions, $[M+K]^+$, $[M+Na]^+$, $[M-K]^-$, and $[M-Na]^-$ ions are also detected with FAB-MS.

Numerous studies have employed FAB-MS to study the structure of large (>400 molecular weight) intact lipids. Jensen et al. (16) utilized FAB-MS to elucidate the structure of the intact molecular species phosphatidylcholine and phosphatidylethanolamine. Kayganich-Harrison and Murphy (17) employed FAB-MS to study the structures of chain-shortened oxidized phosphatidylcholines. We used FAB-MS to identify a new class of sphingolipids (ceramide phosphorylethanolamines) in *Phytophthora infestans* (the causal agent of Late Blight in potatoes, which is the fungal disease that caused the Irish potato famine in the 1850s) (18). We also employed FAB-MS to identify the structure of intact hopanoids (sterol-like pentacyclic triterpenes) in the ethanol-producing bacterium

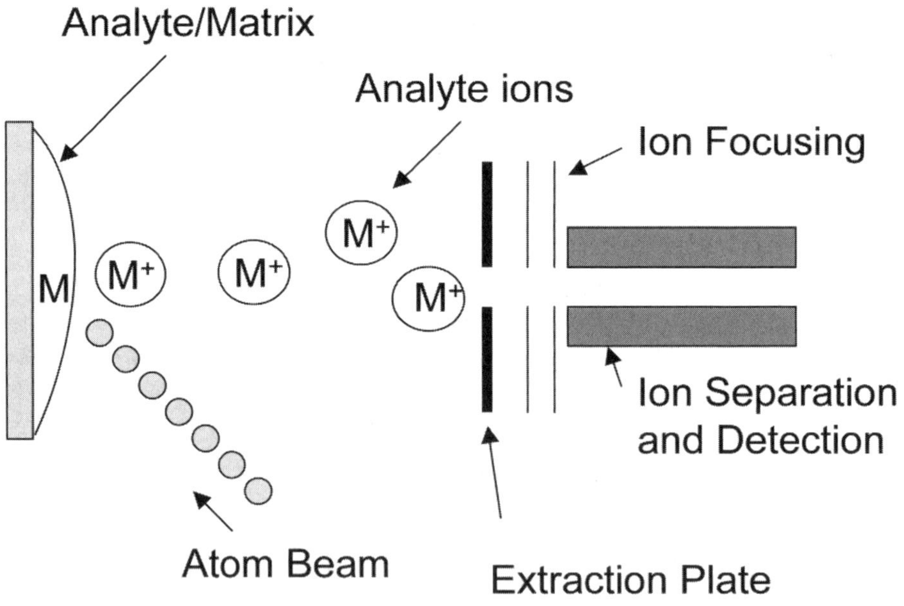

Fig. 5. The mechanism of soft ionization via fast atom bombardment-mass spectrometry (FAB-MS).

Zymomonas mobilis (19). W.W. Christie has a database on his popular Lipid Library website that currently includes 336 references of publications that employ FAB-MS for lipid research (20).

Although FAB-MS has been a very valuable tool for lipid research in the past, it is now being used less frequently; it appears to be slowly being replaced by the four MS methods described below. Reasons for the decline in FAB-MS include the following: the need to purify samples before introduction into the MS; the need for a relatively large sample (2–100 mg) (19); and the fact that other modern MS methods are often more user friendly and can be performed on instruments that are less complex and less expensive.

Soft Ionization via ESI

ESI methods were developed by Fenn and colleagues (21) in the early 1980s at Yale University (Fenn subsequently moved to Virginia Commonwealth University in Richmond, VA and shared the Nobel Prize in Chemistry in 2002). The first commercial units were sold in 1989. These ESI methods were first applied to proteins and peptides (Table 1). The potential of LC-ESI-MS for the identification and analysis of lipids was recognized early by Myer and Kuksis at the University of Toronto, Canada (22). The ESI process takes place at atmospheric pressures (like APCI, APPI, and sometimes MALDI), at which protonated and deprotonated molecules are formed and transferred from the liquid to the gas phase. ESI tends to produce only molecular adduct ions, $[M+X]^+$ (e.g., X = H, Na, K, NH_4), in the positive mode and deprotonated molecular ions, $[M-H]^-$, in the negative mode (Fig. 6). The general lack of fragmentation ions is not a problem for many types of applications, but if additional fragmentation is desirable, collisionally induced dissociation (CID) can be controlled by varying the potential difference between capillary exit and the first skimmer. Several manufacturers are also now marketing electrospray instruments that can produce collisionally induced fragments via coupling two mass analyzers in series (MS-MS) or by employing an ion trap (a technique called MS^n) (the topics of MS-MS and ion traps are described in a later section of this chapter). ESI is ideally suited for the identification of charged lipids such as phospholipids, but it has also proven to be valuable at times for less polar lipids such as triacylglycerols and cholesteryl esters. In 1994, Han and Gross of Washington University in St. Louis, MO introduced a flow injection MS-MS to analyze molecular species of glycerophospholipids, which dispenses with the chromatographic step (23). The first MS step separates a phospholipid class (e.g., phosphatidylcholine or phosphatidylethanolamine) and after collisional dissociation, the second step quantitatively analyzes the individual molecular species for each phospholipid class. The method is now widely adopted for phospholipids, sphingolipids, and plant glycolipids. These new and valuable ESI methods were used as a basis for the new field of lipid profiling that has become known as "lipidomics" (24). The topic of "shotgun lipidomics" using ESI-MS-MS will be covered in

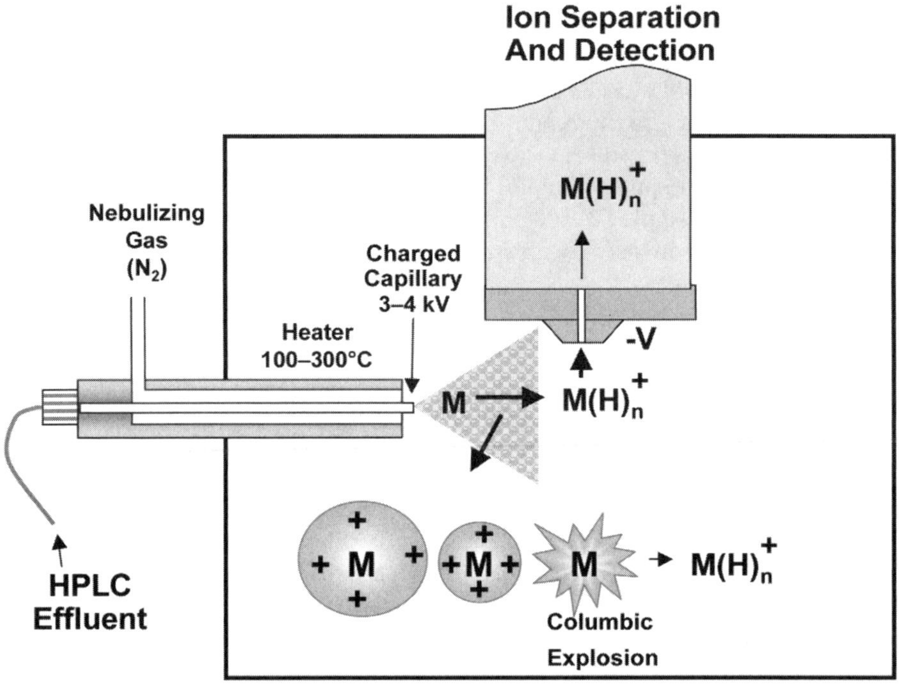

Fig. 6. The mechanism of soft ionization via electrospray-mass spectrometry (ESI-MS). Figure provided by A. Nuñez.

Chapter 3. W.W. Christie's on-line database includes a section on ESI publications for lipid research and it currently includes 387 references (25). With the rapid developments in the field of lipidomics, this number is increasing daily.

Soft Ionization via APCI

APCI instruments have been available commercially since the mid-1980s. As the name implies, "chemical ionization" of analyte molecules in the positive mode occurs by a process in which the corona discharge electrode generates nitrogen ions (N_2^+), which undergo charge transfer to atmospheric water molecules to form hydronium ions (H_3O^+) (26,27) (Fig. 7). The hydronium ions usually act as the primary source for ionization of the analyte molecules. Most APCI instruments operate on the same mass analyzer that is used for ESI. Rozenwell employed reverse phase (RP)-HPLC and APCI to perform quantitative analysis of phytosterols (28). We employed LC-MS-APCI (Fig. 8) to identify intact triacylglycerols (containing almost entirely saturated fatty acids) in food samples (a residue that contained 10%

extractable fat and was probably originally from a stew of goat or lamb) from the funerary banquet of King Midas in Phrygia in 700 B.C. (29). Although APCI is a soft ionization technique, it does cause some fractionation of lipids, as evidenced by the peaks of diacylglycerols derived from collisional dissociation of intact triacylglycerols that were separated by HPLC (Fig. 8). W.W. Christie's website contains a database that currently includes 96 references of publications that employ ESI for lipid research (30).

In 2004 ESA Inc. (Chelmsford, MA) launched a new type of HPLC detector, a Corona™ CAD™ (charged aerosol detector). The nebulization and ionization mechanism of this detector appear to be similar to those employed in APCI (31), except that the total amount of ions is quantified, rather than being separated into ions of different size, as in APCI. One simple way to explain the principle of this detector is to picture Figure 1, with the first step (ionization) linked directly to the third (detection) step. The manufacturers claim that this new detector is much more sensitive than evaporative light-scattering detectors, which have proven to be very valuable for lipid research. This new type of detector has the potential to become a very useful tool for lipid analysts, and the publication of studies evaluating its application in this area is anticipated.

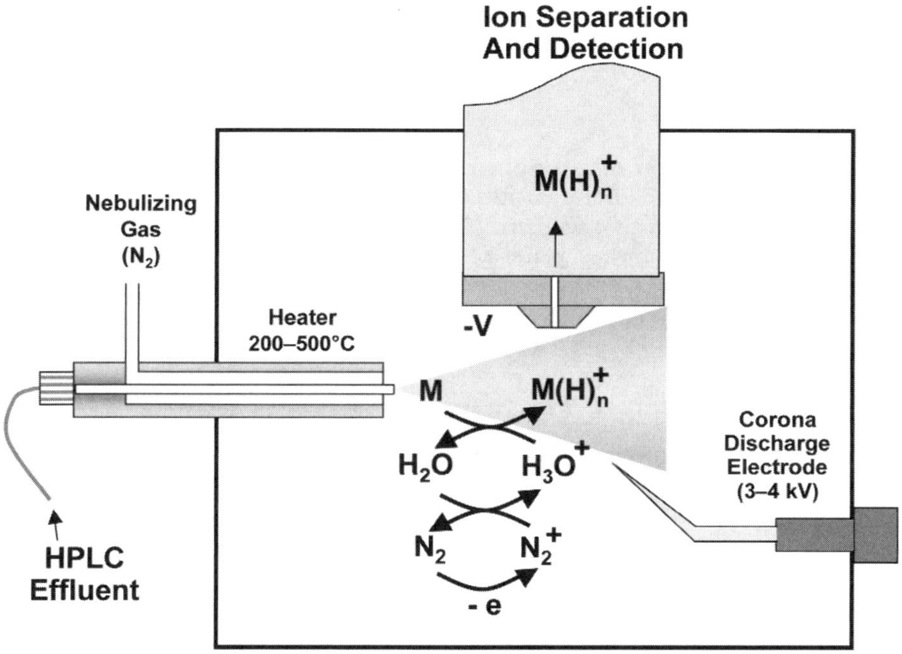

Fig. 7. The mechanism of soft ionization via atmospheric pressure chemical ionization (APCI). Figure provided by A. Nuñez.

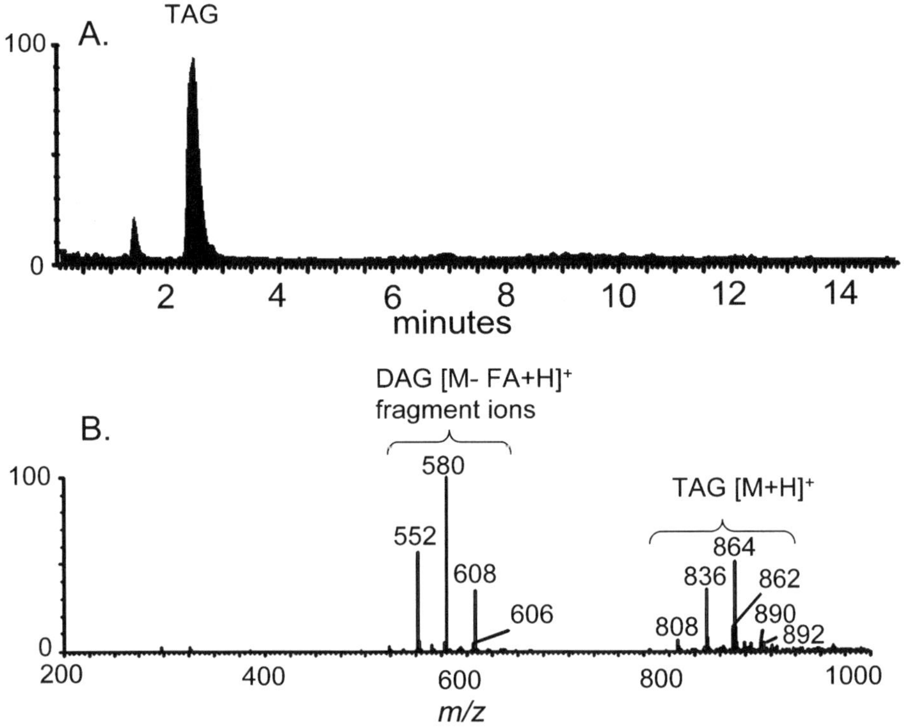

Fig. 8. High performance liquid chromatography-mass spectrometry-atmospheric pressure chemical ionization (HPLC-MS-APCI) total ion chromatogram (A) and mass spectrum (B) of the 2.33-min triacylglycerol peak, providing evidence for the existence of intact triacylglycerols in dried food residues from 700 BC. Abbreviations: TAG, triacylglycerol lipid class; DAGs, diacylglycerol; FA, fatty acid. Reproduced from (29) with permission.

Soft Ionization via APPI

APPI is the newest soft ionization MS method. It was first reported in 2000 by Robb et al. (32,33). The principle of photoionization for GC was introduced in the 1970s as a detection method for GC. The very new field of APPI MS was described in detail in a recent review (34). Like ESI and APCI, the ionization step in APPI occurs at atmospheric pressure. Also, like ESI and APCI, APPI is commonly used as an interface for LC and MS. When used for LC-MS, the mechanism of ionization in APPI (Fig. 9) first involves nebulization of the HPLC effluent (mixture of solvent and analytes). Photons (10 eV) are emitted from a vacuum-UV lamp, and they strike the nebulized analyte molecules and solvent. The exact mechanism of ionization is still being studied, but recent evidence indicates that under the correct conditions $[M+H]^+$ ions are generated by an interaction of the photons with protic solvents (e.g., methanol or isopropanol) (35). This mechanism, which involves indirect ionization via a protic solvent, is called "dopant APPI." There is also evidence that under certain conditions, photons can ionize certain types of analytes directly without the involvement of a solvent (dopant). This latter mechanism is called "direct APPI." Additional studies are warranted to understand the mechanism of APPI and to determine which types of analytes may be ionized directly and which types are ionized indirectly. More information is also required to determine which types and what proportions of solvents are optimal for APPI of lipids.

Several papers employing APPI-MS for the study of lipids and lipid-like metabolites were published by Kostiainen and colleagues in Helsinki, Finland. In

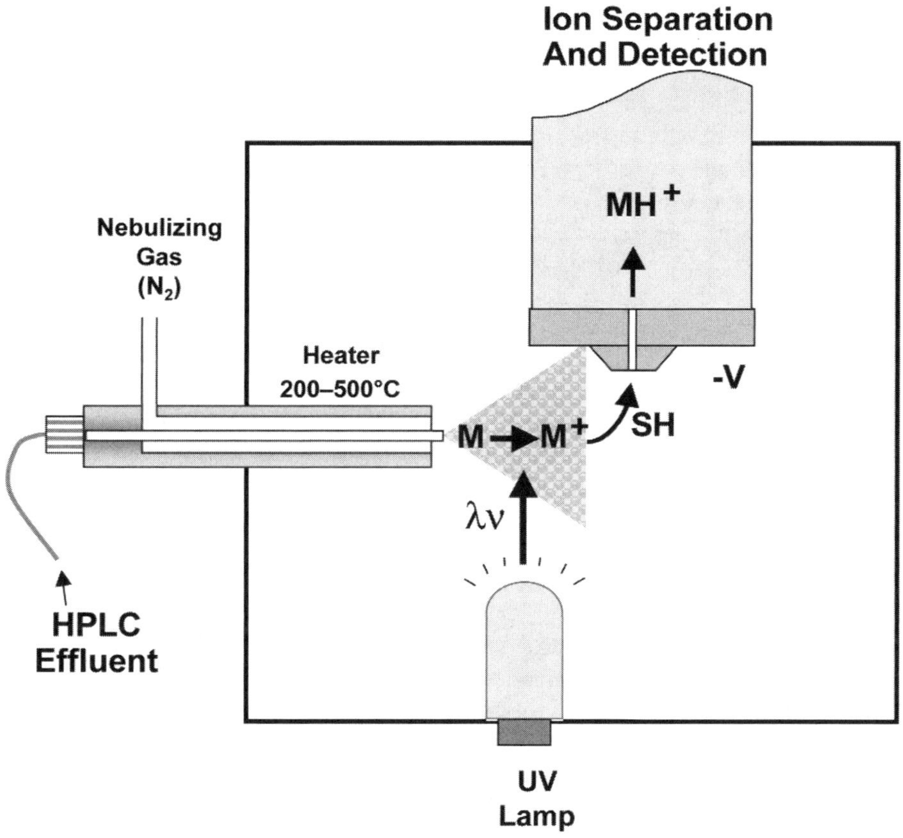

Fig. 9. The mechanism of soft ionization via atmospheric pressure photoionization (APPI). Figure provided by A. Nuñez.

the first study, they compared ESI, APCI, and APPP for the study of flavonoids (36). The limits of detection for catechin with ESI, APCI, and APPI were similar in both the negative ion mode (12, 13, and 11 μM, respectively) and the positive ion mode (46, 36, and 55 μM, respectively). A similar comparative study of the three MS ionization methods for anabolic steroids revealed that ESI was slightly more sensitive than APCI or APPI (37). In the third study, the three MS ionization methods were used to screen 22 drug metabolites in biological samples (38). ESI detected all 22 metabolites, and APCI and APPI detected 12 and 14, respectively.

Although APPI is a very new ionization method, with very few published applications for lipid research, it is a very promising new technique that may someday prove to be extremely valuable for lipid research.

Soft Ionization via Matrix Assisted Laser Desorption Ionization-Time of Flight (MALDI-TOF)

The principles of MALDI were first introduced in 1988 by Tanaka and independently by Hillencamp and Karas [an excellent review on the topic of MALDI-TOF MS for lipid research was recently published by Schiller et al. (39)]. In MALDI instruments, the sample is deposited in a plate with the addition of a UV absorbing matrix. The sample/matrix "spot" is then excited by a pulsing laser (the plate can contain >100 "spots" of samples), causing a charge transfer to the sample, usually a H^+ or Na^+, generating single charged ions (Fig. 10). Most MALDI instruments employ TOF ion analyzer (described in a later section) and are thus called MALDI-TOF; sometimes they have two mass analyzers for collisionally induced dissociation (CID, a principle also discussed in a later section), and the instruments are called MALDI-TOF-TOF. Most MALDI TOF MS instruments are dedicated instruments, but some companies offer them as a combination of MALDI and an API ionization exchangeable source. Some MALDI-TOF instruments require the sample to be placed in a vacuum, whereas for others, the sample can be analyzed at atmospheric pressure (40). Mass analysis by TOF-MS instruments is based on the principle that ions form in the ionization source and, after acceleration by an electrical field, they have approximately the same kinetic energy. Accordingly, mass separation occurs as a function of the speed of the ions moving through the field-free flight tube; lighter ions move more quickly than heavy ones. Fragmentation can be achieved after collision of the ions with an inert gas in a process known as collision-induced dissociation (CID), which with the addition of a quadrupole, allows the MS-MS analysis of selected ions (Q-TOF). A more recent addition to the TOF family is MALDI-TOF-TOF-MS-MS, which consists of two TOF tubes and a CID chamber in between. These systems also do not require preliminary chromatography of the sample. Tanaka was awarded the 2002 Nobel Prize in chemistry for his application of MALDI-TOF to protein profiling (proteomics) (41).

Matrices commonly used with MALDI-TOF instruments consist of low-molecular-weight compounds (in the region of 200–500 Da); sometimes ions derived

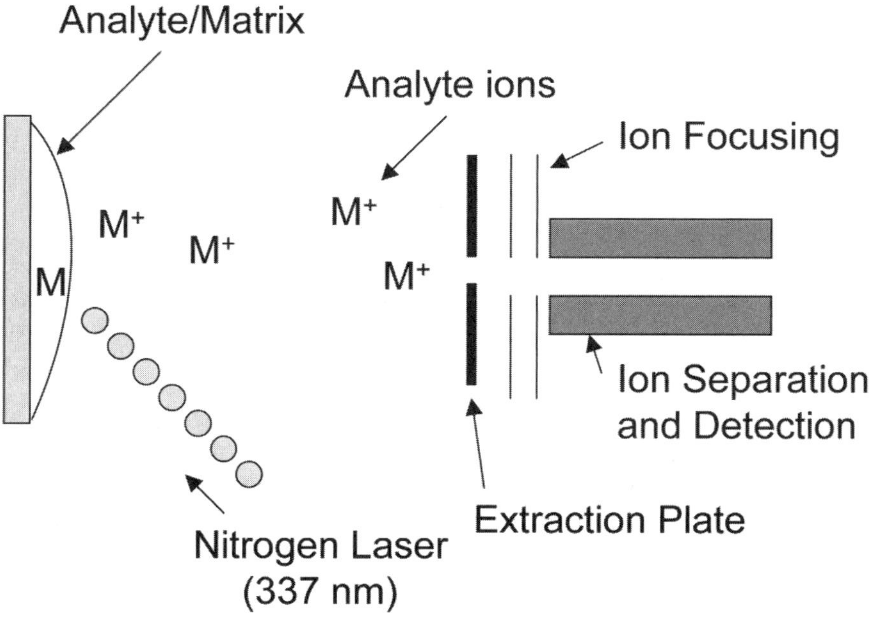

Fig. 10. The mechanism of soft ionization via matrix-assisted laser desorption ionization (MALDI).

from these matrices complicate the interpretation of mass spectra of lipid samples. In recent studies by Ayorinde and colleagues at Howard University (42), the use of meso-tetrakis (pentafluorophenyl) porphyrins as a matrix allowed the analysis of several types of lipids (from free fatty acids to triacylglycerols) without this complication.

MALDI-TOF instruments have radically altered the analysis of peptides and proteins by providing a simple and fast data acquisition tool with significant effect in the development of proteomics (defined as the qualitative and quantitative comparison of proteomes under different conditions to unravel biological processes). MALDI-TOF is a rapid technique that allows high throughput analysis of samples and has many potential applications for screening polar and nonpolar lipids from diverse sources. Some biochemical studies have employed MALDI-TOF to obtain profiling data (lipidomics) of phospholipid molecular species (43). Although ESI is currently used for most lipidomics research, it is likely that in the future, MALDI-TOF will continue to become more popular for lipidomics research. W.W. Christie's on-line database currently includes 123 references of publications that employ MALDI-TOF MS for lipid research (44). Schiller et al. (39) reported that their literature search identified 277 references for MALDI-TOF for lipid research.

High Resolution/Accurate Mass FAB and ESI Methods for the Structural Identification of Lipids

Most mass spectra are obtained with the instruments "tuned" with standards of known mass so that the ions measured are accurate to one decimal place, 0.1 *m/z*. With some mass spectrometers, it is possible to tune the instrument with standards of known mass and obtain a mass spectrum with an "accurate mass" of four decimal places. Examples of structures that were elucidated using this technique are given below.

High-resolution FAB was used to identify hopanoids in *Zymomas mobilis* (19). Two unknowns were separated by normal phase HPLC and their exact masses of 708.5461 $(M+H)^+$ and 708.5444 $(M+H)^+$ were measured, which corresponded to a molecular formula of $C_{41}H_{74}O_8N$. Per-*O*-acetylation of both unknowns increased the molecular weight by an amount equivalents to seven hydroxyl groups. Acid hydrolysis was used to distinguish the two unknowns as bacteriohopanetetrol and bacteriohopanetetrol ether.

High-resolution FAB was used to identify ceramide-phosphorylethanolamine in *Phytophthora infestans* and two other Oomycete species (18). The exact mass of 671.5136 $(M+H)^+$ was measured, which corresponded to a molecular formula of $C_{37}H_{72}O_6N_2P$. Selective hydrolysis and natural remnant magnetizations were used to deduce that this molecular formula was consistent with the structure of ceramide phosphorylethanolamine.

High-resolution ESI was used to identify DFP and CFP (5). An unknown peak was found on the HPLC chromatogram, and its exact mass of 441.1996 $(M+H)^+$ was measured. This mass corresponds to a molecular formula of $C_{24}H_{29}N_2O_6$, which was tentatively identified as ceramide phosphorylethanolamine. This structural identification was confirmed using HPLC-MS-EI (Thermabeam), and the spectra matched that of DFP in the NIST database.

Modern Instrumental Approaches for Mass Separation and Detection

Immediately after ionization (Fig. 1), gas phase ions enter a region of the mass spectrometer known as the mass analyzer. The mass analyzer separates the ions within a selected range of the mass-to-charge (*m/z*) ratios. Ions can be separated by magnetic fields, electric fields, or by measuring the time it takes an ion to travel a fixed distance (Fig. 11).

Quadrupole mass analyzers were developed in the 1950s for electron impact MS (45). Quadrupole mass analyzers contain four precisely parallel rods with a direct current voltage and a superimposed radio frequency (Fig. 11A). By scanning at a preselected radio frequency, one effectively scans a particular mass range. Quadrupoles are the least expensive and most popular mass analyzers for ESI, APCI, and APPI MS.

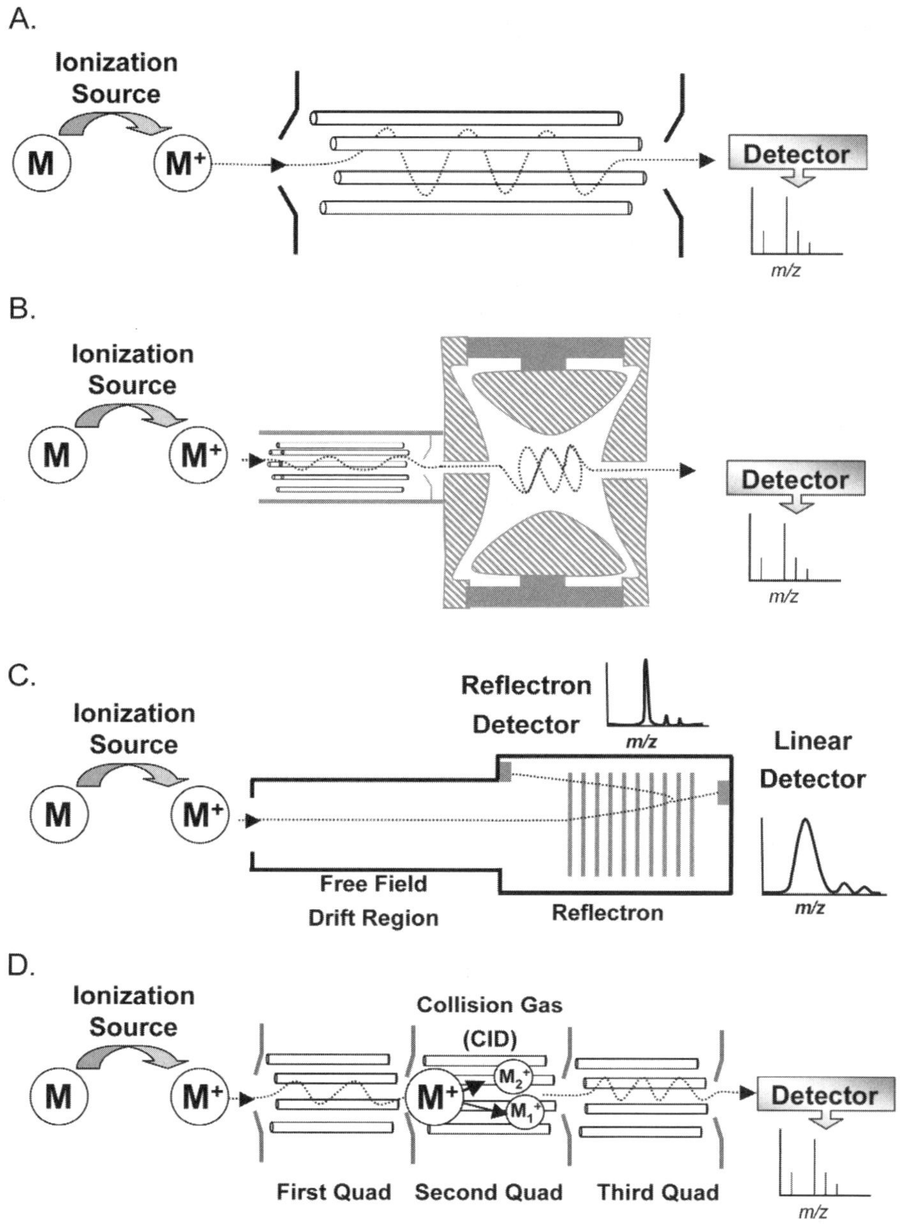

Fig. 11. Methods of mass analysis (ion separation). (A) Quadrupole; (B) ion trap; (C) time of flight; (D) tandem mass spectrometry (MS). Figure provided by A. Nuñez.

Quadrupole ion trap mass analyzers "trap" ions in a radio frequency quadrupole field. It is possible to isolate one ion species by ejecting all others from the trap (Fig. 11B). The isolated ions can subsequently be fragmented by collisional activation and the fragments detected to generate a fragmentation spectrum. Quadrupoles and quadrupole ion traps were both invented by the Nobel Prize winner, Wolfgang Paul. The primary advantage of quadrupole ion traps is that multiple collision-induced dissociations can be performed without the need for multiple analyzers.

The time-of-flight (TOF) analyzer, sometimes called the reflectron time-of-flight (RTOF) analyzer is one of the simplest types of mass analyzers (although not the least expensive) and it is most often used with MALDI. In a TOF analyzer, ions are all accelerated with the same amount of energy. Because the ions have the same energy, yet different mass, they reach the detector at different times (Fig. 11C).

In tandem mass spectrometry (sometimes called MS-MS or MS^n), sample ions are first separated by size in one mass analyzer; then an ion of a particular size is chosen and introduced into a collision cell (Fig. 11D). In the collision cell, the selected ion collides with a collision gas (typically argon or helium) resulting in fragmentation. The resulting fragment ions then enter a second mass analyzer, in which the fragment ions are separated by size and then detected. In MALDI-TOF-TOF, the two mass analyzers are both TOF analyzers.

After ions are separated in the mass analyzer, they enter the ion detector (Fig. 1). The three common strategies for ion detection in modern mass spectrometers include the Faraday cup, the electron multiplier, and the photomultiplier conversion dynode (scintillation counting or Daly detection) (45).

Conclusions and Future Application of Mass Spectrometry to Lipid Chemistry and Biochemistry

The goal of this chapter was to provide a glimpse into some of the advances in mass spectrometry in the last 30 years and how these advances have provided powerful new tools for lipid chemists and biochemists. One of the most promising new fields that has emerged as a result of advances in ESI-MS-MS is that of "lipidomics." I anticipate that MALDI-TOF will soon also be applied to the field of lipidomics and to other areas of lipid research.

Disclaimer

Mention of trade names or commercial products in this publication is solely for the purpose of providing specific information and does not imply recommendation or endorsement by the U.S. Department of Agriculture.

References

1. Anonymous, History of Mass Spectrometry, www.i-mass.com/history.html (accessed January 2006).

2. Moreau, R.A. and A. Nunez, Modern Mass Spectrometry—Four Types of User-Friendly Instruments and Their Applications for Lipid Chemistry and Biochemistry, *INFORM 14:* 536–538 (2003).
3. Christie, W.W., Mass Spectrometry of Fatty Acid Derivatives, www.lipidlibrary.co.uk/mainms.html (accessed January 2005).
4. Christie, W.W., Gas Chromatography—Mass Spectrometry Methods for Structural Analysis of Fatty Acids, *Lipids 33:*343–353 (1998).
5. Moreau, R.A., A. Nunez, and V. Singh, Diferuloylputrescine and *p*-Coumaroylferuloylputrescine, Abundant Polyamine Conjugates in Lipid Extracts of Maize Kernels, *Lipids 36:*839–844 (2001).
6. Nunez, A., Electron Impact Ionization Mass Spectrometry as a Tool for LC/MS Analysis of Lipids and Derivatives, *INFORM 12:*502–509 (2001).
7. Berry, A.M., O.T. Harriot, R.A. Moreau, S.F. Osman, D.R. Benson, and A.D. Jones, Hopanoid Lipids Compose the *Frankia* Vesicle Envelope, Presumptive Barrier of Oxygen Diffusion to Nitrogenase, *Proc. Natl Acad. Sci. 90:*6091–6094 (1993).
8. Berry, A.M., R.A. Moreau, and A.D. Jones, Bacteriohopanetetrol: Abundant Lipid in *Frankia* Cells and in Nitrogen-Fixing Nodule Tissue, *Plant Physiol. 95:*111–115 (1991).
9. Klein, R.A., Mass Spectrometry of the Phosphatidylcholines: Dipalmitoyl, Dioleoyl, and Stearoyl-Oleoyl Glycerylphosphorylcholines, *J. Lipid Res. 12:*123–131 (1971).
10. Klein, R.A., Mass Spectrometry of Phosphatidylcholines: Fragmentation Processes for Dioleyl and Stearoyl-Oleoyl Glycerylphosphorylcholine, *J. Lipid Res. 12:*628–634 (1971).
11. NIST Standard Reference Database 1; NIST/EPA/NIH Mass Spectral Library NIST '02 ASCII Version Reference Database (www.nist.gov/srd/nist1.htm) (accessed January 2006).
12. Wiley Registry of Mass Spectral Data, 7th ed., http://www.wiley.com/WileyCDA/WileyTitle/productCd-0471440973,miniSiteCd-STMDB.html (accessed January 2006).
13. Crawford, C.G., and R.D. Plattner, Ammonia Chemical Ionization Mass Spectrometry of Intact Diacyl Phosphatidylcholine, *J. Lipid Res. 24:*456–460 (1983).
14. Kim, H.Y., and N. Salem, Jr., Liquid Chromatography-Mass Spectrometry of Lipids, *Prog. Lipid Res. 32:*221–245 (1983).
15. Barber, M., R.S. Bordoll, G.J. Elliott, R.D. Sedgwick, and A.N. Tyler, Fast Atom Bombardment Mass Spectrometry, *Anal. Chem. 64:*645A–657A (1982).
16. Jensen, N.J., K.B. Tomer, and M.L. Gross, Fast Atom Bombardment and Tandem Mass Spectrometry of Phosphatidylserine and Phosphatidylcholine, *Lipids 21:*580–588 (1986).
17. Kayganich-Harrison, K.A., and R.C. Murphy, Characterization of Chain-Shortened Oxidized Glycerophosphocholine Lipids Using Fast Atom Bombardment and Tandem Mass Spectrometry, *Anal. Biochem. 221:*16–24 (1994).
18. Moreau, R.A., D.H. Young, P.O. Danis, M.J. Powell, C.J. Quin, K. Beshah, R.A. Slawecki, and R.L. Dilliplane, Identification of Ceramide-Phosphoylethanolamine in Oomycete Plant Pathogens: *Pythium ultimum*, *Phytophthora infestans*, and *Phytophthora capsici*, *Lipids 33:*307–317 (1998).
19. Moreau, R.A., M.J. Powell, S.F. Osman, B.D. Whitaker, W.F. Fett, L. Roth, L., and D.J. O'Brien, Analysis of Intact Hopanoids and Other Lipids from the Bacterium *Zymomonas mobilis* by High-Performance Liquid Chromatography, *Analytical Biochem. 224:*293–301 (1995).

20. Christie, W.W., Literature Survey—Fast-Atom-Bombardment Mass Spectrometry of Lipids, www.lipidlibrary.co.uk/lit_surv/fab_ms.html (accessed January 2006).
21. Whitehouse, C.M., R.N. Dreyer, M. Yamashita, and J.B. Fenn, Electrospray Interface for Liquid Chromatographs and Mass Spectrometers, *Anal. Chem. 57:*675–679 (1985).
22. Myher, J., and A. Kuksis, Electrospray-MS for Lipid Identification, *INFORM 6:*1068–1072 (1995).
23. Han, X., and R.W. Gross, Electrospray Ionization Mass Spectroscopic Analysis of Human Erythrocyte Plasma Membrane Phospholipids, Proc. *Natl. Acad. Sci. USA 91:*10635–10639 (1994).
24. Welti, R., and X. Wang, Lipidomics, *INFORM 14:*607–608 (2003).
25. Christie, W.W., Literature Survey—Electrospray-Ionization Mass Spectrometry of Lipids, www.lipidlibrary.co.uk/lit_surv/elect_ms.html (accessed January 2006).
26. Byrdwell, W.C., APCI-MS for Lipid Analysis, *INFORM 9:*986–996 (1998).
27. Byrdwell, W.C., Atmospheric Pressure Chemical Ionization Mass Spectrometry for Analysis of Lipids, *Lipids 35:*327–346 (2001).
28. Rozenberg, R., N.L. Ruibal-Mendieta, G. Petitjean, P. Cani, D.L. Delacroix, N.M. Delzenne, M. Meurens, J. Quetin-Leclercq, and J.-L. Habib-Jiwan, Phytosterol Analysis and Characterization in Spelt (*Tricitum aestivum* ssp *spelta* L.) and Wheat (*T. aestivum* L.) Lipids by LC/APCI-MS, *J. Cereal Chem. 38:*189–197 (2003).
29. McGovern, P.E., D.L. Glusker, C.W. Beck, E.D. Butrym, R.A. Moreau, A. Nunez, and E. Simpson, The King Midas Funerary Banquet, *Nature 402:*863–864 (1999).
30. Christie, W.W., Literature Survey—Atmospheric-Pressure Chemical Ionization Mass Spectrometry of Lipids, www.lipidlibrary.co.uk/lit_surv/apci_ms.html (accessed January 2006).
31. Dixon, R.W., and D.S. Peterson, Developing and Testing of a Detection Method for Liquid Chromatography Based on Aerosol Charging, *Anal. Chem. 74:*2930–2937 (2002).
32. Robb, D.B., T.R. Covey, and A.P. Bruins, Atmospheric Pressure Photoionization: An Ionization Method for Liquid Chromatography-Mass Spectrometry, *Anal. Chem. 72:*3653–3659 (2000).
33. Robb, D.B., and A.P. Bruins, Atmospheric Pressure Photoionization (APPI): A New Ionization Method for Liquid Chromatography-Mass Spectrometry, Patent No. WO0133605 (2001).
34. Raffaelli, A., and A. Saba, Atmospheric Pressure Photoionization Mass Spectrometry, *Mass Spectrom. Rev. 22:*318–331 (2003).
35. Syage, J.A., Mechanism of $[M + H]^+$ Formation in Photoionization Mass Spectrometry, *J. Am. Soc. Mass. Spectrom. 15:*1521–1533 (2004).
36. Rauha, J-P., H. Vuorela, and R. Kostiainen, Effect of Effluent on the Ionization Efficiency of Flavonoids by Ion Spray, Atmospheric Pressure Chemical Ionization, and Atmospheric Pressure Photoionization Mass Spectrometry, *J. Mass Spectrom. 36:*1269–1280 (2001).
37. Leinonen, A., T. Kuuranne, and R. Kostiainen, Liquid Chromatography/Mass Spectrometry in Anabolic Steroid Analysis—Optimization and Comparison of Three Ionization Techniques: Electrospray Ionization, Atmospheric Pressure Chemical Ionization and Atmospheric Pressure Photoionization, *J. Mass. Spectrom. 37:*693–698 (2002).
38. Keski-Hynnila, H., M. Kurkela, E. Elovaara, L. Antonio, J. Magdalou, L. Luukkanen, J. Taskinen, and R. Kostiainen, Comparison of Electrospray, Atmospheric Pressure Chem-

ical Ionization, and Atmospheric Photoionization in the Identification of Apomorphine, Dobutamine, and Entacapone Phase II Metabolites in Biological Samples, *Anal. Chem. 74:*3449–3457 (2002).

39. Schiller, J., R. Süß, J. Arnhold, B. Fuchs, J. Leßig, M. Müller, M. Petković, H. Spalteholz, O. Zschörnig, and K. Arnold, Matrix-Assisted Laser Desorption and Ionization Time-of-Flight (MALDI-TOF) Mass Spectrometry in Lipid and Phospholipids Research, *Prog. Lipid Res. 43:*449–488 (2004).
40. Moyer, S.C., and R.J. Cotter, Atmospheric Pressure MALDI, *Anal. Chem.*, pp. 469A–476A (2002).
41. Tanaka, K., H. Waki, Y. Ido, S. Akita, Y. Yoshida, and T. Yoshida, Protein and Polymer Analysis up to *m/z* 100,000 by Laser Ionization Time-of-Flight Mass Spectrometry, *Rapid Commun. Mass Spectrum 2*:151–158 (1988).
42. Ayorinde, F.O., Q.L. Keith, and L.W. Wan, Matrix-Assisted Laser Desorption/Ionization Time-of-Flight Mass Spectrometry of Cod Liver Oil and the Effect of Analyte/Matrix Concentration on Signal Intensities. *Rapid Commun. Mass Spectrom. 13:*1762–1769 (1999).
43. Zabrouskov, V., and N.R. Knowles, Lipid Metabolism During Aging of High-a-Linolenate-Phenotype Potato Tubers, *Arch. Biochem. Biophys. 402:*136–148 (2002).
44. Christie, W.W., Literature Survey—Matrix-Assisted Laser Desorption Mass Spectrometry of Lipids, www.lipidlibrary.co.uk/lit_surv/maldi.html (accessed January 2006).
45. Anonymous, Introduction to Mass Spectrometry, http://masspec.scripps.edu/information/intro/ (accessed January 2006).

3 Global Cellular Lipidome Analyses by Shotgun Lipidomics Using Multidimensional Mass Spectrometry

Xianlin Han[a,b] and Richard W. Gross[a,b,c,d]

[a]Division of Bioorganic Chemistry and Molecular Pharmacology, Departments of [b]Medicine, [c]Molecular Biology and Pharmacology, and [d]Chemistry, Washington University School of Medicine, St. Louis, MO 63110, USA

Introduction

Lipidomics is a rapidly expanding research frontier, which gains its utility from quantifying lipids directly from organic solvent extracts of biological tissues and fluids; it does so by integrating many different modern techniques including mass spectrometry (MS) and the separation sciences (1–3). For decades, lipids have been recognized as essential metabolites in cellular function. The roles of lipids in cellular function are complex and include its functions as (i) a barrier to establish appropriate chemical and electrical gradients for cellular organelles; (ii) a matrix to facilitate specific conformations and dynamics for productive protein-protein and lipid-protein interactions; (iii) a reservoir of lipid second messengers to propagate cellular signaling in cell growth, differentiation, death, and response to stimuli; and (iv) a cellular energy depot to supply energy for multiple different cellular functions. The recent emergence of lipidomics, of course, is not due to any changes in the long-standing important roles that lipids play in cellular processes, but rather to modern technological advances in MS coupled with the recent recognition of the role of lipids in many epidemic diseases in industrialized societies, including obesity, atherosclerosis, stroke, hypertension, and diabetes. These disorders are collectively referred to as the "metabolic syndrome" (4). These lipid-related diseases are taking a huge toll in human pain, suffering, and productivity in modern society. Therefore, one long-term goal of lipidomics is to reveal the biochemical mechanisms underlying these diseases, to discover novel biomarkers for the early diagnosis of these diseases, to assist in the discovery of new drugs, and to evaluate drug efficacy.

The first essential step in lipidomics is determination of a total lipid profile (i.e., lipidome) because a cellular lipidome is the metabolic signature of the cell's hormonal, environmental, and nutritional history. Additionally, determination of alterations in the lipidome also provides knowledge about the biophysical state of cellular membranes, differences in lipid pools and turnover rates, changes in cellular energy supply, and the levels of lipid second messengers. Due to the recent development of various soft ionization techniques such as electrospray ionization (ESI), matrix-assisted laser desorption/ionization (MALDI), and atmospheric pressure chemical ionization (APCI), MS has emerged as the preferred physical method for the profiling of the cellular lipidomes [see (1,3,5–11) for recent reviews].

Among these mass spectrometric techniques, ESI-MS is the most prominent and has enjoyed the most success. The advantages of ESI-MS are as follows. First, a complete quantitative analysis of lipid classes, subclasses, and individual molecular species with high efficiency without prior chromatographic separation is feasible. Second, a higher signal-to-noise ratio is present in ESI-MS compared with other traditional MS approaches. Third, ionization efficiency or instrument response factors of individual molecular species in a polar lipid class depends only on the electrical properties of the lipid class and is within experimental error when mass measurement is performed at a low lipid concentration (again with lipids possessing large dipole moments). Fourth, there is a nearly linear relation between the relative intensities of molecular ions and the mass of individual lipids over a wide dynamic range in a low concentration regime for lipids with large dipole moments. In lipids without large dipoles, correction factors and multidimensional methods have been employed with great success. Finally, the reproducibility of sample measurements is excellent (<5% of experimental error). Thus, it is evident that ESI-MS and ESI-MS-MS have become essential tools for measuring cellular lipidomes during cellular perturbations and in disease states [see (1,3,5–9) for recent reviews].

Cellular lipidomes are highly complex and variable at three levels of initial discrimination. First, cellular lipids are quite different among different species, cell types, cellular organelles, membranes, and membrane microdomains (e.g., caveola and/or rafts). Second, the lipidome of each cell type is comprised of different mole percentages (mol%) of specific lipid classes, subclasses, and molecular species that are comprised of different lengths, degrees of unsaturation, different locations of double bonds, and potential branching in aliphatic chains. Finally, the cellular lipidome is dynamic, depending on nutritional conditions, hormonal influences, health status, exercise levels, and many other factors. Thus, tens of thousands of possible lipid molecular species are predictably present in the cellular lipidome at attomole to nanomole concentrations of lipids per milligram of protein. To determine the complexities inherent in each lipidome, multiple chromatographic techniques [e.g., HPLC and/or thin-layer chromatography (TLC)] in combination with different detection methods [e.g., MS, tandem MS, MS^n, Fourier transform (FT)-MS] are employed to resolve different lipid classes and individual molecular constituents in each cell type, fluid, or domain of interest (12–15).

However, it must be recognized that several potential issues must be addressed when significant on-column dwell times are employed and resolution of multiple different component species is incomplete. First, HPLC on silica-based matrices is accompanied by intrapreparative silica-catalyzed hydrolysis of vinyl ether linkages. Second, silica-based surfaces serve as solid phase supports to catalyze α-hydroxy migration (e.g., 1,2-diacylglycerol to 1,3-diacylglycerol, or 2-acyl lysolipids to 1-

acyl lysolipids) (16). Third, appropriate methods for quantitation must be employed because internal standards and each individual molecular species elute from the HPLC matrix with distinct retention times and peak shapes (e.g., differential peak trailing from heterogeneous interactions with the stationary phase) (17–19). In addition, the analysis of individual molecular species by multiple chromatographic steps is time consuming and labor intensive, and errors are propagated from multiple steps and transfers. Therefore, such time-consuming and error-prone procedures are not suitable for the requirements of large-scale studies of lipids (i.e., high-throughput lipidomics). We wish to specifically point out that although expensive and time-consuming chromatographic separations can be avoided through appropriate use of shotgun lipidomics by analysis directly from extracts of biological samples, chromatography or other enrichment approaches are necessary for successful elucidation of the extremely low abundance regime of the lipidome due to sensitivity considerations alone.

Building on the pioneering work of Fenn and colleagues (20), in the late fall of 1991, we and others began employing ESI-MS for phospholipid analysis (21–25). Due to the complex nature of chromatographic techniques as stated above, various methodologies based on direct infusion of lipid extracts without prechromatographic separation were developed and employed by our group and many of our colleagues [e.g., (1,14,21,26–39)]. One successful approach, "shotgun lipidomics," exploits the synergy between the proximal separation of lipid classes in the ion source (i.e., intrasource separation) and subsequent multidimensional MS [see (1,3,9) for recent reviews].

In this chapter, the principles, strategies, and applications of shotgun lipidomics are discussed. Although lipidomics is still in an early stage of development compared with genomics and proteomics, the technology available in lipidomics is already able to provide many new insights into the biological mechanisms of multiple disease states. The development of lipidomics will lead to a new level of understanding in lipid-related diseases through the use of MS to gain fundamental insights into the biological functions of cellular lipids through a systems biology analysis of disease-related alterations in the lipidome.

Intrasource Separation of Lipid Classes in Shotgun Lipidomics

We recognized during our initial lipid analysis (21) that physical processes present in the electrospray ion source using ESI-MS could be utilized to selectively separate differently charged moieties and generate discrete class selective ions under the influence of a high electrical potential (typically ~4 kV). As shown in Figure 1, in the positive-ion mode, an electrospray ion source selectively ionizes cations, whereas the anions are selectively left at the end of the spray capillary (they end up disposed of as waste after oxidation/reduction reactions at the tip of the spray capillary or in the source) if there are both positively and negatively charged moieties present

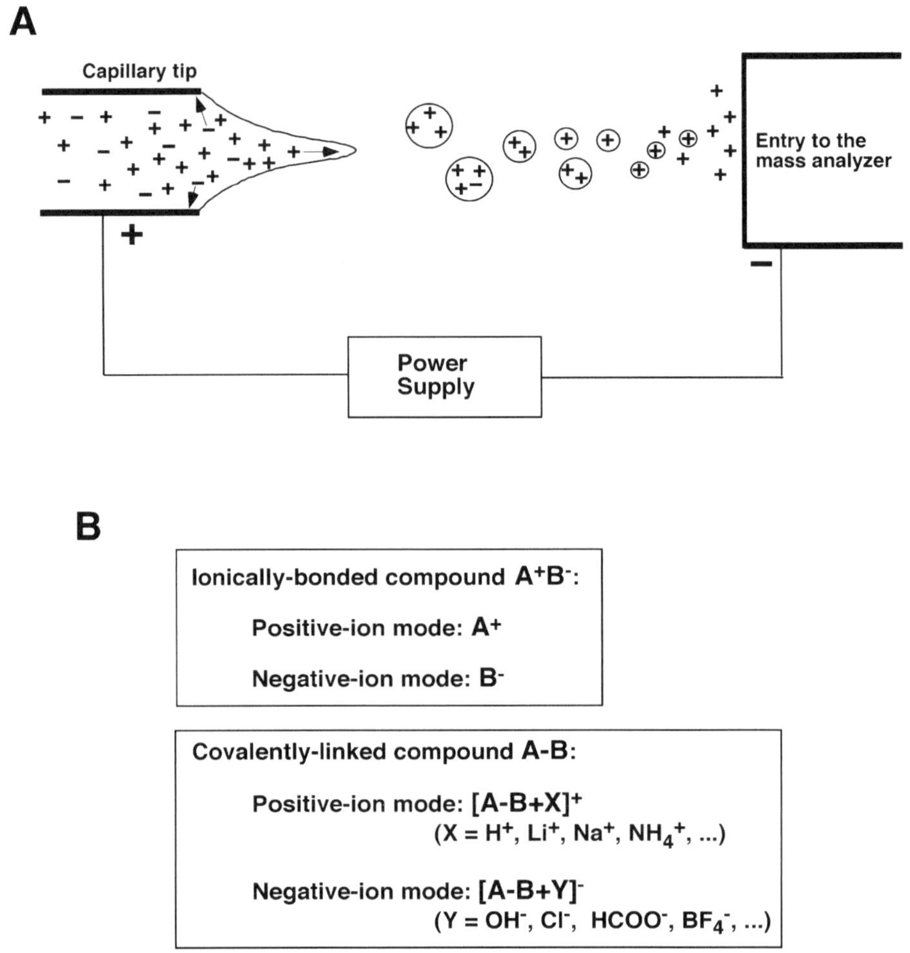

Fig. 1. The principle of electrospray ionization. (A) A schematic diagram of the principle of electrospray ionization in positive-ion mode. (B) The relation of ion formation with the electrical properties of an analyte. The adduct ion X or Y for a covalently linked polar compound depends on the availability of small cation or anion in the solution and the affinity of the adduct ion with the polar compound.

in the infused solution. Similarly, in the negative-ion mode, the ion source selectively ionizes anions and selectively removes the cationic moieties to waste. However, even if the analytes in the infused solution do not carry separatable charges, these compounds can interact with small cation(s) (e.g., H^+, Li^+, Na^+, NH_4^+, K^+) or anion(s) (e.g., OH^-, Cl^-, formate, acetate) (whatever is available in the matrix) to yield adduct ions in the presence of high positive or high negative electric fields, respectively. The ionization efficiencies of these electrically neutral analytes depend on the inherent dipoles of the compounds, the concentration of the small matrix ions, the affinity of the small ions toward the analytes, and the resultant electrochemical properties of the adducts.

On the other hand, although there are tens of thousands of individual lipid molecular species present in the cellular lipidome, these species naturally belong to a much smaller number of lipid classes related by a common head group. Different lipid classes possess different electrical properties, largely depending on the nature of the polar head groups. Based upon their electrical properties, however, one can generally classify lipid classes into three main categories (3). The lipid classes in the first category are those carrying at least one net negative charge under weakly acidic pH conditions (i.e., near pH 5) and are therefore called anionic lipids. Lipid classes in this category include cardiolipin, phosphatidylglycerol, phosphatidylinositol and its polyphosphate derivatives, phosphatidylserine, phosphatidic acid, sulfatide, acyl-CoA, and anionic lysophospholipids. The lipid classes in the second category are those that are electrically neutral under weakly acidic pH conditions, but become negatively charged under alkaline pH conditions. Therefore, they are referred to as weakly anionic lipids. Lipid classes in this category include ethanolamine glycerophospholipid (PE), lysoPE, nonesterified fatty acids and their derivatives, bile acids, and ceramide. The remaining lipid classes belong to the third category, which includes choline glycerophospholipid (PC), lysoPC, sphingomyelin, cerebroside, acylcarnitine, diacylglycerol, triacylglycerol, and cholesterol and its esters. This category of lipid classes is referred to as electrically neutral but polar or polarizable.

Given the physical factors affecting the process of selective ionization of analytes in the electrospray ion source and the different electrical properties of each of the lipid classes, we recognized that the electrospray ion source can be used to resolve lipid classes in a crude lipid extract into different categories based on the intrinsic electrical properties of each lipid class. Such a separation is analogous to the use of an ion-exchange column to separate individual lipid classes [as was employed previously (17)] before mass spectrometry in the initial mass spectrometric investigations (12,13). However, compared with ion-exchange chromatography, intrasource separation is rapid, direct, and reproducible and avoids artifacts inherent in chromatography-based systems. This methodology for lipid class separation is now referred to as intrasource separation of lipids (3,9,38).

A practical strategy for separation of these categories of lipids based on their differential intrinsic electrical properties was developed (1,14) and is illustrated in Figure 2. Specifically, the first category of lipids (i.e., anionic lipids) can be selectively ionized directly in the negative-ion mode and analyzed from diluted lipid extracts by negative-ion ESI-MS. Next, we make the diluted lipid extract solution (which is used in last step) mildly basic by the addition of a small amount of LiOH (or other suitable base); then, the second category of lipids (i.e., weakly anionic lipids) can be analyzed by negative-ion ESI-MS. Most of the other lipid classes

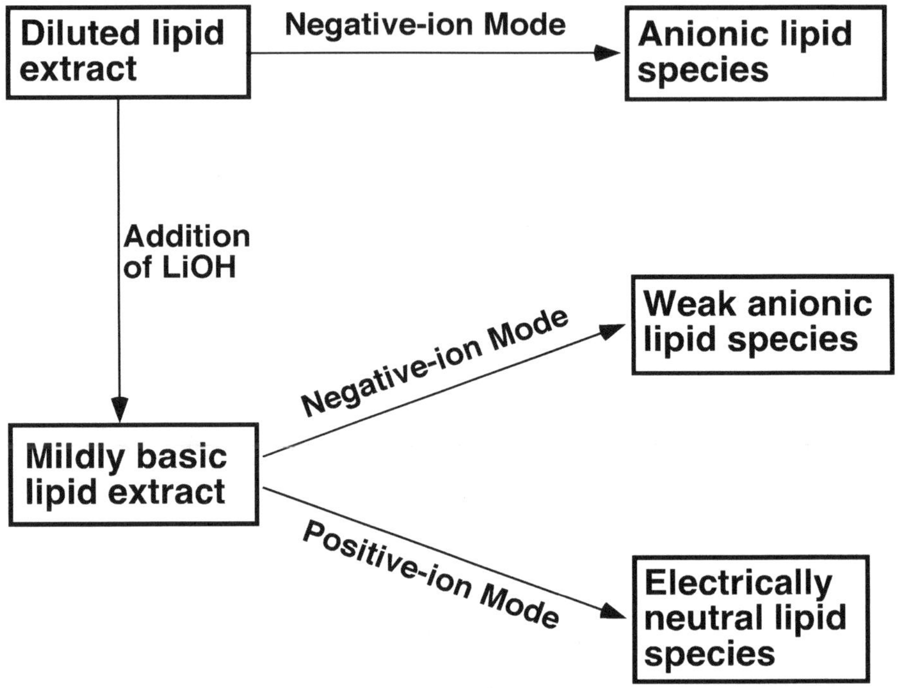

Fig. 2. Schematic diagram of shotgun lipidomics based on intrasource separation directly from a crude extract of a biological sample.

except categories 1 and 2 belong to the third category, which can be analyzed directly from the mildly alkalinized diluted lipid extract (i.e., the solution used in the last step) in positive-ion ESI-MS because lipids in the first and second categories are now anionic under these conditions. Through this approach, a comprehensive series of mass spectra with respect to each of the aforementioned conditions can be obtained for each category of lipids. For example, Figure 3 shows the three corresponding mass spectra from a typical lipid extract of mouse myocardium.

We would like to point out that the abundant PE pseudomolecular ions in the mass spectrum acquired in the negative-ion mode after addition of LiOH largely reflect the higher mass content of PE molecular species compared with individual anionic lipids in the lipid extract because all anionic phospholipid molecular species including PE species possess similar ionization efficiencies under the mildly basic experimental conditions. Therefore, one must consider whether anionic lipids in the extract could overlap with PE molecular species and thus affect their identification and quantitation. In fact, because the mass abundance of PE molecular species is much higher than that of anionic lipid molecular species in most of biological samples (40), and significant overlaps are not present in most cases, the effects of these

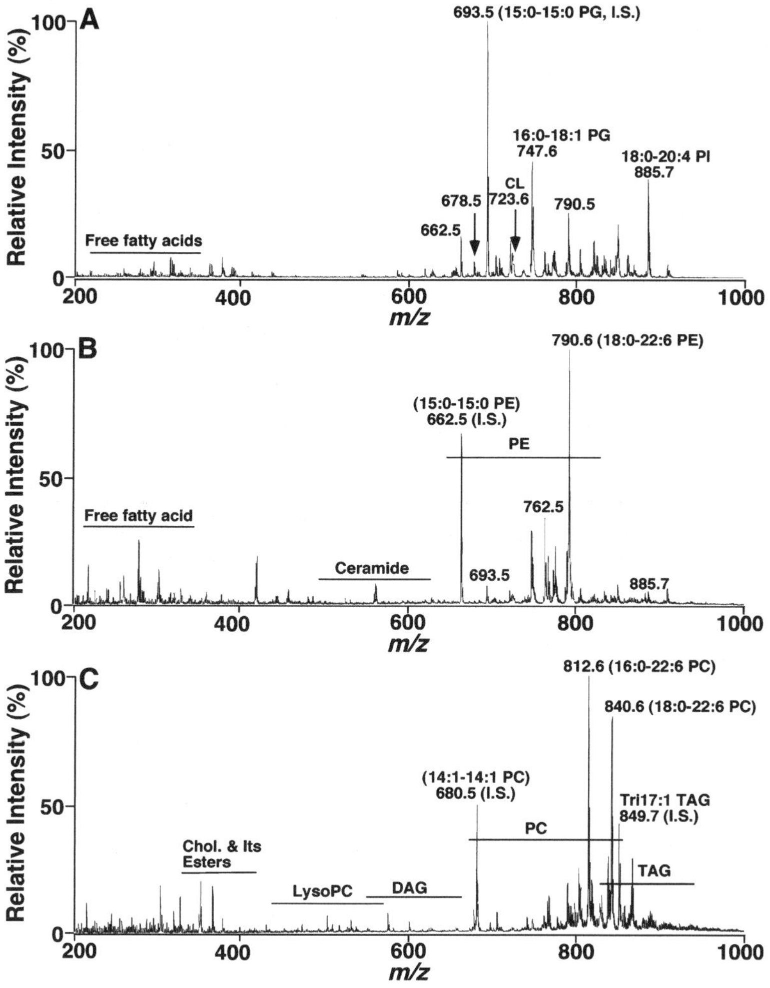

Fig. 3. An example of electrospray ionization (ESI) mass spectra of lipid classes resolved by intrasource separation from a crude lipid extract of a mouse myocardial homogenate. Mouse myocardial lipid extracts were prepared as described previously (59). ESI mass spectra were acquired in negative-ion mode after dilution to a total lipid concentration of ~100 pmol/mL with 1:1 chloroform:methanol (vol/vol) (panel A), acquired in the negative-ion mode from the diluted lipid solution after the addition of 50 nmol LiOH/mg protein (panel B), and acquired in the positive-ion mode from the diluted lipid solution after the addition of 50 nmol LiOH/mg protein (panel C). "I.S." denotes internal standard; "CL" represents doubly charged cardiolipin. All mass spectral traces are displayed after normalization to the base peak in each individual spectrum.

anionic species on the identification and quantitation of PE molecular species is generally negligible. For example, the base peak at *m/z* 693.4 (i.e., 15:0–15:0 phosphatidylglycerol, an internal standard for anionic lipid analysis) in Figure 3A is only ~10% of the intensity of the ion peak at *m/z* 662.5 (i.e., 15:0–15:0 PE, the internal standard for PE analysis) in Figure 3B where a 10:1 molar ratio of PE and phosphatidylglycerol internal standards was used. Thus, the ratio of these peaks largely reflects the selectivity of intrasource separation. Typically, these ratios are utilized to verify the adequacy of intrasource separation as a routine quality control procedure. In addition, molecular species of phosphatidylserine largely form doubly charged molecular ions under the stated alkaline conditions. Thus, the effects of these molecular species on PE analysis are minimal. Moreover, due to the naturally occurring differences in individual PE molecular species and anionic lipid species, only a few of the anionic lipid molecular species in theory, or in practice, actually overlap with PE species, further minimizing errors from this approach (Figs. 3A and 3B). Finally, due to the presence of a different number of nitrogen atoms in PE and most of the classes of anionic lipids, the nitrogen rule (41) can be utilized to identify and deconvolute the partially overlapped anionic lipid species from PE molecular species (if present) by applying correction factors for ^{13}C isotopomer intensities as previously described (3,42,43). In addition, whatever residual small overlaps might be present can be ascertained easily and accounted for after multidimensional analysis.

It should be noted that there are also the chloride adducts of other polar lipids [e.g., PC species (Fig. 3A) and cerebrosides (39)] or other anion adducts (e.g., formate and/or acetate if they are present in the infused solution) in the mass spectrum acquired under condition 1. The abundance of these anion adducts is generally low (after back extraction and dilution); thus, the presence of these adducts does not affect the identification and quantitation of anionic lipids by two-dimensional (2D) MS as discussed in the next sections. In fact, these anionic adducts are very useful for the identification and refinement of the associated polar lipid classes such as PC (38) and cerebrosides (39). In the case of analysis of the second category of lipids, due to both the high mass content of PE species and higher ionization efficiency compared with the anionic adduct formation, the effects of anionic adducts on the identification and quantitation of PE molecular species are truly negligible. For example, ion peaks at *m/z* 708.6 and 690.6, corresponding to the chloride and hydroxy adducts of 15:0–15:0 PtdCho (an internal standard for the quantitation of PC molecular species, which is very abundant in the lipid extract as demonstrated in Fig. 3C) in Figs. 3A and 3B, respectively, are nearly absent or minimal. Moreover, due to the difference of mass distribution between PE molecular species and PC molecular species adducts; it is fortuitous that only a few of the PC molecular species in practice actually overlap with PE species if such PC adducts are present. Therefore, the effects of these adduct ions on PE analyses can be neglected with errors of only 2–3% in PE quantitation in extreme cases.

Finally, we would also like to point out that each ion peak in each of these mass spectra represents at least one and very often more than one lipid molecular species.

Each pseudomolecular ion peak in each mass spectrum may contain nominal isobaric species resulting either from members of the same lipid class or from other class(es) in the category. Although product ion ESI-MS analyses can be performed individually to identify the molecular species underneath each ion peak at this stage, it is labor intensive and time consuming. More effective deconvolution of isobaric species can be accomplished through the use of multidimensional MS with appropriate array analysis as described below.

Identification of Individual Molecular Species of Lipids by 2D MS

Although there are tens of thousands of potential lipid molecular species present in a cellular lipidome, we recognized that most cellular lipids are multiple discrete covalent assemblies of a backbone (e.g., glycerol or sphingosine) with linear combinations of various aliphatic chains (typically 14–22 carbons long containing variable degrees of unsaturation) with (or without) a wide variety of polar head groups. For example, molecular species of lipid classes such as monoacylglycerol (MAG), diacylglycerol (DAG), triacylglycerol (TAG), glycolipid classes, and phospholipid classes all can be expressed by one general structure with three building blocks linked to a glycerol backbone as shown in Figure 4A. In the structure, building blocks I and II can be either a hydrogen or an acyl chain or an aliphatic chain linked by ether or vinyl ether, whereas building block III can vary from a hydrogen or acyl chain in MAG and DAG to an acyl chain in TAG to various sugar ring(s) and their derivatives in glycolipids, in addition to phosphoesters (e.g., phosphocholine, phosphoethanolamine, phosphoglycerol, phosphoserine, and phosphoinositol) in phospholipids. Sphingosine (Fig. 4B) is the most prominent backbone for sphingolipid molecular species. Thus, all of the sphingolipid molecular species containing this backbone can be written by a general structure with two building blocks linked to the sphingosine backbone as shown in Figure 4B. Building block I is either a hydrogen or an acyl chain, whereas building block II is comprised of the various head groups of sphingolipid classes as shown in the box of Figure 4B. The mass content of these two general structures represented plus that of cholesterol and its esters (which can be determined readily by an assay kit) accounts for >95% of the mass in a cellular lipidome in most cases. Collectively, these features make lipidome analysis ideally suited for multidimensional computer array analysis.

Therefore, if one could effectively and unambiguously identify the building blocks presented in each pseudomolecular ion, the complexities in the lipidome could be deconvoluted and readily solved. We recognized that the techniques of neutral loss and precursor-ion scanning could be used to characterize at least one of these building blocks in each individual scanning (Figure 5). Therefore, if one would perform all of the neutral loss and/or precursor-ion scans that characterize all of the potential building blocks of the lipids that are displayed in a mass spectrum acquired under a specific condition, and analyze all potential neutral loss and/or pre-

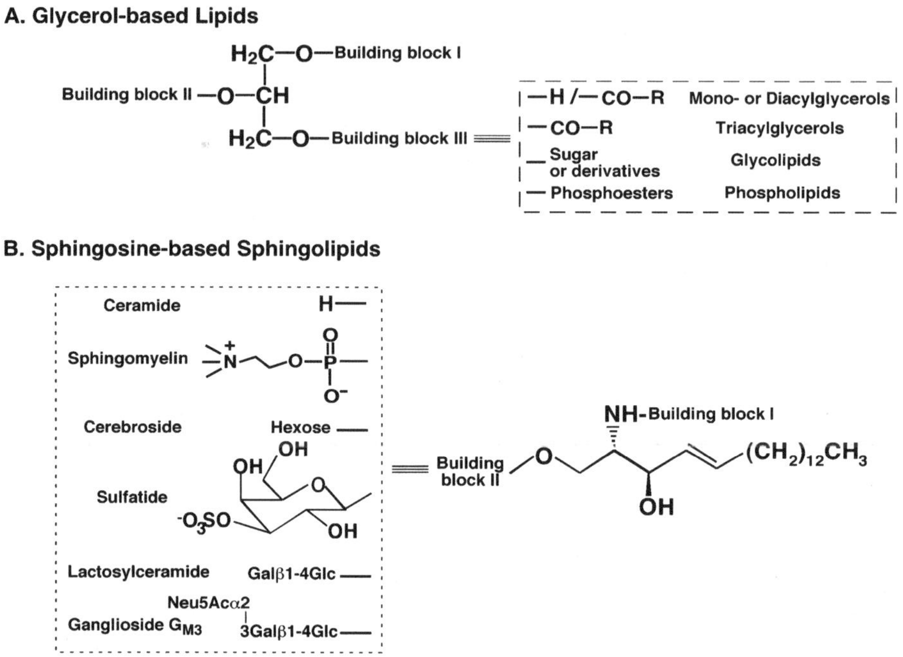

Fig. 4. General structures of glycerol-based and sphingosine-based lipids. (A) General structure of glycerol-based lipids in which three building blocks are attached to the hydroxyl groups of a glycerol backbone. Potential candidates of the building block III and the corresponding lipid classes are listed to the right of the panel. (B) General structure of sphingosine-based sphingolipids. Building block I is a hydrogen or an acyl amide, and building block II is the head group of a sphingolipid class as listed to the left of the panel.

cursor-ion scans from each building block together with the primary ion mass spectrum in an arrayed format, each (pseudo)molecular ion could be identified. Figure 6 shows such an array of these mass spectra acquired from the analysis of mouse myocardial anionic phospholipids. Following this concept, a new technique, called 2D or multidimensional MS, in which other experimental conditions such as ionization conditions (source temperature and spray voltage), fragmentation conditions (collision gas pressure, collision energy, or collision gas), or other modification are included as additional dimensions, was developed recently (1,3,9,38,42).

When 2D MS is used to identify lipid molecular species, the first dimension is comprised of the primary (molecular or pseudomolecular) ions in the *x*-axis of *m/z*, whereas the second dimension is comprised of the individual building blocks (i.e., polar head groups and/or aliphatic chains) of lipids (characterized by either neutral loss scanning or precursor-ion scanning or both) in an axis of mass (in the case of

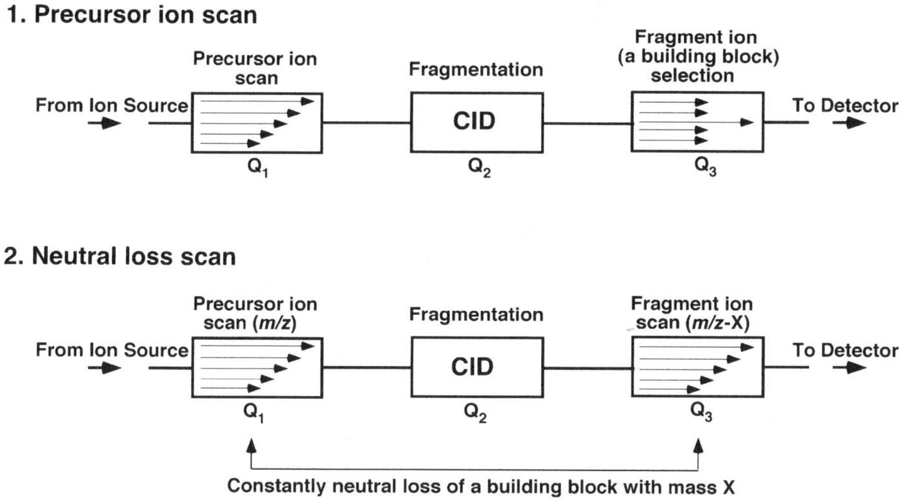

Fig. 5. Schematic diagram of precursor-ion scanning and neutral loss scanning by a triple-quadrupole type instrument used to identify the building blocks of lipids. (A) A schematic diagram of precursor-ion scanning in which one of the building blocks at a time is selectively monitored by the third quadrupole while the first quadrupole is scanned in the mass/charge range of interest. (B) A schematic diagram of neutral loss scanning in which a neutral loss of a mass (corresponding to a characterized building block of interest) between the first and third quadrupoles is selectively determined while the first quadrupole is scanned in the mass/charge range of interest.

neutral loss scanning) or *m/z* (in the case of precursor-ion scanning) (Fig. 6). The 2D mass spectrum exploits array analysis techniques, integrating both the primary ion mass spectrum and associated neutral loss/precursor-ion spectra to determine the molecular composition and amount of a lipid constituent from a single automated platform. This series of arrayed spectra is entirely analogous to 2D nuclear magnetic resonance (NMR) spectroscopy in which the axes are comprised of distinct frequency domains.

Although a 2D mass spectrum includes a collection of tandem mass spectra from neutral loss and/or precursor-ion scanning of lipid molecular ions, 2D MS analysis is totally different from tandem MS analysis. One feature of a 2D mass spectrum is that each imaginary mass spectrum along a vertical line through each *m/z* of the primary ion (see the broken lines in Fig. 6) represents a pseudo-product ion mass spectrum of a precursor ion at the primary ion mass spectrum (first dimension) crossed with the broken line. Therefore, many of the features present in product-ion analysis can be achieved from the 2D MS analysis. For example, regiospecific identification of each individual molecular species (44) and quantitative analysis of isobaric species are two important features of product-ion analyses

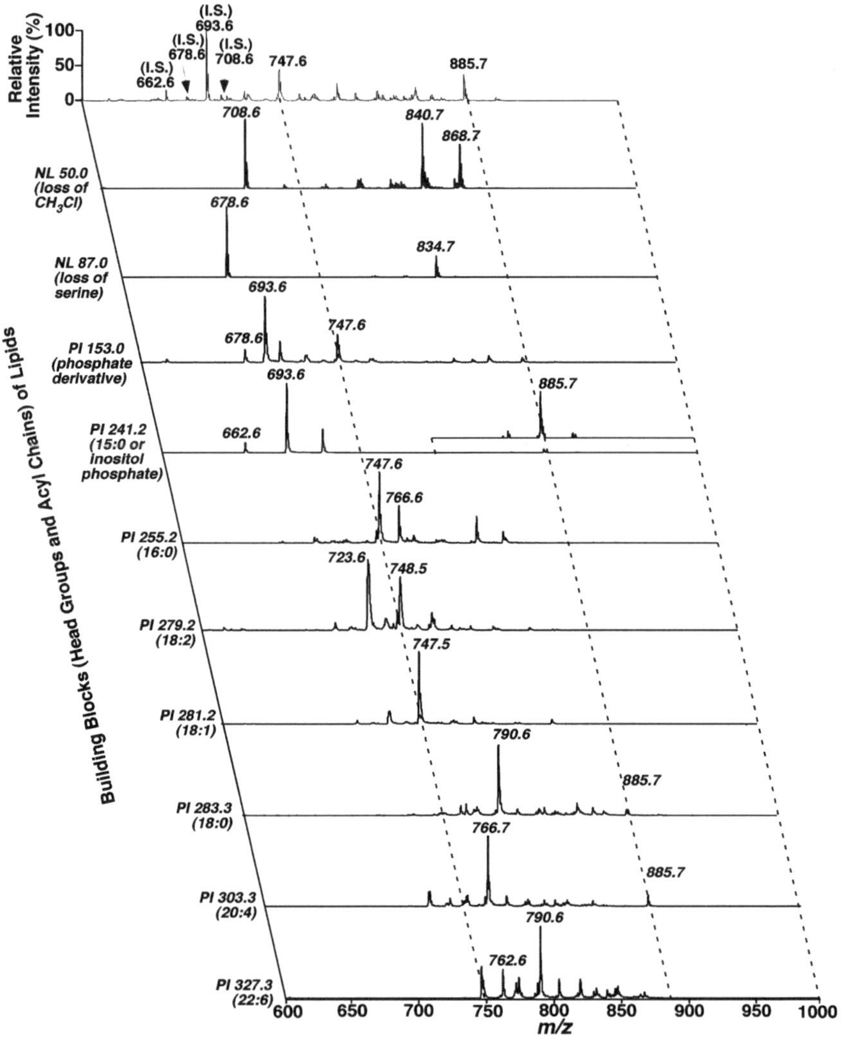

Fig. 6. An example of a two-dimensional electrospray ionization (ESI) mass spectrum of a mouse heart chloroform extract in the negative-ion mode. A conventional ESI mass spectrum was acquired in the negative-ion mode directly from a diluted myocardial lipid extract (i.e., panel A of Figure 2) before analysis of lipid building blocks in the second dimension by precursor-ion (PI) scanning and neutral loss (NL) scanning as indicated. Each mass spectral scan was acquired as described previously (38). "I.S." denotes internal standard; (m:n) indicates an acyl chain containing m carbons and n double bonds. All mass spectral traces were displayed after normalization to the base peak in each individual spectrum.

(among others). The former is identified by comparison of the cross-peak intensities of ions derived from the acyl chain building blocks, whereas the latter is achieved by deconvolution of all cross peaks corresponding to acyl chain building blocks in 2D MS (38,45). Moreover, 2D analysis offers much more information from an arrayed format of both product ions and precursor-ion scans with single n-dimensional representations. Another very important feature of a 2D mass spectrum is the dramatic increase of dynamic range relative to a selected internal standard (see next section). Therefore, quantitation and refinement of low-abundance molecular species with a selected internal standard for each lipid class or even subclass can also be readily achieved by 2D MS analyses (38,39). Most importantly, 2D MS analysis of lipids can be automated, thus representing a high throughput platform for the global analyses of the cellular lipidomes.

Quantitation of Lipid Molecular Species by 2D MS

Generally, quantitation of any compound has to be made by comparison to either an internal or external standard similar to the compound of interest. The former is added during the sample preparation and analyzed at the same time as the sample is analyzed. The latter is analyzed separately but under identical conditions with the sample of interest. A calibration curve is generally established by using the external standard. Both techniques have some advantages and disadvantages. The former is known for its simplicity and accuracy if the internal standard is within a linear dynamic range of the measurement with the sample. However, selection of the internal standard may be very difficult and the dynamic range(s) of the measurement must be predetermined. Utilizing the latter, it is hard to control the measurements being conducted under identical conditions, particularly when multiple steps of sample preparation, separation, and quantitation are involved. Global analyses of the cellular lipidome are such complicated processes that a method using external standards alone is thereby excluded. Therefore, employing internal standards, and related groups of standards, is the best way to quantitate the complex cellular lipidomes.

What kinds of compounds can be used as internal standards for quantitation of cellular lipidomes by MS? Ideally, quantitation of any compound by MS can be accurately made only by comparison of its peak intensity with that of a stable-isotope-incorporated and chemically identical internal standard within a linear dynamic range. However, it is not feasible to use thousands of internal standards for the quantitative analyses of a complex lipidome although quantitation of a few known lipids can be achieved by employing such a method (46). Fortunately, we found and verified experimentally that the electrospray ionization efficiency of lipid molecular species depends predominantly on the electrical properties of the lipid polar head groups in the concentration range of pmol/μL or lower after correction for different ^{13}C isotopomer distributions (3,9,14,21,47). These principles were verified independently by others (19,48,49). This finding laid the foundation for quantitation of a class of lipids that possess an identical polar head group by using a molecular species in the class with reasonable accuracy (~5%).

However, we specifically emphasize that identical ionization efficiency of lipid molecular species in a class is valid only in the low concentration regime; this is not due to the limitation of the linear dynamic range of concentration but to lipid aggregation with some solvents used for lipid analysis by ESI-MS by some investigators in the higher concentration regimes (>50 pmol/μL) with unfavorable solvents. Lipids, unlike other analytes, are unique in terms of their high hydrophobicity. When concentrations of lipids increase, they tend to aggregate to form micelles, even in some organic solvents. The longer the chain length and the higher the degrees of saturation of a lipid species, the lower the critical micellar concentration of the compound. Therefore, molecular species containing short and/or polyunsaturated acyl chains might show higher apparent response factors than those containing long and/or saturated acyl chains at a high lipid concentration (48,50). The maximal concentrations of lipids at which lipid-lipid interactions are small obviously depend on the solvent components used in the infusion solution. For example, we found the maximal concentration of lipids at which lipid aggregation is small is ~1, 10, and 100 pmol/μL in 1:2, 1:1, and 2:1 of chloroform:methanol (vol/vol), respectively. Any solvent system containing water, acetonitrile, or a high percentage of methanol is not favored and should be avoided if possible for lipid analysis by ESI-MS.

The requirement of a linear dynamic range using an internal standard must be classified further because there exist many different measures of dynamic range. One is the dynamic range of concentration in which the quantitative technique is linear. This is the most commonly accepted meaning of the concept of dynamic range in the literature. For the mass spectrometric analysis of lipids, this dynamic range defines the relation between ion counts of a species and the concentration of this species. This linear dynamic range is over 1000-fold in the low concentration regime and has been confirmed by multiple studies (19,21,25,26,48). Another measure of dynamic range is the relative ratio of an internal standard vs. the individual molecular species of interest. Due to the presence of background noise (e.g., chemical noise) and baseline drift (i.e., instrumental stability) in some cases, only an ~100-fold dynamic range (from 0.1 to 10 of the ratio) of this measure can be obtained (39). However, with the help of tandem MS in a 2D MS format, a 1000-fold dynamic range can be achieved through two-stage processing as long as the concentration measures of dynamic range are linear over 1000-fold (38,39). First, the abundant molecular species in a class are quantitated by comparison with a preselected internal standard for the lipid class in the first-dimensional (pseudomolecular ion, primary ion) mass spectrum. Next, these values are used as endogenous standards for ratiometric comparisons to quantitate or refine the mass content of low-abundance individual molecular species from a suitable tandem mass spectrum. By employing this two-step processing, we find that a 1000-fold dynamic range can be readily achieved in almost all cases because background noise is dramatically reduced and various intensity peaks of the same class with highly similar fragmentation kinetics can be found in the primary ion spectra to serve as ratiometric markers for the quantitation of low-abundance molecular species.

A set of endogenous internal standards plus the original exogenous internal standard are generally well distributed in biological samples in terms of different aliphatic chain lengths and degrees of unsaturation. Therefore, these endogenous standards represent better standards than human-selected internal standards for lipid quantitation by tandem MS (29,34,51) in which the overlap of added internal standard ions with endogenous molecular ions must be considered, thereby limiting the candidates that can be selected for exogenous internal standards. One weakness present in 2D MS analysis of lipids to quantitate and/or refine low-abundance molecular species is that the endogenous set of standards is secondary to the original internal standard; thus, the experimental errors of the mass content of these low-abundance molecular species are amplified. However, the total mass content of these low-abundance molecular species typically accounts only for <5 mol% of the entire mass of the class. Therefore, the amplified experimental error for the mass content of these low-abundance species will not substantially affect the accuracy of quantitation for the entire class of lipids, and relative comparisons among these species are quite accurate if approximate endogenous standards are used.

Applications of Shotgun Lipidomics in Biological Studies

Although the principle of global lipidome analysis using intrasource separation has been developed for >10 years (21), shotgun lipidomics using intrasource separation and multidimensional MS has only begun to be more generally appreciated in the last few years. However, the power of this technology is evidenced by its broad applications and the results revealed from studies using this technology [see (1,9,47,52,53) for recent reviews]. At its current state of development, shotgun lipidomics using intrasource separation and multidimensional MS can analyze most of the major lipid classes and many of the minor lipid classes of mammalian cellular lipidomes, hundreds to thousands of lipid molecular species, and >95% of the mass content (including cholesterol mass content, which is readily determined by an assay kit) of a cellular lipidome. The lipid classes in mammalian samples that can be quantitated by shotgun lipidomics include, but are not limited to, choline glycerophospholipid, ethanolamine glycerophospholipid, cardiolipin, phosphatidylglycerol, phosphatidylinositol, phosphatidylserine, phosphatidic acid, sphingomyelin, galactocerebroside, glucocerebroside, sulfatide, triacylglycerol, ceramide, lysoPC, acylcarnitine, acyl-CoA, diacylglycerol, and nonesterified fatty acid.

The first study of shotgun lipidomics by 2D ESI-MS was the quantitation and fingerprinting of mouse myocardial TAG molecular species (42). Because TAG molecular species are comprised of three identical/different acyl chains, which are the building blocks of the TAG species, the 2D mass spectrum for TAG analysis can be constructed by neutral loss scanning of all naturally occurring fatty acids from lithiated TAG molecular ions. From 2D MS analysis, individual isobaric TAG molecular species (which are abundant in TAG in lipid extracts of biological samples) can be readily identified and quantitated. Unlike in polar lipid classes, the polar head group is absent in TAG molecular species. Therefore, ionization efficiency of individual TAG molecular species depends greatly on the acyl chain lengths and the degree of unsaturation in the species. Through establishment of an algorithm that correlates ionization efficiency with acyl chain physical properties (42), quantitative analysis of TAG molecular species can be achieved by shotgun lipidomics using 2D MS [see (1,3,47) for recent reviews]. To date, this approach represents the most sensitive, accurate, and efficient technique for the quantitation of individual TAG molecular species. It was applied extensively in biological, pathological, and pathophysiological studies in the last three years [e.g., (38,45,54–58)].

The newly reported 2D MS analysis of cerebrosides represents another interesting example of lipid analysis by shotgun lipidomics using 2D MS (39). Cerebrosides are polar but electrically neutral molecules; therefore, cerebroside molecular species can be readily analyzed in the positive-ion mode after addition of a small amount of LiOH as discussed above. However, due to their large affinity for chloride, cerebroside molecular species can also be ionized as chlorine adducts in the negative-ion mode without the addition of LiOH. Although the ionization sensitivity of cerebroside is not comparable to that of anionic lipids, the abundant mass content of cerebroside in some biological samples (e.g., brain) allows cerebroside ion abundance to be easily measured in brain extracts. Therefore, the major molecular species of cerebroside can be redundantly quantitated in both positive- and negative-ion modes (39). By exploiting the differential loss of HCl in hydroxy- and nonhydroxy-cerebroside molecular species, the molecular species in these subclasses of cerebroside can be identified (39). The low-abundance molecular species of cerebroside, however, are quantitated in the spectra from neutral loss of either 162.1 or 210.1 u in the positive-ion mode using the determined major molecular species as internal standards in a 2D MS format (39).

Recently, shotgun lipidomics using 2D ESI/MS was used to study lipid storage and metabolism in hormone-induced 3T3-L1 differentiating adipocytes (45). 2D ESI MS analyses demonstrated that unbranched fatty acids containing an odd number of carbons were dramatically accumulated in all major lipid classes in the differentiated adipocytes. Specifically, PC, PE, and TAG contain 15, 23, and 33%, respectively, of molecular species containing odd chain length unbranched fatty acids. These results indicate that rapid α-oxidation of unbranched fatty acids occurs in the adipocytes. Further studies found that the double bonds in odd chain length unbranched fatty acids were located exclusively at the Δ9 position, suggesting the presence of two critical processes in fatty acid handling in adipocyte lipid storage and metabolism (45). First, the absence of Δ8 unsaturated odd chain length fatty acids indicates that α-oxidation cannot occur in monounsaturated fatty acids (e.g., oleic and palmitoleic acids). Second, α-oxidation of saturated fatty acids occurs before Δ9 desaturation.

Very recently, shotgun lipidomics using 2D MS was exploited to investigate energy mobilization during modest caloric deprivation in mice and the mobilization of lipids in this process (59). Remarkably, multiple specific changes in the murine myocardial lipidome were present after only brief periods of food deprivation (4 and 12 h). For example, PC and PE molecular species containing long and polyunsaturated acyl chains were substantially depleted in murine myocardium after 12 h of food deprivation. The lost mass accounted for a total decrease of 39 nmol/mg protein in the pools and represented ~25% of total phospholipid mass and ~20 cal of Gibbs free energy/g wet weight of tissue. Furthermore, no alterations were found in other myocardial phospholipid pools such as phosphatidylserine and phosphatidylinositol after food deprivation. TAG mass was not changed in mouse myocardium during food deprivation, but during 12 h of refeeding, myocardial TAG increased nearly threefold and returned to its basal low levels after 24 h of refeeding. In contrast to the specific changes in lipid pools in murine myocardium, no changes in phospholipid mass were found in skeletal muscle, but a dramatic decrease in skeletal muscle (or skeletal muscle associated) TAG mass was present after 12 h of fasting. These results identify phospholipids as a rapidly mobilizable energy source during modest energy restriction in mouse myocardium, whereas TAG species are the major source of energy reserves in skeletal muscle.

Summary

Shotgun lipidomics, based on intrasource separation, multidimensional MS, and computer-assisted array analysis, is an emerging powerful technology in lipidomics. Through effective intrasource separation of lipid classes based on their intrinsic electrical propensities, analyses of lipids from crude extracts of biological samples can be conducted directly and effectively. Appropriate multidimensional array analysis of lipid (pseudo)molecular ions and their fragments can lead to identification and quantitation of most individual lipid molecular species. Because most biological lipids are linear combinations of aliphatic chains, backbones, and head groups, a rich repertoire of lipid building blocks represents experimental observables that can be reconstructed by computer analysis in conjunction with their pseudomolecular ions to determine the lipid molecular structures comprising the lipidome from the tissue, cell, or fluid of interest directly from its lipid extract. Through this approach, dramatic increases in the accessible dynamic range and discrimination of isobaric molecular species can be achieved without any prior column chromatography or operator-dependent supervision. At its current state of development, shotgun lipidomics can analyze >20 lipid classes, thousands of lipid molecular species, and >95% of the mass content of a cellular lipidome. Thus, understanding the biochemical mechanisms underlying lipid-mediated disease states will be greatly facilitated by the power of shotgun lipidomics.

Acknowledgments

This work was supported by National Institutes of Health grants PO1HL57278 and RO1AG23168 as well as the Neurosciences Education and Research Foundation. The authors are grateful to Dr. Kui Yang, Ms. Hua Cheng, and Ms. Kora Fikes for their help with techniques.

References

1. Han, X., and R.W. Gross, Global Analyses of Cellular Lipidomes Directly from Crude Extracts of Biological Samples by ESI Mass Spectrometry: A Bridge to Lipidomics, *J. Lipid Res. 44*:1071–1079 (2003).
2. Lagarde, M., A. Geloen, M. Record, D. Vance, and F. Spener, Lipidomics Is Emerging [Editorial], *Biochim. Biophys. Acta 1634*:61 (2003).
3. Han, X., and R.W. Gross, Shotgun Lipidomics: Electrospray Ionization Mass Spectrometric Analysis and Quantitation of the Cellular Lipidomes Directly from Crude Extracts of Biological Samples, *Mass Spectrom. Rev. 24*:367–412 (2005).
4. Kenchaiah, S., J.C. Evans, D. Levy, P.W. Wilson, E.J. Benjamin, M.G. Larson, W.B. Kannel, and R.S. Vasan, Obesity and the Risk of Heart Failure, *N. Engl. J. Med. 347*:305–313 (2002).
5. Pulfer, M., and R.C. Murphy, Electrospray Mass Spectrometry of Phospholipids, *Mass Spectrom. Rev. 22*:332–364 (2003).
6. Griffiths, W.J., Tandem Mass Spectrometry in the Study of Fatty Acids, Bile Acids, and Steroids, *Mass Spectrom. Rev. 22*:81–152 (2003).
7. Welti, R., and X. Wang, Lipid Species Profiling: A High-Throughput Approach to Identify Lipid Compositional Changes and Determine the Function of Genes Involved in Lipid Metabolism and Signaling, *Curr. Opin. Plant Biol. 7*:337–344 (2004).
8. Ivanova, P.T., S.B. Milne, J.S. Forrester, and H.A. Brown, LIPID Arrays: New Tools in the Understanding of Membrane Dynamics and Lipid Signaling, *Mol. Interv. 4*:86–96 (2004).
9. Han, X., and R.W. Gross, Shotgun Lipidomics: Multi-Dimensional Mass Spectrometric Analysis of Cellular Lipidomes, *Expert Rev. Proteomics 2*:253–264 (2005).
10. Schiller, J., R. Suss, J. Arnhold, B. Fuchs, J. Lessig, M. Muller, M. Petkovic, H. Spalteholz, O. Zschornig, and K. Arnold, Matrix-Assisted Laser Desorption and Ionization Time-of-Flight (MALDI-TOF) Mass Spectrometry in Lipid and Phospholipid Research, *Prog. Lipid Res. 43*:449–488 (2004).
11. Byrdwell, W.C., Atmospheric Pressure Chemical Ionization Mass Spectrometry for Analysis of Lipids, *Lipids 36*:327–346 (2001).
12. Gross, R.W., High Plasmalogen and Arachidonic Acid Content of Canine Myocardial Sarcolemma: A Fast Atom Bombardment Mass Spectroscopic and Gas Chromatography-Mass Spectroscopic Characterization, *Biochemistry 23*:158–165 (1984).
13. Gross, R.W., Identification of Plasmalogen as the Major Phospholipid Constituent of Cardiac Sarcoplasmic Reticulum, *Biochemistry 24*:1662–1668 (1985).
14. Han, X., R.A. Gubitosi-Klug, B.J. Collins, and R.W. Gross, Alterations in Individual Molecular Species of Human Platelet Phospholipids During Thrombin Stimulation: Electrospray Ionization Mass Spectrometry-Facilitated Identification of the Boundary Conditions for the Magnitude and Selectivity of Thrombin-Induced Platelet Phospholipid Hydrolysis, *Biochemistry 35*:5822–5832 (1996).

15. Byrdwell, W.C., Dual Parallel Liquid Chromatography/Dual Mass Spectrometry (LC2/MS2) of Bovine Brain Total Lipid Extract, *J. Liq. Chromatogr. Relat. Technol. 26*:3147–3181 (2003).
16. Creer, M.H., and R.W. Gross, Separation of Isomeric Lysophospholipids by Reverse Phase HPLC, *Lipids 20*:922–928 (1985).
17. Gross, R.W., and B.E. Sobel, Isocratic High-Performance Liquid Chromatography Separation of Phosphoglycerides and Lysophosphoglycerides, *J. Chromatogr. 197*:79–85 (1980).
18. Taguchi, R., J. Hayakawa, Y. Takeuchi, and M. Ishida, Two-Dimensional Analysis of Phospholipids by Capillary Liquid Chromatography/Electrospray Ionization Mass Spectrometry, *J. Mass Spectrom. 35*:953–966 (2000).
19. DeLong, C.J., P.R.S. Baker, M. Samuel, Z. Cui, and M.J. Thomas, Molecular Species Composition of Rat Liver Phospholipids by ESI-MS/MS: The Effect of Chromatography, *J. Lipid Res. 42*:1959–1968 (2001).
20. Fenn, J.B., M. Mann, C.K. Meng, S.F. Wong, and C.M. Whitehouse, Electrospray Ionization for Mass Spectrometry of Large Biomolecules, *Science 246*:64–71 (1989).
21. Han, X., and R.W. Gross, Electrospray Ionization Mass Spectroscopic Analysis of Human Erythrocyte Plasma Membrane Phospholipids, *Proc. Natl. Acad. Sci. U. S. A. 91*:10635–10639 (1994).
22. Duffin, K.L., J.D. Henion, and J.J. Shieh, Electrospray and Tandem Mass Spectrometric Characterization of Acylglycerol Mixtures That Are Dissolved in Nonpolar Solvents, *Anal. Chem. 63*:1781–1788 (1991).
23. Weintraub, S.T., R.N. Pinckard, and M. Hail, Electrospray Ionization for Analysis of Platelet-Activating Factor, *Rapid Commun. Mass Spectrom. 5*:309–311 (1991).
24. Kerwin, J.L., A.R. Tuininga, and L.H. Ericsson, Identification of Molecular Species of Glycerophospholipids and Sphingomyelin Using Electrospray Mass Spectrometry, *J. Lipid Res. 35*:1102–1114 (1994).
25. Kim, H.Y., T.C. Wang, and Y.C. Ma, Liquid Chromatography/Mass Spectrometry of Phospholipids Using Electrospray Ionization, *Anal. Chem. 66*:3977–3982 (1994).
26. Lehmann, W.D., M. Koester, G. Erben, and D. Keppler, Characterization and Quantification of Rat Bile Phosphatidylcholine by Electrospray-Tandem Mass Spectrometry, *Anal. Biochem. 246*:102–110 (1997).
27. Schneiter, R., B. Brugger, R. Sandhoff, G. Zellnig, A. Leber, M. Lampl, K. Athenstaedt, C. Hrastnik, S. Eder, G. Daum, F. Paltauf, F.T. Wieland, and S.D. Kohlwein, Electrospray Ionization Tandem Mass Spectrometry (ESI-MS/MS) Analysis of the Lipid Molecular Species Composition of Yeast Subcellular Membranes Reveals Acyl Chain-Based Sorting/Remodeling of Distinct Molecular Species En Route to the Plasma Membrane, *J. Cell Biol. 146*:741–754 (1999).
28. Brugger, B., R. Sandhoff, S. Wegehingel, K. Gorgas, J. Malsam, J.B. Helms, W.D. Lehmann, W. Nickel, and F.T. Wieland, Evidence for Segregation of Sphingomyelin and Cholesterol During Formation of COPI-Coated Vesicles, *J. Cell Biol. 151*:507–518 (2000).
29. Blom, T.S., M. Koivusalo, E. Kuismanen, R. Kostiainen, P. Somerharju, and E. Ikonen, Mass Spectrometric Analysis Reveals an Increase in Plasma Membrane Polyunsaturated Phospholipid Species upon Cellular Cholesterol Loading, *Biochemistry 40*:14635–14644 (2001).
30. Duffin, K.L., M.G. Obukowicz, W.J. Salsgiver, D.J. Welsch, C. Shieh, A. Raz, and P. Needleman, Lipid Remodeling in Mouse Liver and Plasma Resulting from Δ6 Fatty Acid Desaturase Inhibition, *Lipids 36*:1203–1208 (2001).

31. Han, X., D.M. Holtzman, and D.W. McKeel, Jr., Plasmalogen Deficiency in Early Alzheimer's Disease Subjects and in Animal Models: Molecular Characterization Using Electrospray Ionization Mass Spectrometry, *J. Neurochem.* *77*:1168–1180 (2001).
32. Williams, S.D., F.F. Hsu, and D.A. Ford, Electrospray Ionization Mass Spectrometry Analyses of Nuclear Membrane Phospholipid Loss After Reperfusion of Ischemic Myocardium, *J. Lipid Res.* *41*:1585–1595 (2000).
33. Ekroos, K., I.V. Chernushevich, K. Simons, and A. Shevchenko, Quantitative Profiling of Phospholipids by Multiple Precursor Ion Scanning on a Hybrid Quadrupole Time-of-Flight Mass Spectrometer, *Anal. Chem.* *74*:941–949 (2002).
34. Welti, R., W. Li, M. Li, Y. Sang, H. Biesiada, H.-E. Zhou, C.B. Rajashekar, T.D. Williams, and X. Wang, Profiling Membrane Lipids in Plant Stress Responses. Role of Phospholipase Dα in Freezing-Induced Lipid Changes in *Arabidopsis*, *J. Biol. Chem.* *277*:31994–32002 (2002).
35. Esch, S.W., T.D. Williams, S. Biswas, A. Chakrabarty, and S.M. Levine, Sphingolipid Profile in the CNS of the Twitcher (Globoid Cell Leukodystrophy) Mouse: A Lipidomics Approach, *Cell. Mol. Biol.* *49*:779–787 (2003).
36. Jain, S., K. Jayasimhulu, and J.F. Clark, Metabolomic Analysis of Molecular Species of Phospholipids from Normotensive and Preeclamptic Human Placenta Electrospray Ionization Mass Spectrometry, *Front. Biosci.* *9*:3167–3175 (2004).
37. Mitchell, T.W., N. Turner, A.J. Hulbert, P.L. Else, J.A. Hawley, J.S. Lee, C.R. Bruce, and S.J. Blanksby, Exercise Alters the Profile of Phospholipid Molecular Species in Rat Skeletal Muscle, *J. Appl. Physiol.* *97*:1823–1829 (2004).
38. Han, X., J. Yang, H. Cheng, H. Ye, and R.W. Gross, Towards Fingerprinting Cellular Lipidomes Directly from Biological Samples by Two-Dimensional Electrospray Ionization Mass Spectrometry, *Anal. Biochem.* *330*:317–331 (2004).
39. Han, X., and H. Cheng, Characterization and Direct Quantitation of Cerebroside Molecular Species from Lipid Extracts by Shotgun Lipidomics, *J. Lipid Res.* *46*:163–175 (2005).
40. Cullis, P.R., D.B. Fenske, and M.J. Hope, Physical Properties and Functional Roles of Lipids in Membranes, in *Biochemistry of Lipids, Lipoproteins and Membranes,* edited by D.E. Vance and J. Vance, Elsevier, Amsterdam, The Netherlands, 1996, 1–33.
41. McLafferty, F.W., and F. Turecek, *Interpretation of Mass Spectra*, 4th ed., University Science Books, Sausalito, California, 1993, p. 371.
42. Han, X., and R.W. Gross, Quantitative Analysis and Molecular Species Fingerprinting of Triacylglyceride Molecular Species Directly from Lipid Extracts of Biological Samples by Electrospray Ionization Tandem Mass Spectrometry, *Anal. Biochem.* *295*:88–100 (2001).
43. Han, X., Characterization and Direct Quantitation of Ceramide Molecular Species from Lipid Extracts of Biological Samples by Electrospray Ionization Tandem Mass Spectrometry, *Anal. Biochem.* *302*:199–212 (2002).
44. Han, X., and R.W. Gross, Structural Determination of Picomole Amounts of Phospholipids Via Electrospray Ionization Tandem Mass Spectrometry, *J. Am. Soc. Mass Spectrom.* *6*:1202–1210 (1995).
45. Su, X., X. Han, J. Yang, D.J. Mancuso, J. Chen, P.E. Bickel, and R.W. Gross, Sequential Ordered Fatty Acid α Oxidation and $\Delta 9$ Desaturation Are Major Determinants of Lipid Storage and Utilization in Differentiating Adipocytes, *Biochemistry* *43*:5033–5044 (2004).
46. Harrison, K.A., K.L. Clay, and R.C. Murphy, Negative Ion Electrospray and Tandem Mass Spectrometric Analysis of Platelet Activating Factor (PAF) (1-Hexadecyl-2-acetylglycerophosphocholine), *J. Mass Spectrom.* *34*:330–335 (1999).

47. Han, X., and R.W. Gross, Specific Lipid Alterations in Alzheimer's Disease and Diabetes: Shotgun Global Cellular Lipidome Analyses by Electrospray Ionization Mass Spectrometry Using Intrasource Separation, in *Functional Lipidomics,* edited by L. Feng and G.D. Prestwich, Marcel Dekker, Inc., New York, 2005, pp. 285–306.
48. Koivusalo, M., P. Haimi, L. Heikinheimo, R. Kostiainen, and P. Somerharju, Quantitative Determination of Phospholipid Compositions by ESI-MS: Effects of Acyl Chain Length, Unsaturation, and Lipid Concentration on Instrument Response, *J. Lipid Res. 42*:663–672 (2001).
49. Hermansson, M., A. Uphoff, R. Kakela, and P. Somerharju, Automated Quantitative Analysis of Complex Lipidomes by Liquid Chromatography/Mass Spectrometry, *Anal. Chem. 77*:2166–2175 (2005).
50. Zacarias, A., D. Bolanowski, and A. Bhatnagar, Comparative Measurements of Multicomponent Phospholipid Mixtures by Electrospray Mass Spectroscopy: Relating Ion Intensity to Concentration, *Anal. Biochem. 308*:152–159 (2002).
51. Brugger, B., G. Erben, R. Sandhoff, F.T. Wieland, and W.D. Lehmann, Quantitative Analysis of Biological Membrane Lipids at the Low Picomole Level by Nano-Electrospray Ionization Tandem Mass Spectrometry, *Proc. Natl. Acad. Sci. U. S. A. 94*:2339–2344 (1997).
52. Han, X., Lipid Alterations in the Earliest Clinically Recognized Stage of Alzheimer's Disease: Implication of the Role of Lipids in the Pathogenesis of Alzheimer's Disease. *Curr. Alz. Res. 2*:65–77 (2005).
53. Gross, R.W., C.M. Jenkins, J. Yang, D.J. Mancuso, and X. Han, Functional Lipidomics: The Roles of Specialized Lipids and Lipid-Protein Interactions in Modulating Neuronal Function, *Gross 77*:52–64 (2005).
54. Finck, B.N., J.J. Lehman, T.C. Leone, M.J. Welch, M.J. Bennett, A. Kovacs, X. Han, R.W. Gross, R. Kozak, G.D. Lopaschuk, and D.P. Kelly, The Cardiac Phenotype Induced by PPARα Overexpression Mimics That Caused by Diabetes Mellitus, *J. Clin. Investig. 109*:121–130 (2002).
55. Finck, B.N., X. Han, M. Courtois, F. Aimond, J.M. Nerbonne, A. Kovacs, R.W. Gross, and D.P. Kelly, A Critical Role for PPARα-Mediated Lipotoxicity in the Pathogenesis of Diabetic Cardiomyopathy: Modulation by Dietary Fat Content, *Proc. Natl. Acad. Sci. U. S. A. 100*:1226–1231 (2003).
56. Listenberger, L.L., X. Han, S.E. Lewis, S. Cases, R.V. Farese, Jr., D.S. Ory, and J.E. Schaffer, Triglyceride Accumulation Protects Against Fatty Acid-Induced Lipotoxicity, *Proc. Natl. Acad. Sci. U. S. A.* 100:3077–3082 (2003).
57. Mancuso, D.J., D.R. Abendschein, C.M. Jenkins, X. Han, J.E. Saffitz, R.B. Schuessler, and R.W. Gross, Cardiac Ischemia Activates Calcium-Independent Phospholipase $A_2\beta$, Precipitating Ventricular Tachyarrhythmias in Transgenic Mice: Rescue of the Lethal Electrophysiologic Phenotype by Mechanism-Based Inhibition, *J. Biol. Chem. 278*:22231–22236 (2003).
58. Newberry, E.P., Y. Xie, S. Kennedy, X. Han, K.K. Buhman, J. Luo, R.W. Gross, and N.O. Davidson, Decreased Hepatic Triglyceride Accumulation and Altered Fatty Acid Uptake in Mice with Deletion of the Liver Fatty Acid-Binding Protein Gene, *J. Biol. Chem. 278*:51664–51672 (2003).
59. Han, X., H. Cheng, D.J. Mancuso, and R.W. Gross, Caloric Restriction Results in Phospholipid Depletion, Membrane Remodeling and Triacylglycerol Accumulation in Murine Myocardium, *Biochemistry 43*:15584–15594 (2004).

4 LC/MS and Chiral Separation

Arnis Kuksis[a] and Yutaka Itabashi[b]

[a]Banting and Best Department of Medical Research, University of Toronto, Toronto, M5G 1L6 Canada; [b]Graduate School of Fisheries Sciences, Hokkaido University, Hakodate, 041-8611, Japan

Introduction

Ever since the successful initial application of electrospray ionization-tandem mass spectrometry (ESI-MS/MS) (1,2) and atmospheric pressure chemical ionization-MS/MS (APCI-MS/MS) (3) to glycerolipid analysis, the impression has grown that prior chromatographic fractionation constitutes an unnecessary complication in MS/MS analysis of lipids. It has been suggested that the results obtained by direct MS/MS analysis of organic extracts avoid many pitfalls associated with multistep sequential chromatographic separations (4). Moreover, rapid profiling methods have further extended the depth of MS/MS analysis of crude lipid extracts. These issues are still under debate by many investigators (5), but the necessity for consideration of isotopomer effects in any quantitative approach is obvious. More recently, the need for rapid profiling of tissue lipids (lipidomics) has further favored (6,7) the view that MS/MS analyses of crude total lipid extracts can provide all the analytical data for tissue comparisons and investigation of various pharmacological and metabolic signals.

Although careful analysis of complex mixtures of natural fats has shown the benefits of prior on-line chromatography for MS/MS analysis (8), an absolute necessity of prior chromatographic separation has been obvious only for analysis of enantiomers and diastereomers. Amongst these chiral chromatography has clearly been the most thoroughly documented and routinely employed. While the enantiomers of low molecular weight hydroxy fatty acids can be effectively resolved by GLC on chiral-phase capillary columns, HPLC on chiral-phase columns is required for enantiomeric resolution of high molecular weight glycerolipids (9,10). The use of chiral-phase HPLC is also advantageous for the analysis of temperature sensitive derivatives of high molecular weight hydroxy fatty acids.

The theoretical principles of chiral-phase resolution of enantiomers have been discussed by Pirkle and Pochapsky (11), while Kuksis and Itabashi (12) and Itabashi (13) have recently reviewed their applicability to the resolution of racemic glycerolipids. It is noted that chiral-phase HPLC resolution of both racemic hydroxy fatty acids and acylglycerols requires a three-point interaction, which is provided by the three different substituents of the prochiral carbon carrying the hydroxyl group or its derivative. In chromatography, a three-point attachment (binding) is not always necessary for chiral discrimination. In many cases, repulsive steric interactions or attractive forces are invoked, usually in combination with one or more bonding interactions, to explain chiral recognition (11). The "three-point-binding" theory that Ogsten proposed to explain the enantiospecific nature of enzymatic reactions differs from the three-point rule in chromatography (11,12). A general method for finding a suitable chiral HPLC column is lacking and the choice of HPLC columns remains empirical.

The present review discusses the applications of chiral-phase HPLC in the MS/MS analysis of enantiomeric hydroxy fatty acids, eicosanoids, acylglycerols and glycerophospholipids.

Materials and Methods

Reagents and Suppliers. Grignard reagent (Aldrich Chemical Co., Milwaukee, WI): freshly diluted 0.5 M ethyl magnesium bromide in dry ethyl ether. (*R*)-(-)- and (*S*)-(+)-1-(1-Naphthyl)ethyl isocyanates (Fluka, Sigma-Aldrich, Canada); *N,N′*-dicyclohexylcarbodiimide (DCC) and *N,N′*-dimethyl-4-aminopyridine (DMAP) (Wako Pure Chemicals, Osaka, Japan); 3,5-dinitrophenyl isocyanate (3,5-DNPI) (Fluka, Buchs, Switzerland and Sumika Chemical Analysis Service, Osaka, Japan); 2-aminoanthracene (96% purity, Aldrich); triphosgene (Aldrich); acetonitrile (dehydrated, Wako Pure Chemicals); 2-propanol; hexane; chloroform; methanol; ethanol; dichloromethane; 1,2-dichloroethane; toluene; trifluoroacetic acid (TFA); pyridine (all HPLC grade, Wako Pure Chemicals); glycerolipid and glycerophospholipid standards (Avanti Polar Lipids, Alabaster, AL, USA); Aldrich-Sigma (St. Louis, MO, USA); Wako Pure Chemicals (Tokyo, Japan).

Normal and Chiral-phase Columns and Suppliers: Normal phase columns: Phenomenex Luna 3 silica columns (100 × 2.0 mm ID), and a guard column (4 × 2.0 mm ID) (Phenomenex, Torrance, CA, USA); Supelcosil (5 μm particles, 250 × 4.6 mm ID) and a guard column, Supelco, Sigma-Aldrich, Canada); Spherisorb silica (3 μm particles, 100 × 4.6 mm ID, Waters, Milford, MA, USA); Polyvinyl alcohol bonded silica column (YMC PVA-SIL, 100 × 3 mm ID, YMC, Kyoto, Japan). Chilar-phase Columns: *N*-(*S*)-2-(4-chlorophenyl)isovaleroyl-D-phenylglycine ionically bonded to γ-aminopropyl silanized silica (OA-2100, 250 × 4.6 mm ID, 5 μm particles); *N*-(*R*)-1-(1-naphthyl)ethylaminocarbonyl-(*S*)-valine (OA-4100, 250 × 4.6 mm ID, 5 μm particles); *N*-(*S*)-1-(1-naphthyl)ethylaminocarbonyl-(*R*)-valine (OA-4100R, 250 × 4.6 mm ID, 5 μm particles), *N*-(*S*)-1-(1-naphthyl)ethylaminocarbonyl-(*S*)-*tert*-leucine (OA-4600, 250 × 4.6 mm ID, 5 μm particles), and *N*-(*R*)-1-(1-naphthyl)ethylaminocarbonyl-(*R*)-*tert*-leucine (OA-4600R), 250 × 4.6 mm ID, 5 μm particles) covalently bonded to γ-aminopropyl silanized silica (Sumika Chemical Analysis Service, Osaka, Japan); (*R*)- and (*S*)-1-(1-naphthyl)ethylamine polymers covalently bonded to 300 Å-wide-pore spherical silica

(YMC-Pack A-K03 and A-L03) (250 × 4.6 mm ID, 5 μm particles, YMC Inc., Kyoto, Japan); cellulose tris(3,5-dimethylphenylcarbamate) (Chiracel OD, 250 × 4.6 mm ID, 5 μm particles) and amylose tris(3,5-dimethylphenyl carbamate) (Chiralpak AD, 250 × 4.6 mm, ID, 5 μm particles) (Daicel Chemical, Tokyo, Japan; Chiral Technologies, Exton, PA, USA); phenylcarbamate-β-cyclodextrin chemically bonded to silica (Chiral CD-Ph, 250 × 2.0 mm ID, 5 μm particles, Shiseido, Tokyo, Japan). Guard Columns: Sumipax Filter PG-ODS (Sumuka Chemical Analysis Service) to be used together with chiral columns.

Derivatization of Enantiomeric Hydroxy Fatty Acids

Racemic hydroxyeicosatetraenoic acids (HETEs) and hydroxyoctadecadienoic acids (HODEs) synthesized by autoxidation of arachidonic acid and linoleic acid, respectively, are reduced with $NaBH_4$ and initially fractionated by reversed-phase HPLC (e.g. a Beckman Ultrasphere ODS 5 mm column, $MeOH/H_2O/HOAc$) and methylated with ethereal diazomethane (13,14). For improved mass spectrometric detection, the methyl esters may be converted into the pentafluorobenzyl (PFB) esters, which improves the sensitivity of response in the electron capture mode (15). The PFB esters are prepared by reaction of the hydroxy fatty acids (nanogram quantities) in CH_3CN (100 μL) with 100 μL of PFB-Br in CH_3CN (1:19, v/v) followed by 100 μL of diisopropylethylamine (DIPE) in CH_3CN (1:9, v/v) and heating the solution at 60°C for 60 min. The solution was allowed to cool, evaporated to dryness under nitrogen at room temperature, and redissolved in 100 μL of hex-ane/EtOH (97:3, v/v) for normal-phase chiral chromatography ready for LC/MS analysis (16). The PFB esters of fatty acid epoxides were prepared using PFB-Br in *N,N*-diisopropylethylamine and purified by reversed-phase HPLC on a Bondapak C_{18} column (3.9 × 300 mm, 10 μm, Waters Associates, Milford, MA) utilizing a linear gradient from 30% H_2O/70% CH_3CN to 100% CH_3CN over 30 min at 1 μL/min. Catalytic hydrogenation was performed as described by Hawkins et al. (17).

Derativization of Enantiomeric Diacylglycerols (DAGs) and Monoacylglycerols (MAGs)

The 3,5-dinitrophenylurethane (3,5-DNPU) derivatives of the DAGs are prepared by reacting 1 mg of free DAGs and about 2 mg of 3,5-DNPI (Sumika Chemical Analysis Service) in 400 μL dry toluene in presence of 40 μL dry pyridine for 1 h at room temperature (18,19). The resulting 3,5-DNPUs are purified by TLC on a silicic acid plate (20 × 20 cm, 0.25 mm thick layer) containing a fluorescence indicator (Kodak, Rochester, NY). The reaction mixture, dissolved in 400 μL of $CHCl_3$, is spotted on the plate and the plate developed up to 15 cm using petroleum ether/1,2-dichloroethane/EtOH (40:10:3, by vol) as the developing solvent. Bands are visualized under UV and the 3,5-DNPU fraction (R_f 0.5–0.6) was scraped off the plate and recovered from the adsorbent by extraction with diethyl ether.

Bis-(3,5-DNPU) derivatives of MAGs [naturally occurring or released from triacylglycerols (TAGs) and DAGs by lipases or chemical degradation] are prepared by dissolving (less than 2 mg) in dry chloroform (2 mL) in the presence of dry pyridine (40 μL) and an excess amount of 3,5-DNPI (10–20 mg) (Sumika Chemical Analysis Service). The reaction is completed at room temperature overnight (18,20).

Preparation of Diastereomeric Derivatives

Naturally occurring free DAGs or DAGs released from TAGs by lipases or chemical degradation (<2 mg) are dissolved in dry toluene (0.3 mL) and (*R*)-(-)-1-(1-naphthyl)ethyl isocyanate (10 μL) and 4-pyrrolidinopyridine (4 mg) are added. The mixture is heated at 50°C overnight (21,22). After evaporation of solvents under dry N_2, the reaction products are dissolved in $MeOH/H_2O$ (95:5, v/v) and applied to Sep-Pak C_{18} column (Waters) and the naphthylethylurethane (NEU) derivatives eluted with acetone (10 mL) (23). In parallel, the (*S*)-(+)-1-(1-naphthyl)ethyl isocyanate is employed and the derivatives resolved.

Bis-(3,5-DNPU) derivatives of phosphatidylglycerol (PtdGro) or lipids containing PtdGro (<2 mg) were prepared by dissolving in dry toluene or dry chloroform (2 mL) in the presence of dry pyridine (40 μL) an excess amount of 3,5-DNPI (10–20 mg). The reaction was completed at room temperature overnight (24). The crude bis-DNPUs were purified by TLC on silica gel HF-plates (20 × 20 cm, 0.5 mm thickness) containing 5% $(NH_4)_2SO_4$, using $CHCl_3/MeOH/H_2O$ (42:8:1, by vol) as the developing solvent. Bands were visualized under UV and the bis-DNPU derivatives (R_f 0.7) were recovered from the adsorbent with $CHCl_3/MeOH$ (2:1, v/v). UV spectra of PtdGro derivatives were recorded at 226 or 254 nm and were essentially the same as those obtained for bis-DNPU derivatives of MAGs.

Scheme 1 shows the structures of derivatization reagents for HPLC separation of enantiomeric glycerolipids. A, 3,5-dinitrophenyl isocyanate; B, 2-anthryl isocyanate; C, phenylethyl isocyanate; D, 1-(1-naphthyl)ethyl isocyanate; E, fluoro-(1-naphthyl)acetic acid; F, 2-*tert*-butyl-2-methyl-1,3-benzodioxole-4-carboxylic acid. The derivatization for enantiomer separation of glycerolipids has been discussed previously (12,13).

Chiral-Phase HPLC of Enantiomeric Hydroxy Fatty Acids

Characteristically, an individual column type is capable of separating a limited range of racemic eicosanoids. Thus, the Chiracel OD column, which is the most versatile of the Chiracel class, will resolve 8-HETE and 12-HETE, whereas 15-HETE is not resolved. Therefore, a battery of chiral columns is required to allow the investigator to deal with a wide variety of fatty acid derivatives (14,25). Likewise, difficulties have been experienced in the chiral-phase resolution of the tosylated hydroxystearate, including its methyl and PFB esters. A baseline separation,

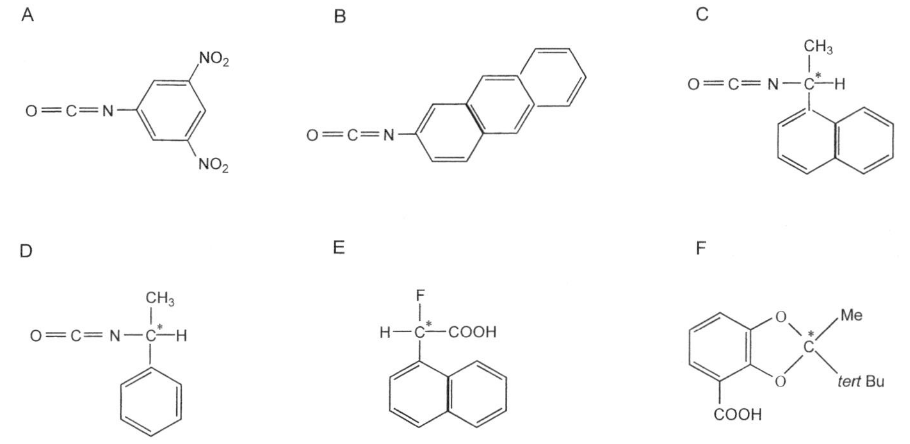

Scheme 1. The structures of derivatization reagents for HPLC separation of enantiomeric glycerolipids. A, 3,5-dinitrophenyl isocyanate; B, 2-anthryl isocyanate; C, phenylethyl isocyanate; D, 1-(1-naphthyl)ethyl isocyanate; E, fluoro-(1-naphthyl)acetic acid; F, 2-*tert*-butyl-2-methyl-1,3-benzodioxole-4-carboxylic acid. The derivatization for enantiomer separation of glycerolipids has been discussed previously (1,13).

however, could be achieved using the methyl ester in combination with Chiralpak AD columns and hexane/EtOH (100:2, v/v) as the solvent (14,25).

Chiral-Phase HPLC of Enantiomeric Epoxides

Hammonds et al. (16) used chiral-phase HPLC for separation of the enantiomers of all four epoxides of arachidonic acid. Despite extensive evaluations of several commercially available chiral-phase HPLC columns, satisfactory resolution of epoxyeicosatetraenoic acids (EETs) was achieved only with Chiralcel OB and OD columns operated in either normal or reversed-phase mode. The epoxides were analyzed as methyl esters or PFB esters. The 5,6-EET-Me enantiomers were resolved on the OB column operated in the presence of EtOH/H_2O mixtures, which, however, tended to degrade the column. Enantiomers of 14,15-EET-Me, 8,9-EET-PFB and 11,12-EET-PFB were separated on normal-phase HPLC using either OB or OD columns. Oliw (26) has reported the separation of epoxides of fatty acids by chiral-phase HPLC on a stationary phase with cellulose carbamate (Chiralcel OC). The resolution was improved by hydrogenation.

Chiral-Phase HPLC of Enantiomeric Glycerolipids

Chiral-phase HPLC of bis-(3,5-DNPU) derivatives of MAGs was carried out with chiral columns (250 × 4.6 mm ID) containing (*R*)- and (*S*)-1-(1-naphthyl)ethylamine polymeric phases covalently bonded to 300 Å-wide-pore spherical silica (5 μm particles, YMC Inc.) (24). The two chiral phases have opposite configuration and would be expected to give opposite order of elution of enantiomers. Usually, several micrograms of the bis-DNPU derivatives dissolved in CH_2Cl_2 was injected using an automatic sample injector and eluted isocratically with a Hewlett-Packard 1050 pump using hexane/CH_2Cl_2/MeOH (60:20:20, by vol) as the mobile phases. Peaks were monitored at 226 nm or 254 nm using a Hewlett-Packard 1050 variable wavelength UV detector. Alternatively, chiral-phase HPLC of bis-DNPU derivatives of MAGs was carried out with chiral columns (250 × 4.6 mm ID, Sumika Chemical Analysis Service) containing *N*-(*S*)-2-(4-chlorophenyl)isovaleroyl-D-phenylglycine ionically bonded to γ-aminopropyl silanized silica (OA-2100, 5 μm particles (18), *N*-(*R*)-1-(1-naphthyl)ethylaminocarbonyl-(*R*)-valine covalently bonded to γ-aminopropyl silanized silica (OA-4100R, 5 μm particles) (20,27) and *N*-(*S*)-1-(1-naphthyl)ethylaminocarbonyl-(*R*)-valine covalently bonded to γ-aminopropyl silanized silica (OA-4100R, 5 μm particles) (27). The OA-4100 and OA-4100R phases have an opposite configuration and would be expected to give the opposite order of elution of enantiomers.

Itabashi and Takagi (18,19) have performed the resolution of enantiomeric *sn*-1,2- and *sn*-2,3-, and the isomeric X-1,3-DAGs as their 3,5-DNPU derivatives by HPLC with a column (250 × 4.6 mm ID) containing (*R*)-1-(1-naphthyl)ethylamine polymeric phase covalently bonded to 300 Å wide pore spherical silica (5 μm particles, YMC-Pack A-KO3, YMC Inc.). The chiral column was eluted isocratically at 28°C using hexane/1,2-dichloroethane/EtOH (40:10:1, by vol) as the mobile phase at a constant flow rate of 0.5–1 mL/min. Usually 10–20 mg of 3,5-DNPUs dissolved in the same solvent as the mobile phase was injected on the column using an automatic sample injector. Peaks were monitored at 226 or 254 nm.

Alternatively, chiral-phase HPLC of enantiomeric *sn*-1,2-DAGs was performed on a Hitachi L-7000 liquid chromatograph equipped with a Hitachi L-4100 UV detector (Hitachi, Tokyo, Japan) (28). Two columns containing (*R*)- and (*S*)-1-(1-naphthyl)ethylamine polymers as the stationary phases (250 × 4.6 mm ID, 5 μm particles, YMC-Pack A-K03 and A-L03, YMC) were used. A guard column, Sumipax Filter PG-ODS (Sumika Chemical Analysis Service), was attached to the inlet of each column. The analysis was done isocratically at 20°C using hexane/CH_2Cl_2/EtOH (80:20:2, by vol) and hexane/CH_2Cl_2/CH_3CN (80:15:5, by vol) as the mobile phases at a constant flow rate of 0.5 mL/min.

Normal Phase HPLC of Diastereomeric DAG Derivatives

The *sn*-1,3-, *sn*-2,3- and *sn*-1,2-DAGs as the diastereomeric derivatives are resolved using in series two Supelcosil LC-Si columns (250 × 4.6 mm ID, 5 μm particles, Supelco-Aldrich) (21–23). The columns are installed in a Waters 550 liquid chromatograph connected to a Waters 990 photodiode array detector. The samples in hexane (10 μL) are injected into the columns and the derivatives are eluted with 0.37% 2-propanol in hexane at a flow rate of 0.7 mL/min, and peaks detected at 280 nm.

Chiral-Phase HPLC of Enantiomeric Glycerophospholipids

Chiral-phase HPLC of bis-(3,5-DNPU) derivatives of PtdGro is carried out with chiral columns (250 × 4.6 mm ID) containing (*R*)- and (*S*)-1-(1-naphthyl)ethylamine polymeric phases covalently bonded to 300 Å-wide-pore spherical silica (5 μm particles, YMC Inc.) (25). The two chiral phases have opposite configuration and would be expected to give the opposite order of elution of enantiomers. Usually, several micrograms of the bis-DNPU derivatives dissolved in dichloromethane are injected using an automatic sample injector and eluted isocratically with a Hewlett-Packard 1050 pump using hexane/CH_2Cl_2/MeOH/TFA (60:20:20:0.2, by vol) for PtdGro and hexane/CH_2Cl_2/MeOH (60:20:20, by vol) for MAGs as the mobile phases. Peaks are monitored at 226 nm or 254 nm using a Hewlett-Packard 1050 variable wavelength UV detector.

The chiral-phase resolution of the bis-(3,5-DNPU) derivatives of enantiomeric PtdGro was improved by performing the separations at 10°C (29).

LC/ESI-MS and MS/MS of Stereoisomeric Lipids

Direct ESI-MS/MS analyses of hydroxy fatty acids have been performed under the general conditions described for phospholipids (2). The samples were infused into the ESI source via a 50 μm I.D. fused silica transfer line using a Harvard Apparatus pump at 2–3 μL/min. Negative-ion ESI-MS and MS/MS used an orifice voltage from 25–100 V.

Conventional ESI- and APCI-MS/MS have limited sensitivity, which makes it difficult to conduct studies in cell culture systems where only trace amounts of non-esterified bioactive lipids are present. The use of electron capture APCI-MS/MS overcomes this problem. More appropriately, LC-MS of chiral hydroxyl fatty acids has been conducted following preparation of the ME or PFB derivatives (14,25). Lee et al. (15) combined normal phase chiral LC/MS on Chiralpak AD-H column (250 × 4.6 mm I.D., 5 μm; Daicel Chemical Industries, Ltd., Tokyo, Japan) with MS conducted on a Thermo Finningan TSQ 7000 triple stage quadrupole instrument equipped with an ion source in negative ion mode. The TSQ 7000 operating conditions were as follows: vaporizer temp., 500°C; heated capillary temp., 230°C; corona discharge needle, set at 16 μA. The sheath gas (nitrogen) and auxiliary gas (nitrogen) pressures were 40 psi and 10 (arbitrary units), respectively. Collision-induced dissociation (CID) was performed using argon as the collision gas at 2.7 m Torr in the second (rf-only) quadrupole. For full-scan and selected reaction monitoring (SRM) analyses, unit resolution was maintained for both precursor and product ions. LC/SRM-MS analysis was conducted using 50 μg of each component as its PFB derivative. Enantiomers and regioisomers of diverse bioactive lipids can be quantified using stable isotope dilution methodology coupled to normal phase chiral chromatography and electron capture APCI-MS/MS.

Originally, the molecular species of chiral DAG were identified by LC/MS using a direct liquid inlet interface (30,31). For LC/MS of the 3,5-DNPU derivatives of DAGs, with direct liquid inlet interface, about 1% of the HPLC effluent was admitted to a Hewlett-Packard Model 5985 quadrupole mass spectrometer via a direct liquid inlet interface (30). Positive CI-spectra were taken every 5 s over the entire chromatogram in the mass range 200–900, and the data were recorded by a computer. Chloride attachment negative CI spectra of the 3,5-DNPU were recorded over a similar mass range. The HPLC solvent served as the reagent gas and the source of chloride for chloride attachment negative CI (31). The stored data were recalled and analyzed with the help of Hewlett-Packard data system (Model HP-1000E) and a graphics terminal (Model HP 2648A).

More recently, the DAG-DNPUs were identified by passing the effluent of the normal phase HPLC column into a Hewlett-Packard 5989A single quadrupole mass spectrometer equipped with a Hewlett-Packard Model 5997A nebulizer-assisted ESI interface (32). N_2 was used as both nebulizing gas (60 psi) and drying gas (60 psi, 270°C). Positive ESI spectra were taken following post-column addition of $CHCl_3$/MeOH/30% NH_4OH (75:24.5:0.5, by vol) at 0.6 mL/min. The capillary exit voltage (CapEx) was 220 V and the mass range scanned was 500–720 (200–1000 in test runs). Capillary, endplate, and cylinder voltages were –4, –3.5, and –5 kV, respectively, in the positive ion mode. The CapEx, which determines the extent of fragmentation by means of CID was varied between +90 and +270 V. Identification of the various unknown species was performed on basis of retention time of standards, averaged full mass spectra, and the fragmentation patterns. The relative proportions of the molecular species of the DAG were calculated from the areas of the peaks obtained by single ion mass chromatograms extracted from the total ion current spectra.

Flow injection/ESI-MS of the derivatized and underivatized PtdGro was carried out in positive- and negative-ion modes with a JEOL JMX-SX102A magnetic instrument (24). Samples of 50 pmol/μL were introduced into the ESI source at a flow rate of 1 mL/min. The capillary skimmer voltage was set at 0 V. The mass spectra were taken in the mass range 100–1500. Normal phase LC/MS was performed by admitting the entire HPLC effluent to a Hewlett-Packard Model 5988B quadrupole mass spectrometer equipped with a nebulizer-assisted ESI interface (HP 59987A). The HPLC separation was accomplished with a Spherisorb 3 mm particle column (100 × 4.5 mm ID) and negative ESI spectra were taken in the mass range of 400–1300 amu.

The diastereomeric DAG naphthylethyl urethanes eluted from the normal phase HPLC column may be detected by UV absorption at 280 nm and collected for a reversed-phase HPLC resolution of the molecular species, and identification and quantification by on-line ESI-MS (33). A reversed-phase Supelcosil C_{18} column (250 × 4.5 mm ID) was used with a linear gradient of 20–50% 2-propanol in MeOH over 30 min at a flow rate of 1 mL/min. Positive ESI spectra were obtained by postcolumn addition of MeOH/0.2 M NH_4OAc (50:50, v/v) at 0.2 mL/min. The CapEx was varied between +90 and +270V. Identification of the various unknown species was performed on the basis of retention time of standards, averaged full mass spectra, and the fragmentation patterns (22). The relative proportions of the molecular species of the DAG were calculated from the areas of the peaks obtained by single ion mass chromatograms extracted from the total ion current spectra. The results obtained by the dual column approach were superior to those obtained by normal phase HPLC/ESI-MS alone (33).

Results and Discussion

Steric Analysis of Monohydroxy Fatty Acids

The first generation of chiral HPLC columns gave only marginal separation of HETE enantiomers (34), and for many compounds derivatization was required to achieve baseline resolution (17). More recent chiral columns, particularly of the Chiralcel class of derivatized cellulose-based supports, have provided much improved resolution of individual hydroxy eicosanoids. Currently, optimal performance is obtained with Chiralpak AD columns using hexane and methanol or ethanol as modifiers (14,25,35), as seen in the resolution of racemic 15-HETE methyl ester on Chiralpak AD column. Figure 1 shows the chiral separation of racemic prostaglandin B_2 methyl esters as obtained on a Chiralpak AD column using (A) hexane/EtOH (90:10, v/v), and (B) hexane/MeOH (95:5, v/v) at 1 mL/min and detection by UV at 280 nm (14). A mixture of prostaglandin B_2 (PGB_2) and 15*R*-PGB_2 methyl esters was widely resolved with retention times of 9.5 min (15*R*) and 15.9 min (15*S*) using a solvent of hexane/EtOH (90:10, v/v). Using hexane/MeOH (95:5, v/v) a further resolution was obtained with retention times of the PGB_2 enantiomers eluted at 23.3 (*R*) and 50.5 min (15*S*), which could not be further improved by use of additional MeOH.

In contrast to results obtained with Chiralcel columns, the Chiralpak AD column gives higher chromatographic efficiency (good peak shapes and 5–10,000 theoretical plates/column). Schneider et al. (14,25) report that with the Chiralpak AD column and EtOH or MeOH as modifier, there are striking improvements in separation with a wide variety of lipoxygenase products and other hydroxy derivatives. Of the different fatty acid derivatives tested by Schneider et al. (14) on the Chiralpak AD column, three compounds showed no enantiomeric separation. Two of the products, (±)-*cis*-12-oxophytodienoic acid and (±)-jasmonic acid methyl esters, are cyclic fatty acid derivatives, lacking a hydroxyl group or other oxygenated chiral substituent. The third compound was methyl 14,15-epoxyeicosa-5,8,11-trienoate, a *cis*-epoxy derivative of arachidonic acid.

Steric analysis of ω-4, ω-3 and ω-2-hydroxy fatty acids can be performed by GLC separation of diastereomeric *S*-phenylpropionic acid derivatives (36). It is also possible to partly resolve the enantiomers of 18-HETEand 19-HETE on chiral-phase HPLC using naphthoyl ester derivatives (37).

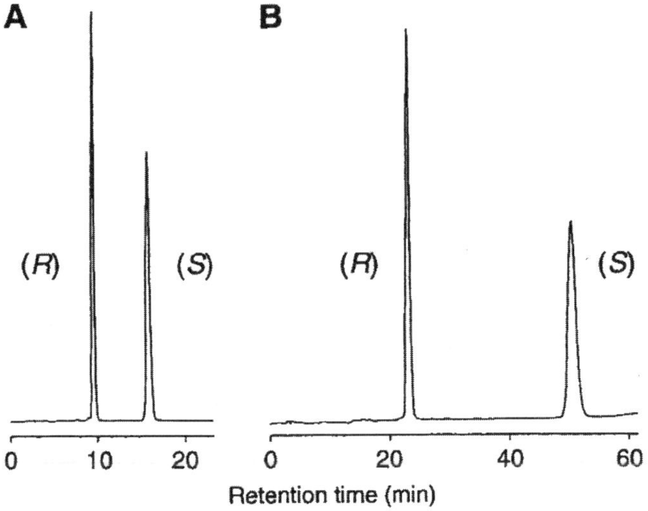

Fig. 1. Chiral separation of prostaglandin B_2 and 15*R*-PGB_2 methyl esters. Racemic PGB_2 methyl ester was analyzed on a Chiralpak AD column (25 cm × 0.46 cm) using (A) hexane/EtOH (90:10, v/v), and (B) hexane/MeOH (95:5, v/v) at 1 mL/min flow rate and detection at UV 280 nm. Racemic prostaglandin B_2 was prepared by mixing PGB_2 with its enantiomer 15*R*-PGB_2, prepared by alkali treatment of 15*R*-PGE_2. *Source:* Schneider et al. (14).

The chirality of the monohydroxyeicosatetraenoic acids in psoriatic skin scales and basophilic leukemia cells was characterized by Baer et al. (38,39). Free hydroxylated fatty acids (9-hydroxy- and 13-hydroxy-linoleic acids, HODD, and 5-, 12-, and 15-hydroxyeicosatetraenoic acids, HETEs) were collected during isocratic runs with MeOH/H_2O/HOAc (80:20:0.1 by vol) on reversed-phase HPLC (ODS-Hypersil columns, 20 cm × 4.6 mm at a flow rate of 0.4 mL/min) and were methylated with ethereal diazomethane and rechromatographed on the same HPLC columns. The methylated fatty acids were collected during the HPLC run, and then rechromatographed on two 25 cm Bakerbond chiral-phase HPLC columns in series (dinitrobenzoylphenyl glycine coupled ionically over aminopropyl residues (J.T. Baker Research Products). The mobile phase mixture for this straight phase HPLC consisted of hexane/2-propanol (1000:15, v/v). Mixing experiments with synthetic standards were carried out for each HETE to confirm exact retention times in the chiral-phase HPLC studies. The hydroxy fatty acid methyl esters resolved by reversed-phase HPLC in the order 12-epi, Δ^6-*trans*-LTB_4, LTB_4, 15-HETE, 12-HETE, 5-HETE, 13-HODD and 9-HODD were collected and subjected separately to chiral-phase HPLC. The identity of the HPLC peaks in the reversed-phase chromatogram was established by cochromatography with standards and by GC/MS of the trimethylsilyl ethers. Each of the hydroxy fatty acid methyl esters was resolved into the *S* and *R*-isomers, with the *S*-isomer being eluted first. The free and esterified 13-HODD acids isolated from the psoriatic scales contained a mixture of the *S/R* stereoisomers, averaging 1.9:1 for free 13-HODD, and 2.4 for the free 9-HODD, which contradicted with the strict *S*-stereospecificity for oxygen insertion exhibited by mammalian lipoxygenase. Incubation of rat basophilic leukemia cells with exogenous arachidonic acid and permeabilizing concentrations of ethanol resulted in the production of 5-, 12-, and 15-HETE. It was demonstrated that the 5-HETE had strict (*S*) stereospecificity while the 12- and 15-HETE were non-racemic mixtures of the stereoisomers with *S/R* ratios averaging 8.6 and 2.2, respectively.

The results suggested that the 15- and 12-HETE acids may be derived from non-lipoxygenase sources. In an attempt to define more accurately the enzymatic origin of these fatty acid derivatives and in assessing their possible role in the pathogenesis of psoriasis, Baer et al. (40) have incubated psoriatic skin scales with radiolabeled arachidonic and linoleic acids and have characterized the monohydroxylated derivatives produced in vitro. The products of incubation with [^{3}H]arachidonic acid were an enantiopure 15(*S*)-[^{3}H]HETE and a nonracemic mixture of the 12-[^{3}H]HETE stereoisomers (*R/S* ratio= 4.5). An enantiopure 13(*S*)-[^{14}C]HODE was produced from [^{14}C]linoleic acid. No radioactive-labeled products were produced with heat-denatured scales. A radiomatic Flo-One A-140 radioactivity detector was used for concurrent measurement of ^{3}H and ^{14}C in the outflow from the HPLC spectrophotometer. Using synthetic standards of racemic 15-HETE and 15(*S*)-[^{3}H]HETE or racemic 12-HETE and 12(*S*)[^{3}H]HETE, it was found that the stereoisomer labeled with tritium had a retention time delayed relative to that of the corresponding unlabeled stereo-isomer, consistent with an isotope effect.

These results provided evidence for two distinct oxygenase activities that are preserved in psoriatic skin scales. One is that of ω-6 oxygenase with strict (*S*) stereospecificity, consistent with the activity of lipoxygenase. This enzyme activity appears to be similar to that of the 15-lipoxygenase which has been described in human keratinocytes. The second activity is that of an arachidonic acid 12(*R*)-oxygenase that has not been observed in normal human epidermis.

In addition to the enzymatic pathways of arachidonic acid metabolism, recent evidence has suggested that during oxidant stress, oxidation of arachidonic acid can occur while still esterified to membrane glycerophospholipids (41–43). The non-enzymatic mechanism of arachidonate oxidation represents an alternative pathway for the generation of biologically active, oxidized arachidonic acid mediators directly or following liberation from the phospholipid (44). Little is known concerning the positional distribution and enantiomeric nature of the non-enzymatic oxidation products of arachidonic acid. Brash et al. (45) reported that the enantiomers of 7-HETE, 10-HETE, and 13-HETE were at least partly resolved by chiral HPLC on Chiralcel OD.

Tandem MS of polyunsaturated hydroxy fatty acids has been extensively investigated by a variety of techniques and the area has been reviewed as part of tandem MS of fatty acids (46).

Nakamura et al. (47) have isolated and characterized murine pulmonary phospholipids and have observed the normal occurrence of 10 isobaric eicosanoids corresponding to incorporation of one oxygen atom into the arachidonate esterified glycerophospholipids. Phospholipids were hydrolyzed to yield the free carboxylic acids prior to reversed and chiral-phase analysis. Reversed-phase HPLC and electrospray tandem MS were used to identify and quantify six monohydroxyeicosatetraenoic acid regioisomers using d_8-HETE as internal standard. Chiral analysis of esterified 15-HETE revealed and *R/S* ratio of 0.98, suggesting operation of a free radical mechanism responsible for generation of this monohydroxy arachidonate phospholipid, and this enantiomeric ratio was 1.10 following treatment of the mouse lung with *tert*-BuOOH. The chiral analysis of 15-HETE was performed following hydrolysis of the phospholipid and isolation by reversed-phase HPLC column (2 × 150 mm, 3 μm C_{18} Ultremex column; Phenomenex). 15-HETE was isolated with 50% B (CH_3CN/MeOH, 65:35, v/v) at a flow rate of 200 μL/min at a retention time of 24.5 min. The collected samples were then methylated with ethereal diazomethane and the methyl esters of 15(*R*)- and 15(*S*)-HETE were separated by normal phase HPLC, which was performed using a chiral column (4.6 × 250 mm, Chiralcel OD; Chiral Technologies, Exton, PA) with hexane/2-propanol (95:5, v/v) at a flow rate of 500 μL/min. The 15(*R*)- and 15(*S*)-HETE methyl esters were collected and hydrolyzed with sodium hydroxide. The recovered fatty acids were hydrogenated by bubbling hydrogen gas through a methanol solution of the sample for 20 min after addition of Rh/Al_2O_3 (1 mg) as catalyst. Samples were than taken to dryness and derivatized as the pentafluorobenzyl (PFB) ester trimethylsilyl ethers as described previously (48). GLC-MS was carried out in the EC negative ion mode using a non-polar 15 m × 0.25 mm DB-1 (J&W Scientific, Folsom, CA) column with 0.25 μm film thickness. The hydroxy fatty acids were determined by monitoring the selected ions at *m/z* 339 and 403, respectively, which corresponded to the loss of PFB radical from 15-hydroxyeicosanoate PFB ester and $[^{18}O_2]$15(*S*)-hydroxyeicosanoate PFB, respectively, at a retention time of 12.5 min.

Bylund et al. (49) studied the oxygenation mechanism of HETE using chiral-phase HPLC with an ion trap mass spectrometer. Incubation of human recombinant cytochrome P_{450}(CYP) 2C9 under ^{18}O gas showed that all HETEs had incorporated ^{18}O to the same degree. Chiral-phase HPLC showed that CYP2C9 formed 15*R*-HETE (72% of the *R* enantiomer), 13*S*-HETE (90%), and 11*R*-HETE (57%). Reversed-phase HPLC-MS analysis revealed that CYP219 oxygenated arachidonic acid to 19-HETE, 14,15-epoxyeicosatrienoic acid (EET), and 8,9-EET as main metabolites. Steric analyses were performed by chiral-phase HPLC after methylation (45). Chiralcel OD-H (5 μm; 250 × 4.6 mm) was used for steric analysis of 13-HETE and Chiralcel OB-H (5 μm; 250 × 4.6 mm) was used for steric analysis of 15-HETE, 11-HETE, 13-HODE, and 9-HODE (45,50). The chiral compounds were eluted with 2% 2-propanol in hexane, at 0.5 mL/min. Specifically, chiral-phase HPLC showed that CYP2C9 formed 13*S*-HETE (90% *S*-enantiomer), 15*R*-HETE (72%), and 11*R*-HETE (57%). A previously reported CYP2C9 also formed

12*R*-THE (85%) and 10-HETE (49). Furthermore, chiral-phase HPLC showed that CYP2C9 formed 13*R*-HODE (90% *R*-enantiomer) and 9*R*-HODE (77%). All mass spectrometric characterization of the hydroxy fatty acids was performed by reversed-phase LC/ESI-MS using MeOH/H_2O/HOAc (80:20:0.01, by vol) at 0.2 mL/min as the mobile phase.

Hamberg (50) incubated linoleic acid with prostaglandin-endoperoxide H synthase-2 (PGHS-2) from ovine placenta and isolated a product consisting of regio- and stereoisomeric hydroxyoctadecadienoic acids (HODE). Analysis by normal phase HPLC followed by chiral-phase HPLC demonstrated that the linoleic acid was preferentially oxygenated at C_9 to produce the following mixture of HODEs: 9(*R*)-HODE (52%), 9(*S*)-HODE (11%), 13(*R*)-HODE (2%), and 13(*S*)-HODE (35%). A parallel incubation of linoleic acid with microsomal prostaglandin-endoperoxide H synthase-1 (PGHS-1) from ovine vesicular gland resulted in a product consisting of 9(*R*)-HODE (73%), 9(*S*)-HODE (9%), 13(*R*)-HODE (1%) and 3(*S*)-HODE (17%). It was concluded that the initial steps of the PGHS-2 and PGHS-1-catalyzed oxygenations proceed with identical stereochemistry and involve stereospecific removal of the pro-*S*-hydrogen from the ω-8-methylene group of the substrate.

Oliw et al. (51) had earlier investigated the NADPH-dependent oxygenation of linoleic acid by liver microsomes of phenobarbial-treated rats and had shown the formation of 11-HODE, 9-HODE, and 13-HODE. Experiments with stereospecifically deuterated linoleic acid at C_{11} showed that 9-HODE (80% *R*) and 13-HODE (85% *R*) were formed by initial abstraction of the pro-*R* and pro-*S* hydrogens at C_{11}, respectively, with subsequent insertion at C_9 or C_{13}. 11-HODE was formed by suprafacial hydrogen abstraction and oxygen insertion.

Suzuki et al. (52) have developed a determination method for human urinary 12-HETE using LC-MS/MS. This method, which includes simple extraction and detection in the selected reaction monitoring mode, allows an accurate determination of 12-HETE. There was a significant sex difference in urinary 12-HETE levels. Chiral analysis of 12-HETE using LC-MS/MS with column switching technique revealed that the major enantiomer was 12(*S*)-HETE. Furthermore, measurements of the urinary level in patients with diabetes mellitus (DM) indicated that there could be difference in production of 12(*S*)-HETE between genders and that 12(*S*)-HETE may play a role in the pathogenesis of DM. Reversed-phase HPLC was performed on a C_{18} Capcell Pak UG120 (Shiseido, Tokyo, Japan: 1.5 × 150 mm, 3 μm) using isocratic elution with CH_3CN/H_2O/HOAc (60:40:0.01, by vol) with a flow rate of 100 μL/min. The column was maintained at 40°C. Column-switching technique was used for chiral analysis of urinary 12-HETE. The 12-HETE was separated using C_{18} Capcell Pak UG120 column (Shiseido, Tokyo, Japan: 1.0 × 75 mm; 3 μm) using isocratic elution with CH_3CN/H_2O/HOAc (50:50:0.02, by vol) at a flow rate of 100 μL/min. The column was maintained at 40°C. The fraction containing 12-HETE was introduced into the Chiral CD-Ph column (Shiseido, Tokyo, Japan; 2.0 × 250 mm, 5 μm) using isocratic elution with

MeOH/H_2O/HOAc (65:35:0.02, by vol) at a flow rate of 100 μL/min. Chiral CD-Ph column was maintained at 15°C. Column effluent was introduced into the mass spectrometer using a fused-silica capillary between 40.0 and 65.0 min after injection. 12-HETE was detected using LC-MS/MS in the SRM mode. The SRM was performed by monitoring the transitions between *m/z* 319 and *m/z* 179 for 12-HETE and between *m/z* 327 and *m/z* 184 for I. S. Collision gas (argon) pressure was 2.0×10^{-13} (mBar). The capillary voltage was –3000 V and the source temperature was 150°C. Cone voltage was –25 V with collision energy of 15 eV. Peak areas and the calibration curve were obtained using the MassLynx program (Micromass, Manchester, UK).

The method was applied to specify the enzymatic origin of urinary 12-HETE by analyzing the chirality of urinary 12-HETE in healthy volunteers and patients with DM using LC-MS/MS with the column switching technique described above. Maintenance of phenylcarbamate-β-cyclodextrin bonded chiral column at 15°C made it possible to separate each 12-HETE enantiomers. Chiral analysis of 12-HETE revealed that the major enantiomer in human urine was 12(*S*)-HETE, indicating that human urinary 12-HETE was produced by the 12(*S*)-lipoxygenase pathway. Figure 2 shows chiral-phase LC/ESI-MS analysis of 12-HETE in human urine, which indicates that the major enantiomer is 12(*S*)-HETE. In this study, 12(*R*)-HETE was not detected in any of the urine samples. As formation of the 12(*S*)-enantiomer can be accounted for by the 12(S)-lipoxygenase while it is believed that 12(*R*)-HETE is produced by cytochrome P450 (53) or 12(*R*)-lipoxygenase pathways (54), the data of Suzuki et al. (52) indicate that human urinary 12-HETE was produced by the 12(*S*)-lipoxygenase pathway.

Bayer et al. (55) have reported a simple and rapid enantioselective determination of the fatty acid derivatives 13(*R,S*)-HODE, 9(*R,S*)-HODE and 12(*R,S*)-HETE, using HPLC with Chiralpak AD as the chiral selector and ESI-MS. A Chiralpak AD column with amylose tris-(3,5-dimethylphenylcarbamate) coated on silica-gel as the chiral stationary phase (Chiral Technologies Europe, Illkirch, France; 250 × 4.6 mm ID, 20 μm) was used. MS/MS fragmentation was performed on a quadrupole ion trap mass spectrometer LCQ (Thermo Finnigan, San Jose, CA, USA) equipped with a nanoelectrospray ion source (Proxeon Biosynthems A/S, Odense, Denmark). Approximately 3 μL of the analyte solution (samples: solutions used for enantioselective HPLC were vaporized under nitrogen stream and taken up in 50 μL ethanol; reference compounds: 0.2 μg/mL in ethanol) were loaded into laboratory-made gold-coated glass capillaries and negative ions generated at a spray voltage between –900 to 1000 V. The temperature of the heated transfer-capillary was 180°C, the capillary voltage was –30 V, the tube lends –15 V. Spectra were averaged over 10–20 scans. The relative collision energy for fragmentation was set between 20–28%. Figure 3 shows the enantioseparation of 12-HETE fraction, 13-HODE fraction and 9-HODE fraction, using Chiralpak AD column and hexane/EtOH/MeOH/HOAc (93:4:3:0.1, by vol) as the eluting solvent at a flow rate of 1 mL/min. (55). In case of 12-HETE, the use of methanol as alcohol modifi-

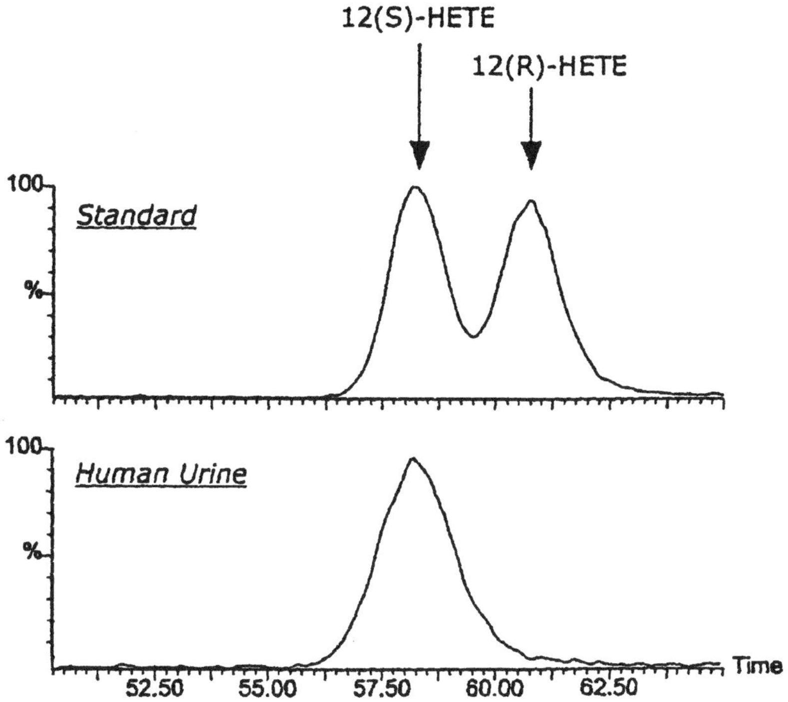

Fig. 2. Chiral analysis of 12-HETE in human urine using LC-MS/MS with column switching technique. The phenylcarbamate-β-cyclodextrin bonded chiral column (CD-Ph column (250 × 2.0 mm I.D., 5 μm, Shiseido, Tokyo,) was eluted isocratically at 15°C with MeOH/H_2O/HOAc (65:35:0.2, by vol) at a flow rate of 100 μL/min. The 12-HETE was extracted from urine using solid-phase extraction cartridge. *Source:* Suzuki et al. (52).

er was mandatory. The analyzed fractions had been collected from a sample extract using LiChrospher Si-60 column. ESI-MS with MS/MS fragmentation was performed on a quadrupole ion trap spectrometer LCQ (Thermo Finnigan, San Jose, CA, USA) equipped with a nanoelectrospray ion source (Proxeon Biosystems A/S, Odense, Denmark). The enantiomer distribution of 12-HETE and 9-HODE in psoriatic skin scales of untreated patients showed no significant differences, while samples of patients under systematic treatment exhibited a lower predominance of 12(*S*)-HODE than samples of untreated patients. The results were in good accordance with data obtained previously (52).

Hiratsuka et al. (56) have reported the chiral-phase HPLC separation of (*R*)- and (*S*)-hydroxynonenal (HNE) enantiomers. HPLC was performed on a chiral column (Chiralcel OB, 10 μm pore size, 4.6 × 250 mm, Daicel Chemical Industries, Tokyo, Japan) in 2% (v/v) 2-propanol/hexane (0.8 mL/min). The (*S*)-HNE and (*R*)-

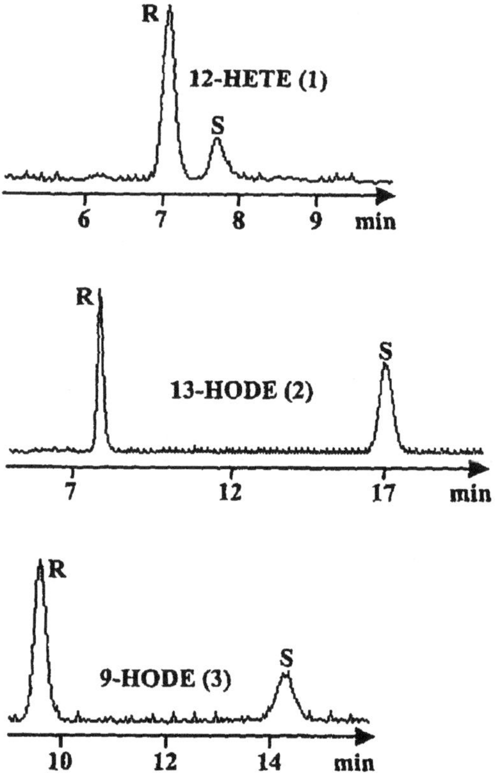

Fig. 3. Enantioseparation of 12-HETE, 13-HODE and 9-HODE fractions isolated from skin scales and chromatographed using Chiralpak AD column with amylase tris-(3,5-dimethylphenylcarbamate) coated on silica gel 250 mm × 4.6 mm I.D., 10 μm; Chiral Technol. Europe, Illkirch, France). The column was eluted with hexane/EtOH/MeOH/HOAc (93:4:3:0.1, by vol) at a flow rate of 1 mL/min. The fractions were collected from a solvent extract using Lichrospher Si-60 column (250 × 4 mm I.D., 5 μm; Merck, Darmstadt, Germany). The column was eluted with n-hexane/2-propanol/HOAc (100:2:0.1, by vol) at a flow rate of 1 mL/min. *Source:* Bayer et al. (55).

HNE were eluted at 17.8 and 22.4 min, respectively. The authors have shown that the enzyme glyceraldehydes-3-phosphate dehydrogenase was irreversibly and (*S*)-selectively inactivated by the enantiomers of racemic 4-hydroxy-2(*E*)-nonenal.

Steric Analysis of Dihydroxy Fatty Acids

Vicinal *threo*-dihydroxyeicosatrienoic acid (DiHETrE), formed by hydrolysis of epoxyeicosatrienoic acid (EpETrE), has been resolved by chiral HPLC as methyl ester derivatives using Chiralcel OC and Chiralcel OD columns (Diacel) (57). Resolution was improved by hydrogenation. The PFB ester derivatives of the hydrogenated vicinal diols yielded better separation on the Chiralcel CD column (Diacel) and the elution times were shortened. The *R,R*-enantiomers of the dihydroxyeicosanoic acids (DHEA) eluted before the *S,S*-enantiomers. These results have not been extended to other diols, e.g. to *threo*-17,18-dihydroxyeicosanoic and to *erythro*-17,18-DHEA (58).

Knothe et al. (59) have made *erythro/threo*-assignments of the two hydrogenated dihydroxy isomers obtained from oleic acid in the spectra of methyl 9,10-dihydroxyoctadecanoate. These authors distinguished the *erythro* and *threo* forms by differences in the chemical shifts of the carbons a to the hydroxy-bearing carbons. Takagi (60) showed that vic-dihydroxy acid diastereomers of 9,10-dihydroxyoctadecanoic acid can be resolved as the 3,5-DNPU into four enantiomers by HPLC on Sumichiral OA-4500 (25 cm × 4.6 mm I.D.) with hexane/$CHCl_3$/MeOH/TFA (65:20:15:1, by vol) at 1 mL/min.

Steric Analysis of Epoxy Fatty Acids

Chiral-phase HPLC for separation of the enantiomers of all four epoxides of arachidonic acids has been described by Hammonds et al. (16), using Chiralcel OB and Chiralcel OD columns (Diacel). The epoxides were analyzed as methyl esters or as PFB esters. Figure 4 shows the elution profiles for 8,9- and 11,12-EET-PFB enantiomers. Resolution was achieved for individual racemates injected on Chiralcel OD column by developing with 0.08% and 0.15% 2-propanol in hexane. After collection and solvent evaporation, the identity of each chiral pair was confirmed by GC/MS.

Enantiomers of *cis-trans* conjugated HETE and other *cis-trans* conjugated hydroxy fatty acids can be separated by chiral HPLC on (*R*)-(-)-*N*-(3,5-dinitrobenzoyl)-α-phenylglycine columns (17). The separation is improved after derivatization of the hydroxyl groups (34). Brash et al. (35) reported that the enantiomers of 7-HETE, 10-HETE, and 13-HETE were at least partly resolved by chiral-phase HPLC on Chiralcel OD.

Tandem MS of polyunsaturated hydroxy fatty acids has been extensively investigated by a variety of techniques and the area has been reviewed as part of tandem MS of fatty acids (46). Cellulose derivatized with triphenylcarbamate (Chiralcel OC, Diacel), for example, resolved the enantiomers of some epoxides of arachidonic linoleic, and α-linolenic acids, but the enantiomers of the ω-3 epoxides of eicosapentaenoic or linolenic acid were not separated.

Steric Analysis of MAGs

Takagi and Ando (61) proposed a stereospecific analysis of TAGs by chiral-phase HPLC of isomeric MAGs as the di-(3,5-DNPUs). The 1-MAGs prepared from TAGs by partial Grignard degradation were resolved into the *sn*-1- and *sn*-3-

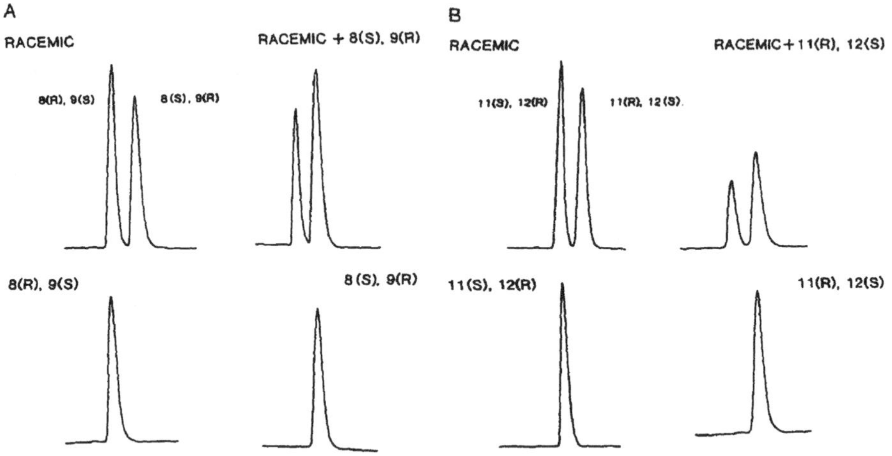

Fig. 4. Separation of 8,9- and 11,12-EET-PFB enantiomers on Chiralcel OD column. Samples of racemic 8,9-[^{14}C]- and 11,12-[1-^{14}C]- and of 8(*S*),9(*R*)-, 8(*R*)9(*S*)-, 11(*S*),12(*R*)-, and 11(*R*),12(*S*)-EET-PFB were individually injected onto the column (250 × 4.6 mm I.D., J. T. Baker Chemical Co., Phillipsburg, NJ). The absorbance of column eluants was monitored 210 nm. (A) Elution profiles for racemic 8,9-, 8(*R*),9(*S*)-, and 8(*S*),9(*R*)-EET-PFB and for co-injected mixture of racemic 8,9- and 8(*S*),9(*R*)-EET-PFB. Resolution was achieved with a solvent mixture of 0.08% 2-propanol in hexane at 2.0 mL/min. (B) Elution profiles for racemic 11,12-, 11(*S*),12(*R*)-, and 11(*R*),12(*S*)-EET-PFB and for a co-injected mixture of racemic 11,12- and 11(*R*),12(*S*)-EET-PFB. Resolution was achieved with a solvent mixture of 0.15% 2-propanol in hexane at 1 mL/min. *Source:* Hammonds et al. (16).

MAGs on a chiral column (Sumichiral OA-4100). The method, which permits a clear separation of the enantiomeric MAGs, has been extensively applied in the stereospecific analyses of polyunsaturated fats and oils (20). Ando et al. (62) recently reported further improvement in the method by subjecting the isomeric and enantiomeric *sn*-1-, *sn*-2- and *sn*-3-MAG derivatives to chromatography on two chiral columns. A column of Sumichiral OA-4100 was used for isolation of *sn*-1- plus *sn*-2-MAG and *sn*-3-MAG fractions, and then one of Sumichiral OA-4000 for separation of *sn*-1-MAG and *sn*-2-MAG fractions. Assignments of each fatty acid to the *sn*-1-, *sn*-2- and *sn*-3-positions were obtained from the peak area ratio of each MAG molecular species relative to mononononadecanoylglycerol, formed from the internal standard, and the fatty acid composition of each position was calculated on the basis of the assignments. The method has been widely employed in stereospecific analyses of TAGs (62–65). Takagi (66) has reported improved resolution of isomeric and enantiomeric MAGs and monoalkylglycerols as the 3,5-DNPUs by chiral-phase HPLC on Sumipax OA-4100 or YMC A-K03 columns by lowering column temperature and reducing the flow rate. Since the monoalkylglycerols were also resolved into enantiomers, it should also be possible to subject alkylacylglycerols to stereospecific analyses using this methodology.

Itabashi et al. (67) have reported a chiral-phase HPLC resolution of *sn*-1,2- and *sn*-2,3-isopropylidene glycerols as the 3,5-DNPUs. The separations were performed on two different stationary phases of opposite configuration: *N*-(*S*)-1-(1-naphthyl)ethylaminocarbonyl-(*S*)-*tert*-leucine (OA-4600) and *N*-(*R*)-1-(1-naphthylethylaminocarbonyl-(*R*)-*tert*-leucine (OA-4600R). This caused a reversal in the order of enantiomer elution. Complete enantiomer separation, which permitted an accurate estimation of the optical purity of each enantiomer, was easily achieved on both columns (each 250 × 4.6 mm I.D.) within 15 min after an injection by an isocratic elution with hexane/CH_2Cl_2/EtOH (40:12:3, by vol) at a flow rate of 0.5 mL/min.

Steric Analysis of DAGs

A chiral-phase HPLC resolution of the *sn*-1,2(2,3)-DAGs is based on the conversion of the DAGs into the 3,5-DNPUs by reaction with the 3,5-DNPI and the separation of the enantiomeric urethanes on a chiral stationary phase consisting of bonded (*R*)-(+)-1-(1-naphthyl)ethylamine (19). The DAG derivatives are resolved by an isocratic elution with hexane/1,2-dichloroethane/EtOH 40:10:1 (by vol) as the mobile phase, with the *sn*-1,2-enantiomers emerging ahead of the *sn*-2,3-enantiomers. The method has been utilized extensively in stereospecific analyses of the DAG moieties derived from natural TAGs by random degradation with the Grignard reagent (67–72). Figure 5 shows the total and fragment ion current profiles of the DNPU-derivatives of *rac*-1,2-DAGs derived from corn oil TAGs as obtained by chiral-phase LC/NCI (negative chemical ionization)-MS with chloride attachment (31). The detection of the DAG-DNPUs as the chloride attachment fragment ions in the negative ion mode increased the sensitivity of analysis about 100-fold over the detection in the positive ion mode.

The DNPU groups can be removed from DAGs by silolysis with trichlorosilane-triethylamine in dry toluene (68). The *N*-silylated DNPU group can be removed by exposure to water, which regenerates the pure enantiomers. The pure isomers can be recovered and the molecular species of individual enantiomers quantified by GLC with flame ionization following trimethylsilylation of the free DAGs.

Itabashi et al. (28) have shown that under certain conditions the chiral-phase resolution of enantiomers can be combined with the resolution of the reverse (regio) isomers. Figure 6 shows the chiral-phase resolution of the reverse isomers and enantiomers of 1,2-DAGs containing polyunsaturated fatty acyl chains as 3,5-DNPUs on (*R*)-1-(1-naphthyl)ethylamine polymer (A-K03). Each mixture of the reverse isomers of *rac*-1,2-DAGs was analyzed separately. The peak identities are shown in the figure. An opposite order of elution of the enantiomers and reverse isomers was obtained on (*S*)-1-(1-naphthyl)ethylamine polymer (A-L03).

Okabe et al. (73) have reported a highly sensitive method for the separation and detection of enantiomeric and regioisomeric DAGs as 2-anthrylurethanes by chiral-phase HPLC with fluorescence detection. The derivatives were obtained by

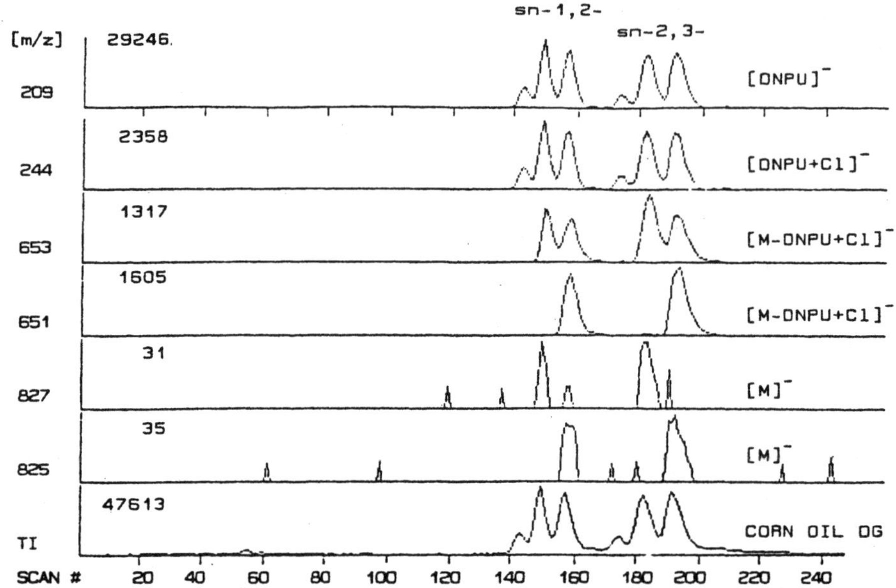

Fig. 5. Total negative chemical ionization (NCI) current and fragment ion profiles of *rac*-1,2-DAG DNPU derived from corn oil TAGs as obtained by chiral-phase LC/NCI-MS: *m/z* 653, 18:1/18:2; *m/z* 651, 18:2/18:2; *m/z* 827, 18:1/18:2 DNPU; *m/z* 825, 18:2/18:2 DNPU. LC-MS conditions: Chiral-phase column (25 cm × 4.6 mm I.D.) containing (*R*)-naphthylethylamine polymer bonded to 300A wide pore spherical silica (YMC-Pack A-K03, YMC Inc., Kyoto, Japan); Solvent system, linear gradient of Solvent A (isooctane/*tert*-butyl methyl ether/2-propanol/CH_3CN, 80:10:5:5, by vol) containing 1% CH_2Cl_2 (20%) to Solvent B (hexane/CH_2Cl_2/EtOH (40:10:1, by vol) (Solvent B) (80%). Ion source pressure, 0.7 torr; temp., 150°C. *Source:* Marai et al. (32).

reaction of DAG and 2-anthryl isocyanate for 3 h at 25°C. The isocyanates were obtained by reaction of 2-aminoanthracene and triphosgene in acetonitrile at 100°C. Satisfactory resolution of the enantiomers and regioisomers was achieved on a (*R*)-1-(1-naphthyl)ethylamine polymeric phase, using a mixture of a hexane/Cl_2CH_2/EtOH (150:10:1, by vol) as the mobile phase. The formation of various hydrogen bonding, dipole-dipole stacking and charge transfer complexes between the urethane derivatives and the stationary phase was thought to contribute to the enantiomer separation. The detection limit of 2-anthrylurethanes was 1 fmol when the signal-to-noise ratio was 3:1. Under the working conditions, the *sn*-1,2-enantiomers eluted ahead of the *sn*-2,3-enantiomers and both were preceded by the X-1,3-regioisomers. The chiral-phase HPLC separation was based on the degree of unsaturation of the DAGs as 2-anthrylurethanes. *sn*-2,3-Distearoylglycerol overlapped with *sn*-1,2-dioleoylglycerol.

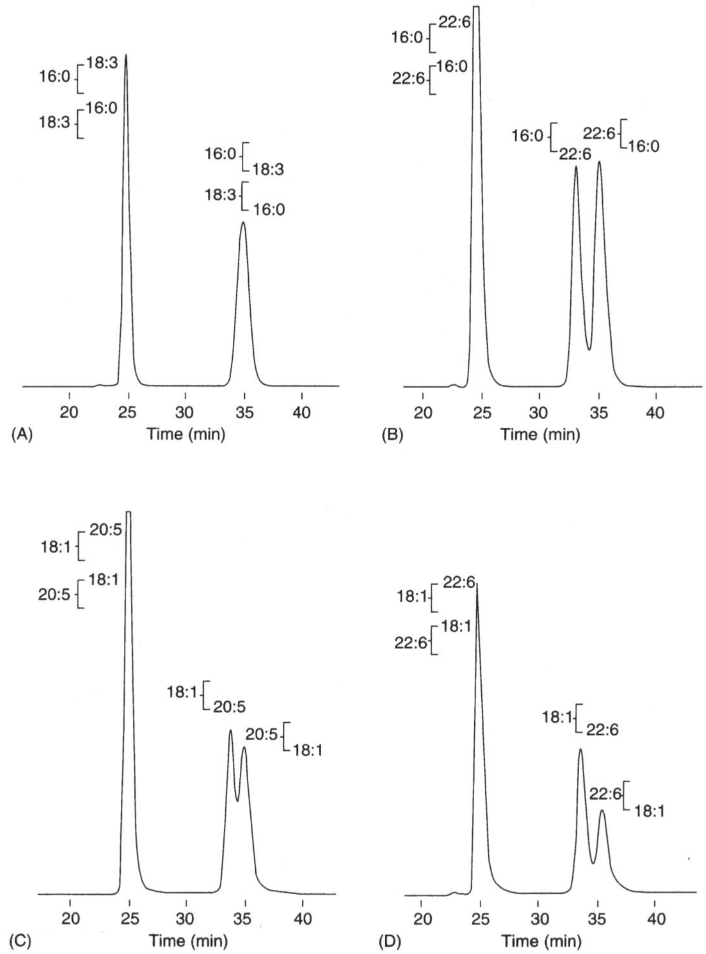

Fig. 6. Chiral-phase HPLC resolution of the reverse isomers and enantiomers of 1,2-DAGs containing polyunsaturated acyl chains as 3,5-DNPUs on (*R*)-1-(1-naphthyl)ethylamine (A-K03). Peaks (left to right): (A) 1-linolenoyl-2-palmitoyl plus 1-palmitoyl-2-linolenoyl-*sn*-glycerol and 2-palmitoyl-3-linolenoyl plus 2-linolenoyl-3-palmitoyl-*sn*-glycerol; (B) 1-docosahexaenoyl-2-palmitoyl plus 1-palmitoyl-2-docosahexaenoyl-*sn*-glycerol, 2-palmitoyl-3-docosahexaenoyl and 2-docosahexaenoyl-3-palmitoyl-*sn*-glycerol; (C) 1-eicosapentaenoyl-2-oleoyl plus 1-oleoyl-2-eicosapentaenoyl-*sn*-glycerol, 2-oleoyl-3-eicosapentaenoyl and 2-eicosapentaenoyl-3-oleoyl-*sn*-glycerol; (D) 1-docosahexaenoyl-3-oleoyl plus 1-oleoyl-2-docosahexaenoyl-*sn*-glycerol, 2-oleoyl-3-docosahexaenoyl and 2-docosahexaenoyl-3-oleoyl-*sn*-glycerol. Mobile phase: hexane/CH_2Cl_2/EtOH (80:20:2, by vol). Peak quantification by UV (254 nm). *Source:* Itabashi et al. (28).

Piyatheerawong et al. (74) have recently reported an analytical method in which enantiomeric *sn*-1,2(2,3)-DAGs can be directly separated without any derivatization. The method is based on Chiralcel OD TM technology, which allows separation of underivatized *sn*-1,2- and *sn*-2,3-DAGs, provided a conventional silica gel column is used in tandem to separate an X-1,3-DAG from *sn*-1,2(2,3)-DAGs before they enter into the chiral stationary phase column. There is no need to isolate the fraction of *sn*-1,2(2,3)-DAG in advance. For this purpose, a silica gel column (Ultrasphere 5 μm column, 250 × 4.6 mm ID, Bachman Coulter Inc., Fullerton, CA, USA) was connected in series with a chiral stationary phase column (cellulose-tris-3,5-dimethylphenylcarbamate)-impregnated silica, 250 × 4.6 mm ID, Daicel Chemical Co., Tokyo, Japan). The mobile phase was hexane-2-propanol (300:7, v/v) at a flow rate of 1.0 mL/min using an HPLC pump (Model 880-PU, Jasco, Tokyo, Japan) at 25°C. A 5 μL aliquot of the sample (containing 10–70 mg of 1,2- and/or 2,3-DAGs dissolved in hexane) was injected using a 10 μL injection loop. The peaks were detected using ELSD (Shimadzu Corporation, Kyoto, Japan). An authentic mixture of 1,3-dicapryloyl and *rac*-1,2-dicapryloylglycerols subjected to the tandem column HPLC yielded the X-1,3-, and *sn*-2,3- and *sn*-1,2-DAGs at 26, 32 and 34 min, respectively. Comparable results were obtained for isomeric mixtures of dioleoylglycerols. The new method was used for the evaluation of the stereospecificity of action of immobilized *Candida antarctica* lipase B and *Rhizomucor miehei* lipase, which possessed opposite stereospecificity. Although enantiomeric separation without derivatization is convenient and may prevent isomerization during preparation of 3,5-DNPUs, it does not exclude isomerization of DAGs in the free form. Free DAGs are known to undergo isomerization even in a solid state (75,76).

Takahashi et al. (27) utilized chiral-phase HPLC for the determination of the stereochemical configuration of the glycerol moieties in glycoglycerolipids. For this purpose DAGs and MAGs were released from the glycosyldi- and MAGs, respectively, by periodate oxidation followed by hydrazinolysis. The released DAGs and MAGs were converted into the 3,5-DNPU and bis-(3,5-DNPU) derivatives, respectively, using standard procedures. The derivatives were separated on two chiral-phases of opposite configuration, (*R*)- and (*S*)-1-(1-naphthyl)ethylamine polymers (YMC A-KO3 and A-LO3) for DAGs and *N*-(*R*)-1-(1-naphthyl)ethylaminocarbonyl-(*S*)-valine (OA-4100) and *N*-(*S*)-1-(1-naphthyl)ethylaminocarbonyl-(*R*)-valine (OA-4100R) for MAGs. Using the above method, the authors determined the glycerol configuration in the glycosyldiacylglycerols (monogalactosyl, digalactosyl-, and sulfoquinovosyldiacylglycerols) and glycosylmonoacylglycerols (monogalactosyl-, digalactosyl, and sulfoquinovosylmonoacylglycerols) isolated from spinach leaves and the coralline red alga *Corallina pilulifera*. The results clearly showed that the glycerol moieties in all the glycoglycerolipids examined have an *S*-configuration (*sn*-1,2-DAGs and *sn*-1-MAGs).

Laakso and Christie (77) and Christie et al. (21) have revised an earlier proposed (78) method of resolution of racemic DAGs as diastereomers. This method depends on the conversion of the enantiomeric *sn*-1,2(2,3)-DAGs into either diastereomeric

NEUs or PEUs (phenylethyluretanes) using optically active naphthylethyl or phenylethyl isocyanate, respectively, followed by separation of the diastereomers on a conventional silica gel column. The method has been utilized for stereospecific analyses of TAGs (77–80). Ågren and Kuksis (22,23) and Kuksis et al. (33) have recently reported improved conditions for the resolution and identification of the diastereomeric DAGs and have combined the method of resolution with on-line MS. Ågren and Kuksis (22) obtained almost complete separation of the *sn*-1,2- and *sn*-2,3-DAGs as their (*R*)- and (*S*)-(+)-1-(1-naphthyl)ethylurethane derivatives using Supelcosil LC-Si columns (5 μm, 250 × 4.5 mm I.D.) and an isocratic elution with 0.37% 2-propanol in hexane at a flow rate of 0.7 mL/min. However, a part of 38 and 40 acyl carbon *sn*-1,2-DAG eluted concurrently with *sn*-2,3-DAG. For example. 16:1/22:6 *sn*-1,2-DAG was resolved into two peaks (indicating a separation on the basis of fatty acid location in DAG), and the first one eluted with the *sn*-2,3-DAG fraction. The overlap with the corresponding *sn*-2,3-DAG species only in a few cases and these values could be corrected using results from the runs of (*S*)-form NEU derivatives (23). The diastereomeric NEU derivatives of *sn*-1,2- and *sn*-2,3-DAGs were identified by a single quadrupole mass spectrometer (Hewlett-Packard Model 5989A, Hewlett-Packard, Palo Alto, CA). The NEU derivatives resolved on two normal phase silica gel columns (Supelcosil LC-Si, 5 μm, 25 cm × 4.6 mm ID, Supelco Inc., Bellefonte, PA) combined in series by 0.37% 2-propanol in hexane were admitted to the mass spectrometer at 0.7 mL/min. Positive chemical ionization was obtained by post-column addition of $CHCl_3$/MeOH/30% NH_4OH (75:24.5:0.5, by vol) at 0.6 mL/min. The capillary exit voltage was 220 V and the mass range scanned was *m/z* 500-720. Figure 7 illustrates an LC/ESI-MS analysis of *sn*-2,3-DAGs and *sn*-1,2-DAGs derived from rat plasma VLDL TAGs. The figure shows the total positive ion current profile along with the single ion mass chromatograms of the major molecular species indicated by the protonated nominal masses and carbon number/double bond number (22).

Normal phase HPLC separation of enantiomeric DAGs on achiral silica gel columns has been carried out for their diastereomers prepared with other chiral UV or fluorescent labeling reagents, such as phenylethyl isocyanate (81), fluoro-(1-naphthyl)acetic acid (82), (*S*)-(+)-2-*tert*-butyl-2-methyl-1,3-benzodioxole-4-carboxylic acid (83) and potentially others (84).

Steric Analysis of Glycerophospholipids

Enantiomeric glycerophospholipids may arise due to the presence of a chiral center in the glycerol, the polyhydroxy headgroup, or in a hydroxy fatty acid esterified to the glycerol backbone. Chiral-phase HPLC for the resolution enantiomeric glycerophospholipids was first used by Itabashi and Kuksis (24) during a reassessment of the stereochemical configuration of natural phosphatidylglycerols (PtdGros). For this purpose, the synthetic and natural PtdGro were converted to their bis-(3,5-DNPU) derivatives, which were separated by HPLC using two columns having chiral phases of opposite configuration, (*R*)- and (*S*)-1-(1-naphthyl)ethylamine poly-

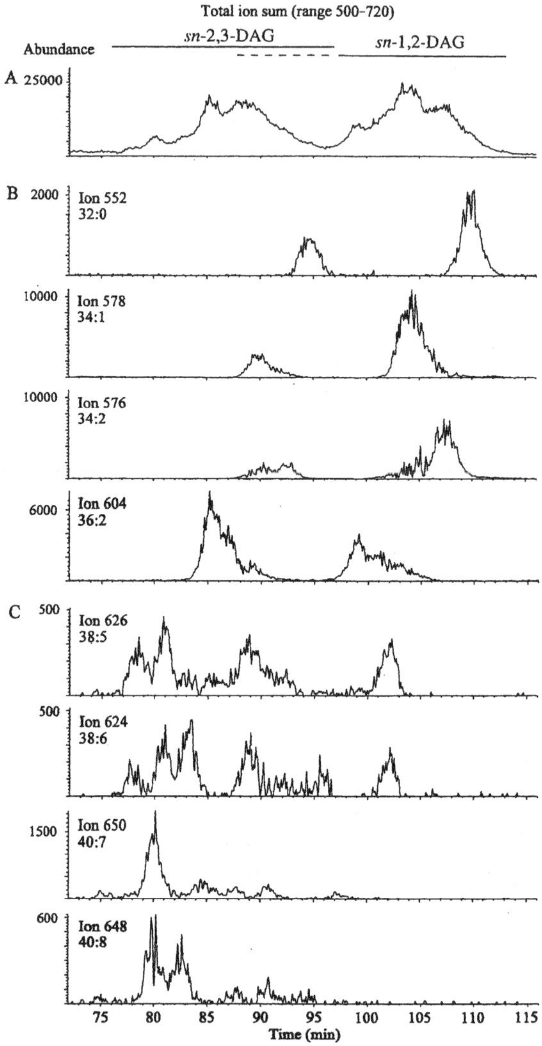

Fig. 7. Chiral-phase LC/ESI-MS analysis of *sn*-2,3-DAG (70–96 min) and sn-1,2-DAG (96–115 min, with some exceptions) of rat plasma VLDL TAG. (A) Total positive ion chromatogram (ion sum from *m/z* 500 to 720); (B) and (C), single ion mass chromatograms of selected molecular species indicated by protonated nominal mass and carbon number/number of double bonds. HPLC conditions: two normal phase silica gel columns (Supelcosil LC-SI, 5μm, 25 cm × 4.6 mm ID, Supelco Inc.) in series were used and 0.37% isopropanol in hexane was used as mobile phase at a flow of 0.7 mL/min. Positive ionization obtained with postcolumn addition of $CHCl_3$/MeOH/30% NH_4OH (75:24.5:0.5, by vol) at 0.6 mL/min. *Source:* Ågren et al. (22).

mers. The molecular species were identified by on-line negative-ion ESI-MS. Absolute configurations of the resolved peaks were assigned by comparison with the elution order of the corresponding *sn*-1(3)-MAG enantiomers as bis-(3,5-DNPU) derivatives on the same column. They demonstrated that chiral-phase HPLC and negative-ion ESI-MS provided direct and unambiguous information about the configuration, identity, and quantity of molecular species in natural and synthetic PtdGros. Subsequently, Sato et al. (85) have employed this method for the determination of the stereochemical configuration of PtdGros prepared by bacterial phospholipase D-catalyzed transphosphatidylation from PtdCho and glycerol. The resulting PtdGros were converted into bis-(3,5-DNPU) derivatives, which were separated on a column containing (*R*)-1-(1-naphthyl)ethylamine polymer. All chromatograms obtained for the bacterial phospholipase D products showed two clearly resolved peaks of different proportions of diastereomers ranging from 30–40% 1,2-diacyl-*sn*-glycero-3-phospho-1′-*sn*-glycerol (*R,S*-configuration) and 60–70% 1,2-diacyl-*sn*-glycero-3-phospho-3′-*sn*-glycerol (*R,R*-configuration). The asymmetric proportions were absent from the diastereomers prepared with cabbage or peanut phospholipase D, which gave equimolar mixtures of *R,S*- and *R,R*-diastereomers. It was concluded that bacterial phospholipase D catalyzed transphosphatidylation proceeds to a large extent stereospecifically, showing a preference for the *sn*-3′-position of the glycerol molecule. Figure 8 demonstrates a marked effect of temperature on the stereoselectivity of phospholipase D from *Streptomyces septatus* toward the two primary hydroxyl groups of glycerol in the transphosphatidylation reaction of PtdCho to PtdGro (29). The proportion of the *R,R*-isomer produced with phospholipase D from *Streptomyces* TH-2 and *Actinomadura* sp. increased gradually from 50% *R,R* and 50% *R,S* at 50–60°C to 70% *R,R* and 30% *R,S* at 0°C. On the basis of these studies, Sato et al. (86) developed a simple method for synthesizing diastereomerically pure PtdGros, namely 1,2-diacyl-*sn*-glycero-3-phospho-3′-*sn*-glycerol (*R,R*-configuration) and 1,2-diacyl-*sn*-glycero-3-phospho-1′-*sn*-glycerol (*R,S*-configuration). For this purpose, diastereomeric 1,2-*O*-isopropylidene PtdGro were prepared from 1,2-diacyl-*sn*-glycero-3-phosphocholine and enantiomeric 1,2-*O*-isopropylidene glycerols by transphosphatidylation with phospholipse D from *Actinomadura* in acetate buffered diethyl ether at 30°C. The isopropylidine protective group was removed by heating the diastereomeric isopropylidine PtdGro at 100°C in trimethylborate in the presence of boric acid (75).

Itabashi and Kuksis (87) investigated the separation of enantiomeric and diastereomeric bis(monoacylglycerol)phosphate (BMP) on chiral-phase HPLC. For this purpose, the BMP stereoisomers were prepared by lipase (*Rhizopus niveus*) hydrolysis of the corresponding bis-phosphatidic acids synthesized chemically from PtdGros of known configuration, which permitted the assignment of the absolute configurations of the resolved BMP peaks. The synthetic BMP isomers were converted to their bis-(3,5-DNPU) derivatives, which were separated by HPLC using two columns (250 × 4.6 mm I.D.) having chiral phases of opposite configurations, *N*-[(*S*)-1-(1-naphthyl)ethylaminocarbonyl]-L-*tert*-leucine (OA-

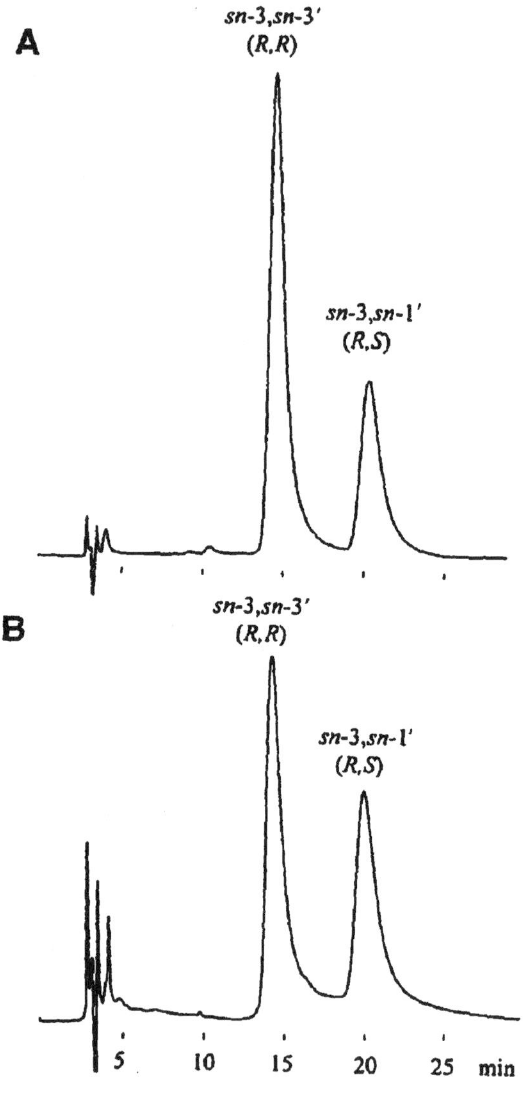

Fig. 8. Chiral-phase HPLC profiles of the bis-(3,5-DNPU) derivatives of the PtdGros generated from 1,2-dioleoyl-*sn*-glycero-3-phosphocholine and glycerol by transphosphatidylation with phospholipase D from *Streptomyces septatus* TH-2. (A) 0°C; (B) 60°C. *sn*-3, *sn*-1′, 1,2-dioleoyl-*sn*-glycero-3-phospho-1′-*sn*-glycerol (*R,S* configuration); *sn*-3, *sn*-3′, 1,2-dileoyl-*sn*-glycero-3-phospho- 3′-*sn*-glycerol (*R,R* configuration). Chiral-phase HPLC was done on A-KO3 (25 cm × 4.6 mm ID, YMC, Kyoto, Japan) using hexane/CH_2Cl_2/MeOH/TFA (60:20:20:0.2, by vol) as the mobile phase at a flow rate of 1 mL/min. Column temp., 10°C; detection, 254 nm UV. *Source:* Sato et al. (29).

4600) and *N*-[(*R*)-1-(1-naphthyl)ethylaminocarbonyl]-D-*tert*-leucine (OA-4600R). The analysis was done isocratically using hexane/CH_2Cl_2/MeOH/TFA (70:20:10:0.2, by vol) and peaks were detected under UV (254 nm). Complete resolution was achieved for the enantiomeric 2-palmitoyl-*sn*-glycerol-3-phospho-3'-(2'-palmitoyl)-*sn*-glycerol (*R,R*-configuration on the chiral centers) and 2-palmitoyl-*sn*-glycerol-1-phospho-1′-(2′-palmitoyl)-*sn*-glycerol (*S,S*), and the corresponding meso compound (*R,S* or *S,R*) eluted between the *R,R*- and *S,S* -enantiomers (Figure 9). The elution order of the *R,R*- and *S,S*- enantiomers reversed on the two columns. The negative ESI-MS spectra of the derivatives, which were essentially the same for each peak resolved on the chiral phases, gave a prominent deprotonated molecule $[M-H]^-$. Joutti and Renkonen (88) had described a procedure for stereoanalysis of radiochemically labeled glycerophospholipids on the basis of labeled α-glycerophosphate, which retains its original configuration when liberated upon alkaline hydrolysis of the lipids.

Finally, Itabashi et al. (89) have reported the chiral-phase resolution of diastereomeric phosphatidyl-D-serine (D-PtdSer) and L-serine (L-PtdSer) following derivatization with 3,5-DNPI and purification by normal phase HPLC on a polyvinyl alcohol-bonded silica column (YMC PVA-SIL, 100 × 3 mm I.D.). Complete resolution of the diastereomeric L-PtdSer and D-PtdSer was achieved by chromatography on two chiral stationary phases of opposite configuration, (*R*)- and (*S*)-1-(1-naphthyl)ethylamine polymers (YMC A-KO3 and A-LO3 columns, each 250 × 4.6 mm ID), using hexane/CH_2Cl_2/MeOH/TFA (70:20:10:0.2, by vol) as the mobile phase. The diastereomers eluted with a reversal in the elution order on the two columns.

Figure 10 shows the resolution of synthetic L-PtdSer and D-PtdSer as 3,5-dinitrophenyl-urea derivatives by chiral-phase HPLC on A-L03 and A-K03, respectively. The mobile phase in both instances was hexane/CH_2Cl_2/MeOH/TFA (70:20:10:0.2, by vol) with a flow rate of 1 mL/min at 10°C. The peak elution was monitored by UV (254 nm) and identification confirmed by ESI-MS. The diastereomeric 1,2-diacyl-*sn*-glycero-3-phospho-L-serine (L-PtdSer) and 1,2-diacyl-*sn*-glycero-3-phospho-D-serine (D-PtdSer) were prepared from L- and D-serine, respectively, by phospholipase D (*Streptomyces*)-catalyzed transphosphatidylation of 1,2-diacyl-*sn*-glycero-3-phosphocholine. For chiral-phase HPLC resolution, the PtdSers were converted into the 3,5-dinitrophenylurea by reaction with 3,5-DNPI and purified by normal phase HPLC on a polyvinyl alcohol bonded silica column (YMC PVA-SIL, 100 × 3 mm I.D.). The purified urea derivatives were introduced into a chiral column by a column-switching method. Complete resolution of the diastereomeric L- and D-PtdSer derivatives was achieved on two chiral stationary phases of opposite configuration, (*R*)- and (*S*)-1-(1-naphthyl)ethylamine polymers (YMC A-K03 and A-L03 columns, each 250 × 4.6 mm I.D.), using hexane/CH_2Cl_2/MeOH/TFA (70:20:10:0.1, by vol). Negative ESI-MS gave the same spectra for the urea derivatives of the diastereomeric L-PtdSer and D-PtdSer, yielding a prominent deprotonated molecule $[M-H]^-$. Formation of various hydrogen bonds, dipole-dipole stacks, and charge transfer complexes

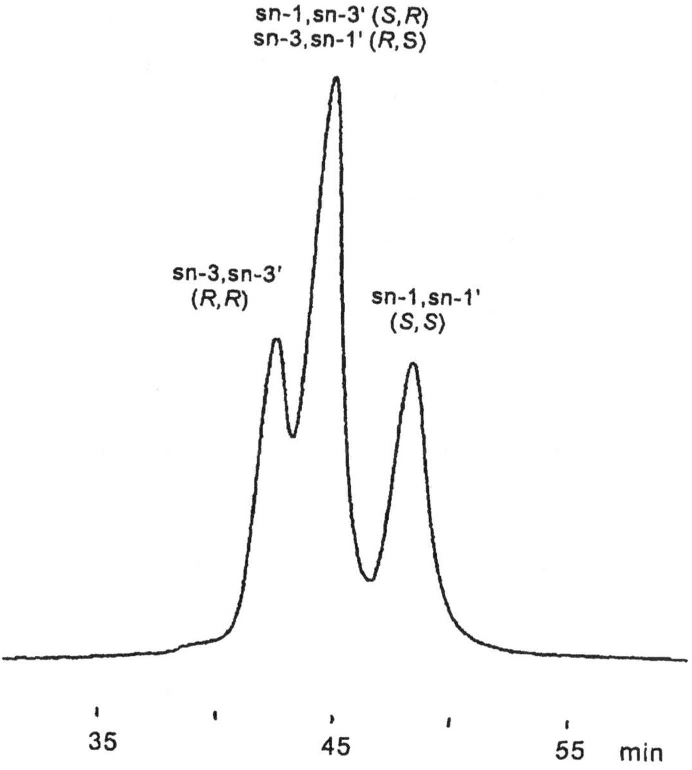

Fig. 9. Chiral-phase HPLC resolution of the enantiomeric and diastereomeric bis-(3,5-DNPU) derivatives of synthetic bis(monoacylglycerol)phosphates (2-palmitoyl-*rac*-glycero-1-phospho-1′-(2′-palmitoyl)-*rac*-glycerol). Peak identification: *sn*-3,*sn*-3′, 2-palmitoyl-*sn*-glycero-3-phospho-3′-(2′-palmitoyl)-*sn*-glycerol (*R,R* configuration); *sn*-1,*sn*-1′, 2-palmitoyl-*sn*-glycero-1-phospho- 1′-(2′-palmitoyl)-*sn*- glycerol (*S,S* configuration); *sn*-1,*sn*-3′, 2-palmitoyl-*sn*-glycero-1-phospho-3′-(2-palmitoyl)-*sn*-glycerol (*S,R* configuration); *sn*-3,*sn*-1′, 2-palmitoyl-*sn*-glycero-3-phospho-1′-(2′-palmitoyl)-*sn*-glycerol (*R,S* configuration). The *sn*-1,*sn*-3' (*S,R*) isomer is a meso compound and identical to *sn*-3,*sn*-1′ (*R,S*) isomer. Column: (*S*)-1-(1-naphthyl)ethylaminocarbonyl-(*S*)-*tert*-leucine (Sumichiral OA-4600, 250 × 4.6 mm I.D.); Mobile phase: hexane/CH_2Cl_2/MeOH/TFA (70:20:10:0.2, by vol) at 0.5 mL/min. Column temp., 10°C; Peaks were detected by UV (254 nm). *Source:* Itabashi and Kuksis (87).

between the solutes and the chiral stationary phases was thought to be responsible for the diastereomer separation (90). Using this method, the PtdSer isolated from rat brain and mackerel flesh clearly indicated that the serine moiety in both samples had the L-configuration.

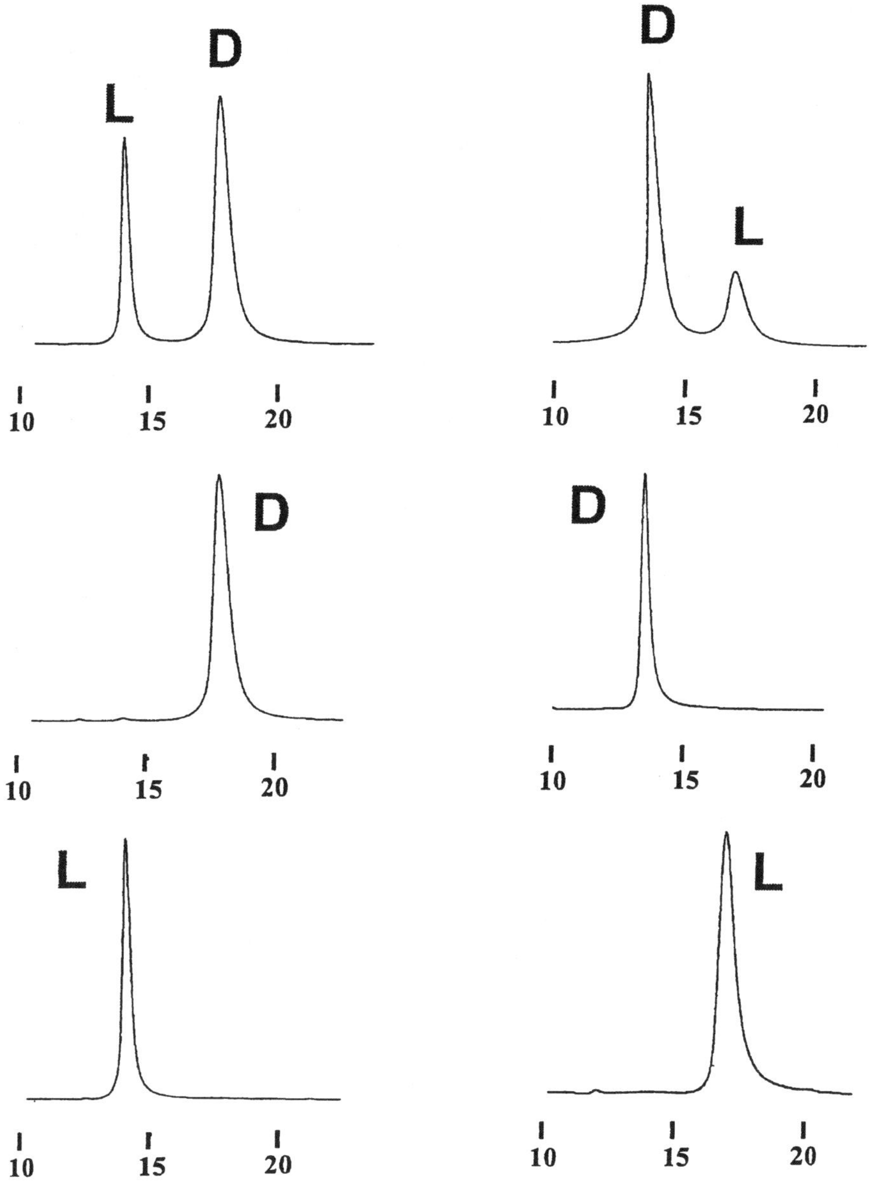

Fig. 10. Chiral-phase HPLC separation of the 3,5-dinitrophenylurea derivatives of synthetic dipalmitoyl GroPSer(D) and dipalmitoyl GroPSer(L). *Left Panel*, A-L03 column; *Right Panel*, A-K03 column. Mobile phase: hexane/CH_2Cl_2/MeOH/TFA (70:20:10:0.2, by vol) at 1 mL/min (UV detection, 254 nm). Column temp., 10°C. *Source:* Itabashi et al. (89).

Summary

This review shows that the chiral-phase HPLC separations of synthetic MAG and DAG demonstrated by Itabashi and Takagi in the 1980s can be readily extended to the resolution of naturally occurring enantiomeric MAGs and DAGs and to acylglycerols derived from complex lipids. A combination of chiral-phase HPLC with on-line MS permits the identification of the molecular species of the enantiomers and reverse isomers. A practical consequence of this development has been an accurate determination of the stereospecific positional placement of fatty acids in all glycerolipids. The development of improved chiral-phase columns along with optimized chromatographic elution conditions has permitted the further extension of chiral-phase LC/MS the determination of the molecular species of enantiomers of intact glycerophospholipids, such as phosphatidylserine, phosphatidylglycerol and bis(lysophosphatidic acid), which has allowed a reinvestigation of the structural configuration of natural polyglycerol phosphates previously assessed by laborious classical methodology only. Clearly, the new methodology is ready for meaningful practical application in metabolic and lipidomic studies of which, thus far, there have been few.

Acknowledgments

These studies were supported in part by various Grants-in-Aid for Scientific Research from the Ministry of Education, Sciences, Sports and Culture of Japan, and by Research Fellowships of the Japan Society for the Promotion of Science for Young Scientists. The work in Canada was supported by the Heart and Stroke Foundation of Ontario, Toronto, Canada and the Medical Research Council of Canada, Ottawa, Ontario.

References

1. Han, X., and R.W. Gross, Electrospray Ionization Mass Spectroscopic Analysis of Human Erythrocyte Plasma Membrane Phospholipids, *Proc. Natl. Acad. Sci. USA* *91*:10635–10639 (1994).
2. Kerwin, J.L., A.R. Tuininga, and L.H. Ericsson, Identification of Molecular Species of Glycerophospholipids and Sphingomyelin Using Electrospray Mass Spectrometry, *J. Lipid Res.* *35*:1102–1114 (1994).
3. Byrdwell, W.C., and W.E. Neff, Qualitative and Quantitative Analysis of Triacylglycerols Using Atmospheric-Pressure Chemical Ionization Mass Spectrometry, in *New Techniques and Applications in Lipid Analysis* (R.E. McDonald and M.M. Mossoba, editors), AOCS Press, Champaign, IL, pp. 45–80 (1997).
4. Han, X., R.A. Gubitosi-Klug, B.J. Collins, and R.W. Gross, Alterations in Individual Molecular Species of Human Platelet Phospholipids During Thrombin Stimulation: Electrospray-Ionization Mass Spectrometry-Facilitated Identification of the Boundary Conditions for the Magnitude and Selectivity of Thrombin-Induced Platelet Phospholipid Hydrolysis, *Biochemistry* *35*:5822–5832 (1996).
5. Koivusalo, M., P. Haimi, L. Heikinheimo, R. Kostiainen, and P. Somerharju, Quantitative Determination of Phospholipid Compositions by ESI-MS: Effects of Acyl Chain Length, Unsaturation, and Lipid Concentration on Instrument Response, *J. Lipid Res.* *42*:663–672 (2001).

6. Han, X., J. Yang, H. Cheng, H. Ye, and R.W. Gross, Toward Fingerprinting Cellular Lipidomes Directly from Biological Samples by Two-Dimensional Electrospray Ionization Mass Spectrometry, *Anal. Biochem. 330*:317–331 (2004).
7. Han, X., and R.W. Gross, Shotgun Lipidomics: Electrospray Ionization Mass Spectrometric Analysis and Quantitation of Cellular Lipidomes Directly from Crude Extracts of Biological Samples, *Mass Spectrom. Rev. 24*:367–412 (2005).
8. Karlsson, A.Å., P. Michelsen, Å. Larsen, and G. Odham, Normal Phase Liquid Chromatography Class Separation and Species Determination of Phospholipids Utilizing Electrospray Mass Spectrometry/Tandem Mass Spectrometry, *Rapid Commun. Mass Spectrom. 10*:775–780 (1996).
9. Takagi, T., Chromatographic Resolution of Chiral Lipid Derivatives, *Progr. Lipid Res. 29*:277–298 (1990).
10. Kuksis, A., Analysis of Positional Isomers of Glycerolipids by Non-Enzymatic Methods, in *Advances in Lipid Methodology–Three* (W.W. Christie, editor), The Oily Press, Dundee, Scotland, pp. 1–36 (1996).
11. Pirkle, W.H., T.C. Pochapsky, Considerations of Chiral Recognition Relevant to the Liquid Chromatographic Separation of Enantiomers, *Chem. Rev. 89*:347–362 (1989).
12. Kuksis, A. and Y. Itabashi, Regio- and Stereospecific Analysis of Glycerolipids, *Methods 36*:172–185 (2005).
13. Itabashi, Y., Chiral Phase HPLC of Glycerolipids, in *HPLC of Acyl Lipids* (J.T. Lin and T.A. McKeon, editors), HNB Publishing, New York, NY, pp. 167–197 (2005).
14. Schneider, C., W.E. Boeglin and A.R. Brash, Enantiomeric Separation of Hydroxy Eicosanoids by Chiral Column Chromatography: Effect of the Alcohol Modifier, *Anal. Biochem. 287*:186–189 (2000).
15. Lee, S.H., M.V. Williams, R.N. DuBois and I.A. Blair, Targeted Lipidomics Using Electron Capture Atmospheric Pressure Chemical Ionization Mass Spectrometry, *Rapid Commun. Mass Spectrom. 17*:2168–2176 (2003).
16. Hammonds, T.D., I.A. Blair, J.R. Falck, and J.H. Capdevila, Resolution of Epoxyeicosatrienoate Enantiomers by Chiral Phase Chromatography, *Anal. Biochem. 182*:300–303 (1989).
17. Hawkins, D.J., H. Kühn, E.H. Petty, and A.R. Brash, Resolution of Enantiomers of Hydroxyeicosatetraenoate Derivatives by Chiral Phase High-Pressure Liquid Chromatography, *Anal. Biochem. 173*:456–462 (1988)
18. Itabashi, Y., and T. Takagi, High Performance Liquid Chromatographic Separation of Monoacylglycerol Enantiomers on a Chiral Stationary Phase. *Lipids 21*, 413–416 (1986).
19. Itabashi, Y., and T. Takagi, High Performance Liquid Chromatographic Separation of Diacylglycerol Enantiomers on a Chiral Stationary Phase, *J. Chromatogr. 402*:257–264 (1987).
20. Ando, Y., and T. Takagi, Micromethod for Stereospecific Analysis of Triacyl-*sn*-Glycerols by Chiral-Phase High Performance Liquid Chromatography, *J. Am. Oil Chem. Soc. 70*:1047–1049 (1993).
21. Christie, W.W., B. Nikolova-Damyanova, P. Laakso, and B. Herslof, Stereospecific Analysis of Triacyl-*sn*-Glycerols via Resolution of Diastereomeric Diacylglycerol Derivatives by High Performance Liquid Chromatography on Silica, *J. Am. Oil Chem. Soc. 68*:695–701 (1991).
22. Ågren, J.J., and A. Kuksis, Analysis of Diastereomeric DAG Naphthylethylurethanes by Normal-Phase HPLC with On-Line Electrospray MS. *Lipids 37,* 613–619 (2002).

23. Ågren, J.J., A. Ravandi, A. Kuksis, and G. Steiner, Structural and Compositional Changes in Very Low Density Lipoprotein Triacylglycerols During Basal Lipolysis, *Eur. J. Biochem. 269*:6223–6232 (2002).
24. Itabashi, Y., and A. Kuksis, Reassessment of Stereochemical Configuration of Natural Phosphatidylglycerols by Chiral-Phase High-Performance Liquid Chromatography and Electrospray Mass Spectrometry, *Anal. Biochem. 254*:49–56 (1997).
25. Schneider, C., W.E. Boeglin, S. Lai, J.K. Cha, and A.R. Brash, Synthesis and Applications of Stereospecifically ^{3}H-Labeled Arachidonic Acids as Mechanistic Probes for Lipoxygenase and Cyclooxygenase Catalysis, *Anal. Biochem. 284*:125–135 (2000).
26. Oliw, E.H., Separation of Epoxides of Fatty Acids by Chiral Phase HPLC on a Stationary Phase with Cellulose Carbamate (Chiralcel OC), *J. Chromatogr. 583*:231–235 (1992).
27. Takahashi, T., Y. Itabashi, M. Suzuki, and A. Kuksis, A Simple Method for the Determination of Stereochemical Configuration of the Glycerol Moieties in Glycoglycerolipids by Chiral Phase High Performance Liquid Chromatography, *Lipids 36*:741–748 (2001); Itabashi, Y., A. Kuksis, L. Marai, and T. Takagi, HPLC Resolution of Diacylglycerol Moieties of Natural Triacylglycerols on a Chiral Phase Consisting of Bonded (*R*)-(+)-1-(1-naphthyl)ethylamine, *J. Lipid Res. 31*:1711–1717 (1990).
28. Itabashi, Y., J.J. Myher, and A. Kuksis, High Performance Liquid Chromatographic Resolution of Reverse Isomers of 1,2-Diacyl-*rac*-Glycerols as 3,5-Dinitrophenylurethanes, *J. Chromatogr. A 893*:261–279 (2000).
29. Sato, R., Y. Itabashi, A. Suzuki, T. Hatanaka, and A. Kuksis, Effect of Temperature on the Selectivity of Phospholipase D Toward Glycerol in Transphosphatidylation of Phosphatidylcholine to Phosphatidylglycerol, *Lipids 39*:1019–1023 (2004).
30. Itabashi, Y., L. Marai, and A. Kuksis, Identification of Natural Diacylglycerols as the 3,5-Dinitrophenylurethanes by Chiral Phase Liquid Chromatography with Mass Spectrometry, *Lipids 26*:951–956 (1991).
31. Marai, L., A. Kuksis, J.J. Myher, and Y. Itabashi, Liquid chromatography chloride attachment negative chemical ionization mass spectrometry of diacylglycerol dinitrophenylurethanes, *Biol. Mass Spectrom. 21*:541–547 (1992).
32. Myher, J.J., A. Kuksis, and P.W. Park, Stereospecific Analysis of Docosahexaenoic Acid-Rich Triacyglycerols by Chiral-Phase HPLC with On-Line Electrospray Mass Spectrometry, in *New Techniques and Applications in Lipid Analysis* (R.E. McDonald and M.M. Mossoba, editors), AOCS Press, Champaign, IL, pp. 100–120 (1997).
33. Kuksis, A., J. Park, and J.J. Agren, Improved Stereochemical Analysis of Triacylglycerols (TAGs) as Diastereomeric Diacylglycerol (DAG) Naphthylethylurethanes (NEUs) by Normal and Reversed Phase LC/ESI-MS, in *Abstracts, 95th AOCS Annual Meeting & Expo*, Cincinnati, Ohio, May 9–12, 2004, page 3.
34. Kühn, H., R. Wiesner, V.Z. Lankin, A. Nekrasov, L. Aler, and T. Schewe, Analysis of the Stereochemistry of Lipoxygenase-Derived Hydroxypolyenoic Fatty Acids by Chiral Phase High Pressure Liquid Chromatography, *Anal. Biochem. 160*:24–34 (1987).
35. Brash, A., W. Boeglin, G. Coffa, Z. Yu, and C. Schneider, Chiral HPLC Resolution of Eicosanoids and Other Oxylipins, in *Abstracts, 95th AOCS Annual Meeting & Expo*, Cincinnati, Ohio, May 9–12, 2004, page 3.
36. Hammarstrom, S. and M. Hamberg, Steric Analysis of 3-, ω4-, ω3-, and ω2-Hydroxy Acids and Various Alkanols by Gas-Liquid Chromatography, *Anal. Biochem. 52*:169–179 (1973).

37. Oliw, E.H., Resolution of Enantiomers of 19-Hydroxyeicosatetraenoate and 18-Hydroxyeicosatetraenoate by Chiral Phase High-Performance Liquid Chromatography of Naphthoyl Ester Derivatives, *J. Chromatogr. 526*:525–529 (1990).
38. Baer, A.N., P.B. Castello, and F.A. Green, Characterization of the Chirality of the Monohydroxy-eicosatetraenoic Acids Produced by Rat Basophilic Leukemia Cells, *Biochem. Biophys. Res. Commun. 169*:332–338 (1990).
39. Baer, A.N., P.B. Costello, and F.A. Green, Free and Esterified 13(R,S)-Hydroxyoctadecadienoic Acids: Principal Oxygenase Products in Psoriatic Skin Scales, *J. Lipid Res. 31*:125–130 (1990).
40. Baer, A.N., P.B. Costello, and F.A. Green, Stereospecificity of the Products of the Fatty Acid Oxygenases Derived from Psoriatic Scales, *J. Lipid Res. 32*:341–347 (1991).
41. Morrow, J.D., J.A. Awad, H.J. Boss, I.A. Blair, and L.J. Roberts II, Non-Cyclooxygenase-Derived Prostanoids (F_2-Isoprostanes) are Formed in situ on Phospholipids, *Proc. Natl. Acad. Sci. USA 89*:10721–10725 (1992).
42. Kayganich-Harrison, K.A., D.M. Rose, R.C. Murphy, J.D. Morrow and L.J. Roberts, Collision-Induced Dissociation of F_2-Isoprostane-Containing Phospholipids, *J. Lipid Res. 34*:1229–1235 (1993).
43. Morrow, J.D., and L.J. Roberts, The Isoprostanes: Unique Bioactive Products of Lipid Peroxidation, *Progr. Lipid Res. 36*:1–21 (1997); Morrow, J.D., and L.J. Roberts II, The Isoprostanes: Current Knowledge and Directions for Future Research, *Biochem. Pharmacol. 51*:1–9 (1996).
44. Khaselev, N., and R.C. Murphy, Structural Characterization of Oxidized Phospholipid Products Derived from Arachidonate-Containing Plasmenyl Glycerophosphocholine, *J. Lipid Res. 41*:564–572 (2000).
45. Brash, A.R., W.E. Boeglin, J.H. Capdevila, S. Yeola, and I.A. Blair, 7-HETE, 10-HETE, and 13-HETE are Major Products of NADPH-Dependent Arachidonoic Acid Metabolism in Rat Liver Microsomes: Analysis of Their Stereochemistry, and Stereochemistry of Their Acid-Catalyzed Rearrangement, *Arch. Biochem. Biophys. 321*:485–492 (1995).
46. Griffiths, W.J., Tandem Mass Spectrometry in the Study of Fatty Acids, Bile Acids, and Steroids, *Mass Spectrom Revs. 22*:81–152 (2003).
47. Nakamura, T., P.M. Henson, and R.C. Murphy, Occurrence of Oxidized Metabolites of Arachidonic Acid Esterified to Phospholipids in Murine Lung Tissue, *Anal. Biochem. 262*:23–32 (1998).
48. Wheelan, P., J.A. Zirolli, and R.C. Murphy, Low Energy Fast Atom Bombardment Tandem Mass Spectrometry of Monohydroxy Substituted Unsaturated Fatty Acids, *Biol. Mass Spectrom. 22*:465–473 (1993).
49. Bylund, J., J. Ericsson, and E.H. Oliw, Analysis of Cytochrome P_{450} Metabolites of Arachidonic and Linoleic Acids by Liquid Chromatography-Mass Spectrometry with Ion Trap MS^2, *Anal. Biochem. 265*:55–68 (1998).
50. Hamberg, M., Stereochemistry of Oxygenation of Linoleic Acid Catalyzed by Prostaglandin-Endoperoxide H Synthases-2, *Arch. Biochem. Biophys. 349*:376–380 (1998).
51. Oliw, E.H., D. Brodowsky, L. Hornsten, and M. Hamberg, Bis-allylic Hydroxylation of Polyunsaturated Fatty Acids by Hepatic Monooxygenase and Its Relation to the Enzymatic and Non-Enzymatic Formation of Conjugated Hydroxy Fatty Acids, *Arch. Biochem. Biophys. 300*:434–439 (1993).

52. Suzuki, N., T. Hishinuma, T. Saga, J. Sato, T. Toyota, J. Goto, and M. Mizugaki, Determination of Urinary 12(S)-Hydroxyeicosatetraenoic Acid by Liquid Chromatography-Tandem Mass Spectrometry With Column-Switching Technique: Sex Difference in Healthy Volunteers and Patients With Diabetes Mellitus, *J. Chromatogr. B 783*:383–389 (2003).
53. Capdevila, J., P. Yadagiri, S. Manna, and J.R. Falck, Absolute Configuration of the Hydroxyeicosatetraenoic Acids (HETEs) Formed During Catalytic Oxygenation of Arachidonic Acid by Microsomal Cytochrome P-450, *Biochem. Biophys. Res. Commun. 141*:1007–1011 (1986).
54. Boeglin, W.E., R.B. Kim, and A.R. Brash, A 12*R*-Lipoxygenase in Human Skin: Mechanistic Evidence, Molecular Cloning, and Expression, *Proc. Natl. Acad. Sci. USA 95*:6744–6749 (1998).
55. Bayer, M., A. Mosandi, and D. Thaci, Improved Enantioselective Analysis of Polyunsaturated Hydroxy Fatty Acids in Psoriatic Skin Scales Using High Performance Liquid Chromatography, *J. Chromatogr. B 819*:323–328 (2005).
56. Hiratsuka, A., K. Hirose, H. Saito, and T. Watabe, 4-Hydroxy-2(E)-Nonenal Enantiomers: (*S*)-Selective Inactivation of Glyceraldehyde-3-Phosphate Dehydrogenase and Detoxification by Rat Glutathione S-Transferase A4-4, *Biochem. J. 349*:729–735 (2000).
57. Capdevila, J.H., S. Wei, A. Kumar, J. Kobayasi, J.R. Snapper, D.C. Zeldin, R.K. Bhatt, and J.R. Falck, Resolution of Dihydroxyeicosanoates and of Dihydroxyeicosatrienoates by Chiral Phase Chromatography, *Anal. Biochem. 207*:236–240 (1992).
58. Oliw, E.H., J. Bylund, and C. Herman, Bisallylic Hydroxylation and Epoxidation of Polyunsaturated Fatty Acids by Cytochrome P_{450}, *Lipids 31*:1003–1021 (1996).
59. Knothe, G., D. Weisleder, M.O. Bagby, and R.E. Peterson, Hydroxy Fatty Acids Through Hydroxylation of Oleic Acid With Selenium Dioxide/*tert*-Butylhydroperoxide, *J. Am. Oil Chem. Soc. 70*:401–404 (1993).
60. Takagi, T., Resolution of *vic*-Dihydroxy Acid Diastereomers to Four Enantiomers by High-Performance Liquid Chromatography, *J. Am. Oil Chem. Soc. 71*:547–548 (1994).
61. Takagi, T., and Y. Ando, Stereospecific Analysis of Triacyl-*sn*-Glycerols by Chiral High-Performance Liquid Chromatography, *Lipids 26*:542–547 (1991).
62. Ando, Y., M. Stake, and Y. Takahashi, Reinvestigation of Positional Distribution of Fatty Acids in Docosahexanoic Acid-Rich Fish Oil Triacyl-*sn*-Glycerols, *Lipids 35*:579–582 (2000).
63. Ando, Y., T. Ota, Y. Matsuhira, and K. Yazawa, Stereospecific Analysis of Triacyl-*sn*-Glycerols in Docosahexaenoic Acid-Rich Fish Oils, *J. Am. Oil Chem. Soc. 73*:483–487 (1996).
64. Ando, Y., K. Nishimura, N. Aoyanagi, and T. Takagi. Stereospecific Analysis of Fish Oil Triacyl-*sn*-Glycerols, *J. Am. Oil Chem. Soc. 69*:417–424 (1992).
65. Taylor, D.C., S.L. MacKenzie, A.R. McCurdy, P.B.E. McVetty, E.M. Giblin, E.W. Pass, S.J. Stone, R. Scarth, S.R. Rimmer, and M.D. Pickard, Stereospecific Analyses of Seed Triacylglycerols from High-Erucic Acid *Brasssicaceae*: Detection of Erucic Acid at the *sn*-2-Position in *Brassica oleracea* L. Genotypes, *J. Am. Oil Chem. Soc. 71*:163–167 (1994).
66. Takagi, T., Resolution of Chiral Lipid Components Into Enantiomers by Chiral-Phase HPLC, in *Abstracts, 95th AOCS Annual Meeting & Expo*, Cincinnati, Ohio, May 9–12, 2004, p. 2.
67. Itabashi, Y., H. Fujishima, and R. Sato, Chiral Phase High-Performance Liquid Chromatographic Separation of Enantiomeric 1,2- and 2,3-O-Isopropylidine-*sn*-Glycerols as 3,5-Dinitrophenylurethanes, *J. Oleo Sci. 53*:405–412 (2004).

68. Itabashi, Y., A. Kuksis, and J.J. Myher, Determination of Molecular Species of Enantiomeric Diacylglycerols by Chiral Phase HPLC on Polar Capillary GLC, *J. Lipid Res.* *31*:2119–2126 (1990).
69. Itabashi, Y., J.J. Myher, and A. Kuksis, Determination of Positional Distribution of Short-Chain Fatty Acids in Bovine Milk Fat on Chiral Columns, *J. Am. Oil Chem. Soc.* *70*:1177–1181 (1993).
70. Myher, J.J., A. Kuksis, K. Geher, P.W. Park, and D.A. Diersen, Stereospecific Analysis of Triacylglycerols Rich in Long-Chain Polyunsaturated Fatty Acids, *Lipids* *31*:207–215 (1996).
71. Yang, L-Y., A. Kuksis, J.J. Myher, and G. Steiner, Contribution of *De Novo* Fatty Acid Synthesis to Very Low Density Lipoprotein Triacylglycerols: Evidence from Mass Isotopomer Distribution Analysis of Fatty Acids Synthesized from [$^2H_6$]Ethanol, *J. Lipid Res.* *37*:262–274 (1996).
72. Lehner, R., A. Kuksis, and Y. Itabashi, Stereospecificity of Monoacylglycerol and Diacylglycerol Acyltransferases from Intestine as Determined by Chiral Phase High Performance Liquid Chromatography, *Lipids 28*:29–34 (1993).
73. Okabe, H., Y. Itabashi, T. Ota, and A. Kuksis, Highly Sensitive Method for the Separation of Enantiomeric and Regioisomeric Diacylglycerols as 2-Anthrylurethanes by Chiral-Phase High-Performance Liquid Chromatography with Fluorescence Detection, *J. Chromatogr. A 829*:81–89 (1998).
74. Piyatheerawong, W., Y. Iwasaki, and T. Yamane, Direct Separation of Regio- and Enantiomeric Isomers of Diacylglcerols by a Tandem Column High-Performance Liquid Chromatography, *J. Chromatogr. A 1068*:243–248 (2005).
75. Buchnea, D., Stereospecific Synthesis of Enantiomeric Acylglyerols, in *Handbook of Lipid Research*, Vol. 1: *Fatty Acids and Glycerides* (A. Kuksis, editor) Plenum Press, 1978, pp. 233–287.
76. Kodali, D.R., A. Tercyak, D.A. Fahey, and D.M. Small, Acyl Migration in 1,2-Dipalmitoyl-*sn*-Glycerol, *Chem. Phys. Lipids 52*:163–170 (1990).
77. Laakso, P., and W.W. Christie, Chromatographic Resolution of Chiral Diacylglycerol Derivatives: Potential in the Stereospecific Analysis of Triacylglycerols, *Lipids 25*:349–353 (1990).
78. Michelsen, P., E. Aronsson, G. Odham, and B. Åkesson, Diastereomeric Separations of Natural Glycero Derivatives as Their 1-(1-Naphthyl)ethyl Carbamates by High-Performance Liquid Chromatography, *J. Chromatogr. 350*:417–426 (1985).
79. Santinelli, F., P. Damiani, and W.W. Christie, The Triacylglcerol Structure of Olive Oil Determined by Silver Ion High Performance Liquid Chromatography in Combination with Stereospecific Analysis, *J. Am. Oil Chem. Soc. 69*:552–556 (1992).
80. Damiani, P., F. Santinelli, M.S. Simonetti, M. Castellini, and M. Rosi, Comparison Between Two Procedures for Stereospecific Analysis of Triacylglycerols from Vegetable Oils–I: Olive Oil, *J. Am. Oil Chem. Soc. 71*:1157–1162 (1994).
81. Rogalska, E., S. Ransac, and R. Verger, Stereoselectivity of Lipases. II. Stereoselective Hydrolysis of Triglycerides by Gastric and Pancreatic Lipases, *J. Biol. Chem.* *265*:20271–20276 (1990).
82. Sonnet, P.E., J.E. Oliver, R.M. Waters, G. King, and S. Panicker, A Potential Chiral Derivatizing Agent for 1,2-Diglycerides, *Chem. Phys. Lipids 78*:203–208 (1995).
83. Kim, J.H., Y. Nishida, H. Ohrui, and H. Meguro, A Highly Sensitive HPLC Method to Determine the Absolute Configuration of Glycosyl Diacylglycerols Using a Fluorescent Chiral Derivatizing Reagent, *J. Carbohydr. Chem. 14*:889–893 (1995).

84. Toyo'oka, T., Y-M. Liu, N. Hanioka, H. Jinno, M. Ando and K. Imai, Resolution of Enantiomers of Alcohols and Amines by High-Performance Liquid Chromatography After Derivatization with a Novel Fluorescent Chiral Reagent, *J. Chromatogr. A 675*:79–88 (1994).
85. Sato, R., Y. Itabashi, T. Hatanaka, A. Kuksis, Asymmetric *In Vitro* Synthesis of Diastereomeric Phosphatidylglycerols from Phosphatidylcholine and Glycerol by Bacterial Phospholipase D, *Lipids 39*:1030–1018 (2004).
86. Sato, R., Y. Itabashi, H. Fujishima, H. Okuyama, and A. Kuksis, Simple Synthesis of Diastereomerically Pure Phosphatidylglycerols by Phospholipase D-Catalyzed Transphoshatidylation, *Lipids 39*:1025–1030 (2004).
87. Itabashi, Y., and A. Kuksis, Chiral Phase HPLC Separation of *bis*-(monoacylglycerol)phosphate Stereoisomers, in Abstracts, *International Symposium on High Performance Liquid Phase Separations and Related Techniques (HPLC Kyoto)*, Kyoto, September 11–14, 2001, pp. 72–73.
88. Joutti, A. and O. Renkonen, The Stereoconfiguration of *bis*-(Monoacylglycero)phosphate Synthesized *In Vitro* in Lysosomes of Rat Liver: Comparison With the Natural Lipid, *J. Lipid Res. 20*:840–847 (1979).
89. Itabashi, Y., T. Ebata, A. Hatanaka, and A. Kuksis, Chiral Phase HPLC Resolution of Diastereomeric Phosphatidyl-D-Serine and L-Serine, in *Abstracts, 95th AOCS Annual Meeting and Expo*, Cincinnati, May 9–12, pp. 3–4 (2004).
90. Okamoto, Y., M. Kawashima, and K. Hatada, XI. Controlled Chiral Recognition of Cellulose Triphenylcarbamate Derivatives Supported on Silica Gel, *J. Chromatogr. 363*:173–186 (1986).

5 LC/MS and Lipid Oxidation

Arnis Kuksis[a] and Olli Sjovall[b]

[a]Banting and Best Department of Medical Research, University of Toronto, Toronto, M5G 1L6 Canada; [b]Department of Biochemistry and Food Chemistry, University of Turku, Finland, FIN-20014

Introduction

The successful initial demonstration of the suitability of electrospray ionization-tandem mass spectrometry (ESI-MS/MS) (1,2) and atmospheric pressure chemical ionization-MS/MS (APCI-MS/MS) (3) for glycerolipid analysis, and numerous subsequent applications, have left the impression that prior chromatographic fractionation constitutes an unnecessary complication in MS/MS analysis of lipids. In fact, it has been claimed (4,5) that specific losses of molecular species can occur when chromatography is used, so that direct analysis of intact lipids may provide greater accuracy in quantitative analysis. Although this view has been challenged by others (6,7), who have pointed out that the comparisons may have been made to inferior chromatographic analyses, direct MS/MS analysis of lipids is currently widely employed, despite occasional evidence of lipid class interference (8). Recently, the need for rapid profiling of tissue lipids (lipidomics) has further favored (9,10) the view that MS/MS analyses of crude total lipid extracts can provide all the analytical data for tissue comparisons and investigation of various pharmacological and metabolic signals.

Although an absolute need of prior chromatographic separation is obvious only for analysis of enantiomers and diastereomers of hydroxy fatty acids and oxidized glycerolipids, detailed analysis of mixtures of natural fats also show the benefits of prior chromatography for on-line MS/MS identification and quantification of complex natural lipids. While the low molecular weight oxo-lipids can be effectively resolved by GLC on either polar or non-polar columns, HPLC on normal or reversed-phase columns is necessary for resolution of high molecular weight glycerolipids prior to MS/MS analyses (11–13). The present review discusses recent developments in the application of both normal and reversed-phase HPLC to facilitate MS/MS of oxo-fatty acids and oxidized cholesteryl esters, triacylglycerols (TAGs) and glycerophospholipids. The use of chiral phase HPLC for LC/MS analyses of enantiomeric lipids has been reviewed elsewhere (14,15).

Materials and Methods

Reagents and Suppliers. Tert-butylhydroperoxide (TBHP) (Sigma Chemical Co., St. Louis, MO); 2,4-dinitrophenylhydrazine (DNPH) (Aldrich Chemical Co., Milwaukee, WI); 2-aminoanthracene (96% purity); Triphosgene (Aldrich); acetonitrile (dehydrated); 2-propanol; *n*-hexane; chloroform; methanol; ethanol; dichloromethane; 1,2-dichloroethane; toluene; trifluoroacetic acid (TFA); pyridine (all HPLC grade, Wako Pure Chemicals, Tokyo, Japan).

Normal and Reversed-Phase Columns and Suppliers. Normal phase: Phenomenex Luna 3 silica columns (100 × 2.0 mm ID, Phenomenex, Torrence, CA) and a guard column (4 mm × 2.0 mm ID, Phenomenex, Torrence, CA); Supelcosil (5 µm particles, 250 × 4.6 mm ID, Supelco, Sigma-Aldrich, Canada); Spherisorb silica (3 µm particles, 100 × 4.6 mm ID, Waters, Milford, MA, USA); Polyvinyl alcohol bonded silica column (YMC PVA-SIL, 100 × 3 mm ID, YMC, Kyoto, Japan).

Derivatization of Hydroxy Fatty Acids and Esters. The hydroxyeicosatetraenoic acids (HETEs), dihydroxyeicosatetraenoic acids (DHETEs) and hydroxyoctadecanoic acids (HODEs) synthesized by autoxidation of arachidonic acid or linoleic acid, respectively, are reduced with $NaBH_3$ and initially fractionated by reversed-phase HPLC (e.g. a Beckman Ultrasphere ODS 5 mm column, $MeOH/H_2O/HOAc$) and methylated with ethereal diazomethane (16). For improved mass spectrometric detection, the methyl esters may be converted into the pentafluorobenzyl (PFB) esters, thus increasing the sensitivity of response in the electron capture mode. The PFB esters may be prepared as described by Lee et al. (17). Picomolar quantities of the acids in CH_3CN (100 mL) were reacted with 100 mL of PFB bromide in CH_3CN (1:19, v/v) followed by 100 mL of diisopropylethanolamine in CH_3CN (1:9, v/v) and the solution heated at 60°C for 60 min. The reaction mixture was allowed to cool, evaporated to dryness under N_2 at room temp., and redissolved in 100 mL of hexane/EtOH (97:3, v/v) for normal phase LC/MS analysis. The epoxides were analyzed as methyl or PFB esters. The resolution was improved by hydrogenation. The PFB esters of fatty acid epoxides are prepared as described by Hammonds et al. (18).

The trimethylsilyl (TMS) and *tert*-butyldimethylsilyl (TBDMS) ethers and oximes were prepared as suggested by the manufacturers of the reagents (Supelco Catalogue). Dinitrophenylhydrazone (DNPH) derivatives of the core aldehydes were prepared as described by Esterbauer and Cheeseman (19). Chemical reductions of hydroperoxides and ozonides were performed by standard procedures (20,21).

HPLC and LC/ESI-MS/MS of Lipids. This section summarizes the general methodolgy employed in LC-MS and LC-MS/MS of the oxygenated fatty acids, steryl esters and glycerolipids. More detailed description of the methods is given under specific applications as well as in legends to figures.

Hydroperoxy and hydroxy fatty acid esters were resolved on either normal or reversed-phase HPLC columns as required. Nakamura et al. (22,23) used reversed-phase HPLC and ESI-MS/MS to identify and quantify the monohydroxyeicosatetraenoic (EET) acid regioisomers using d_8-HETE as internal standard. The organic acids were taken up in 100 μl water containing 0.05% (v/v) AcOH with pH adjusted to 5.7 using NH_4OH (mobile phase A) and the sample filtered (0.2 μm) and injected (50 μL) onto reversed-phase HPLC column (1 × 150 mm, 5 μm ODS Prodigy column, Phenomenex, Torrance, CA) and the column eluted at a flow rate of 50 μL/min. A linear gradient starting from 20% A to 100% B (CH_3N/MeOH, 65:35, v/v) over 15 min was employed and the column effluent introduced into the mass spectrometer through a 0.5 × 50 μm ID fused silica capillary. MS was performed on a Sciex API III triple quadrupole mass spectrometer (PE-Sciex, Thornhill, Ontario) in the negative ion mode using ultra high purity air as nebulizing gas to prevent glow discharge at the ion spray needle. Argon was used in the second quadrupole collision cell.

Bylund et al. (24,25) used reversed-phase LC-MS with an ion trap mass spectrometer to resolve the epoxy and hydroxy derivatives of arachidonic and linoleic acids. A reversed-phase column (5 μm, 250 × 2 mm; Chromasil 5 C_{18} 100 A, Phenomenex) and a guard column (Opti-Gard, 1 mm C_{18}) were used. The HPLC column was eluted with MeOH/H_2O/HOAc (80:20:0.01, by vol) at 0.2 mL/min. The effluent was admitted to an ion trap mass spectrometer (LCQ, Finnigan MAT, San Jose, CA) and subjected to APCI and ESI (24). Negative ions were monitored and $PGF_{1\alpha}$ was used for tuning. Leukotriene B_4 ω-side chain hydroxylation products were separated on a RP-LC Symmetry C_{18} (4.6 × 250 mm) column (Waters) by a gradient 50:50:0.1 (by vol). MeOH/H_2O/HOAc to 100:0.1 (v/v) MeOH/HOAc over 25 min) at 1 mL/min. The identity of the metabolites was confirmed by LC/MS (Agilent 1100 LC/MSD trap), where it was subjected to ESI (25). Negative ions were monitored by full scan (*m/z* 100–400) and subjected to MS/MS (*m/z* 347→ 367 → full scan).

Piazza et al. (26) used reversed-phase HPLC in combination with APCI-MS for the identification of isomeric 1,2- and 1,4-diol derivatives of conjugated linoleic acid. For this purpose, two Symmetry 3, 5 μm C_{18} columns (150 × 2.1 mm and 50 × 2.1 mm) (Waters, Milford, MA) connected in series were used. The mobile phase composition and the linear gradient used were 0–5 min H_2O/CH_3CN (40:60, v/v); 5–30 min H_2O/CH_3CN (40:60, v/v) to CH_3CN (100%); 30–54 min CH_3CN (100%) at a flow rate of 0.25 mL/min. The 1,4-diol isomers were resolved into two isomers by normal-phase HPLC on a Dynamax Macro HPLC Si column (250 × 10 mm) (Rainin, Woburn, MA). The mobile phase composition and the gradient used were 0–6 min C_3H_7OH/hexane (5:95, v/v); 6–18 min C_3H_7OH/hexane (13:87, v/v); 18–35 min C_3H_7OH/hexane (5:95, v/v). Epoxide methyl esters were separated by normal phase (Dynamax Macro HPLC Si column). The mobile phase composition and gradient used were 0–30 min C_3H_7OH/hexane (2:98, v/v) to C_3H_7O/hexane (9:91, v/v) with a flow rate of 1.5 mL/min. Alternatively, epoxide methyl esters were separated by reversed-phase HPLC using Adsorbosphere C_{18} 5μ column (250 × 10 mm) (Alltech, Deerfield, IL). The mobile phase composition and gradient used were 0–5 min H_2O/CH_3CN (15:85, v/v); 5–30 min H_2O/CH_3CN (15:85, v/v) to CH_3CN (100%); 30–35 min CH_3CN (100%). After separation the products were characterized by EI-MS (Thermabeam Mass Detector; Waters) and APCI-MS (Micromass ZMD; Waters). When using APCI-MS, the reversed-phase gradient contained 0.1% HOOH, and the APCI-MS detector was set to scan the mass range of *m/z* 150–550 at 400 amu/s.

Havrilla et al. (27) have reported the separation and analysis of cholesteryl ester hydroperoxides with coordination (Ag^+) ion spray LC-MS (CIS-MS). The CIS-MS method provides the most useful approach to analysis of cholesteryl linoleate hydroperoxides. The method makes use of the fact that silver ions form readily detected Ag^+ adducts with peroxides and hydroperoxides, and these give ions at $[M+107]^+$ and $[M+109]^+$, which undergo specific fragmentations typical of hydroperoxides and cyclic peroxides. By addition of $AgBF_4$ postcolumn, silver ion CIS-MS/MS can be coupled to normal phase HPLC. Typically, autoxidation of cholesteryl linoleate yields carbon 9 and carbon 13 substituted compounds, which can be readily distinguished by chromatography on narrow bore silica columns at 150 μL/min of mobile phase (0.35% 2-propanol in hexane). SRM utilizes the differences in Hock fragmentation of the hydroperoxides to differentiate the eluting regioisomers: *m/z* 787 to 647 (C_9 core aldehyde) and *m/z* = 787 to 687 (C_{13} core aldehyde).

For this purpose the peroxidic acid oxidation products were isolated from the unoxidized lipid by semipreparative chromatography (0.5% 2-propanol in hexane, 10 mL/min). The oxidized fractions of cholesteryl linoleate (10–25 min) and cholesteryl arachidonate (10–35 min) were collected, concentrated, and analyzed by LC-CIS/MS. For HPLC sample introduction, a Hewlett-Packard 1090 HPLC system was used. For cholesteryl linoleate and cholesteryl arachidonate hydroperoxide analysis, normal phase HPLC sample introduction was carried out using two tandem Beckman Ultrasphere narrow bore 5 μm silica columns (2.0 mm × 25 cm) operated in isocratic mode with 0.35% 2-propanol in hexane. For analysis of cyclic peroxide mixtures, sample introduction was carried out using a single Beckman Ultrasphere narrow bore 5 μm silica (2.0 mm × 25 cm) operated in isocratic mode with 1.0% 2-propanol in hexane. An Upchurch PEEK high pressure mixing tee was connected next in a series for the postcolumn addition of the silver salts. The silver tetraborate ($AgBF_4$) solution (0.25 mM in 2-propanol) was added via Harvard Apparatus (Cambridge, MA) syringe pump at a flow rate of 75 μL/min. A section of PEEK tubing (1.04 m, 0.25 mm ID) allowed time for the complexation of the silver to the lipid while delivering effluent to the mass spectrometer. A Rheodyne 7725 injector was fitted with a 00 μL PEEK loop for 20–50 μL sample injections. The CIS-MS was performed using a Finnigan TSQ-7000 (San Jose, CA) triple quadrupole mass spectrometer equipped with a standard APCI-ESI ionization source outfitted with 100 μL deactivated fused Si capillary (27).

Earlier, Kamido et al. (28) used reversed-phase HPLC with direct liquid inlet electron capture MS (DLI/EC-MS) for the analysis of cholesteryl ester core aldehydes following conversion into the 2,4-DNPHs. The HPLC separation was performed on Supelco LC-18 column (250 mm × 4.6 mm, ID) (Supelco, Mississauga, Ontario) with a linear gradient of 30–90% CH_3CH_2CN in CH_3CN in 30 min.

Herrera et al. (29) resolved the cholesteryl linoleate ozonides by reversed-phase HPLC using Supelcosil LC-18 columns (250 × 4.6 mm, ID, 5 μm) (Supelco), which were eluted with a linear gradient of 20–80% 2-propanol in MeOH in 40 min. The effluent along with a post-column flow of 0.2 mL/min of NH_4Ac in 2-propanol was admitted via an ESI interface to a Hewlett-Packard Model 5985 quadrupole mass spectrometer. The spectra were recorded in the positive ion mode in the mass range of 300–1200 with an ESI source exit voltage (CapEx) set at 120 V, which yielded characteristic masses for the parent and fragment ions. SIM spectra were extracted from the total ion current profile as needed for peak identification. In specific instances collision induced dissociation was obtained by rerunning the sample at CapEx 300V.

Sjovall et al. (30) resolved reference hydroperoxy, hydroxyl and epoxy TAGs by reversed-phase HPLC with on-line identification by ESI-MS. The reversed-phase HPLC was performed on Supelcosil LC-18 column (250 × 4.6 mm ID) (Supelco, Mississauga, Ontario) using a linear gradient of 20–80% 2-propanol in MeOH (0.85 mL/min) in 30 min. The instrument was a Hewlett-Packard (Palo Alto, CA) Model 1090 liquid chromatograph interfaced with a nebulizer-assisted ESI source connected to a Hewlett-Packard Model 5989 A quadrupole mass spectrometer. The ionization (capillary exit) voltage of this instrument was set at 100 V but could be increased to 300 V, to obtain fragment ions from any clearly resolved components (pseudo MS/MS). For improved ionization, 1% NH_4OH in 2-propanol was added postcolumn at a flow rate of 0.15 mL/min. The unoxidized and oxidized TAGs were identified as $[M+NH_4]^+$ or $[M+Na]^+$ in positive ion mode, whereas the DNPH derivatives of TAG core aldehydes were detected as $[M-H]^-$ ions in negative ion mode. Sjovall et al. (31) have shown that same experimental conditions are suitable for the analysis and identification of the TBHP oxidation products of unsaturated TAGs, including the TBHP adducts. Sjovall et al. (32) have subsequently shown, that the LC/ES-MS is equally well suited for the isolation and identification of the TAG core aldehydes following their isolation as the DNPH derivatives. Earlier, Ravandi et al. (33) used reversed-phase HPLC with on-line thermospray ionization (TSI) MS (LC/TSI-MS) to separate and identify the core aldehydes prepared from synthetic TAGs by ozonization and reduction with phenylphosphine. The core aldehydes were analyzed as the DNPH derivatives.

Myher et al. (12) and Ravandi et al. (20) employed normal phase LC/ESI-MS for separation of both native and oxidized phospholipids on a Spherisorb (3 μm, 100 mm × 4.6 mm ID, Altech, Guelph, Ontario) column. The column was eluted with a linear gradient of 100% A ($CHCl_3/MeOH/NH_4OH$, 80:19.5:0.5, by vol) to 100% B ($CHCl_3/MeOH/H_2O/NH_4OH$, 60:34.5:5.5:0.5, by vol) in 14 min, then at 100% B for 10 min. The system resolved the common phospholipid classes on basis of overall polarity, but caused only a partial resolution of the corresponding natural and oxidized species. The oxygenated glycerophospholipids, including ozonides, hydroperoxides, hydroxides, and core aldehydes gave protonated and sodiated molecular ions, which were seen as the ozonization products without fragmentation. Ravandi et al. (34) had previously used normal phase HPLC with on-line ESI-MS to separate the hydroperoxides, hydroxides and core aldehydes of intact PtdCho obtained by copper oxidation. The LC/ESI-MS was performed by splitting the HPLC flow 1:50, which resulted in admission of 20 μL/min to a Hewlett-Packard model 5988B quadrupole mass spectrometer equipped with a nebulizer-assisted ESI interface. Both positive and negative ion spectra were taken in the mass range 100 to 1100 amu. Capillary exit voltage was set at 120 V, but it could be increased to 300 V for fragmention. Oxidized standard phospholipids were used as standards.

Kamido et al. (21,35) resolved and identified the core aldehydes of human LDL phosphatidylcholine following their extraction with acidified $CHCl_3$/MeOH containing 2,4-DNPH and dephosphorylation with phospholipase C. The DNPH derivatives of the DAGs were separated by reversed-phase HPLC on a Supelcosil LC-18 column (250 × 4.6 mm ID) using a linear gradient of 30–90% CH_3CH_2CN in CH_3CN as the eluting solvent. About 1% of the column effluent was admitted to a Hewlett-Packard Model 5985B quadrupole mass spectrometer via a DIL interface. NICI (electron capture) mass spectra were taken every 5 s over the entire chromatogram.

Khaselev and Murphy (36) used on-line reversed-phase LC-MS to analyze the oxidation products of 1-O-hexadec-1′-enyl-arachidonoyl GroPCho by hydroxyl radical generated from Cu(II)/H_2O_2. On-line analysis of the oxidized mixtures of plasmalogen and different diacyl GroPCho species was carried out using a linear greadient from 100% mobile phase A [MeOH/H_2O/CH_3CN (60:20:20, by vol)] to 100% mobile phase B (1 mM methanolic NH_4Ac) over 40 min followed by isocratic elution with 100% B for 20 min. The HPLC was operated at a flow rate of 50 μl/min, using a Columbus C_{18} column (1 × 150 mm, 5μ; Phenomenex). MS analyses of the phospholipids were performed in the negative ion mode using MRM, monitoring specific transitions of the phospholipid negative ion molecular ion species $[M-15]^-$ that are collisionally activated to the carboxylate anions from the fatty acids esterified at either *sn*-1, *sn*-2-, or both. The transition *m/z* 750→303 was used to detect unoxidized 16:0p/20:4-GroPCho and *m/z* 766→303 for unoxidized 16:0a/20:4-GroPCho, where *p* and *a* represent alkyl and alkenyl 16:0, respectively. Molecular weight alone did not define each potential oxidized or unoxidized GroPCho species.

Results and Discussion

Hydroxy and Hydroperoxy Fatty Acids

MS/MS of polyunsaturated hydroxy fatty acids has been extensively investigated by a variety of techniques and the area has been extensively reviewed (37). Current methods for the study of hydroxy fatty acids involve the use of GC-MS with electron impact ionization (EI) or negative chemical ionization (NCI), which are 2–3 orders of magnitude more sensitive than positive LC/ESI-MS. MacPherson et al. (38) reported the use of HPLC with ESI-MS to identify the arachidonic acid metabolites in *Limulus amoebocytes*. ESI-MS is ideal for readily ionizable species containing an acidic function, but instability of free hydroxy fatty acids may lead to polymerization. Therefore, derivatization of hydroxy fatty acids is essential. Santiago-Vazquez et al. (39) have described the use of LC/APCI-MS for the analysis of algal hydroxy fatty acid methyl esters. Figure 1 shows an LC/APCI-MS analysis of an organic extract of a cell free incubation mixture of *Euglena gracilis*. Four peaks are discerned with retention times similar to those of commercial stan-

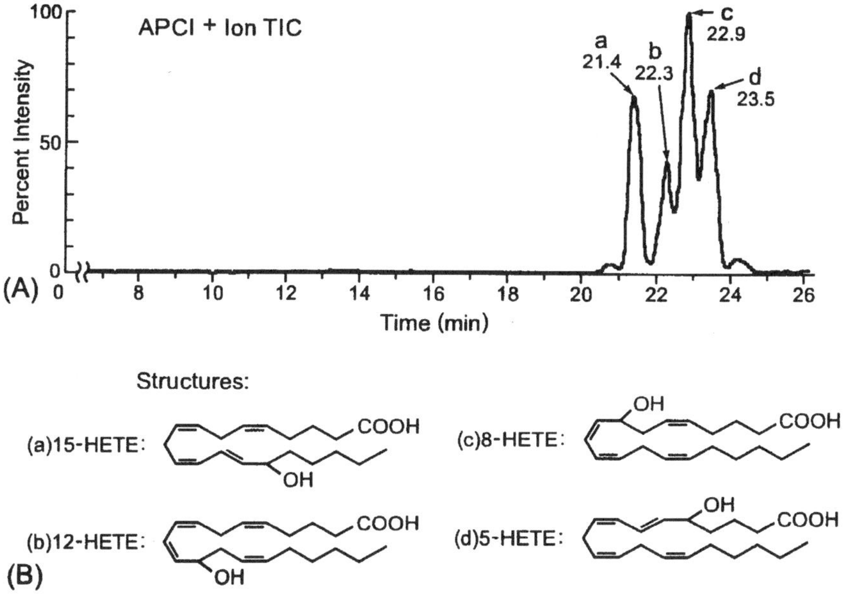

Fig. 1. Reversed-phase LC/APCI-MS profile of four HETEs produced by *E. gracilis'* cell free extract incubated with arachidonic acid. (A) Total ion current profile. The peaks are identified as follows: (a) 15-HETE ME; (b) 12-HETE ME; (c) 8-HETE ME and (d) 5-HETE ME. (B) Structures of the metabolites from *E. gracilis*. Samples were analyzed by LC/APCI-MS in positive ion mode (Michrom BioResources UMA HPLC system, Auburn, CA) coupled to a VG Platform II Mass Spectrometer (Micromass, England). The samples (1–5 μg of the crude extract) were separated on a C_{18} column (4.6 mm × 250 mm, 5 μm particle size, 100 A pore size, Varian, Walnut Creek, CA) using optimized mobile phase gradient starting at CH_3CN/H_2O (50:50, v/v) going to CH_3CN/H_2O (95:5, v/v) in 30 min and then run isocratically for another 30 min at 1 mL/min. *Source*: Santiago-Vazquez et al. (39).

dards and with *m/z* 317, which corresponds to the major ion in many of the arachidoinic acid derived hydroxy fatty acids. The peaks were identified as the methyl (ME) esters of 15- HETE ME (21.4 min), 12-HETE ME (22.3 min), 8-HETE ME (22.9 min) and 5-HETE ME (23.5 min). A detection limit of 20 pg/μl per injection was determined for 5-HETE ME based on signal to noise ratio of the *m/z* 317 ion, which corresponds to the loss of a hydroxyl group [M-17].

Pruzanski et al. (40) employed LC/ESI-MS in the negative ion mode to identify the peroxy fatty acids released by group IIA secretory phospholipase A_2 from the autoxidized PtdCho of HDL and acute phase HDL. Mono- and di-hydroperoxy linoleic and arachidonic acids were recognized as early emerging peaks from a normal phase HPLC column eluted with a gradient of $CHCl_3$/MeOH/NH_4OH. Figure 2 shows the total negative ion current profile along with SIMs for the mono- and dihydroperoxides and hydroxides of 18:2 and 20:4 acids as obtained for acute phase HDL after 2 h of incubation with group IIA secretary phospholipase A_2. The single ion mass chromatograms were extracted from the total ion current profile obtained by normal phase LC/ESI-MS for the total lipid extract of acute phase HDL. The oxygenated fatty acids were eluted as separate peaks following the corresponding non-oxygenated fatty acids. The ion abundance recorded on the left side of the mass chromatogram indicates that the monohydroperoxide and monohydroxide of 18:2 predominated.

Schneider et al. (16) utilized reversed-phase HPLC for the analysis of the autoxidation products 13*S*-hydroperoxy-9*Z*,11*E*-octadecadienoic (13*S*-HPODE) and 9*S*-hydroperoxy-10*E*,12*Z*-octadecadienoic (9*S*-HPODE) acids. Reversed-phase HPLC [Waters Symmetry C18 5 μm, 0.46 × 25 cm, CH_3CN/H_2O/HOAc (60:40:0.01, by vol)] at a flow rate of 1 mL/min. was used. The column effluent was monitored with an HP 1040 diode array detector.

Oliw et al. (41) used normal phase HPLC with ion-trap mass spectrometer to analyze the hydroperoxides of the main metabolites of oleic, linoleic, α-linolenic and γ-linolenic acids, which are formed by manganese lipoxygenase (Mn-LO) and linoleate diol synthase (LDS). MS^3 analysis of hydroperoxides and MS^2 analysis of dihydroxy- and monohydroxy metabolites yielded many fragments with information on the position of oxygenated carbons. M-LO oxygenated C-11 and C-13 of 18:2*n*6, 18:3*n*3, and 18:3*n*6 in a ratio of approximately 1:1-3 at high substrate concentrations. 8-Hydroxy-9(10)epoxystearate was identified as a novel metabolite of LDS and oleic aid.

Bylund et al. (24,25) have used reversed-phase LC-MS with an ion trap mass spectrometer to resolve the metabolites of arachidonic and linoleic acids, including the epoxides, produced by human recombinant cytochrome P_{450} (CYP) enzymes. The CYP_{2C9} converted arachidonic acid, octadeuterated arachidonic, and linoleic acids to epoxides. Figure 3 shows a reversed-phase LC/ESI-MS analysis of DHETs, HETEs, and EETs formed by cytochrome P_{450} (CYP_{2C9}) oxidation along with the mass chromatograms characterizing the individual oxygenated fatty acid species (24). Earlier, an extensive separation of the epoxides of arachidonic acids,

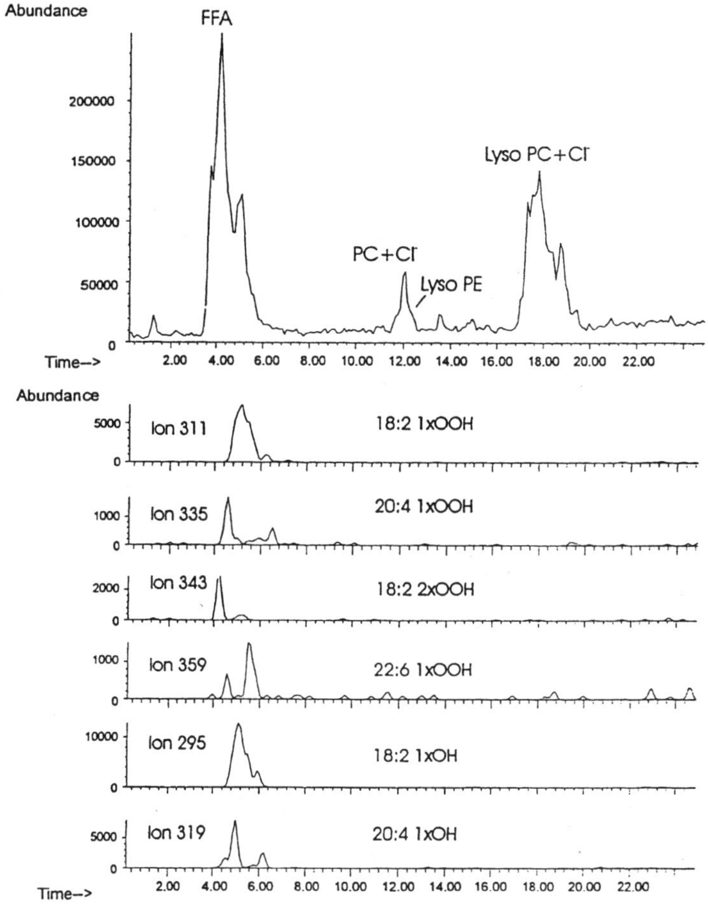

Fig. 2. Selected ion mass chromatograms of oxo-fatty acids released from acute phase HDL after 2 hr hydrolysis with secretory group IIA phospholipase A_2 as obtained by normal phase LC/ESI-MS. *Upper Panel,* total negative ion current profile; *Lower Panel, S*ingle negative ion mass chromatograms. Peak identification is given in the figures. The peak doublets and multiplets are due to separation of positional and geometric isomers of oxygenated fatty acids. Column: Spherosorb 3 μ (100 mm × 4.6 mm ID, Alltech Associates, Deerfield, IL); solvent: linear gradient of 100% solvent A to 100% Solvent B in 14 min, then Solvent B for 10 min. Solvent A: $CHCl_3$/MeOH/30% NH_4OH 80:19.5:0.5, by vol); Solvent B: $CHCl_3$/MeOH/H_2O/NH_4OH, 60:34:5.5:0.5, by vol). Instrumentation: Hewlett-Packard Model 1060 Liquid Chromatograph interfaced with Hewlett-Packard Model 5988B quadrupole mass spectrometer equipped with a nebulizer assisted ESI interface. Capillary exit voltage was set at 120 V, with electron multiplier at 1795 V. *Source:* Pruzanski et al. (40).

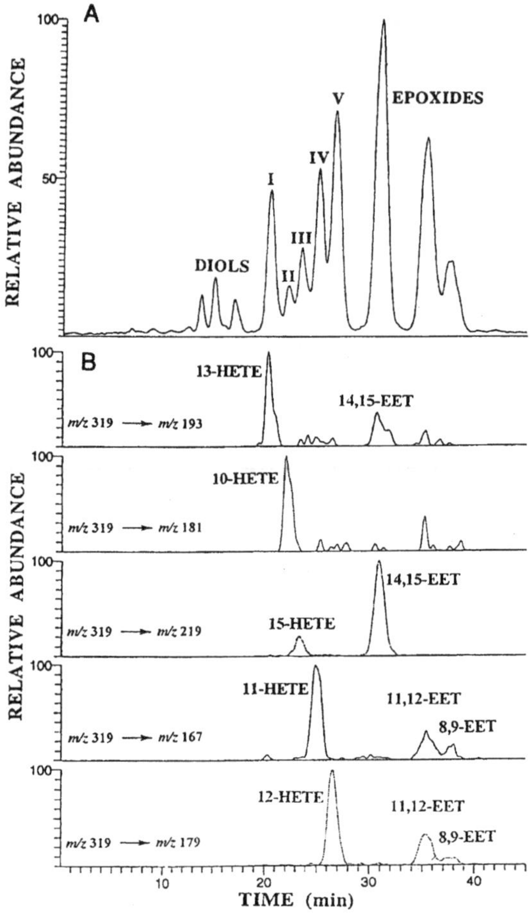

Fig. 3. LC/ESI-MS analysis of DHETs, HETEs, and EETs formed by cytochrome P_{450} oxidase (CYP2C9). (A) Mass chromatogram of the carboxylate anions showing the combined relative intensities of *m/z* 337 (DHETs) and *m/z* 319 (HETEs and EETs). Peak I contained 13-HETE and 19-HETE, peak II 10-HETE, peak III 15-HETE, peak IV 11-HETE, and peak V 12-HETE. (B) Detection of 13-HETE (*m/z* 193), 10-HETE (*m/z* 181), 15-HETE (*m/z* 219), 11-HETE (*m/z* 167), and 12-HETE (*m/z* 179) by monitoring selected ions formed by MS/MS 319 to full-scan. Column: Chromasil 5 C_{18} 10 A (Phenomenex) (5 µm, 250 × 2 mm, ID) and a guard column Opti-Gard, 1 mm C_{18}). The reversed-phase HPLC column was eluted with MeOH/H_2O/HOAc (80:20:0.01, by vol) at 0.2 mL/min. Column effluent was admitted to an ion trap mass spectrometer (LCQ, Finnigan MAT, San Jose, CA) and subjected to APCI or ESI. For ESI the capillary temperature was set at 230°C. The collision energy was set at 25–30%. Negative ioins were monitored and $PGF_{1\alpha}$ (Upjohn) was used for tuning. *Source:* Bylund et al. (24).

including chiral isomers, had been obtained by Hammonds et al. (18) using Chiralcel OB and Chiralcel OD columns in combination with MS/MS. The epoxides were analyzed as methyl esters or as PFB esters.

Prostaglandins and Isoprostanes

In this chapter the LC/MS analysis of prostaglandins is discussed together with the LC/MS of the isomeric isoprostanes. Current methods for the study of prostaglandins and eicosanoids in general involve GC-MS with EI or NCI. Waugh and Murphy (42) reported LC/MS analyses of four regioisomers of F_2-isoprostanes formed by free radical oxidation of arachidonic acid. Eight synthetic $PGF_{2\alpha}$ isomers were found separable by reversed-phase HPLC with a gradient of CH_3CN/MeOH in HOAc. The epimeric prostaglandins were detected as they eluted from the HPLC column by LC/ESI-MS/MS with collision induced dissociation of the carboxylate anions and a loss of C_2H_4O (44 u) from the 1,3-diol cyclopentane ring. Elution of these synthetic compounds occurred between 12 and 17 min: 8-epi-$PGF_{2\alpha}$ eluted in the earliest peak and $PGF_{2\alpha}$ eluted last. The elution of these two species was marked by the deuterium-labeled internal standards [$^2H_4$]8-epi-$PGF_{2\alpha}$ and [$^2H_4$]$PGF_{2\alpha}$.

Waugh et al. (43) separated F_2-isoprostanes by initial reversed-phase HPLC and detected them using ESI-MS with the characteristic loss of 44 u (C_2H_4O) from the common 1,3-diol cyclopropane ring found in these eicosanoids. CID of the carboxylate anions from the separated F_2-isoprostanes formed abundant ions characteristic for regioisomers of Type I (*m/z* 115), Type III (*m/z* 127), and Type IV (*m/z* 193), which made possible characterization of these three family sub-types by LC-MS/MS. Separation of F_2-isoprostane free acids was carried out on a Phenomenex Ultracarb ODS column (5 μm, 25 cm × 4.6 mm ID, Torrance, CA) at 1.0 mL/min with a linear gradient starting with 25% solvent B (CH_3N/MeOH, 95:5, v/v) and increasing to 30% B in 50 min. Mobile phase system A consisted of 0.05% acetic acid adjusted to pH 5.7 with NH_4OH. For LC/MS/MS a postcolumn splitter yielding 15 mL/min flow into the mass spectrometer was employed. In some experiments, [$^2H_4$]$PGF_{2\alpha}$ and [$^2H_4$]8-epi-$PGF_{2\alpha}$ (10 μg each) were added to an aliquot of the F_2-isoprostane extract. On-line HPLC analysis was done with a Sciex API-III+ triple quadrupole mass spectrometer (Perkin-Elmer Sciex, Thornhill, Ontario) (mass chromatogram not shown). For some experiments, the first quadrupole of the mass spectrometer was set to transmit *m/z* 353 and 357 and the third quadrupole was set to transmit *m/z* 309 and 313 to monitor the loss of 44 u from the carboxylate anions (MRM), while in other experiments, the first quadrupole was set to transmit *m/z* 353, and the third quadrupole was scanned from *m/z* 50 to 360 to collect all CID ions of the F_2-isoprostane carboxylate anion. For ESI analyses, the curtain gas flow was 1.2 L/min N_2, with nebulizer pressure at 38 psi. Figure 4 shows a reversed-phase HPLC separation of F_2-isoprostanes extracted from the liver of a rat treated with CCl_4, analyzed by direct ESI-MS (43).

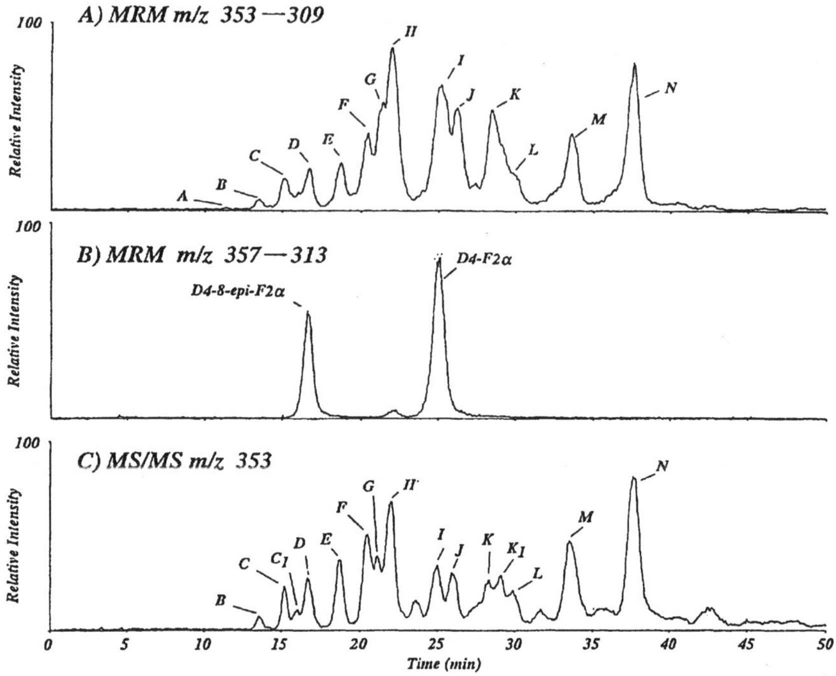

Fig. 4. Reversed-phase LC/ESI-MS separation and identification of F_2-isoprostanes extracted from the liver of a rat treated with CCl_4. (A) Multiple reaction monitoring (MRM) of the loss of 44 u from the molecular anion of the F_2-isoprostanes at *m/z* 353. (B) Multiple reaction monitoring of the loss of 44 u of the molecular anion of D_4-8-epi-$PGF_{2\alpha}$ and D_4-$PGF_{2\alpha}$ at *m/z* 357. (C) Reconstructed total ion current of all product ions following collision induced decomposition and tandem mass spectrometry of carboxylate anion of F_2-isoprostanes, *m/z* 353. HPLC conditions: Phenomenex Ultracarb ODS 5 μm 4.6 mm × 25 cm column (Torrence, CA) at 1.0 mL/min with a linear gradient starting with 25% solvent B (CH_3CN/MeOH, 95:5, v/v) increasing to 30% B in 50 min. Mobile phase A consisted of 0.05% AcOH adjusted to pH 5.7 with NH_4OH. For LC-MS/MS a postcolumn splitter yielding 15 μL/min flow into the mass spectrometer was employed. $[^2H_4]PGF_{2\alpha}$ and $[^2H_4]$8-epi-$PGF_{2\alpha}$ (10 ng each) were added to an aliquot of the F_2-isoprostane extracts in some experiments. The on-line HPLC analysis was carried out with a Sciex API-III+ Triple quadrupole mass spectrometer (Perkin-Elmer Sciex, Thornhill, Toronto, Ontario). *Source:* Waugh et al. (43).

Subsequently, Lawson et al. (44) have developed an LC-MS/MS method for the identification in human urine of two more isomers, 8,12-iso-$iPF_{2\alpha}$-VI and 5-epi-8,12-iso-$iPF_{2\alpha}$-VI, using synthetic standards. HPLC was performed on an ABI 140B solvent delivery system (Perkin-Elmer) utilizing a Hypersil BDS C_{18} column

(23 × 150 mm) packed with 3 μm particles with 130 A pore size (Phenomenex Inc., Torrance, CA). The flow rate was 200 mL/min. The column effluent was split, with 25% going to the ESI source and the remainder to waste. The elution gradient was programmed from 25 to 35% mobile phase B in 20 min, held 20 min, and then programmed to 90% mobile phase B in mobile phase A. Mobile phase A consisted of deionized distilled water, with the pH adjusted to 5.7 with HOAc. Mobile phase (B) was 95% CH_3CN and 5% MeOH. Samples were injected by a Hewlett-Packard Series 1100 autosampler. Figure 5 shows the specificity of LC-MS/MS for F_2-iP analysis with minimal sample preparation (45). A, specificity for class VI F_2-iPs. 10 ng each of $iPF_{2\alpha}$-III, -IV, -V, and VI were analyzed by monitoring four product ions of *m/z* 353. *Upper panel,* the sum of three non-specific product ions at *m/z* 273, 291 and 309; *Lower panel,* the product ion at *m/z* 115, demonstrating specificity for F_2-iPs of class VI; B, The spectrum of class VI F_2 isoprostanes in urine. 10 ng of synthetic tetradeuterated $iPF_{2\alpha}$-VI was added to 1 mL of urine and subjected to reversed-phase solid phase extraction. (A) the product ion at *m/z* 357 to m/z 115, specific for tetradeuterated isoprostanes of class VI; (B) the product ion at *m/z* 353 to *m/z* 115, specific for class VI isoprostanes. $iPF_{2\alpha}$-VI, at a retention time of 11 min and 45 s, is the most abundant F_2 isoprostane previously quantified. Peaks I and II are each present at approximately five times the concentration of $iPF_{2\alpha}$-VI in this sample, which is representative of all samples observed to date. Lawson et al. (45) have published a mini-review on the formation and analysis of isoprostanes, and their use as indices of lipid peroxidation with emphasis on the usefulness of the specificity of LC-MS/MS for analysis of sample with minimal sample preparation. It was suggested that in the future development of even more sensitive and specific methodology, advantage may be taken of the unique ability of class VI F_2-iPs to form a cyclic lactone and hence be easily separated from other F_2-iP classes.

Mallat et al. (46) studied F_2-isoprostanes in human carotid atherosclerosis using on-line reversed-phase LC-MS/MS. The LC-MS/MS quantitative assay for isoprostanes was based on determining the ratio of the abundance of the ion transition (*m/z* 357→313) for the deuterated $PGF_{2\alpha}$ internal standard divided into the area under the curve for the corresponding transition (*m/z* 353→309) observed for the family of isoprostanes. For the isoprostanes, multiple HPLC peaks were readily observed, owing to the presence of multiple isoprostane family members.

Waddington et al. (47) have reported isoprostane derivatives among the fatty acid peroxidation products in human atherosclerotic plaques. Chiral phase HPLC was used to separate the methyl esters of 9- and 13-HODE, 15-HETE, and 11-HETE derivatives. For this purpose the methyl esters were prepared by methylation with diazomethane and subjected to separation using Chiracel OB (250 × 4.6 mm ID) column (Diacel Chemical Industries, Ltd., Japan) and isocratic elution with hexane/2-propanol (95:5, v/v). Absorbance was measured at 235 nm. Kozak et al. (48) used LC/ESI-MS with Zorbax RX-C_{18} narrow bore column (15 cm × 2.1 mm, 5 μm) interfaced to a Finnigan TSQ-7000 triple quadrupole mass spectrome-

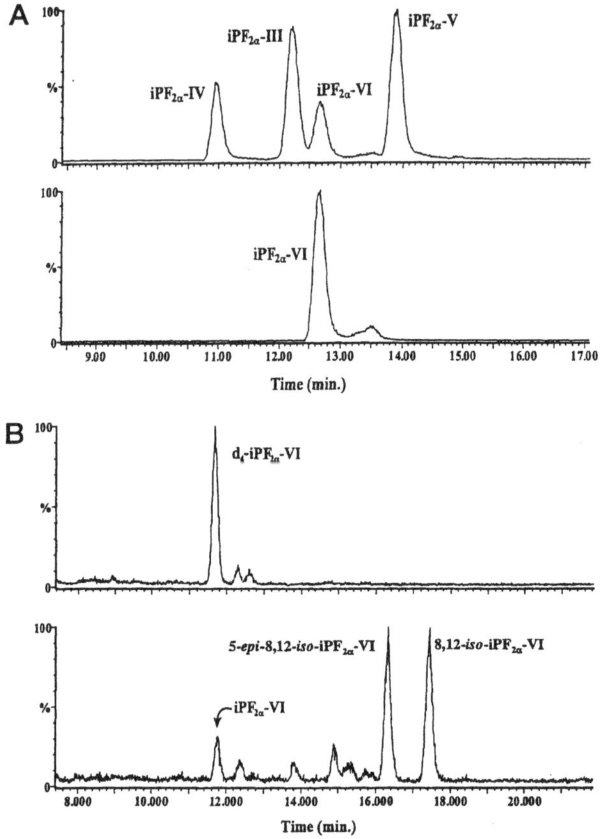

Fig. 5. An LC-MS/MS method specific for class VI F_2-iPs. 10 ng each of $iPF_{2\alpha}$-III, $iPF_{2\alpha}$-IV, $iPF_{2\alpha}$-V and $ipF_{2\alpha}$-VI were analyzed by LC-MS/MS, monitoring four product ions of *m/z* 353. A, ability of LC/MS/MS to resolve isomeric classes of F_2-iPs. *(Upper panel)* The sum of three non-specific product ions at *m/z* 273, 291, and 309; *(Lower panel)* The product ion at *m/z* 115, demonstrating specificity for F_2-iPs of class VI. B, class VI iPs in urine. One mL of human urine was spiked with $[^2H_4]iPF_{2\alpha}$-VI, extracted by reversed phase SPE, and analyzed by LC/MS/MS. *(Upper panel)*, product ion specific for $[^2H_4]$iPs of class VI (*m/z* 357115). *(Lower panel)*, product ion specific for iPs of class (*m/z* 353→115). HPLC conditions: Hypersil BDS C_{18} column (2 × 150 mm) packed with 3 μm particles with 130 A bore size (Phenomenex, Torrance, CA). The flow rate was 100 μL/min. The column effluent was split, with 25% going to ESI source and the remainder to waste. The mobile phase was programmed from 25% to 35% mobile phase B in 20 min, held for 20 min, and then programmed to 90% mobile phase B. Mobile phase A consisted of deionized distilled water, with the pH adjusted to 5.7 with HOAc. Mobile phase B was CH_3CN/MeOH (95:5, v/v). *Source:* Lawson et al. (44).

ter to demonstrate that COX-2 mediated oxygenation of 2-arachidonoylglycerol provides the novel lipid, PGH_2 glycerol ester (PGH_2-Gro), in vivo and in cultured macrophages. Sodiated analytes were eluted with increasing concentrations of CH_3CN in 0.001% aqueous NaOAc. Two primary product masses were observed at *m/z* 449 and *m/z* 417, which corresponded to the sodiated molecular ions of the glyceryl esters of PGH_2, PGE_2, or PGD_2 and 11- or 15-(HETE), respectively.

Roberts et al. (49) have reported the formation of isoprostane-like compounds, termed neuroprostanes, during peroxidation of 22:5n3 and 22:6n3. The formation of the C_{22} neuroprostanes is similar to the formation of isoprostanes from 20:4n6 and proceeds via generation of highly unstable endoperoxide intermediates. There is evidence which shows that the neuroprostanes like isoprostanes are formed in significant amounts in vitro and in vivo from the free radical-catalyzed peroxidation of the 22:6n3-containing phospholipids and are presumably released in the free form by a phospholipase. Reich et al. (50) isolated esterified D_4/E_4-neuroprostanes from rat and human brain and hydrolyzed them by reaction with *A. mellifera* venom phospholipase A_2 for subsequent analysis in the free form. The D_4/E_4-neuroprostanes were purified by HPLC and analyzed by LC/ESI-MS/MS. HPLC was performed on a 25 cm × 4.6 mm Econosil C-18 column (5 μm particles) using an isocratic solvent system of $H_2O/CH_3CN/HOAc$ (68:32:0.1, by vol) at a flow rate of 1 mL/min. D_4/E_4-neuroprostanes eluted at a retention volume of 25–55 mL. Fractions were then analyzed by LC/ESI-MS/MS in NCI using either a 15 cm × 2.1 mm Econosil C-18 column or a 15 cm × 1 mm Zorbax C-18 column. The solvent system employed in each case was a gradient consisting of 20 mM $NH_4Ac/CH_3CN/HOAc$ (90:10:0.1, by vol) to 20 mM $NH_4Ac/CH_3CN/HOAc$ (10:90:0.1, by vol) over the course of 10 min at a flow rate of 0.2 μL/min or 50 μL/min., respectively. Eicosanoids with E/D-type prostane rings are unstable and dehydrate to cyclopentenone-containing compounds possessing A-type and J-type prostane rings, respectively. Fam et al. (51), therefore, explored the A_4/J_4-neuroprostanes from dehydration of E_4/D_4-neuroprostanes and demonstrated that oxidation of 22:6 indeed yielded the anticipated compounds. Esterified A_4/J_4-neuroprostanes in glycerophospholipids were hydrolyzed using saponification with KOH, following treatment with methoxyamine HCl in $CHCl_3$/MeOH (2:1, v/v) for 1 h at room temperature. The A_4/J_4 neuroprostanes were conjugated with the excess of glutathione (GSH) and the adducts purified using a C_{18} Sep-Pak cartridge preconditioned with CH_3CN and 0.1 M NH_4Ac (pH 3.4) and eluted with 10 mL of 95% EtOH and identified by LC-MS. HPLC was carried out using MAGIC 2002 LC system (Michrom BioResources, Auburn, CA) operating in the isocratic mode with the mobile phase of $H_2O/(CH_3)_2CO/HOAc$ (77:22:9:0.1, by vol), and compounds were separated on an Eclipse XDB-C_{18} column (2.1 × 50 mm, 5 μm particle size, Agilent, Palo Alto, CA) at a flow rate of 75 μL/min. The compounds were identified by on-line ESI-MS (51). Fam et al. (51) have attempted to demonstrate the formation of the cyclopentenone neuroprostanes in vivo. Putative A_4/J_4-neuroprostanes were analyzed as free compounds following chemical hydrolysis of the compounds esterified in brain lipid extracts from normal rats. Figure 6 shows a representative ion current chromatogram from a human brain sample (51). The two O-methyloxime isomers of the [$^2H_4$]PGA_2 internal standard are present in the *m/z* 438 chromatogram. The chromatographic peaks in the lower *m/z* 438 ion current chromatogram represent the O-methyloxime isomers of the internal standard [$^2H_4$]PGA_2. In the upper *m/z* 458 chromatogram are a series of peaks that have molecular masses and retention times expected for A_4/J_4-neuroprostanes. The pattern of peaks representing A- and J-ring neuroprostanes was very similar to that obtained from the oxidation of DHA in vitro. Youssef et al. (52) have recently measured brain levels of F_2-isoprostanes and F_4-neuroprostanes and their precursors as sensitive and reliable indicators of oxidative injury. Since comparable levels were observed in all animal age groups, it was concluded that ageing is not accompanied by enhanced brain susceptibility to oxidative stress.

Brame et al. (53) have reported that isoprostane endoperoxide intermediates also undergo rearrangements to form highly reactive γ-ketoaldehyde levuglandin-like compounds. Bernoud-Hubac et al. (54) have proposed to name these nonenzymatically generated γ-ketoaldehydes isoketals to distinguish them from levuglandins formed by rearrangement of the cyclooxygenase endoperoxide intermediate, PGH_2. Utilizing LC-MS analyses, Bernoud-Hubac et al. (54) found that during in vitro oxidation of 22:6n3, neuroketals were formed in greater abundance than the isoketals during oxidation of 20:4n6. The neuroketals were analyzed by HPLC as the lysyl adducts formed during oxidation of 22:6n3 in the presence of lysine (see also below under Glycerophospholipid Hydroperoxides, Isoprostanes and Core Aldehydes).

Steryl Ester Hydroxides, Hydroperoxides, Isoprostanes and Core Aldehydes

Steryl esters represent a high proportion of bound polyunsaturated fatty acids, which are subject to peroxidation. Because the sterol ring also becomes oxidized, chemical degradation may be necessary for structural identification of the oxidation products even when MS/MS is preceded by chromatography. Furthermore, direct analysis of the hydroperoxides or the corresponding alcohols by MS is complicated by the fact that most MS ionization methods result in loss of water from the parent ion.

With EI-MS, steryl esters rarely give detectable molecular ions. Ions representing loss of the fatty acyl moiety are seen with model compounds, but these are of limited value with unknowns. Nonetheless, valuable information can be obtained with some samples. With ammonia as the reagent gas, a good quasi-molecular ion $[M+NH_4]^+$ and ions diagnostic for both sterol and fatty acid moieties are obtained (29,35). Both C_9 and C_{13}-hydroperoxides of linoleate were readily separated and detected by LC/MS (35).

Kenar et al. (55) have identified and quantified the regioisomeric hydroperoxides of cholesteryl linoleate in oxidized human lipoproteins, while Mashima et al.

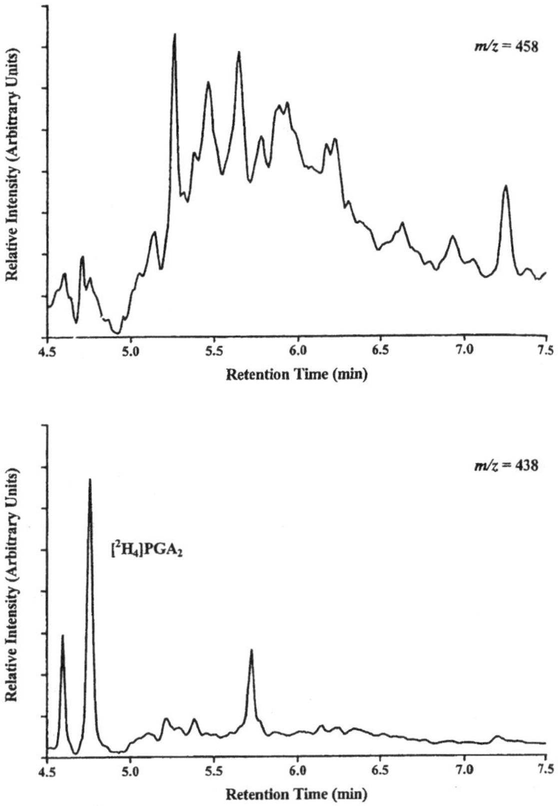

Fig. 6. Selected ion current chromatogram obtained from the analysis of A_4/J_4-NPs esterified in post-mortem human temporal lobe brain tissue. The series of peaks shown in the *m/z* 458 chromatogram represent the putative A_4/J_4-NPs, and the two peaks in the *m/z* 438 ion current chromatogram represent the *syn*- and *anti*-O-methyloxime isomers of the $[^2H_4]PGA_2$ internal standard. The calculated amount of A_4/J_4-NPs present was 74 ng/g of brain tissue. Compounds generated by in vitro oxidation of DHA were purified by normal phase HPLC (Econosil SI column, 25 cm × 4.6 mm ID, 5 μ particle size) using an isocratic solvent system (hexane/2-propanol/HOAc, 97:3:0.1, by vol) at 1 mL/min. A_4/J_4 eluted in with retention time 13–37 mL. The Fraction was then analyzed by LC/ESI-MS/MS in the negative ion mode using a 5 cm × 2.1 mm Zorbax C-18 column (Agilent Technologies, Palo Alto, CA). The solvent system was a gradient consisting of 5 mM $NH_4Ac/CH_3CN/HOAc$ (90:10:0.1, by vol) to 5 mM $NH_4Ac/CH_3CN/HOAc$ (10:90:0.1, by vol) over the course of 10 min at a flow rate of 200 μL/min. The voltage of the capillary was 20.0 V, the capillary temperature 200°C. Collision induced dissociation of molecular ions of putative A_4/J_4-NPs in these fractions were scanned from *m/z* 50 to 400. The mass spectrometer was a Finnigan TSQ 7000 instrument (San Jose, CA). *Source:* Fam et al. (51).

(56) determined the regioisomeric distribution of the hydroperoxides and hydroxides of cholesteryl linoleate in human plasma.

Havrilla et al. (27) have identified the six hydroperoxides anticipated to be formed from cholesteryl arachidonate. The various peaks are characterized by SRM analysis of the parent $107Ag^+$ adduct at *m/z* 811, which gives fragments that identify the position of substitution on the arachidonate chain. The SRM fragments result from cleavage of the cholesterol (Ch) fatty acid ester bond with characteristic fragments derived from Hock fragmentation of the arachidonate. Thus Ch-15-HPETE gives a fragment at *m/z* 343. The 11- and 12-hydroperoxides of cholesteryl arachidonate give identical Hock fragmentation ions, and shows the sum of signals detected from the SRM for *m/z* 303. Ch-8-HPETE and Ch-9-HPETE both give the same Hock fragmentation product and Panel C shows the SRM for this ion at *m/z* = 263, and the corresponding fragmentation of the Ch-5-HPETE gives *m/z* = 327 (Figure 7).

Yin et al. (57) have used LC/MS and GC/MS techniques to demonstrate the formation of isoprostane bicyclic endoperoxides from the autoxidation of cholesteryl arachidonate. All four possible types (I–IV) of bycyclic endoperoxides were found starting from different regioisomeric hydroperoxides of cholesteryl arachidonate. The stereochemistry of Type IV bicyclic endoperoxides was determined by conversion to PFB and TMS derivatives and comparison to synthetic standards by GC/MS. All eight possible diastereomers of the derivatized isoprostanes were observed and were separated by GLC. The bicyclic endoperoxides with the two alkyl chains *syn* on the cyclopentane ring were formed preferentially to those with ante configuration. Substantiual amounts of the *ante*-substituted isoprostanes, including $PGF_{2\alpha}$ were also observed in the mixture.

Earlier workers (28,35) used LC/MS to demonstrate the formation of cholesteryl ester core aldehydes during peroxidation of plasma lipoproteins. To analyze core aldehydes from cholesteryl esters, lipids were isolated from plasma by extraction with acidified $CHCl_3$/MeOH containing 2,4-dinitrophenylhydrazine (DNPH), and the DNPH derivatives formed were resolved by reversed-phase HPLC for identification by on-line MS (21,28). The major aldehydes from peroxidation of cholesteryl esters were found to be the 9-oxononanoyl, 8-oxooctanoyl, and 5-oxovaleroyl esters of cholesterol and 7-ketocholesterol. Hoppe et al. (58) employed reversed-phase LC/ESI-MS to isolate and identify the core aldehydes and acids formed by peroxidation of cholesteryl linoleate by mouse peritoneal macrophages and in human atheroma. The presence of C_5 and C_9 core aldehydes of cholesterol and oxy-cholesterol were identified as the DNPH derivatives as described by Kamido et al. (35).

Kawai et al. (59) measured cholesteryl ester core aldehydes in peroxidized LDL by HPLC of their DNPH derivatives on a Davelosil ODS-MG-5 column (4.6 × 250 mm) with a linear gradient elution from 2-propanol/MeOH (2:8, v/v) to 2-propanol/MeOH (8:2, v/v) for 45 min at a flow rate of 0.8 mL/min. The elution profiles were monitored by absorbance at 370 nm. The peaks were identified by LC-MS with a Jasco PlatformII-LC instrument, 0.5% NH_4OH was added to the mobile phase. The binding of cholesteryl ester core aldehydes to LDL protein was

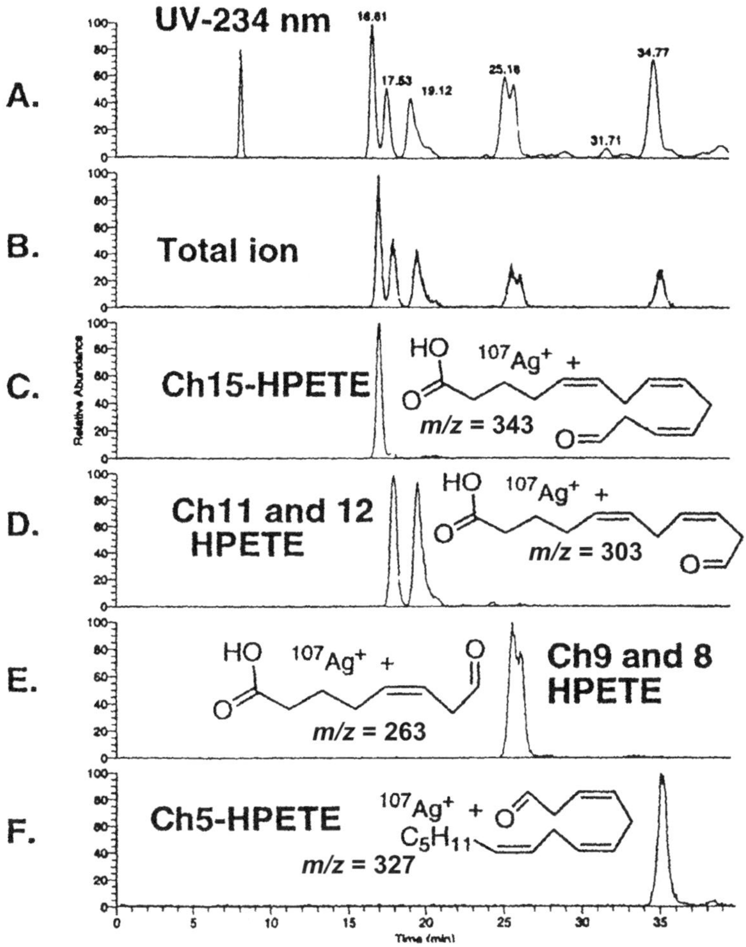

Fig. 7. HPLC/CIS-MS of cholesteryl arachidonate oxidation mixture formed by co-oxidation with cyclohexadiene. (A) UV detection at 234 nm; (B) LC/CIS-MS total ion current; (C) LC/CIS-MS in SRM mode for *m/z* =787 to 343; (D) LC/CIS-MS in SRM mode for *m/z* 787 to 303; (E) LC/CIS-MS in SRM mode for *m/z* 787 to 263; (F) LC/CIS-MS in SRM mode for *m/z* 787 to 327. HPLC conditions: two tandem Beckman Ultrasphere narrow-bore 5 μm silica columns (2.0 mm × 25 cm) operated in isocratic mode with 0.35% 2-propanol in hexane. The flow rate was 150 μL/min. Column effluent was passed through an Applied Biosystems 785 A programmable absorbance UV detector, with detection at 234 nm. An Upchurch PEEK high pressure mixing tee was connected next in series for the postcolumn addition of the silver salts. The silver tetrafluoroborate ($AgBF_4$) solution (0.25 mM in 2-propanol) was added via a Harvard Apparatus (Cambridge, MA) syringe pump at a flow rate of 75 μL/min. *Source:* Havrilla et al. (27).

demonstrated by monoclonal antibodies directed against 9-oxononanoyl cholesterol modified protein.

Unsaturated cholesteryl esters are readily ozonized and several laboratories have used LC/ESI-MS to investigate the ozonization products. Herrera et al. (29) have suggested tentative identities for the products of ozonization of cholesteryl linoleate in a non-participating solvent (hexane). Figure 8 shows the total positive ion current profile of the ozonization products of cholesteryl linoleate as obtained by reversed-phase LC/ESI-MS along with the total mass spectrum recorded over the entire elution range. The total ion current profile suggests the presence of five major peaks representing two to three isomers each. This is reflected in the full mass spectrum, which also contains five major ions (*m/z* 795, *m/z* 811, *m/z* 911, *m/z* 934 and *m/z* 999). Figure 9 shows the single ion mass chromatograms for these ions as extracted from the total ion current profile by the computer (29). The earliest eluted peak (*m/z* 795) corresponds to the ammonium adduct of (5-oxo-5,6-secocholestan-6-al)-3β-linoleate diozonide, *m/z* 777 $[795\text{-}18]^{+\cdot}$. The ion at *m/z* 811 may be attributed to the ammonium adduct of 1,2,4-trioxolane cholesterol-3β-linoleate diozonide, *m/z* 793 $[811\text{-}18]^{+}$. The ion at *m/z* 911 was attributed to an ammonium adduct of (5,6ξ-epidioxy-6ξ-B-homo-6-oxacholestan)-3β-linoleate diozonide to which a 6-carbon alkyl group (84 mass units) has been attached at C_6 of the opened sterol ring *mz* 893 $[911\text{-}18]^{+}$. The ion at *m/z* 934 was attributed to (5,6ξ-epidioxy-6ξ-B-homo-6-oxacholestan)-3β-linoleate diozonide with a monounsaturated 7-carbon monoalkene (96 mass units) as the alkoxy group, *m/z* 916 $[934\text{-}18]^{+}$. Similarly, the ion at *m/z* 999 was attributed to an ammonium adduct of (5,6ξ-epidioxy-6ξ-B-homo-6-oxacholestan)-3β-linoleate diozonide with a monounsaturated 9-carbon monoozonide (171 mass units) as the alkoxy group at C_6, *m/z* 981 $[999\text{-}18]^{+}$. All three of the latter ions exceeded the mass $[793\text{+}18]^{+}$ permissible under full ozonization in non-participating solvents. It should be noted that three highest molecular weight ozonides exceed the molecular weight of the fully ozonized cholesteryl linoleate. Clearly, the sterol ring must have become opened and added alkyl groups at C_6 as observed for the ozonization of cholesterol in participating solvents, when the primary product is believed to be 5-hydroperoxy-B-homo-6-oxa-cholestane-3β,7-diol (60). However, similar ozonization structures have been shown to arise during ozonization of more concentrated solutions in non-participating solvents. Pulfer et al. (61) have investigated by LC-MS/MS the unstable hydroperoxy bis-hemiacetal product obtained during ozonolysis of cholesterol with water as the participating solvent and have characterized it as a hydroperoxy, hydroxyl hemiacetal derivative, which is consistent with the NMR and X-ray crystallographic studies carried out with the more stable methyl ether derivative (60).

Diradylglycerol Hydroxides, Hydroperoxides, Isoprostanes and Core Aldehydes

Diradylglycerol hydroxides, hydroperoxides and core aldehydes were first resolved by LC-MS as an aid in the identification of the corresponding oxygenated glyc-

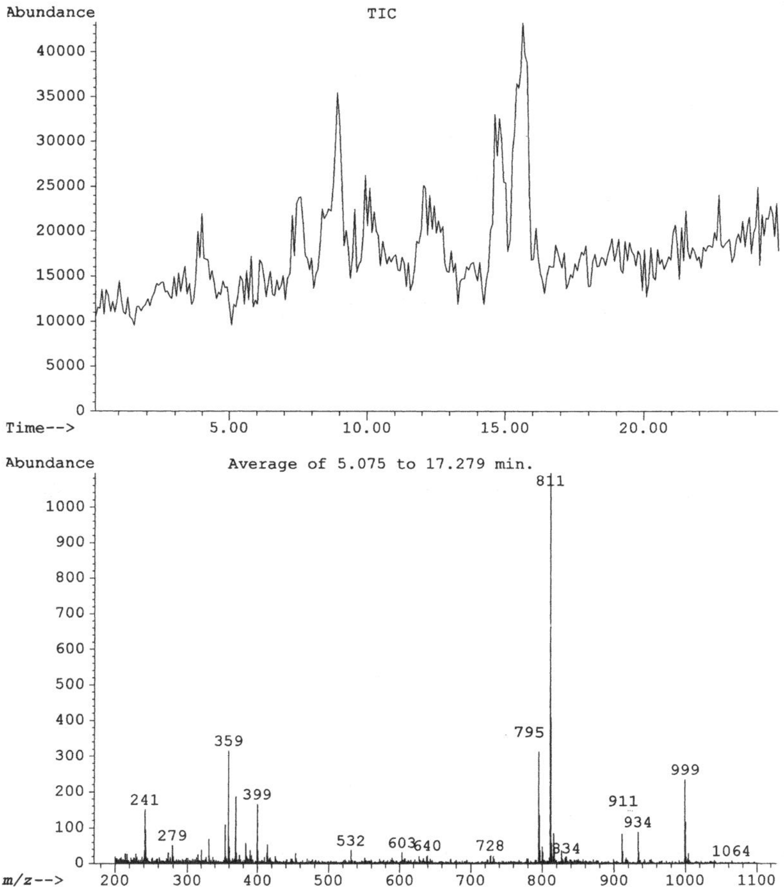

Fig. 8. Reversed-phase LC/ESI-MS analysis of the major products of ozonization of cholesteryl linoleate. (A) Total positive ion current profile of the oxidation products as obtained by LC/ESI-MS. (B) Full mass spectrum (*m/z* 200–1100) recorded over the elution range of the ozonides. HPLC/ESI-MS conditions: columns, Supelcosil LC-18 (250 × 4.6 mm ID); mobile phase, linear gradient of 20–80% 2-propanol in MeOH over 30 min. ESI-MS, positive-ion mode over the mass range 200 to 1100 Da; exit voltage set to 120 V; postcolumn addition of 0.5% NH_4OH in MeOH at 0.3 mL/min. *Source*: Herrera et al. (29).

erophospholipids (21,28,35). Aliquots of the lipid extracts of the DNPH-treated total copper-catalyzed peroxidation mixture of lipoprotein glycerophospholipids were digested with phospholipase C (*Bacillus cereus*) and the released DNPH derivatives of the diradylglycerol core aldehydes were purified by TLC using $CHCl_3$/MeOH (95:5, v/v) as the developing solvent, which resolved the aldehydes

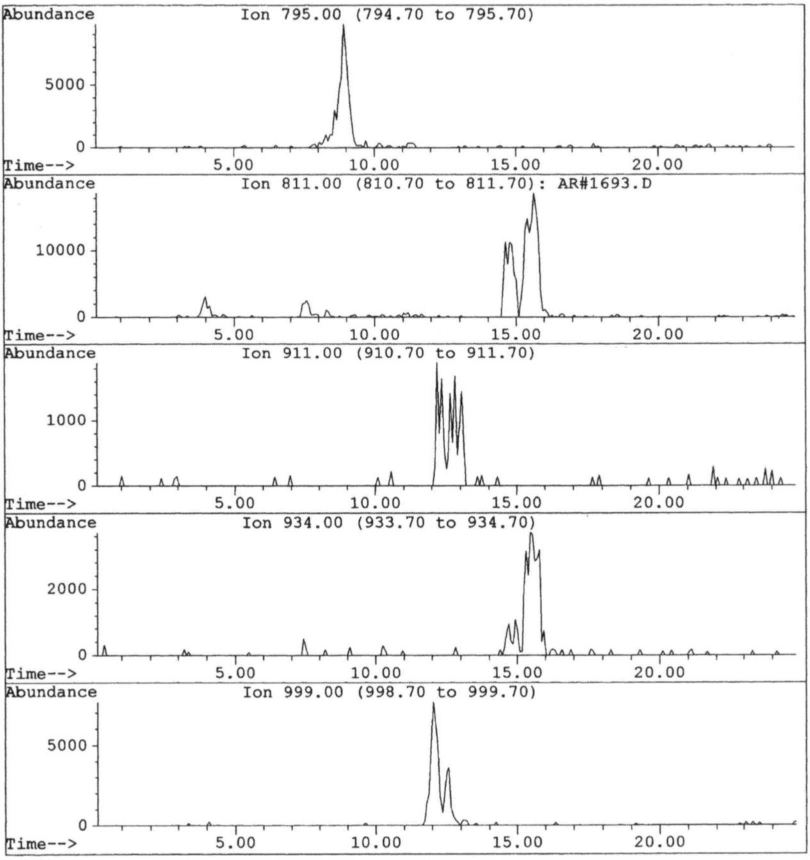

Fig. 9. Single ion mass chromatograms of the major products of ozonization of cholesteryl linoleate. Tentative structures are assigned as follows: *m/z* 795, (5-oxo-5,6-secocholestan-6-al)-3β-linoleate diozonide $[776+18]^+$; *m/z* 811, 1,2,4-trioxolane cholesterol-3β-linoleate diozonide, $[792+18]^+$; *m/z* 911, (5,6ξ-epidioxy-6ξ-B-homo-6-oxacholestan)-3β-linoleate diozonide to which a 6-carbon alkyl group has been attached at C_6 $[892+18]^+$; *m/z* 934, (5,6ξ-epidioxy-6ξ-B-homo-6-oxacholestan)-3β-linoleate diozonide with a monounsaturated 7-carbon alkyl group at C_6, $[916+18]^+$; *m/z* 999, (5,6ξ-epidioxy-6ξ-B-homo-6-oxacholestan)-3β-linoleate diozonide with a 7-carbon monounsaturated alkyl group at C_6 , $[980+18]^+$. HPLC/ESI-MS conditions were as given in Figure 8. *Source:* Herrera et al. (29).

from any hydroxides and hydroperoxides. The isolated DNPH derivatives of the aldehydes were separated by reversed-phase HPLC on a Supelcosil LC-18 column (250 × 4.6 mm ID) using a linear gradient of 30–90% propionitrile in acetonitrile as the eluting solvent. The peaks were monitored at 358 nm and the components were quantified using the DNPH derivatives of appropriate core aldehyde standards. About 1% of the HPLC column effluent was admitted to a Hewlett-Packard Model 5985 quadrupole mass spectrometer via direct liquid inlet (DLI) interface and NICI (electron capture) mass spectra were taken over the entire chromatogram in the mass range 200–900. Single ion plots were obtained by recalling the data stored in the computer.

The diradylglycerol hydroperoxides (after $NaBH_3$ reduction) and hydroxides were analyzed by GC/MS following trimethylsilylation on non-polar capillary columns (21) or by reversed-phase LC/ESI-MS following acetylation as described for the analysis of monoradylglycerols (62).

TAG Hydroxides, Hydroperoxides, Isoprostanes and Core Aldehydes

Reversed-phase HPLC is a most useful technique for resolving oxidized TAGs, with oxygenated compounds eluting ahead of non-oxygenated ones. The greater the number of hydroperoxide groups, the lower the retention time. MS has been an invaluable aid in characterization of both oxidized and normal TAGs, and coupling to HPLC increases its usefulness even further.

Kuksis et al. (11) used reversed-phase HPLC to identify mixed hydroperoxide and core aldehyde derivatives of synthetic and natural TAGs following reaction with the *tert*-BOOH/Fe^{2+} reagent. The oxidation products were extracted with $CHCl_3$/MeOH, reacted with DNPH, and resolved by normal phase TLC into fractions depending on the number of hydrazone derivatives formed. The individual TLC bands were examined by reversed-phase HPLC with on-line TSI-MS. The HPLC separations were performed on an HP-1090 liquid chromatograph equipped with a reversed-phase C-18 Supelcosil column (4.6 × 250 mm), which was eluted with a gradient of 20–80% 2-propanol in MeOH over 30 min. A solution of 0.2 M NH_4Ac/MeOH (1:1, v/v) was added to the postcolumn flow at a rate of 0.2 mL/min.

Kusaka et al. (63) were the first to apply LC/APCI-MS to the analysis of TAG hydroperoxides. Tey determined the composition of the hydroperoxides of *rac*-1-stearoyl-2-oleoyl-3-linoleoylglycerols. HPLC was performed with a reversed-phase C18 column. The hydroperoxidized synthetic TAGs gave three peaks, two of which were identified as hydroperoxides on the basis of the main fragment ions $[M-H_2O+H]^+$, $[M-H_2O+H]^+$ and $[M-R_1(R_3)COOH-H_2O+H]^+$.

Kozak et al. (48) identified the COX-2 oxidation products of 2-arachidonoylglycerol as various prostaglandin esters (see above).

Sjovall et al. (30) used reversed-phase HPLC with on-line ESI-MS for the separation and identification of synthetic hydroperoxy, hydroxy and epoxy derivatives, which were prepared as an aid in identification of peroxidized natural TAGs. Later, Sjovall et al. (31) resolved reference hydroperoxy, hydroxyl and epoxy TAGs by reversed-phase HPLC with on-line identification by ESI-MS. Figure 10 compares the reversed-phase HPLC/ELSD and LC/ESI-MS profiles of the oxidation products of 18:0/18:0/18:2 following a 45-min exposure at 37°C to 7.8 M TBHP. The elution profile detected by ELSD is similar to that recorded for the total positive ion current by MS. The mass spectrum averaged over the acylglycerol elution times is relatively simple when considering the presence of both ammonia and sodium adducts and suggests that the complex patterns result from resolution of isomers. The reversed-phase HPLC was performed on Supelcosil LC-18 column (250 × 4.6 mm ID) (Supelco, Mississauga, Ontario) using a linear gradient of 20–80% 2-propanol in MeOH (0.85 mL/min) in 30 min. The instrument was a Hewlett-Packard (Palo Alto, CA) Model 1090 liquid chromatograph interfaced with a nebulizer-assisted ESI source connected to a Hewlett-Packard Model 5989 A quadrupole mass spectrometer. The ionization (capillary exit) voltage of this instrument was set at 100 V but could be increased to 300 V, to obtain fragment ions from any clearly resoloved components (pseudo MS/MS). For improved ionization, 1% NH_4OH in 2-propanol was added postcolumn at a flow rate of 0.15 mL/min. The unoxidized and oxidized TAGs were identified as $[M+NH_4]^+$ or $[M+Na]^+$ in positive ion mode, whereas the DNPH derivatives of TAG core aldehydes were detected as $[M-H]^-$ ions in negative ion mode. Sjovall et al. (31) have recently shown that same experimental conditions are suitable for the analysis and identification of the *tert*-BOOH oxidation products of unsaturated TAGs, including the TBHP adducts. Sjovall et al. (32) have subsequently shown that the LC/ESI-MS is equally well suited for the isolation and identification of the TAG core aldehydes following their isolation as DNPH derivatives.

Sjovall et al. (64) reported the use of reversed-phase HPLC to investigate the formation of TAG core aldehydes during rapid oxidation of corn and sunflower oils with *tert*-BOOH/ Fe^{2+}. The core aldehydes were isolated as DNPH derivatives by preparative TLC and identified by reversed-phase HPLC with on-line ESI-MS and by reference standards. Using the reversed-phase HPLC system developed earlier (30), a total of 113 species of TAG core aldehydes were specifically identified, accounting for 32–53% of the DNPH-reactive material of high molecular weight, representing 25–33% of the total oxidation products. The major core aldehyde species (50–60% of total TAG core aldehydes) were the mono[9-oxo]nonanoyl- and mono[12-oxo]-9,10-epoxy dodecenoyl- or [12-oxo]-9-hydroxy-10,11-dodecenoyl-DAGs. Figure 11 shows the total negative ion elution profile of the 2,4-DNPH derivatives of oxidized sunflower oil TAGs and full mass spectrum averaged over the acylglycerol elution range (64). The various peaks are eluted as clusters of oxidation products. In order to appear in cluster, the oxo-TAG must possess at least one core aldehyde or ketone group to react with the DNPH. The various core aldehydes were eventually identified by combined TLC, HPLC and LC/MS techniques.

Sjovall et al. (64,65) have since used the above reversed-phase HPLC system to demonstrate significant formation of core aldehydes during rapid oxidation of corn and sunflower oils by *tert*-BOOH/Fe^{2+} reagent (65) and during autoxidation

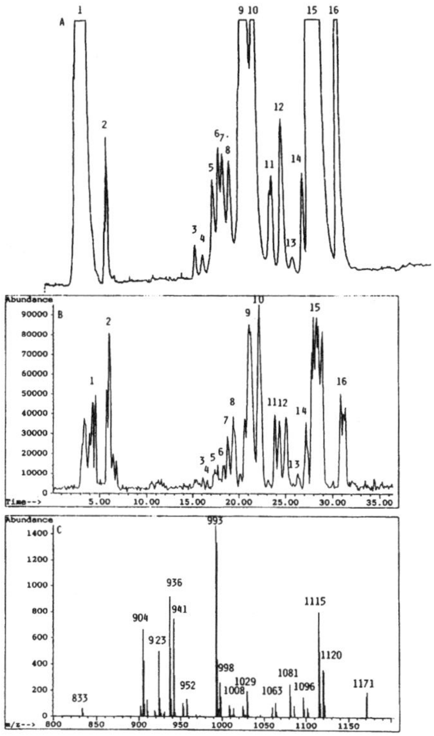

Fig. 10. Comparison of reversed-phase HPLC/ELSD (A) and HPLC/ESI-MS (B) profiles of the oxidation products of 18:0/18:018:2 following a 45 min exposure at 37°C to 7.8 M *tert*-butylhydroperoxide along with the full mass specrum (C) averaged over the elution time (13.587–33.362 min) of the oxoacylglycerol derivatives. Major peaks were identified as follows: Peak 6, 18:0/18:0/18:1, OOH, O-O, TBHP (*m/z* 1058); Peak 7, 18:0/18:0/18:2, OOH, epoxy (*m/z* 952); Peak 8, 18:0/18:0/18:2, OOH, TBHP (*m/z* 1024); Peak 9, 18:0/18:0/18:2, OOH (*m/z* 936); Peak 10, 18:0/18:0/18:1, O-O, diHTBHP (*m/z* 1114); Peak 11, 18:0/18:0/18:3, epoxy (*m/z* 918); Peak 14, 18:0/18:0/18:2, diTBHP (*m/z* 1080); Peak 15, 18:0/18:0/18:2, TBHP (*m/z* 992) Peak 15, 18:0/18:0/18:1 triTBHP (*m/z* 1170); and Peak 16, 1:0/18:0/1:2 (*m/z* 904). HPLC/ELSD conditions: reversed-phase HPLC on a Supelcosil LC-18 column (250 × 4.6 mm ID) using a linear gradient of 20–80% 2-propanol in MeOH (0.85 mL/mn) in 30 min (Hewlett-Packard Model 1050 Liquid Chromatograph coupled to a Varex ELSD II Light Scattering Detector (Varex, MD, USA) using nitrogen as nebulizing gas and an evaporation temperature of 85°C. LC/ESI-MS was performed using a Hewlett-Packard Model 1090 Liquid chromatograph interfaced with a nebulizer assisted ESI source connected to a Hewlett-Packard Model 5989 A quadrupole mass spectrometer. The ionization (capillary exit) voltage of the instrument was set at 170 V. The HPLC conditions were the same as those for HPLC/ELSD. *Source:* Sjovall et al. (31).

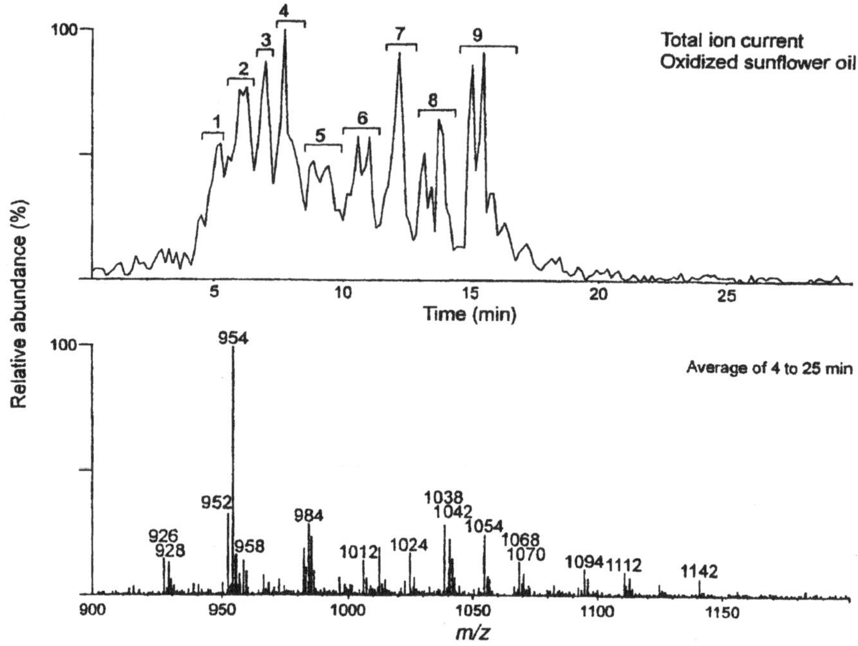

Fig. 11. Total negative ion elution profile of the 2,4-dinitrophenylhydrazine (DNPH) derivatives of oxidized sunflower oil triacylglycerols and full mass spectrum averaged over the acylglycerol elution range. Peaks are numbered as clusters in the elution profile and identified on the basis of subsequent analyses as follows: Cluster 1, trialdehydes; Cluster 2, core aldehyde trihydroxides or epoxydihydroxides; Cluster 3, core aldehyde dihydroxides or epoxyhydroxides; Cluster 4, core aldehyde hydroperoxides or epoxyhydroxides; Cluster 5, core aldehyde hydroperoxides; Cluster 6, TBHP adducts of core aldehyde epoxides; Cluster 7, TBHP adducts of core aldehydes; Cluster 8, core aldehydes; Cluster 9, core aldehydes. LC/ESI-MS conditions were as given in Figure 10. *Source:* Sjovall et al. (64).

of sunflower oil (64). Figure 12 shows the total negative ion current ESI-MS profile of the DNPH derivatives of core aldehydes in a 12-day autoxidation sample of sunflower seed oil (A) along with the full mass spectrum of averaged over the entire elution range (65). The HPLC peaks are labeled by the masses of the major ions. The tentative identification of the ions is given in (B).

The presence of TAG core aldehydes among the peroxidation products of vegetable oils has been demonstrated by Neff and Byrdwell (66) and Byrdwell and Neff (13,67,68) using APCI and reversed-phase LC/ESI/MS with NH_4COOH added as electrolyte. The reversed-phase HPLC was performed on two Inertsil ODS-3 columns (25 cm × 4.6 mm, 5 μm) used in series. The gradient program for the separation of the oxidation products of TAGs was as follows: initial-

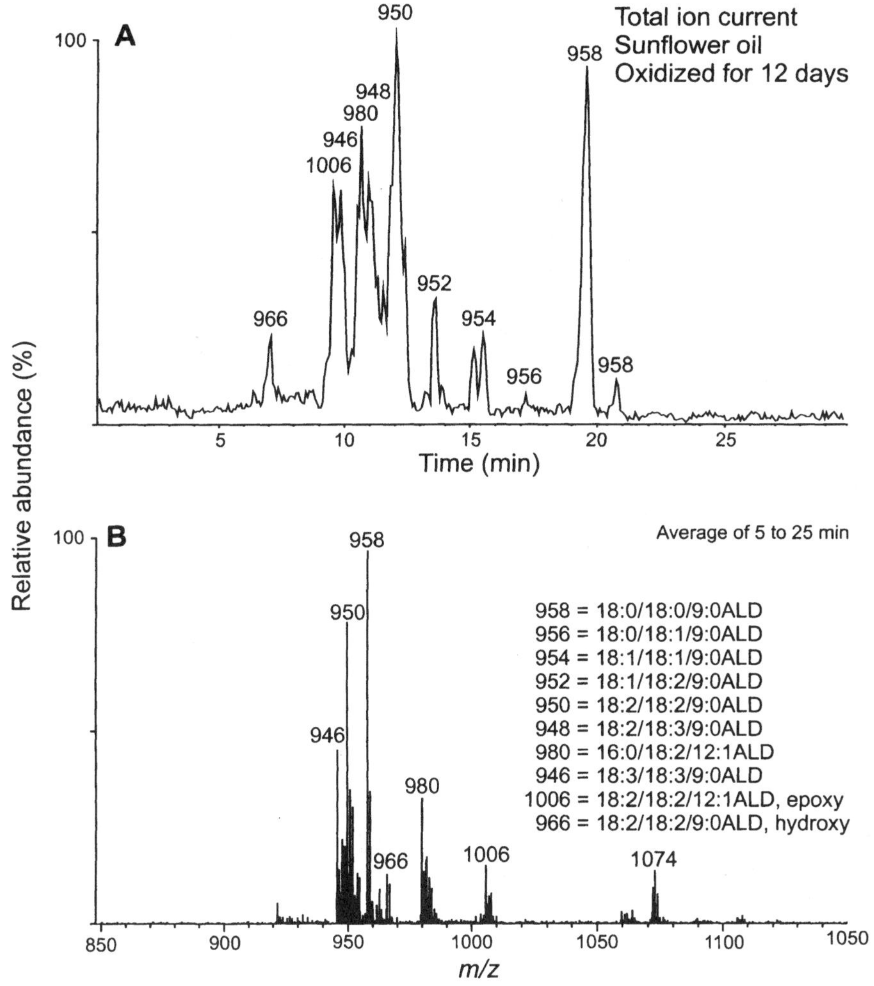

Fig. 12. Total negative ion current profile of DNPH derivatives of core aldehydes in a 12-day autoxidation sample of sunflower seed oil (A) and a full mass spectrum (B) averaged over the core aldehyde elution range (5–25 min). The HPLC peaks are labeled with the masses of the ions and the ion identities are given in B. LC/ESI-MS conditions were as given in Figure 10. *Source:* Sjovall et al. (65).

$MeCl_2/CH_3CN$ (25:75, v/v), held for 20 min; linear from 20 to 50 min to CH_2Cl_2/CH_3CN (30:70, v/v); linear from 50 to 85 min to CH_2Cl_2/CH_3CN (70:30, v/v); recycled to original conditions from 85 to 99 min. The flow rate was 0.8 mL/min. The column effluent was split to deliver the eluates to the mass spectrometer (LCQ Deca ion trap instrument, ThermoFinnigan) and to an ELSD system

(Varex). These researchers detected several hydroperoxide, epoxide and ketone structures. Among the chain shortened derivatives of trioleoylglycerol were found species formed by fragmentation between carbons 9 and 10, the location of the double bond. The earlier eluted products appeared to represent chain-shortened species in which an oxygen functional group remained on the TAG. These molecules could be either core aldehydes or epoxides. The fragmentation patterns of these two classes of isobaric molecules could not be distinguished without preparation of the DNPU derivatives.

Ravandi et al. (20) used reductive ozonolysis to prepare synthetic TAGs containing one, two and three core aldehyde groups per TAG molecule. These core aldehydes were isolated by TLC and their structures were confirmed by HPLC with on-line ESI-MS. Single ion mass chromatograms and full mass spectra showed the presence of TAGs containing one and two aldehyde groups as the DNPH derivatives. The HPLC separations were obtained using a gradient of 20–80% 2-propanol/MeOH system described earlier (11).

Bauer-Plank and Steenhorst-Slikkerveer (69) reported the separation of TAG hydroperoxides in vegetable oils by non-aqueous reversed-phase HPLC with UV detection, while Steenhorst-Slikkerveer et al. (70) used ESI-MS in combination with HPLC for the analysis of core aldehydes, core acids and the mono- and di-oxygenated oxidation products of standard TAGs. This resulted in a separation into classes of TAG oxidation products such as epoxy-, oxo-, hydroperoxy-, hydroxy- and "2.5 glycerides," which were identified by SIM. LC-MS was performed on an HP 1100 LC/MS, which was operated using a binary high pressure pump, an auto-injector, a thermostated column and a diode-array detector. Downstream of the diode array detector, 0.15 mM NaI in EtOH/MeOH 1:1 (v/v) was added at 100 μL/min to the eluant flow by means of another HPLC pump. The HPLC separations were performed on a Waters Diol Column (200 × 3 mm ID). The mobile phase gradient started with hexane and proceeded to a mixture of hexane/methyl *tert*-butyl ether (MTBE) or hexane/PA. When these non-volatile lipid oxidation products were analyzed by reversed-phase HPLC, the various species present were separated according to class: [OOH-TAG, hydroxy-TAG (OH-TAG), epoxy-TAG, etc.] and size [OOH-18:2/18:2/18:2, OOH18:1/18:2/18:2, OOH-18:1/18:1/18:2, etc.].

Kurvinen et al. (71) prepared and characterized the core aldehydes of MAGs as the DNPH derivatives using the LC/ESI-MS methods employed by Sjovall et al. (64,65) to identify the core aldehydes of TAGs.

More recently, Suomela et al. (72,73) have used reversed-phase HPLC in combination with ESI-MS/MS for the detection and identification of oxidized TAGs in human LDL and in the diet and plasma lipoproteins of pigs. The HPLC system consisted of a Hitachi (Tokyo, Japan) L-6200 Intelligent Pump with a Discovery HS C_{18} column (250 × 4.6 mm ID, Supelco Inc., Bellefonte, PA). The column was eluted with a linear gradient of 20–80% 2-propanol in MeOH over 20 min at a flow rate of 0.85 mL/min. Fifteen percent of the effluent was (0.13 mL/min) led to a Finnigan MAT TSQ 700 triple quadrupole mass spectrometer

(Finnigan, San Jose, CA) equipped with nebulizer-assisted ESI interface. As a sheath liquid, 1% NH_4OH in MeOH was added at a flow rate of 3 mL/min to improve ionization. For the detection of TAG core aldehydes, part of the TAG mixture was reacted overnight at 6°C with 2,4-DNPH prior to HPLC/ELSD (Sedex 75, SEDERE, Alfortville, France) and LC/MS analysis.

Glycerophospholipid Hydroperoxides, Isoprostanes and Core Aldehydes

Coupling of MS methods with chromatography provides much more structural information, including double bond location and configuration. The method has proven indispensable for the identification of oxygenated glycerophospholipids. Several groups of investigators have used TSI-MS (21,34,35) and ESI-MS (12,33,34) in combination with normal phase HPLC for the identification of the major hydroperoxides and core aldehydes of PtdCho from oxidized plasma lipoproteins and atheroma tissues.

Myher et al. (12) combined a normal phase HPLC system with on-line ESI-MS for sensitive detection of oxygenated glycerophospholipids. The oxygenated glycerophospholipids, including ozonides, hydroperoxides, hydroxides, and core aldehydes gave protonated and sodiated molecular ions, which were seen as the ionization products without fragmentation. Such pseudomolecular ions had not been previously observed for the oxygenated phospholipids using other LC-MS interfaces (e.g. thermospray, particle beam). The ionization was so soft that non-covalent dimers could be detected for deuterated PtdChos. The method has been applied to the determination of lipid ester ozonides and core aldehydes in liposomes and plasma lipoproteins (21,33,34), and for identification of glycated aminophospholipids from red cells and plasma of diabetics (74). It has been further utilized for the identification and quantification of the peroxynitrite oxidation products of the PtdChos of HDL (75,76). Figure 13 illustrates the LC/ESI-MS analysis of HDL phospholipids following oxidation with SIN-1 for 6 h (75). In other applications the above LC/ESI-MS methods have been used to follow the metabolic fate of the lipid hydroperoxides and core aldehydes in cell cultures (77,78) and in atherosclerotic plaques (79). Adachi et al. (80) have used LC/ESI-MS to demonstrate that the mean plasma levels of PtdCho-OOH are 55.1 ± 30.4 pmol/mL and 16.3 ± 9.9 pmol/mL depending on the exact solvent system employed in the analysis. The use of a Finepak SIL NH_2 column with 2-propanol/MeOH/H_2O as the mobile phase gave a higher resolution than did an LC-18-DB column with MeOH containing 0.01% triethylamine. It was concluded that the plasma levels of hydroperoxy PtdCho commonly reported may be overestimated.

Spickett et al. (81) have described the use of reversed-phase HPLC with positive-ionization ESI-MS for the detection of phospholipid oxidation in oxidatively stressed cells. LC-MS was carried out with Shimadzu LC-10 system and a Phenomex Luna C_8 column (5 μm RP-Sct, 150 × 1 mm ID). This column was operated

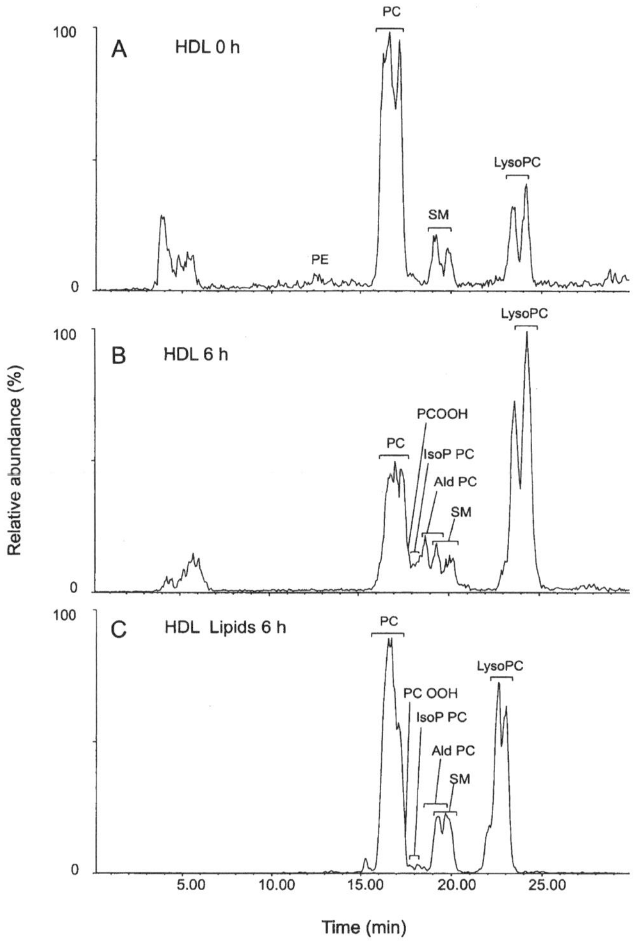

Fig. 13. LC/ESI-MS analysis of HDL phospholipids following oxidation with SIN-1. A, control HDL; B, HDL oxidized for 6 h; C, HDL liposomes oxidized for 6 h. Eluted peaks were identified by relative retention times of standards and the molecular masses of the components as follows: PE, PtdEtn; PC, PtdCho; S, sphingomyelin; LysoPC, lyso PtdCho; PC-OOH, PtdCho hydroperoxides; isoP PC, isoprostane GroPChos; AldPC, PtdCho core aldehydes. LC/ESI-MS analysis was performed using a normal phase silicic acid column (4.6 × 250 mm, Alltech Associates, Deerfield, IL) in a Hewlett-Packard Model 1050 liquid chromatograph, connected to a Hewlett-Packard Model 5989A Quadrupole Mass Spectrometer, equipped with a nebulizer-assisted ESI interface. The column was eluted with a linear gradient of 10% A ($CHCl_3$/MeOH/30% NH_4OH (80:19.5:0.5, by vol) to 100% B ($CHCl_3$/MeOH/H_2O/30% NH_4OH (60:34:5.5:0.5, by vol) for 14 min, followed by 100% B for 10 min, at a flow rate of 1 mL/min. *Source:* Ahmed et al. (75).

at a flow rate of 100 μL/min with an isocratic solvent system of MeOH/hexane/0.1 M NH_4Ac (71:5:7, by vol). The eluent from the column was split 1:1 between the MS and waste. Peak top spectra were collected usually between *m/z* 700 and *m/z* 1000 with a sweep time of 2 s. This may have prevented from detecting any TBHP adducts [see Sjovall et al. (65)]. Figure 14 shows the results for phospholipid vesicles treated with *tert*-BOOH and Fe^{2+} overnight in presence of air (81). The TIC chromatogram shows a variety of oxidized phospholipids eluting in the first 3–10 min, and the native phospholipids eluting at 12 min onwards. In the right hand panel, the RICs at *m/z* 842, 874 and 906 show the appearance of products corresponding to the addition of one-three dioxygens to stearoylarachidonoyl GroPCho (*m/z* 810). The tris-hydroperoxide eluted earlier at 3.2 min, followed by bis-hydroperoxides at 4–6 min with the mono-hydroperoxides eluting last at 8.6 and 9.76 min. For both the mono- and bis-hydroperoxides several components were partially resolved, which were attributed to isomers. In the left hand panel, the ions at *m/z* 790 (4.9 min) and *m/z* 818 (6.7 min) were assigned to the monohydroperoxides of $C_{34:2}$ (native mass 758) and $C_{36:2}$ (native mass 786). Each of the components assigned to hydroperoxides showed two source-induced breakdown products with losses of 18 and 34 mass units, corresponding to dehydration or the loss of H_2O_2.

Cardiolipin is an anionic phospholipid in mitochondrial and bacterial membranes and has recently been reported in human plasma lipoproteins. Because of its high unsaturation, it is anticipated that it would be readily peroxidized. Bergqvist and Kuksis (82) reported the formation of cardiolipin hydroperoxides and core aldehydes during treatment of commercial bovine heart cardiolipin with *tert*-BOOH. The hydroperoxides were resolved and identified by normal phase LC/ESI-MS using the methodology previously employed for the analysis of other oxo-phospholipids (12,20). Figure 15 shows the total ion negative ion profile of the oxidized cardiolipin sample along with the full mass spectrum recorded over the phospholipids elution range (A) and the $[M-1]^-$ ion plots (B) for the major hydroperoxide species of bovine cardiolipin (82). The major species were identified as LLLL-OOH (*m/z* 1480); LLLO-OOH (*m/z* 1482); LLLL-di-OOH (*m/z* 1512); LLLO-di-OOH (*m/z* 1514); LLLL-tri-OOH (*m/z* 1544); and LLLL-tetra-OOH (*m/z* 1576). The molecular species of the oxygenated CL were identified on the basis of the molecular masses provided by MS, the knowledge of the fatty acid composition of the CL, and the relative HPLC retention times. In presence of Fe^{2+} these peroxides yielded small amounts of the corresponding core aldehydes and core acids. Schlame et al. (83) reported the resolution of cardiolipin into 12 molecular species, including oxidized species produced by treatment of cardiolipin with air, cytochrome C, or Cu^{2+}/*tert*-butyhydroperoxide. Kriska et al. (84) have proposed a simple method for the selective determination of phospholipids hydroperoxide families, including cardiolipin hydroperoxides, in complex populations. The method is based on phospholipid peroxide separation by nomal phase high performance TLC (HPTLC) followed by spray detection with N,N,"N" –tetramethyl-p-phenylenediamine (TPD) and densitometric scanning of the purple bands. The

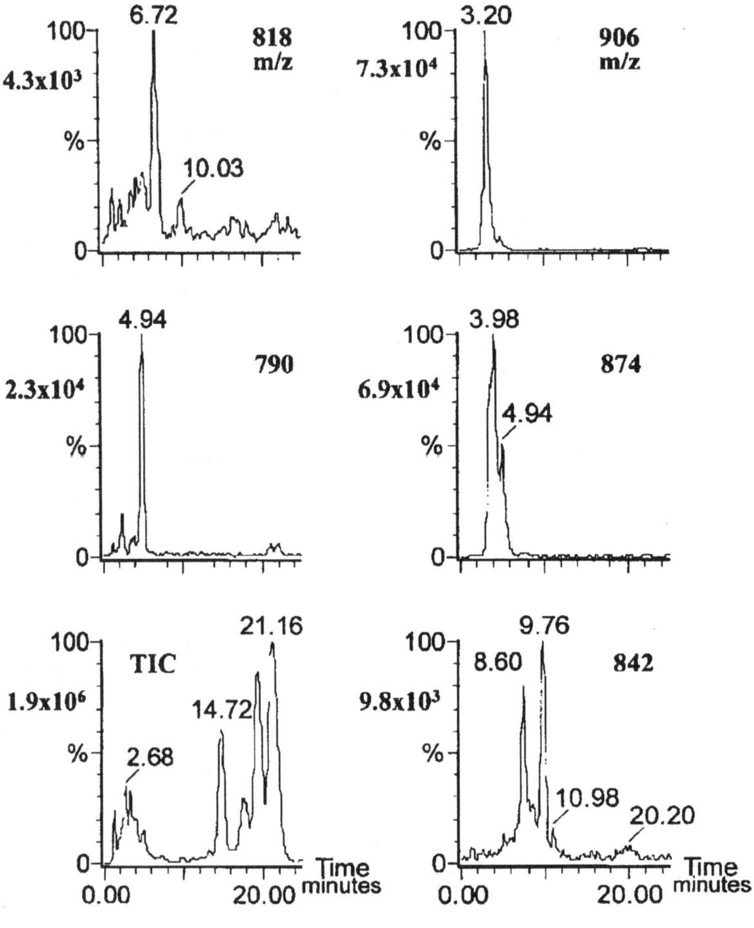

Fig. 14. Reversed-phase LC/ESI-MS of oxidatively modified PtdCho. TIC, Total ion current profile of phospholipid vesicles treated with *tert*-butylhydroperoxide and Fe^{2+}; the oxidized PtdChos are eluted in the first 3–10 min; RIC, Reconstituted ion chromatograms of five different masses. *Right-hand panel* shows the RICs at *m/z* 842, 874 and 906 corresponding to the addition of one to three dioxygens to $C_{38:4}$ GroPCho (stearoylarachidonoyl GroPCho, *m/z* 810). *Left-hand panel* shows the ions at *m/z* 790 and 818 assigned to the monohydroperoxides of $C_{34:2}$ (native mass 758) and $C_{36:2}$ (native mass 786). LC/ESI-MS was performed using Phenomenex Luna column with a flow rate of 0.1 mL/min. Peak-top data were collected between *m/z* 650–950 with a sweep of 2 s. RICs were generated using MassLynx software and are non-smoothed. 100% of the vertical axes corresponds to the ion current intensity indicated to the Y axis of the chromatogram. *Source:* Spickett et al. (81).

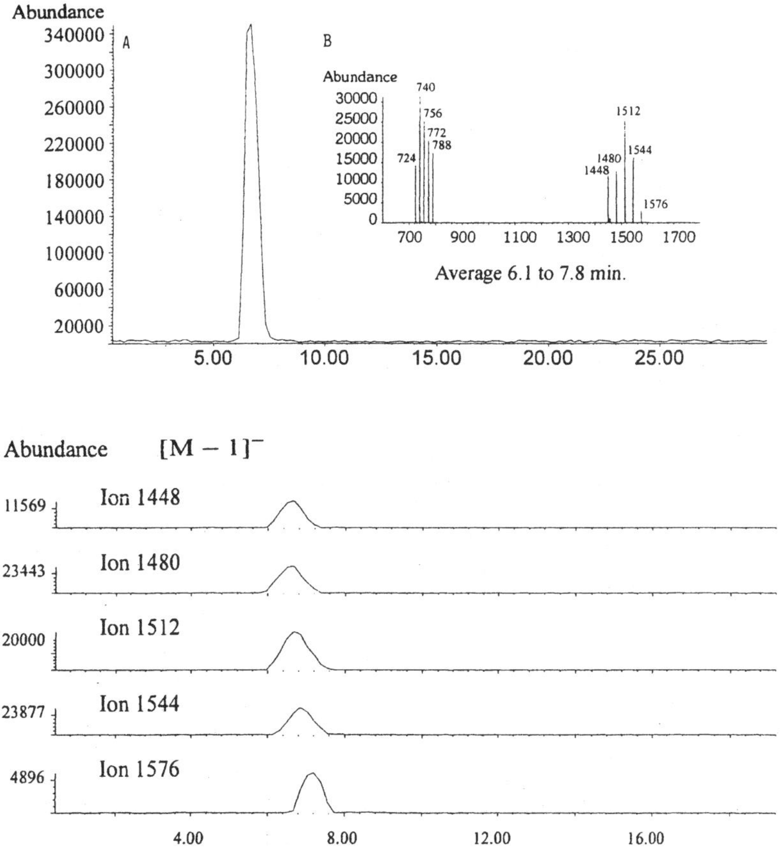

Fig. 15. Normal phase LC/ESI-MS analysis of commercial bovine heart cardiolipin after 1 h treatment with *tert*-butylhydroperoxide and Fe^{2+} in taurocholic acid. *Upper Panel:* (A) total negative ion current; (B) full mass spectrum averaged over the entire HPLC peak. Peak identification: *m/z* 1448, tetralinoleoyl-; *m/z* 1480, tetralinoleoyl-OOH-; *m/z* 1482, trilinoleoyloleoyl-OOH-; *m/z* 1512, tetralinoleoyl-$(OOH)_2$-; *m/z* 1514, trilinoleoyloleoyl-$(OOH)_2$-, *m/z* 1544, tetralinoleoyl-$(OOH)_3$-; and *m/z* 1576, tetralinoleoyl-$(OOH)_4$-GroPGroPGro. *Lower Panel:* Single negative ion mass chromatograms of major species of cardiolipin hydroperoxides. Peak identification is as given in *Upper Panel.* Normal phase LC/ESI-MS was done with a Spherisorb 3μ column (100 × 4.6 mm ID, Alltech Associates) installed in a Hewlett-Packard Model 1050 Liquid Chromatograph connected to a Hewlett-Packard Model 5988 Quadrupole mass spectrometer equipped with a nebulizer assisted ESI interface. The column was eluted with gradient of 100% Solvent A ($CHCl_3$/MeOH/30% NH_4OH (80:19.5:0.5, by vol) to 100% Solvent B ($CHCl_3$/MeOH/H_2O/30% NH_4OH (60:34:5.5:0.5, by vol) in 14 min, then 100% B for 10 min. The flow rate was 1 mL/min. *Source:* Bergqvist and Kuksis (82).

HPTLC-TPD bands presumed to represent CL were verified by MS. For MS analysis, the sample was dissolved in hexane/EtOH (4:1, v/v), mixed with 200-fold molar excess of terthiophene matrix, and subjected to MALDI-TOF analysis using a Voyager-DE PRIO Biospectrometry Workstation (PerSeptive Biosystems, Foster City, CA). Measurements were made in the negative ion reflector mode.

As an alternative, a new method in which silver ion adducts of peroxides and hydroperoxides are subjected to MS has been applied to the structural analysis of oxidized phospholipids by Milne and Porter (85). By addition of $AgBF_4$ postcolumn, silver ion CIS-MS was coupled to reversed-phase HPLC to give definitive structural information. A 24 h oxidation of 1-stearoyl-2-arachidonoyl-*sn*-glycero-3-phosphocholine with 10 mol% 2,2,5,7,8-pentamethyl-6-chromanol yielded six major components on reversed-phase LC-MS (Discovery C_{18} analytical column, Supelco; MeOH/H_2O, 95:5, v/v, mobile phase). The components were identified as 15-HPETE GroPCho (*m/z* 948 to 665); 11-HPETE GroPCho (*m/z* 948 to 625); 12-HEPETE GroPCho (*m/z* 948 to 820); 8-HPETE GroPCho (*m/z* 948 to 585); 9-HPETE GroPCho (m/z 948 to 780) and 5-HPETE GroPCho (*m/z* 948 to 545), in order of increasing retention time.

Isoprostanes and prostaglandin-like hydroxy-polyunsaturated fatty acids are formed via free radical oxidation reactions of arachidonoyl groups esterified to membrane glycerophospholipids (42,43,86). Morrow et al. (86) demonstrated that on normal phase HPLC, the F_2-isoprostane containing phospholipids exhibited much greater polarity than the non-oxidized phospholipids. Normal phase HPLC analysis of lipid extracts from livers of CCl_4-treated rats was performed on a 25 cm × 4.6 m Econosil SI column with 5 mm particles (Alltech Associates), using an isocratic solvent system of hexane/2-propanol/H_2O (4:6:1, by vol) at a flow rate of 1 mL/min. UV absorbance was monitored continuously at 205 nm. Aliquots of fractions eluted were collected, subjected to alkaline hydrolysis and the free F_2-isoprostanes identified and quantified by GC. Kayganich-Harrison et al. (87) confirmed that the F_2-isoprostane is esterified to the GroPCho backbone by CID (collision induced dissociation) of $[M\text{-}CH_3]^-$ ions from oxidized phospholipids. CID of the $[M\text{-}CH_2CHN(H_3)_3]^-$ ion revealed that F_2-isoprostanes were primarily esterified to the *sn*-2-position of the PtdCho. The isoprostane formation is a facile process, which has been subsequently found to take place in vivo (45) and includes the formation of the D_2/E_2 isoprostanes. Normal phase HPLC analysis of lipid extracts was performed on a 25 cm × 4.6 mm ID Econosil SI column, Alltech) with 5 μm particles using an isocratic solvent system of hexane/2-propanol/H_2O (4:6:1, by vol) at a flow rate of 1 mL/min. UV absorbance was monitored continuously at 205 nm. Aliquots of fractions were subjected to hydrolysis by bee venom phospholipase A_2 and then analyzed for free D_2/E_2-isoprostanes. Hall and Murphy (88) used normal phase HPLC with ESI-MS to separate phospholipid classes and analyze the distribution of the major polyunsaturated fatty acyl groups and corresponding oxidation products from red blood cells treated with an oxidizing agent. Subsequently, each of the various oxidized acyl moieties was identified by tandem MS techniques.

Specifically, Nakamura et al. (22,23) have isolated and characterized murine pulmonary phospholipids and have observed the normal occurrence of 10 isobaric eicosanoids corresponding to incorporation of one oxygen atom into the arachidonate containing glycerophospholipids. Phospholipids were hydrolyzed to yield the free carboxylic acids prior to reversed and chiral phase analysis. Reversed-phase LC/ESI-MS/MS was used to identify and quantify six monohydroxyeicosatetraenoic (HETE) acid regioisomers using d_8-HETE as internal standard and an HPLC column (2 × 150 mm, 3μ C18 Ultremex column; Phenomenex). 15-HETE was isolated with 50% B at a flow rate of 200 μL/min at a retention time of 24.5 μL/min. The collected samples were then methylated with ethereal diazomethane and the methyl esters of 15(*R*)- and 15(*S*)-HETE were separated by normal phase HPLC, wich was performed using a chiral column (4.6 × 250 mm, Chiralcel OD; Chiral Technologies, Exton, PA) with hexane/2-propanol (95:5, v/v) at a flow rate of 500 μL/min. The 15(*R*)- and 15(*S*)-HETE methyl esters were collected and hydrolyzed with sodium hydroxide. The recovered fatty acids were hydrogenated by bubbling hydrogen gas through a methanol solution of the sample for 20 min after addition of Rh/Al_2O_3 (1 mg) as catalyst. Samples were then taken to dryness and derivatized as the PFB ester TMS ethers. GLC-MS was carried out in the EC negative ion mode using a non-polar 15 m × 0.25 mm DB-1 (J&W Scientific, Folsom, CA) column with 0.25 μm film thickness. The hydroxy fatty acids were determined by monitoring the selected ions a *m/z* 339 and *m/z* 403, respectively, which corresponded to the loss of PFB radical from 15-hydroxyeicosanoate PFB ester and $[^{18}O_2]$15(S)-hydroxyeicosanoate PFB, respectively, at a retention time of 12.5 min.

Pruzanski et al. (89) used a previously developed LC/ESI-MS method (12,20) to demonstrate the presence of isoprostane GroPho in both normal and acute phase HDL. The major ions corresponded to iso-PGE_2epoxyGroPCho (*m/z* 828), iso-PGE_2/D_2GroPCho (*m/z* 830), and isoPGF_2GroPCho (*m/z* 832). The isoPGE_2/D_2GroPCho (*m/z* 858) and iso-PGF_2GroPCho (*m/z* 860) species were minor and appeared erratically in the acute phase HDL samples. The method also permitted the identification of the PtdCho isoprostanes generated by oxidation of HDL with peroxynitrite (75–78). Figure 16 shows the single positive ion mass chromatograms of the PtdCho isoprostanes accumulated in HDL after 2 h of oxidation with SIN-1 (75). The major species were identified as the epoxy-isoprostane GroPChos and isoprostane GroPChos. Jerlich et al. (90) have reported the formation of PtdCho isoprostanes in stimulated phagocytes. Specifically, the formation of monohydroperoxides (*m/z* 814) and bis-hydroperoxides (*m/z* 846) of palmitoyl/arachidonoyl GroPCho were observed. However, the major oxidized product occurred at *m/z* 828, and was identified as 1-palmitoyl-2-(5,6-epoxyisoprostane E_2)-*sn*-glycero-3-phosphocholine. LC-MS was carried out using a Phenomenex Luna C_8 reversed-phase column (5 mm RP-Sect, 150 mm × 1 mm ID) with an isocratic solvent system of MeOH/hexane/0.1 NH_4Ac (71:5:7, by vol) at a flow rate of 0.1 mL/min. The phospholipid oxidation products in oxidatively stressed cells were detected in positive ESI-MS as described by Spickett et al. (81).

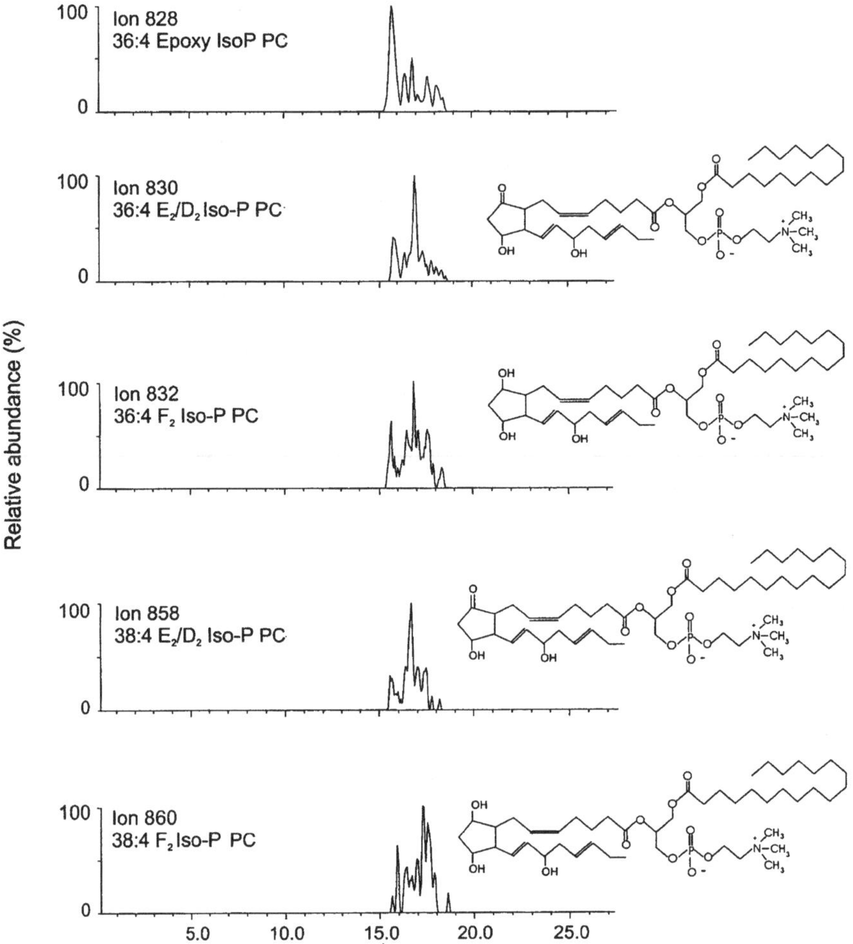

Fig. 16. Single positive ion mass chromatograms of the PtdCho isoprostanes accumulated in HDL after 2 h of oxidation with SIN-1. Peaks are identified and structural formulas given in the figure. Epoxy Iso PC, epoxy-isoprostane GroPChos; IsoP PC, isoprostane GroPChos; 36:4 and 38:4 represent total number of fatty acyl carbons and number of double bonds. LC/ESI-MS conditions were as given in Figure 13. *Source:* Ahmed et al. (75).

Salomon et al. (91) postulated and subsequently confirmed the hypothesis that isolevuglandins are generated upon non-enzymatic phospholipid peroxidation through rearrangement of isoprostane endoperoxides. The eight stereoisomers of LGE_2 are referred to collectively as isoLGE_2 (Scheme 1). Structural isomers of

LGE_2 are also produced as mixtures of eight possible stereoisomers. The terms "isoketal or IsoK" were coined by Bernoud-Hubac et al. (54) as alternatives to the isoLG nomenclature to distinguish them from levuglandins formed by rearrangement of the cyclooxygenase endoperoxide intermediate, PGH_2. Salomon (92) points out that such a distinction is erroneous because the exact same levuglandin molecules, LGE_2 and LGD_2, are generated by both the cyclooxygenase and isoprostane pathways. The difference between the pathways is that stereo and structural isomers are cogenerated with LGE_2 and LGD_2 in free radical-induced autoxidation of arachidonates.

Lipid hydroperoxides, the primary products of lipid peroxidation, are degraded into aldehydes, the secondary products of oxidation. When polyunsaturated glycerophospholipids and cholesteryl esters are oxidized, both water-soluble short-chain aliphatic aldehydes and lipid-soluble core aldehydes (aldehydes still bound to parent molecules) are generated in stoichiometric amounts. Kamido et al. (21,28,35) used LC/TSI-MS to resolve and identify GroPCho core aldehydes generated by OsO_4 oxidation of egg yolk PtdCho followed by $NaIO_4$ cleavage. Both 16:0/9:0ALD and 18:0/9:0ALD GroPCho were generated. The core aldehydes were identified as the hydrazones of the DAGs released by phospholipase C. Later, Kuksis et al. (11), Myher et al. (12) and Ravandi et al. (20) used reversed-phase HPLC for direct resolution of the core aldehydes generated by chemical oxidation of PtdCho from various sources. Khaselev and Murphy (36) and Kayganich-Harrison and Murphy (93) reported a detailed characterization of chain-shortened oxidized GroPCho lipids using FAB-MS and tandem MS.

More recently, Watson et al. (94) used flow injection MS/MS but Watson et al. (95) used LC/ESI-MS for structural identification of the core aldehydes and epoxides of glycerophospholipids in minimally oxidized LDL. The core aldehydes of ether bond-containing glycerophospholipids have also been isolated and characterized. Khaselev and Murphy (36) employed HPLC in their investigation of the oxidation of 20:4*n*6-containing plasmenyl GroPCho. The plasmenyl PCho was oxidized by AAPH. Several major as well as minor GroPCho species were observed when 1-O-hexadecyl-1′-enyl-2-20:4-GroPCho was oxidized for 3 h in the presence of AAPH [2,2′-azo-bis-(2-amidinopropane)hydrochloride]: 1-lyso/20:4 GroPCho (19.3 min); 16:0a/20:4 GroPCho (21.1 min); 16:0p/5:0(carbonyl) GroPCho (27.3 min); 16:0p/5HPETE-GroPCho (40.2 min); 16:0p/5ETE-GroPCho (42.8 min); 16:0p-OH/20:4 GroPCho (44.6 min); and 16:0p/20:4 GroPCho (starting material) (491.1 min). Thus, the reaction products included 1,2-diacyl lipid, a lysophospholipid, oxidation products involving the *sn*-1-position alone, oxidation products involving the *sn*-2-position alone, chain-shortened ω-aldehyde radyl substituents (core aldehydes) as well as products which were oxidized both at the *sn*-1- and *sn*-2-positions.

Kamido et al. (96) reported the identification of core aldehydes of alkyl glycerophospholipids in atheromas. The alkyl 16:0-5:0ALD, 18:0-5:0ALD and alkyl 16:0-9:0ALD and 18:0-9:0ALD along with the corresponding acyl 16:0 and 18:0

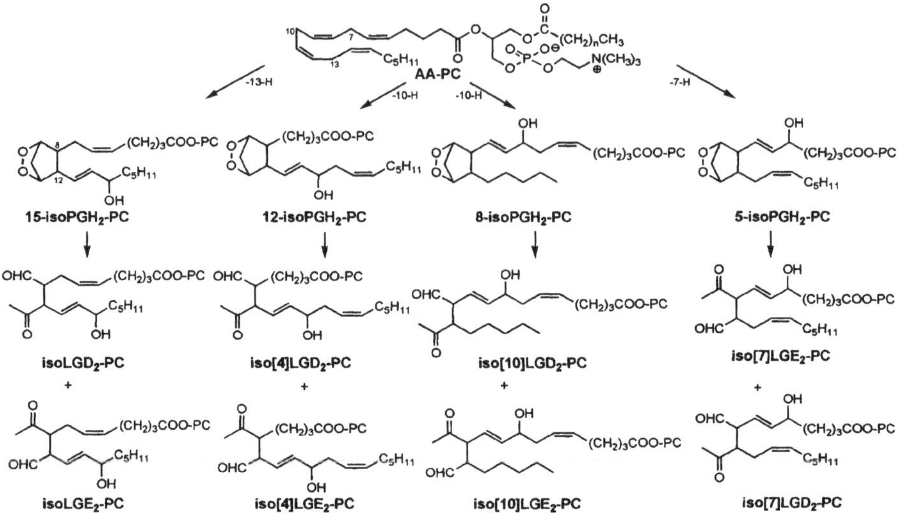

Scheme 1. Free radical-induced cyclooxygenation of arachidonoyl phosphatidylcholine (AA-PC) produces four structurally isomeric endoperoxides that rearrange to give eight structurally isomeric isoLG phospholipids that are each mixtures of eight stereoisomers. *Source:* Salomon (92).

core aldehydes were identified by reversed-phase LC/ESI-MS as the DNPH derivatives of the dialkylglycerol derivatives released by phospholipase C from the corresponding glycerophospholipids from human atheroma (see above under Diradylglycerol Hydroxides, Hydroperoxides and Core Aldehydes).

Due to the high natural abundance and ready oxidation, the linoleate and arachidonate esters of GroPCho constitute the main sources of the 9-oxononanoyl and 5-oxopentanoyl core aldehydes, respectively, isolated and characterized by these methods. Figure 17 shows an ample selection of the common aldehydes resulting from peroxidation of polyunsaturated PtdCho as obtained by Ahmed et al. (75) during peroxynitrite oxidation of HDL. The core aldehydes were analyzed in the underivatized form using previously established methods (12,20).

Subbanagounder et al. (97) demonstrated the presence of γ-hydroxy-α,β-unsaturated aldehydic phospholipids, 1-palmitoyl-2-(5-hydroxy-8-oxooct-6-enoyl)-*sn*-glycero-3-phosphocholine in oxidized palmitoylarachidonoyl GroPCho by LC/ESI-MS/MS, derivatization and chemical synthesis. Flow injection ESI-MS analyses were performed as described by Watson et al. (94). Reversed-phase LC/ESI-MS was performed by employing a C_8 column (Betasil, 2 × 250 mm) at a flow rate of 0.4 mL/min under the LC/MS conditions of Watson et al. (95). LC-MS/MS analysis in both positive and negative ion modes was performed employing the above chromatographic conditions. Derivatization of phospholipids with $NaBH_3$ and methoxylamine was performed under standard conditions. The hydrox-

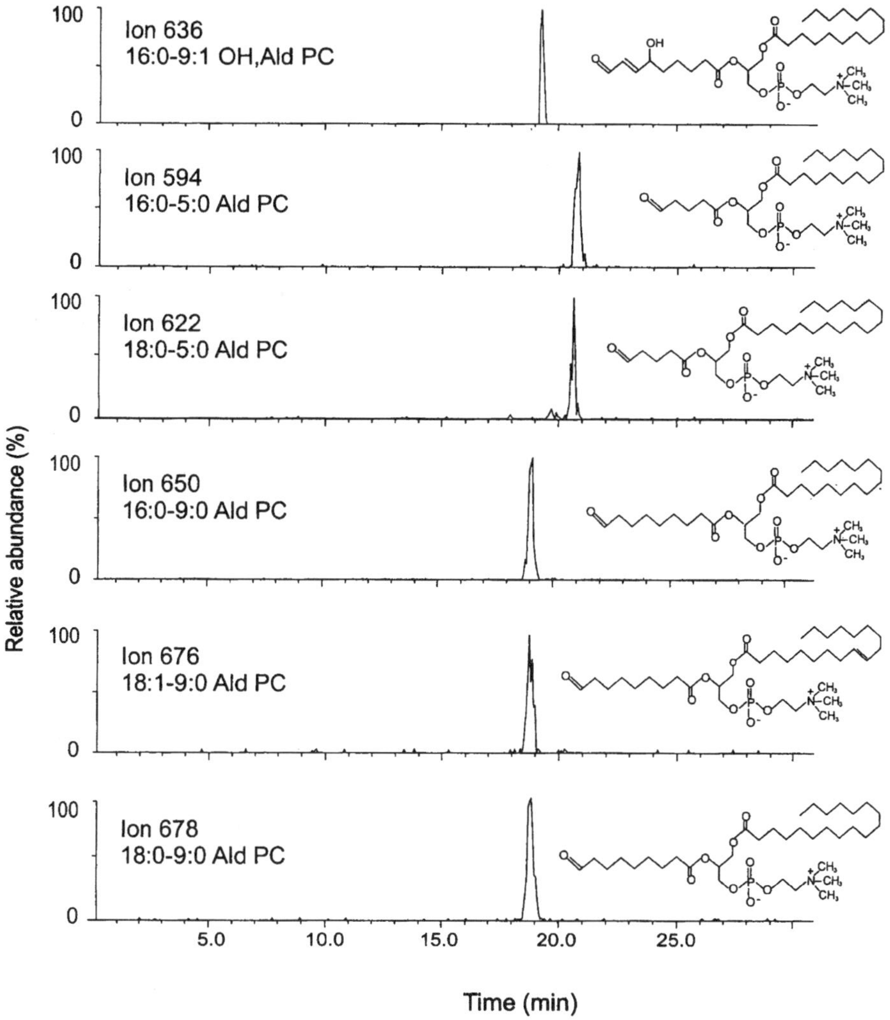

Fig. 17. Single positive ion mass chromatograms of the PtdCho core aldehydes (19–21.5 min) derived from a 6 h oxidation of HDL with SIN-1. Peaks are identified in the figure by carbon and double bond number and by proposed chemical structures. *OH, Ald PC,* hydroxylated aldehydic GroPCho; *Ald PC*, aldehydic GroPCho. LC/ESI-MS conditions were as given in Figure 13. *Source:* Ahmed et al. (75).

yl alkenal was found to accumulate with autoxidation and represented up to 2% of oxidized PtdCho.

Podrez et al. (98) have reported the characterization of a structurally conserved family of oxidized PtdCho that serve as novel high affinity ligand for cells stably transfected with CD36, mediating recognition of multiple oxidized forms of LDL. Specifically, four major structurally-related phospholipids with macrophage scavenger receptor CD36 binding activity were identified from oxidized 1-hexadecanoyl-2-eicosatetra-5′,8′,11′,14′-enoyl-GroPCho and four corresponding structural analogs with CD36 binding activity were identified from oxidized 1-hexadecanoyl-2-octadecadi-9′,12′-enoyl-GroPCho. Each was synthetically prepared, its structure confirmed by multilamellar NMR and high resolution MS. LC/ESI-MS/MS studies demonstrated that oxidized PtdCho are formed during LDL oxidation by multiple distinct pathways. LDL modified by myeloperoxidase (MPO)-generated nitrating intermediates (NO_2-LDL) was formed by incubating LDL (0.2 mg of protein/mL) at 37°C in 50 mM sodium phosphate, pH 7.0, 100 μM DTPA, 30 nM MPO, 100 μg/mL glucose, 20 ng/mL glucose oxidase, and 0.5 mM $NaNO_2$ for 8 h. The lipids were resolved using a ternary ($CH_3CN/MeOH/H_2O$) gradient generated by a Waters 600 E multi-solvent system HPLC, and monitored using ELSD. Lipids, both free and derivatized, were resolved on a Luna C_{18} column (250 mm × 4.6 mm, ID) (5 μm, Phenomenex, Torrance, CA). A discontinuous gradient was obtained by mixing solvent A ($MeOH/H_2O$, 85:15 (v/v) with solvent B (MeOH). MS analyses were performed on a Quatro II triple-quadrupole mass spectrometer (Micromass, Inc.) equipped with an ESI probe and interfaced with an HP 1100 HPLC. Quantification of the various oxidized PtdCho species was performed using LC/ESI-MS/MS in positive ion mode using MRM. HCOOH (0.1%) was included in the mobile phase. The maximum CD36 binding activity exhibited by the oxidized species of PtdCho contained an *sn*-2-acyl group that possessed a terminal ω-hydroxy (or oxo)-α,β-unsaturated carbonyl. The phospholipid hydroxyalkenals were isolated by LC/ESI/MS/MS as described by Podrez et al. (98). The phospholipid hydroxyalkenals derived from arachidonoyl and linoleoyl GroPCho were found to be enriched in fresh atherosclerotic lesions from the aorta of Watanabe heritable hyperlipidemic rabbits (98). Kamido et al. (96) and Ravandi et al. (79), however, failed to specifically recognize the hydroxyalkenal GroPChos in various extracts of human atherosclerotic lesions, while the [5-oxo]valeroyl- and [9-oxo]nonanoyl-esters of the palmitoyl and stearoyl lysoGroPCho could readily be seen. It should be noted, however, that the newly identified mono hydroxides of [8-oxo]octanoyl GroPChos possess the same mass as the known [9-oxo]nonanoyl GroPCho identified earlier. These compounds either had not been formed under the autoxidation conditions of the human atheromas or had decomposed to the stable final saturated core aldehydes during preparation and storage of the lipid extracts as shown (98). However, Hoff et al. (99) using the LC/ESI-MS/MS methods of Watson et al. (94,95) were able to demonstrate the presence of the hydroxyalkenal derivatives of PtdCho (5-hydroxy-8-oxo-6-octenoate and 9-hydroxy-12-oxo-10-dodecenoate derived from arachidonate and linoleate, respectively) in fresh human atherosclerotic lesions in the unbound form as well as covalent adducts with proteins.

Glycerophospholipid core aldehydes have also been found in plant tissues. Ou et al. (100) reported the normal phase LC/ESI-MS total ion profile of oxidized soybean PtdCho along with the $[M+1]^+$ or $[M+Na]^+$ mass chromatograms of the major core aldehyde species. The total ion current profile showed a major peak for the unoxidized PtdCho (PC) along with smaller peaks for the core aldehydes (PC-ALD) and core acids (PC-ACID). The major core aldehyde species are 16:0/9:0ALD (*m/z* 650); 18:0/9:0ALD (*m/z* 678); 16:0/13:1ALD (*m/z* 704), 18:1/9:0ALD (*m/z* 676) and the sodiated 16:0/9:0ALD (*m/z* 672), as expected from the known composition of the molecular species of soybean PtdCho. The PC-ACID peak was made up of the ω-carboxy homologues corresponding to the core aldehydes, as indicated by their chromatographic migration and the presence of extra oxygen (16 mass units) in the molecule.

Finally, Itabe et al. (101) combined TLC with HPLC and FAB-MS to purify and establish the structure of the PtdCho core aldehydes prepared by the osmium tetroxide method previously described by Kamido et al. (21). ^{14}C-Labeled and non-labeled 16:0/18:1 and16:0/18:2, and 16:0/20:4 GroPCho yielded the 16:0/9:0ALD and 16:0/5:0ALD, respectively, which were purified by TLC and resolved by reversed-phase HPLC (Lichrosorb RP-18, Merck) using a gradient of Solvent A (MeOH, CH_3CN/H_2O (616:264:120, by vol) and Solvent B ($MeOH/CH_3CN$, 70:30, v/v). The structures of the synthetic compounds were confirmed by FAB-MS spectra recorded for the $[M+H]^+$ and $[M+Na]^+$ ions.

Summary

This brief review shows that incorporation of a preliminary chromatographic step in the mass spectrometric strategy of analysis of oxygenated fatty acids, steryl esters, TAGs and glycerophospholipids constitutes an essential development in the identification and quantification of complex oxo-lipids. While an on-line combination of the two analytical systems is not an absolute necessity, it does facilitate both identification and quantification by fully accounting for all components present at all times. The HPLC/MS combinations discussed here have been deliberately confined to the normal and reversed-phase separations; the absolute necessity of chiral phase HPLC for the resolution of enantiomers should not be overlooked as pointed out in a parallel chapter in this book. In conclusion, it is satisfying to note that the curiosity-driven methodology developed for the isolation and identification of the oxo-lipids, including the core aldehydes, has now found application in medical and pharmacological research and that some of the oxo-lipids identified may have biological importance. Though extremely powerful, the HPLC/MS approaches have not been widely adopted up to now because of the relatively high cost of the equipment and the expense involved in its operation and maintenance.

Acknowledgments

These studies were supported in part by the Heart and Stroke Foundation of Ontario, Toronto, Ontario and the Medical Research Council of Canada, Ottawa, Ontario, Canada and the Finnish Food Research Foundation, Helsinki, Finland.

References

1. Han, X. and R.W. Gross, Electrospray Ionization Mass Spectroscopic Analysis of Human Erythrocyte Plasma Membrane Phospholipids. *Proc. Natl. Acad. Sci. USA 91,* 10635–10639 (1994).
2. Kerwin, J.L., A.R. Tuninga and L.H. Ericsson, Identification of Molecular Species of Glycerophospholipids and Sphingomyelin Using Electrospray Mass Spectrometry. *J. Lipid Res. 35,* 1102–1114 (1994).
3. Byrdwell, W.C. and W.E. Neff, Qualitative and Quantitative Analysis of Triacylglycerols Using Atmospheric-Pressure Chemical Ionization Mass Spectrometry. In *New Techniques and Applications in Lipid Analysis* (R. E. McDonald and M. M. Mossoba, editors), AOCS Press, Champaign, IL, pp. 45–80 (1997).
4. Han, X.L, R.A. Gubitosi-Klug, B.J. Collins and R.W. Gross, Alterations in Individual Molecular Species of Human Platelet Phospholipids During Thrombin Stimulation: Electrospray-Ionization Mass Spectrometry-Facilitated Identification of the Boundary Conditions for the Magnitude and Selectivity of Thrombin-Induced Platelet Phospholipid Hydrolysis. *Biochemistry 35,* 5822–5832 (1996).
5. DeLong, C.J., P.R.S. Baker, M. Samuel, Z. Cui, and M.J. Thomas, Molecular Species Composition of Rat Liver Phospholipids by ESI-MS/MS: The Effect of Chromatography. *J. Lipid Res. 42,* 1959–1968 (2001).
6. Koivusalo, M., P. Haimi, L. Heikinheimo, R. Kostiainen, and P. Somerharju, Quantitative Determination of Phospholipid Composition by ESI-MS: Effects of Acyl Chain Length, Unsaturation, and Lipid Concentration on Instrument Response. *J. Lipid Res. 42,* 663–672 (2001).
7. Kuksis, A., Lipids. In *Chromatography,* 6th Edition, *Journal of Chromatography* Library, Vol. 69B (E. Heftmann, editor), Elsevier, Amsterdam, 2004, pp. 739–837.
8. Schiller, J., R. Süß, M. Petkovic, O. Zschornig, and K. Arnold, Negative-Ion Matrix-Assisted Laser Desorption And Ionization Time-of-Flight Mass Spectra of Complex Phospholipid Mixtures in the Presence of Phosphatidylcholine: A Cautionary Note on Peak Assignment. *Anal. Biochem. 309,* 311–314 (2002).
9. Han, X., J. Yang, H. Cheng, H. Ye, R. W. Gross, Toward Fingerprinting Cellular Lipidomes Directly from Biological Samples by Two-Dimensional Electrospray Mass Spectrometry. *Anal. Biochem. 330,* 317–331 (2004).
10. Han, X. and R.W. Gross, Global Analyses of Cellular Lipidomes Directly from Crude Extracts of Biological Samples by ESI Mass Spectrometry: A Bridge to Lipidomics. *J. Lipid Res. 44,* 1071–1079 (2003).
11. Kuksis, A., J.J. Myher, L. Marai, and K. Geher, Analyses of Hydroperoxides and Core Aldehydes of Triacylglycerols. In *Proceedings of 17th Nordic Lipid Symposium* (Y. Malkki, ed.), Lipidforum, Bergen, Norway, pp. 230–238 (1993).
12. Myher, J.J., A. Kuksis, A. Ravandi and N. Cocks, Normal-Phase Liquid Chromatography/Mass Spectrometry for Sensitive Detection of Oxygenated Glycerophospholipids. *INFORM 5,* 478–479 (1994).
13. Byrdwell, W.C. and W.E. Neff, Autoxidation Products of Normal and Genetically Modified Canola Varieties Determined Using Liquid Chromatography with Mass Spectrometric Detection. *J. Chromatogr. A 905,* 85–102 (2001).
14. Kuksis, A. and Y. Itabashi, Regio- and Stereo-Specific Analysis of Glycerolipids. *Methods* 36, 172–185 (2005).
15. Kuksis, A. and Y. Itabashi, LC-MS and Chiral Separation. In *New Techniques and*

Applications in Lipid Analysis and Lipidomics (M. M. Mossoba, J. K. G. Kramer, J. T. Brenna and R. E. McDonald, editors), AOCS Press, 2005, in press.
16. Schneider, C., W.E. Boeglin, S. Lai, J.K. Cha and A.R. Brash, Synthesis and Applications of Stereospecifically ^{3}H-Labeled Arachidonic Acids as Mechanistic Probes for Lipoxygenase and Cyclooxygenase Catalysis. *Anal. Biochem. 284*, 125–135 (2000).
17. Lee, S. H., M.V. Williams, R.N. DuBois and I.A. Blair, Targeted Lipidomics Using Electron Capture Atmospheric Pressure Chemical Ionization Mass Spectrometry. *Rapid Commun. Mass Spectrom. 17,* 2168–2176 (2003).
18. Hammonds, T.D., I.A. Blair, J.R. Falck, and J.H. Capdevila, Resolution of Epoxyeicosatrienoate Enantiomers by Chiral Phase Chromatography. *Anal. Biochem. 182,* 300–303 (1989).
19. Esterbauer, H. and K.H. Cheeseman, Determination of Aldehydic Peroxidation Products: Malonaldehyde and 4-Hydroxynonenal. *Methods Enzymol. 186*, 407–421 (1990).
20. Ravandi, A., A. Kuksis, J. J. Myher and L. Marai, Determination of Lipid Ester Ozonides and Core Aldehydes by High-Performance Liquid Chromatography with On-Line Mass Spectrometry. *J. Biochem. Biophys. Methods 30*, 271–285 (1994).
21. Kamido, H., A. Kuksis, L. Marai, J.J. Myher and H. Pang, Preparation, Chromatography and Mass Spectrometry of Cholesteryl Ester and Glycerolipid-Bound Aldehydes. *Lipids 27*, 645–650 (1992).
22. Nakamura, T., D.L. Bratton, and R.C. Murphy, Analysis of Epoxyeicosatrienoic Acids Esterified to Phospholipids in Human Red Blood Cells by Electrospray Tandem Mass Spectrometry. *J. Mass Spectrom. 32,* 888–896 (1997).
23. Nakamura, T., P.M. Henson, and R. C. Murphy, Occurrence of Oxidized Metabolites of Arachidonic Acid Esterified to Phospholipids in Murine Lung Tissue. *Anal. Biochem. 262,* 23–32 (1998).
24. Bylund, J., J. Ericsson and E.H. Oliw, Analysis of Cytochrome P_{450} Metabolites of Arachidonic and Linoleic Acids by Liquid Chromatography-Mass Spectrometry with Ion Trap MS^2. *Anal. Biochem. 265*, 55–68 (1998).
25. Bylund, J., A. G. Harder, K. G. Maier, R. J. Roman and D. R. Harder, Leukotriene B_4 ω-Side Chain Hydroxylation by CYP4F5 and CYP4F6. *Arch. Biochem. Biophys. 412*, 34–41 (2003).
26. Piazza, G.J., A. Nunez, and T.A. Foglia, Isolation of Unsaturated Diols After Oxidation of Conjugated Linoleic Acid with Peroxygenase. *Lipids 38*, 255–261 (2003).
27. Havrilla, C.M., D.L. Hachey, and N.A. Porter, Coordination (Ag^+) Ion Spray-Mass Spectrometry of Peroxidation Products of Cholesterol Linoleate and Cholesterol Arachidonate: High Performance Liquid Chromatography—Mass Spectrometry Analysis of Peroxide Products from Polyunsaturated Lipid Autoxidation. *J. Am. Chem. Soc. 122,* 8041–8055 (2000).
28. Kamido, H., A. Kuksis, L. Marai and J.J. Myher, Identification of Core Aldehydes Among in vitro Peroxidation Products of Cholesteryl Esters. *Lipids 28*, 331–336 (1993).
29. Herrera, D., A. Ravandi and A. Kuksis, Reversed-phase Liquid Chromatography-Mass Spectropmetry of Some Ozonization Products of Cholesteryl Linoleate. In Abstracts, *94th AOCS Annual Meeting & Expo*, Kansas City, MO, May 4–7, 2003, page 10.
30. Sjovall, O., A. Kuksis, L. Marai and J. J. Myher, Elution Factors of Synthetic Oxotriacylglycerols as an Aid in Identification of Peroxidized Natural Triacylglycerols by Reversed-Phase High Performance Liquid Chromatography with Electrospray Mass Spectrometry. *Lipids 32*, 1211–1218 (1997).

31. Sjovall, O., A. Kuksis and H. Kallio, Reversed-Phase High-Performance Liquid Chromatographic Separation of *tert*-Butyl Hydroperoxide Oxidation Products of Unsaturated Triacylglycerols. *J. Chromatogr. A 905*, 119–132 (2001).
32. Sjovall, O., A. Kuksis and H. Kallio, Analysis of Molecular Species of Peroxide Adducts of Triacylglycerols Following Treatment of Corn Oil with *tert*-Butyl Hydroperoxide. *Lipids 36,* 1347–1356 (2001).
33. Ravandi, A., A. Kuksis, and N.A. Shaikh, Glucosylated Glycerophosphoethanolamines Are the Major LDL Glycation Products and Increase LDL Susceptibility to Oxidation. *Arteriscler. Thromb. Vasc. Biol. 20*, 467–477 (2000).
34. Ravandi, A., A. Kuksis, N.A. Shaikh, and G. Jackowski, Preparation of Schiff Base Adducts of Phosphatidylcholine Core Aldehydes and Aminophospholipids, Amino Acids and Myoglobin. *Lipids 32*, 989–1001 (1997).
35. Kamido, H., A. Kuksis, L. Marai and J.J. Myher, Lipid Ester-Bound Aldehydes Among Copper-Oxidized Peroxidation Products of Human Plasma Lipoproteins. *J. Lipid Res. 36,* 1876–1886 (1995).
36. Khaselev, N. and R.C. Murphy, Peroxidation of Arachidonate Containing Plasmenyl Glycerophosphocholine: Facile Oxidation of Esterified Arachidonate at Carbon-5. *Free Radical Biol. & Med. 29*, 620–632 (2000).
37. Griffiths, W.J., Tandem Mass Spectrometry in the Study of Fatty Acids, Bile Acids, and Steroids. *Mass Spectrom. Revs. 22*, 81–152 (2003).
38. MacPherson, J.C., J.G. Pavlovich and R.S. Jacobs, Biosynthesis of Arachidonic Acid Metabolites in *Limulus polyphemus* Amebocytes: Analysis by Liquid Chromatography-Elecrospray Ionization Mass Spectrometry. *Biochim. Biophys. Acta 1303*, 127–136 (1996).
39. Santiago-Vazquez, L.Z, L.D. Mydlarz, J.G. Pavlovich and R.S. Jacobs, Identification of Hydroxy Fatty Acids by Liquid Chromatography-Atmospheric Pressure Chemical Ionization Mass Spectroscopy in *Euglena gracilis*. *J. Chromator. B 803*, 233–236 (2004).
40. Pruzanski, W., E. Stefanski, F.C. de Beer, M.C. de Beer, P. Vadas, A. Ravandi and A. Kuksis, Lipoproteins Are Substrates for Human Secretory Group IIA Phospholipase A_2: Preferential Hydrolysis of Acute Phase HDL. *J. Lipid Res. 39*, 2150–2160 (1998).
41. Oliw, E. H., C. Su, T. Skogstrom and G. Benthin, Analysis of Novel Hydroperoxides and Other Metabolites of Oleic, Linoleic, and Linolenic Acids by Liquid Chromatography-Mass Spectrometry with Ion Trap MS^n. *Lipids 33*, 843–852 (1998).
42. Waugh, R.J. and R.C. Murphy, Mass Spectrometric Analysis of Four Regioisomers of F_2-Isoprostanes Formed by Free Rradical Oxidation of Arachidonic Acid. *J. Amer. Soc. Mass Spectrom. 7,* 490–499 (1996).
43. Waugh, R.J., J.D. Morrow, L.J. Roberts, and R.C. Murphy, Identification and Relative Quantification of F_2-Isoprostane Regioisomers Formed in vivo in the Rat. *Free Radic. Biol. Med. 23*, 943–954 (1997).
44. Lawson, J.A., H. Li, J. Rokach, M. Adiyaman, S-W. Hwang, S. P. Khanapure, and G. A. Fitzgerald, Identification of Two Major F_2 Isoprostanes, 8,12-iso- and 5-epi-8,12-Isoprostane $F_{2\alpha}$-VI, in Human Urine. *J. Biol. Chem. 273*, 29295–29301 (1998).
45. Lawson, J.A., J. Rokach, and G.A. Fitzgerald, Isoprostanes: Formation, Analysis and Use as Indices of Lipid Peroxidation in vivo. *J. Biol. Chem. 274*, 24441–24444 (1999).
46. Mallat, Z., T. Nakamura, J. Chan, G. Leseche, A. Tedgui, J. Maclouf and R.C. Murphy, The Relationship of Hydroxyeicosatetraenoic Acids and F_2-Isoprostanes to Plaque Instability in Human Carotid Atherosclerosis. *J. Clin. Invest. 103*, 421–427 (1999).

47. Waddington, E., K. Sienuarine, I. Puddey and K. Croft, Identification and Quantification of Unique Fatty Acid Oxidation Products in Human Atherosclerotic Plaque Using High Performance Liquid Chromatography. *Anal. Biochem. 292,* 234–244 (2001).
48. Kozak, K.R., S.W. Rowlinson, and L.J. Marnett, Oxygenation of the Endocannabinoid. 2-Arachidonoylglycerol, to Glyceryl Prostaglandins by Cyclooxygenase-2. *J. Biol. Chem. 275,* 33744–33749 (2000).
49. Roberts, L.J., T.J. Montine, W.E.R. Markesbery, A.R. Tapper, P. Hardy, S. Chemtob, W.D. Dettbarn, and J.D. Morrow, Formation of Isoprostane-like Compounds (Neuroprostanes) in vivo from Docosahexaenoic Acid. *J. Biol. Chem. 273,* 13605–13612 (1998).
50. Reich, E.E., W. E. Zackert, C. J. Brame, Y. Chen, L.J. Roberts, II, D. L. Hachey, T. J. N. Montine, and J. D. Morrow, Formation of Novel D-Ring and E-Ring Isoprostane-like Compounds (D_4/E_4-Neuroprostanes) in vivo from Docosahexaenoic Acid. *Biochemistry 39,* 2376–2383 (2000).
51. Fam, S.S., L.J. Murphey, E.S. Terry, W.E. Zackert, Y. Chen, L. Gao, S. Pandalai, G. L. Milne, L.J. Roberts, N.A. Porter, et al., Formation of Highly Reactive A-Ring and J-Ring Isoprostane-like Compounds (A_4/J_4-Neuroprostanes) in vivo from Docosahexaenoic Acid. *J. Biol. Chem. 277,* 36076–36084 (2002).
52. Yousef, J.A., L.S. Birnbaum, L.L. Swift, J.D. Morrow, and M.Z. Badr, Age-Independent, Grey Matter-Localized, Brain-Enhanced Oxidative Stress in Male Fischer 344 Rats: Brain Levels of F_2-Isoprostanes and F_4-Neuroprostanes. *Free Rad. Biol. Med. 34,* 1631–1635 (2003).
53. Brame, C.J., R.G. Salomon, J.D. Morrow, and L.J. Roberts II, Identification of Extremely Reactive γ-Ketoaldehydes (isolevuglandins) as Products of the Isoprostane Pathway and Characterization of their Lysyl Protein Adducts. *J. Biol. Chem. 274,* 13139–13146 (1999).
54. Bernoud-Hubac, N., S.S. Davies, O. Boutaud, T.J. Montine, and L.J. Roberts II, Formation of Highly Reactive γ-Ketoaldehydes (Neuroketals) as Products of the Neuroprostane Pathway. *J. Biol. Chem. 276,* 30964–30970 (2001).
55. Kenar, J.A., C.M. Havrilla, N.A. Porter, J.R. Guyton, S.A. Brown, K.F. Klemp and E. Selinger, Identification and Quantification of the Regioisomeric Cholesteryl Linoleate Hydroperoxides in Oxidized Human Low Density Lipoprotein and High Density Lipoprotein, *Chem. Res. Toxicol. 9,* 737–744 (1996).
56. Mashima, R., K. Onodera, and Y. Yamamoto, Regioisomeric Distribution of Cholesteryl Linoleate Hydroperoxides and Hydroxides in Plasma from Healthy Humans Provides Evidence for Free Radical-Mediated Lipid Peroxidation in vivo. *J. Lipid Res. 41,* 109–115 (2000).
57. Yin, H., C. M. Havrilla, J.D. Morrow, and N.A. Porter, Formation of Isoprostane Bicyclic Endoperoxides from Autoxidation of Cholesteryl Arachidonate. *J. Am. Chem. Soc. 124,* 7745–7754 (2002).
58. Hoppe, G., A. Ravandi, D. Herrera, A. Kuksis, and H.F. Hoff, Oxidation Products of Cholesteryl Linoleate Are Resistant to Hydrolysis in Macrophages, Form Complexes with Proteins, and Are Present in Human Atherosclerotic Lesions. *J. Lipid Res. 38,* 1347–1360 (1997).
59. Kawai, Y., A. Saito, N. Shibata, M. Kobayashi, S. Yamada, T. Osawa, and K. Uchida, Covalent Binding of Oxidized Cholesteryl Esters to Protein. *J. Biol. Chemistry. 278,* 21040–21049 (2003)

60. Paryzek, Z. and U. Rychlewska, Ozonolysis of Cholesterol and Other Δ^5-Steroids in the Presence of Alcohols: A Revised Mechanism and Hydroperoxide Structure of the Solvent-Participated Product, Confirmed by X-Ray Analysis. *J. Chem. Soc. Perkin Trans. 2,* 2313–2318 (1997).
61. Pulfer, M.K., K. Harrison, and R.V. Murphy, Direct Electrospray Tandem Mass Spectrometry of the Unstable Hydroperoxy Bishemiacetal Product Derived from Cholesterol Ozonolysis. *J. Am. Soc. Mass Spectrom. 15,* 194–202 (2004).
62. Hartvigssen, K., A. Ravandi, K. Bukhave, G. Hølmer and A. Kuksis, Regiospecific Analysis of Neutral Ether Lipids by Liquid Chromatography/Electrospray Ionization/Single Quadrupole Mass Spectrometry: Validation with Synthetic Compounds. *J. Mass Spectrom. 36,* 1116–1124 (2001).
63. Kusaka, T., S. Ishihara, M. Sakaida, A. Mifune, Y. Nakano, K. Tsuda, M. Ikeda and H. Nakano, Composition Analysis of Normal Plant Triacylglycerols and Hydroperoxidized *rac*-1-Stearoyl-2-Oleoyl-3-Linoleoyl-*sn*-Glycerols by Liquid Chromatography-Atmospheric Pressure Chemical Ionization Mass Spectrometry. *J. Chromatogr. A 730,* 1–7 (1996).
64. Sjovall, O., A. Kuksis, and H. Kallio, Formation of Triacylglycerol Core Aldehydes During Rapid Oxidation of Corn and Sunflower Oils with *tert*-Butylhydroperoxide/Fe^{2+}. *Lipids 37,* 81–94 (2002).
65. Sjovall, O., A. Kuksis and H. Kallio, Tentative Identification and Quantification of TAG Core Aldehydes as Dinitrophenylhydrazones in Autoxidized Sunflower Seed Oil Using Reversed-Phase HPLC with Electrospray Ionization MS. *Lipids 38,* 1179–1190 (2003).
66. Neff, W.E. and W.C. Byrdwell, Characterization of Model Triacylglycerol (Triolein, Trilinolein and trilinolenin) Autoxidation Products via High-Performance Liquid Chromatography Coupled with Atmospheric Pressure Chemical Ionization Mass Spectrometry. *J. Chromatogr. A 818,* 269–286 (1998).
67. Byrdwell, W.C., and W.E. Neff, Dual Parallel Electrospray Ionization and Atmospheric Pressure Chemical Ionization Mass Spectrometry (MS), MS/MS and MS/MS/MS for the Analysis of Triacylglycerols and Triacylglycerol Oxidation Products. *Rapid. Commun. Mass Spectrom. 16,* 300–319 (2002).
68. Byrdwell, W.C., and W.E. Neff, Non-Volatile Products of Triolein Produced at Frying Temperatures Characterized Using Liquid Chromatography with On-Line Mass Spectrometric Detection. *J. Chromatogr. A 852,* 417–432 (1999).
69. Bauer-Plank, C., and L. Steenhorst-Slikkerveer, Analysis of Triacylglycerol Hydroperoxides in Vegetable Oils by Nonaqueous Reversed-phase High Performance Liquid Chromatography with Ultraviolet Detection. *J. Am. Oil Chem. Soc. 77,* 477–482 (2000).
70. Steenhorst-Slikkerveer, L., X. Louter, H.-G. Jansen, and C. Bauer-Plank, Analysis of Nonvolatile Lipid Oxidation Products in Vegetable Ooils by Normal-Phase High Performance Liquid Chromatography with Mass Spectrometric Detection. *JAOCS 77,* 837–845 (2000).
71. Kurvinen, J.-P., A. Kuksis, A. Ravandi, O. Sjovall and H. Kallio, Rapid Complexing of Oxoacylglycerols with Amino Acids, Peptides and Aminophospholipids. *Lipids 34,* 299–305 (1999).
72. Suomela, J.-P., M. Ahotupa, O. Sjovall, J.-P. Kurvinen, and H. Kallio, New Approach to the Analysis of Oxidized Triacylglycerols in Lipoproteins. *Lipids 39,* 507–512 (2004).
73. Suomela, J.-P., M. Ahotupa, O. Sjovall, J.-P. Kurvinen and H. Kallio, Diet and Lipoprotein Oxidation: Analysis of Oxidized Triacylglycerols in Pig Lipoproteins. *Lipids 39*, 639–647 (2004).

74. Ravandi, A., A. Kuksis and N.A. Shaikh, Glycated Phosphatidylethanolamine Promotes Macrophage Uptake of Low Density Lipoprotein and Accumulation of Cholesteryl Esters and Triacylglycerols. *J. Biol. Chem. 274*, 16494–16500 (1999).
75. Ahmed, Z., A. Ravandi, G.F. Maguire, A. Emili, D. Draganov, B.N. La Du, A. Kuksis and P. W. Connelly, Apolipoprotein A-I Promotes the Formation of Phosphatidylcholine Core Aldehydes that are Hydrolyzed by Paraoxonase (PON-1) During High Density Lipoprotein Oxidation with a Peroxynitrite Donor. *J. Biol. Chem. 276,* 24473–24481 (2001).
76. Ahmed, Z., A. Ravandi, G.F. Maguire, A. Emili, D. Draganov, B.N. La Du, A. Kuksis and P.W. Connelly, Multiple Substrates for Paraoxonase-1 During Oxidation of Phosphatidylcholine by Peroxynitrite. *Biochem. Biophys. Res. Commun. 290,* 391–396 (2002).
77. Ahmed, Z., S. Babaei, G.F. Maguire, D. Draganov, A. Kuksis, B.N. La Du, and P.W. Connelly, Paraoxonase-1 Reduces Monocyte Chemotaxis and Adhesion to Endothelial Cells Due to Oxidation of Palmitoyl Linoleloyl Glycerophosphocholine. *Cardiovasc. Res. 57,* 225–231 (2003).
78. Ahmed, Z., A. Ravandi, G. F. Maguire, A. Kuksis and P. W. Connelly, Formation of Apoprotein AI-Phosphatidylcholine Core Aldehyde Schiff Base Adducts Promotes Uptake by THP-1 Macrophages. *Cardiovasc. Res. 58*, 712–720 (2003).
79. Ravandi, A., S. Babaei, R. Leung, J.C. Monge, G. Hoppe, H. Hoff, H. Kamido and A. Kuksis, Phospholipids and Oxophospholipids in Atherosclerotic Plaques at Different Stages of Plaque Development, *Lipids 39*, 97–109 (2004).
80. Adachi, J., N. Yoshioka, R. Funae, Y. Nagasaki, T. Naito and Y. Ueno, Phosphatidylcholine Hydroperoxide Levels in Human Plasma Are Lower Than Previously Reported. *Lipids 39*, 891–896 (2004).
81. Spickett, C.M., N. Rennie, H. Winter, L. Zambonin, L. Landi, A. Jerlich, R.J. Schaur, and A.R. Pitt, Detection of Phospholipod Oxidation in Oxidatively Stressed Cells by Reversed-Phase HPLC Coupled With Positive-Ionization Electroscopy MS. *Biochem. J. 355*, 449–457 (2001) .
82. Bergqvist, M. and A. Kuksis, Liquid Chromatography with On-Line Electrospray Mass Spectrometry of Oxidized Diphosphatidylglycerol. In *New Techniques and Applications in Lipid Analysis* (R. E. McDonald and M. M. Mossoba, editors), AOCS Press, Champaign, IL, pp. 81–99 (1997).
83. Schlame, M., I. Haller, L.R. Sammaritano, and T.J. Blanck, Effect of Cardiolipin Oxidation on Solid-Phase Immunoassay for Antiphospholipid Antibodies. *Throm. Haemost. 86,* 1475–1482 (2001).
84. Kriska, T., W. Korytowski, and A.W. Girotti, Role of Mitochondrial Cardiolipin Peroxidation in Apoptotic Photokilling of 5-Aminoevulinate-Treated Tumor Cells. *Arch. Biochem. Biophys. 433*, 435–446 (2005).
85. Milne, G.L. and Porter, N.A., Separation and Identification of Phospholipid Peroxidation Products. *Lipids 36,* 1265-1275 (2001).
86. Morrow, J. D., J.A. Awad, H. J. Hollis, I.A. Blair, and L.J. Roberts II, Non-Cyclooxygenase-Derived Prostanoids (F_2-Isoprostanes) are Formed in situ on Phospholipids, *Proc. Natl. Acad. Sci. USA 89*, 10721–10725 (1992).
87. Kayganich-Harrison, K.A., D.M. Rose, R.C. Murphy, J.D. Morrow and L.J. Roberts, Collision Induced Dissociation of F_2-Isoprostanes-Containing Phospholipids. *J. Lipid Res. 34,* 1229–1235 (1993).

88. Hall, L.M. and R.C. Murphy, Analysis of Stable Oxidized Molecular Species of Glycerophospholipids Following Treatment of Red Blood Cell Ghosts with t-Butylhydroperoxide. *Anal. Biochem. 258*, 184–194 (1998).
89. Pruzanski, W., E. Stefanski, F.C. de Beer, M.C. de Beer, A. Ravandi and A. Kuksis, Comparative Analysis of Lipid Composition of Normal and Acute-Phase High Density Lipoproteins. *J. Lipid Res. 41*, 1035–1047 (2000).
90. Jerlich, A., R.J. Schaur, A.R. Pitt, and C.M. Spickett, The Formation of Phosphatidylcholine Oxidation Products by Stimulated Phagocytes. *Free Radical Research 37,* 645–653 (2003).
91. Salomon, R.G., G. Subbanagounder, U. Singh, J. O'Neil, and H. F. Hoff, Oxidation of Low-Density Lipoproteins Produces Levuglandin-Protein Adducts. *Chem. Res. Toxicol. 10*, 750–759 (1997).
92. Salomon, R.G., Distinguishing Levuglandins Produced Through the Cyclooxygenase and Isoprostane Pathways. *Chem. Phys. Lipids 134*, 1–20 (2005).
93. Kayganich-Harrison, K. A. and R. C. Murphy, Characterization of Chain-Shortened Oxidized Glycerophosphocholine Lipids Using Fast Atom Bombardment and Tandem Mass Spectrometry, *Anal. Biochem. 221*, 16–24 (1994).
94. Watson, A.D., N. Leitinger, M. Navab, F.F. Faull, S.Horkko, J.L. Witzum, D. Schwenke, R.G. Salomon, W. Sha, G. Subbanagounder, et al., Structural Identification by Mass Spectrometry of Oxidized Phospholipids in Minimally Oxidized Low Density Lipoprotein that Induce Monocyte/Endothelial Interactions and Evidence for Their Presence in vivo. *J. Biol. Chem. 272*, 13597–13607 (1997).
95. Watson, A.D., G. Subbanagounder, D.S. Welsbie, K.F. Faull, M. Navab, M.E. Jung, A.M. Fogelman and J.A. Berliner, Structural Identification of a Novel Pro-Inflammatory Epoxyisoprostane Phospholipid in Mildly Oxidized Low Density Lipoprotein. *J. Biol. Chem. 274,* 24787–24798 (1999).
96. Kamido, H., H. Eguchi, H. Ikeda, T. Imaizumi, K. Yamana, K. Hartvigsen, A. Ravandi, and A. Kuksis, Core Aldehydes of Alkyl Glycerophosphocholines in Atheroma Induce Platelet Aggregation and Inhibit Endothelium-Dependent Arterial Relaxation. *J. Lipid Res. 43,* 158–166 (2002).
97. Subbanagounder, G., Y. Deng, C. Borromeo, A.N. Dooley, J.A. Berliner and R.G. Salomon, Hydroxyl Alkenal Phospholipids Regulate Inflammatory Functions of Endothelial Cells. *Vasc. Phamacol. 38*, 201–209 (2002).
98. Podrez, E.A., E. Poliakov, Z. Shen, R. Zhang, Y. Deng, M. Sun, P.J. Finton, L. Shan, B. Gugiu, P.L. Fox, H.F. Hoff, R.G. Salomon and S.L. Hazen, Identification of a Novel Family of Oxidized Phospholipids that Serve as Ligands for the Macrophage Scavenger Receptor CD36, *J. Biol. Chem. 277*, 38503–38516 (2002) .
99. Hoff, H.F., J. O'Neil, Z. Wu, G. Hoppe and R.L. Salomon, Phospholipid Hydroxyalkenals. Biological and Chemical Properties of Specific Oxidized Lipids Present in Atherosclerotic Lesions. *Arterioscler. Thromb. Vasc. Biol. 23*, 275–282 (2003).
100. Ou, Z., A. Ogamo, L. Guo, and Y. Konda, Identification And Quantification of Choline Glycerophospholipids that Contain Aldehyde Residues by Fluorometric High Performance Liquid Chromatography. *Anal. Biochem. 227*, 289–294 (1995).
101. Itabe, H., K. Suzuki, R. Hosova, M. Mori, Y. Higashi, Y. Fujimoto, and T. Takano, Preparation of Radioactive Aldehyde-Containing Phosphatidylcholine. *Anal. Biochem. 285,* 151–155 (2000).

6 Structural Analysis of Unsaturated Fatty Acid Methyl Ester Isomers with Acetonitrile Covalent-Adduct Chemical Ionization (CACI) Tandem Mass Spectrometry

J. Thomas Brenna

Cornell University, Savage Hall, Ithaca, New York 14853

Introduction

Unsaturated fatty acids (UFAs) and polyunsaturated fatty acids (PUFAs) are carboxylic acid–terminated, linear hydrocarbon chains with double bonds interspersed in specific positions and geometries along the chain. UFAs in mammalian tissues are overwhelmingly abundant in specific configurations. Double bonds are all-*cis*, and when more than one double bond is present (i.e., for PUFAs), they are methylene-interrupted. PUFAs of this nature are referred to as *homoallylic*, or *homoconjugated*, as shown in Figure 1. The number of carbons in each UFA is even, and the homoallylic arrangement of double bonds in PUFA predominates in mammals.

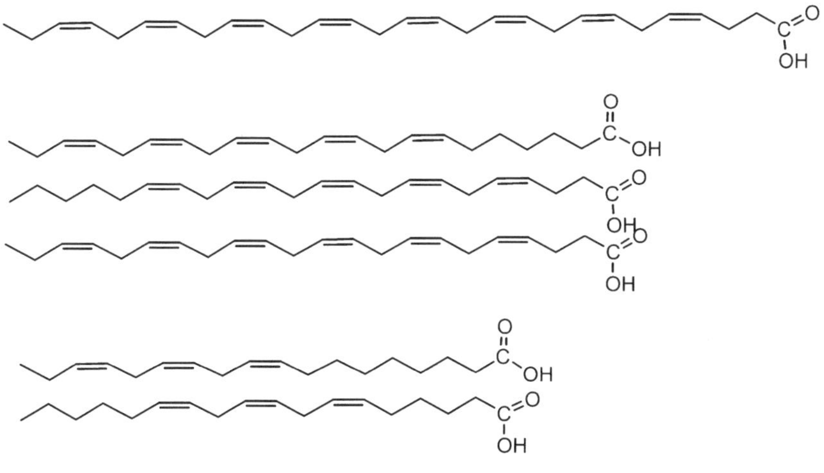

Fig. 1. Six homoallylic (homoconjugated) polyunsaturated fatty acids (PUFAs), including two sets of metabolically important isomers. Top: 28:8n-3; middle: 22:6n-3 and isomeric 22:5 long-chain polyunsaturates; bottom: isomeric 18:3s.

Important classes of UFAs consumed in foods worldwide, and particularly in the United States, have come under scrutiny in the last decade. The *trans* fatty acids, often referred to as simply "*trans*," are formed by partial catalytic hydrogenation of commodity oil sources such as cottonseed, soybean, and palm. A driving factor for this procedure is economic; the least expensive, high-quality edible oils available at a particular time can be modified to have any desired melting point and therefore be useful in many processed and finished food products. Partial hydrogenation results in geometric and positional isomerization of the remaining double bonds in the fatty acid (FA). In PUFAs, conjugated and *trans* double bonds are more stable thermodynamically than *cis* and methylene-interrupted double bonds; thus conjugated systems are formed abundantly from homoallylic PUFAs. Epidemiologic studies suggest that these *trans* UFAs are proatherogenic, more so than many saturated fatty acids (SFAs) (e.g., 16:0) (1), and contribute to the epidemic of type II diabetes (2).

A second source of conjugated FA is the dairy cow. Ruminant milk is rich in conjugated dienes because of the action of rumen bacteria and the cow's own physiology. Studies over the past decade showed in model organisms that conjugated linoleic acids (CLAs) have potent bioactivities that are exquisitely sensitive to the double bond position (3). The most commonly cited examples are *c*9,*t*11-CLA, which has potent anticarcinogenic activity, and *t*10,*c*12,-CLA, which inhibits the synthesis of milk fat and is generally an antiobesity agent (4,5). At least 19 other CLAs were identified in various bovine milk fats, many with unknown bioactivities (structures of 13 of these are shown in Fig. 2). The bioactivities of *trans* FAs with ≥3 double bonds are not known. In general, CLAs, and conjugated PUFAs with ≥3 double bonds are present at <1% of total FA; therefore, very sensitive and specific methods are required for detection and structure elucidation. The general term for PUFA in which at least 2 double bonds are not separated by a methylene group is *non-methylene-interrupted PUFA* (NMI-PUFA).

Conventional Methods for Characterization of UFA Double Bond Position

Gas Chromatography (GC), GC-Mass Spectrometry (MS). Since the 1960s, FA structure and quantitative analysis were done by lipid extraction using organic solvents, followed by conversion of all FA to fatty acid methyl esters (FAME) regardless of the original lipid class by saponification/methylation or transmethylation (6). It is widely appreciated that GC can yield very high resolution of FAME mixtures; however, a common misconception is that this is because FAME are more volatile than the corresponding FA. Although greater volatility is a factor, it is more important that FAME are aprotic and thus do not hydrogen bond to the fused silica support, which causes excessive peak broadening and tailing.

When a lipid class is desired, separations by thin layer chromatography (TLC) or high-performance liquid chromatography (HPLC) are done, fractions collected, and FAME prepared. FAME are then analyzed by GC with a flame ionization

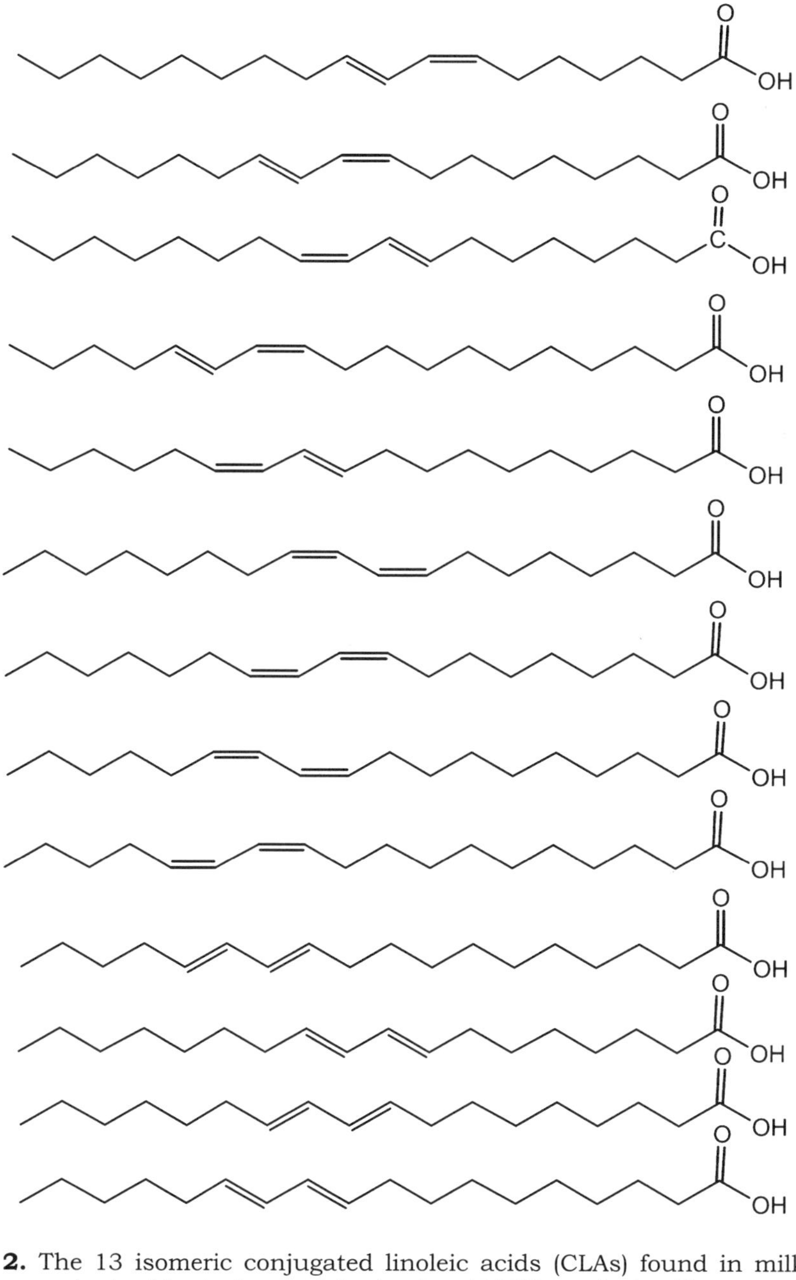

Fig. 2. The 13 isomeric conjugated linoleic acids (CLAs) found in milk fat using covalent-adduct chemical ionization (CACI) analysis. Chromatogram presented in Figure 9.

detector (FID). FAME chromatography provides very high resolution and near baseline separation of more than 100 FAME in a single chromatogram is possible. Assignment of structure is routinely done by comparison of retention times, which cannot be considered a definitive technique, and in any case, does not apply to unknowns for which there are no standards. The classical method for determining double bond position is by chemical treatment of purified FA by ozonolysis; this produces small fragments for subsequent GC analysis.

The increasingly wide availability of MS has fueled an increase in the use of GC-MS for identification. The structural information available in a mass spectrum depends in part on the ionization method used and the extent of fragmentation. The most common gas phase ionization methods are electron impact (EI) and chemical ionization (CI). In EI, electrons are accelerated to a kinetic energy of ~70 eV (thermal energy at room temperature is ~0.07 eV) and collide with the gas phase analyte to eject an electron and yield a positive ion; this also imparts a variable amount of additional energy depending on whether the collision is head-on (low-impact parameter) or glancing (high-impact parameter). The additional energy causes the nascent ion to dissociate into a neutral fragment and an ionized fragment, where the latter is observed. EI of FAME generally does not yield fragments that are useful for assignment of double bond position. This is because the positive charge, which tends to drive dissociation of ions, is not strongly localized on the nascent ion. In addition, EI results in extensive fragmentation that is not easily predicted or rationalized. Molecular ions, defined as the analyte missing a single electron, are weak; thus, molecular weight is not easily determined with EI for higher molecular weight FAME.

After its introduction in the 1960s, CI rapidly evolved to mean the transfer of a proton (H^+) to an analyte from a reagent gas present in excess in the gas phase. Reagent gases are typically small hydrocarbons (methane, isobutane) that are admitted to an ion source at 1000-fold excess compared with the analyte. Electrons emitted from a filament cause numerous cascading reactions that result in abundant protonated reagent species, which donate protons to analytes with greater proton affinity. Common reagent ions [CH_5^+, $CH_2^+(CH_3)_3$] have lower proton affinity than most analytes and thus are efficient ionization sources. The hallmark of CI MS is greatly reduced fragmentation compared with EI because of the lower amount of excess energy deposited into the nascent ion during the ionization event. Molecular ions are often observed, but again, fragments that reveal the double bond position are not reliably observed.

GC/MS and Derivatives Other than Methyl Esters. The inability of EI and CI of FAME to produce reliable fragments indicative of double bond position has long been known to lie in the diffuse distribution of charge on the analyte ion (7). It is now routine to derivatize with one of a few different reagents to localize the charge and produce reproducible charge-remote fragmentation [CRF, (8)]. Common acylation groups are picolinyl esters (9) and dimethyloxazoline (DMOX) derivatives. These derivatives result in ions with strong charge localizing groups. Statistically driven fragmentation results in a peak representing breakage of each carbon-carbon bond, single or double, in the chain, thus, the double bond position can be read off from the mass intervals of either 14 (CH_2) or 13 (CH), with the latter appearing consecutively. Mass spectra of these derivatives and an explanation of their interpretation are available on the web (10).

There are numerous disadvantages to these techniques. GC analysis of FAME remains the method of choice for quantitative analysis of FA profiles; thus, most real samples have been converted to FAME. Esterification with alternative reagents requires additional wet chemistry and is usually not routine. They are particularly inconvenient for low-abundance FA or for PUFA with a complex double bond structure. Chromatography phases and temperature programs, and therefore chromatographic resolution, are not as well developed for alternative derivatives as they are for FAMEs. Other esters are necessarily of higher molecular weight and can present volatility problems for high molecular weight FA. Reagents used to form other esters are sufficiently harsh that double bond rearrangement is possible. Finally, because FAME are the usual form in which FA are analyzed, unknown peaks are nearly always detected in the analysis of FAME mixtures, not as picolinyl or DMOX esters. Derivatization of an aliquot of the same sample to make picolinyl or DMOX esters necessarily involves new GC conditions, and chromatographic peaks for picolinyl or DMOX derivatives generally do not elute from the column in the identical order as the FAME. It can therefore be difficult to match similar peaks in FAME and picolinyl or DMOX chromatograms. For all of these reasons, methods that permit direct analysis of FAME are highly desirable.

Liquid Introduction/Atmospheric Pressure Chemical Ionization (APCI)-Electrospray Ionization (ESI). Comprehensive reviews on liquid-source tandem MS of glycerolipids appeared in 2003 (11,12). ESI introduction is now routine for glycerolipid regiospecific analysis, e.g., (13). When lithium is included with the analyte, collisional activation of lithiated and dilithiated phospholipid (PL) results in fragments for fatty acyl groups with intensities corresponding to the *sn*-1 or *sn*-2 positions. No double bond positional information is obtained in these spectra. It might be possible to accomplish simultaneous regiospecific and double bond positional analysis with CRF and tandem MS. This was accomplished for FAME using high-energy collisional dissociation in tandem magnetic sector MS (14). However, triple quadrupole MS and quadrupole time-of-flight (TOF) in routine use for tandem MS do not generate collisions in the keV range, which appear to be required to produce these fragments; thus, no double bond positional information is obtained from instruments now in wide use.

Why a High Sensitivity Method Is Essential. Many thousands of mammalian FA profiles published since 1960 show that the major components of most mammalian tissues and fluids are saturates and monounsaturates. However, the most metabolically active FAs are the PUFAs (2 or more double bonds) and NMI-PUFAs, which are most often present at <5% of all FAs, and often <1%. In addition, with few exceptions, most PUFAs, including NMI-PUFAs, exist primarily as components of several glycerolipid classes usually with several different FAs occupying the complementary position (*sn*-1, *sn*-2) on the glycerol molecule. All told, each PUFA may be present on 10 or more distinct molecular species, thereby splitting their concentrations among these various forms and further lowering their molecular concentrations by an order of magnitude or more. Each species is likely to have distinct biochemical properties. It is these individual molecular species that are the targets of lipidomic analysis.

CACI: Covalent Adduct Chemical Ionization

We noted above that the term "chemical ionization" refers specifically to proton attachment, which is generally an electrostatic interaction of a proton with a nucleophilic site on a molecule, such as a lone electron pair. Although gas phase ion-molecule chemical reactions involving covalent bond making/breaking are the subject of hundreds of papers in the chemical literature, and although such reactions have been studied in the context of analysis, no techniques that we are aware of have emerged that are based on this approach. A review of this literature suggests that the main shortcoming of these approaches is that they were performed with single-stage MS instruments incapable of MS-MS, and thus cannot isolate the relatively low-intensity reaction products from single-stage mass spectra dominated by unimolecular decomposition products. All covalent modifications to analytes continue to be performed by wet chemistry before MS analysis.

Since 1999, we published a series of papers demonstrating a purely mass spectrometric technique for double bond localization in FAME that relies on gas phase derivatization (15–20). This method is performed under mass spectrometer conditions that are accurately described as CI, using acetonitrile as the reagent ion. However, because it relies upon a very different chemical principle than proton-attachment CI, the name is deceiving. Because of the success of this approach, and for clarity in distinguishing this sort of CI from conventional CI, we introduce a general acronym here to describe analytical methods that rely upon an ion-molecule reaction followed by mass spectral analysis: Covalent-Adduct Chemical Ionization = CACI (with the "CI" pronounced as in "cappuccino," as "chee": "catchy"). This term applies to the use of reagent ions beyond acetonitrile, and can be used generally to refer to methods in which gas phase derivatization reactions create covalent adducts that are subsequently collisionally activated to produce diagnostically useful ions. Here we restrict its use to acetonitrile CACI for PUFA FAME analysis.

Acetonitrile CACI for FAME Analysis

Three-dimensional (3D), internal ionization ion traps facilitate the use of low vapor pressure reagent gases for CI because of the high concentration of reagent ions that can be maintained in the trapping region compared with the open, nontrapping fields in the CI source of quadrupole or magnetic sector instruments. All of our studies were performed in 3D internal ionization traps, either a Varian Saturn 3 or a Saturn 2000. When acetonitrile (CH_3CN, *m/z* 41) is maintained under CI conditions, the mass spectrum appears as is shown in Figure 3. Aside from the expected dominant ions at *m/z* 42 (CH_4CN^+), *m/z* 40 (CH_2CN^+), an additional ion appears at *m/z* 54. Oldham (21) showed that this ion results from the reaction of CH_3CN with itself to transfer a CH_3 to yield (1-methyleneimino)-1-ethenylium (MIE), with the structure $CH_2{=}C{=}N^+{=}CH_2$ according to the reaction scheme in Figure 4.

When neutral FAME are introduced into the CI reaction mixture, a series of ions results in the MS-1 spectrum shown in Figure 5 for methyl arachidonate (20:4n-6). The expected pseudomolecular ion resulting from proton attachment is observed ("MH") along with three other ions that appear to be characteristic of FAME. At highest mass is the [M+54] ion from the covalent addition of MIE with the double bonds of FAME. Other ions correspond to the loss of 32 (methanol) and 50 (methanol + water) from MH, and a low intensity but analytically useful ion is found for the loss of 32 from [M+54]. The intensity of these ions provides useful information on double bond geometry as discussed below. In acetonitrile CACI, the [M+54] ion is the relevant covalent adduct to be isolated and collisionally activated to yield an MS-2 spectrum.

Homoallylic FAME [M+54] ions collisionally dissociate to form two ions that are normally of greatest intensity in the MS-2 spectrum. Studies with homoallylic FAME (15,16) reveal systematic fragmentations that can be rationalized on the basis of neutral analyte structure as shown in Figure 6. Fragmentation can be either vinylic

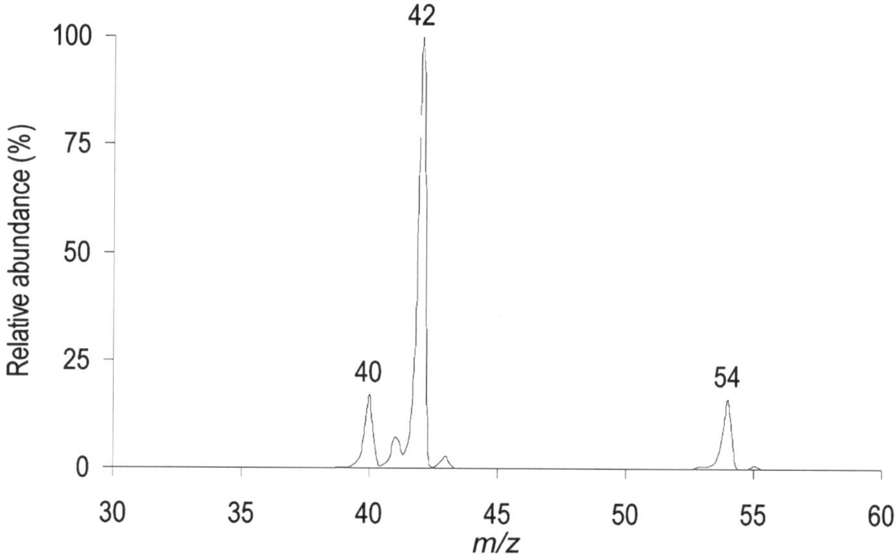

Fig. 3. Mass spectrum of CH_3CN under chemical ionization (CI) conditions.

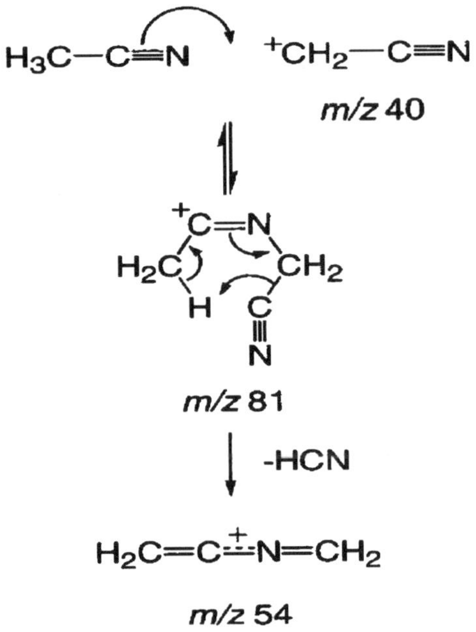

Fig. 4. Formation mechanism for (1-methyleneimino)-1-ethenylium (MIE). *Source*: Ref. 21.

or allylic to a double bond, and an H can be transferred either to or from the neutral fragment. The precise combination is defined according to the number of double bonds in the homoallylic system. Monoenes fragment at positions allylic to the double bond site, and a proton is transferred to the ion. Dienes fragment at positions vinylic to

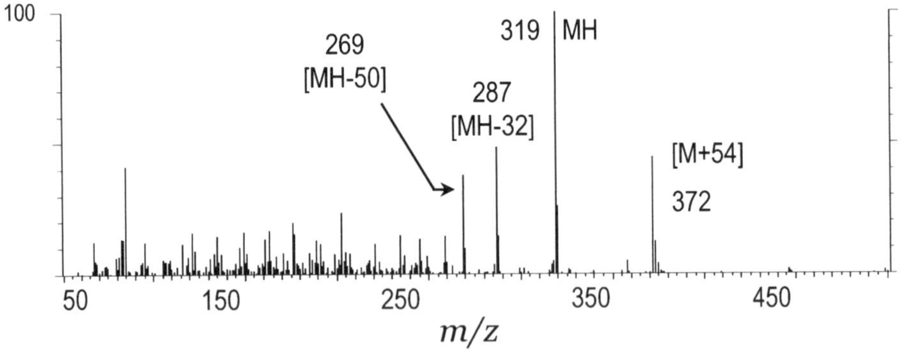

Fig. 5. Single-stage mass spectrometry (MS) of arachidonic acid (20:4n-6) upon covalent-adduct chemical ionization (CACI) (CH_3CN CI). Four ions characteristic of fatty acid methyl esters (FAME), including losses of CH_3OH and CH_3OH+H_2O from the MH, were observed.

CIMS/MS of Homoallylic FAME (Methylene Interrupted)

- Monoenes:
  - allylic fragmentation; higher unsaturation vinylic.
- Monoenes and dienes:
  - H transfer to ion; higher unsaturation H transfer to neutral.

CK Van Pelt, JT Brenna (1999) *Analytical Chemistry* 71: 1981.

NJ Oldham, A Svatoš (1999) *Rapid Comm MS* 13: 331.

CK Van Pelt, BK Carpenter, JT Brenna (1999) *Journal of the American Society for Mass Spectrometry* 10 (12): 1253.

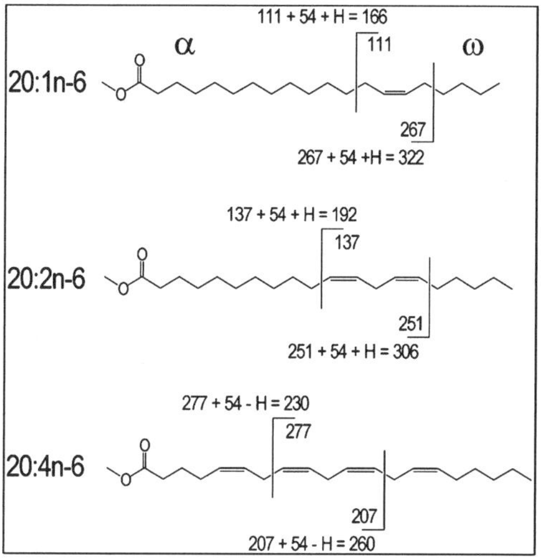

Fig. 6. Systematics of diagnostic ion production.

the double bond system and transfer a proton to the ion. Trienes and higher unsaturates fragment within the double bond system and transfer a proton to the neutral.

As a result of these highly reproducible dissociations, the two dominant MS-2 products completely define the position of double bonds in homoallylic FAME. We introduced the terms α and ω diagnostic ions, based on whether the α (C2) or ω (methyl) carbon is part of the observed fragment, as also shown in Figure 6. Tables of MS-2 diagnostic ions that permit the double bond structure of homoallylics to be read off from the masses of the diagnostic ions are now available (18).

A straightforward illustration of the technique is the analysis of monoene present in a sample of partially hydrogenated vegetable oil. The [M+54] ion for Me18:1, appearing at *m/z* 350, is isolated and fragmented. Figure 7 (upper trace) shows a reconstructed ion chromatogram from all MS-MS ions during elution from the GC column. Although a high resolution 100-m column (CP-Sil88) routinely used for *trans* analysis is employed, there is no resolution of the individual isomers. The lower plots are reconstructed ion plots for the diagnostic ions expected for each designated 18:1. Each isomer is baseline resolved, with the *trans* isomers eluting before the *cis* isomers. A distribution of both *trans* and *cis* isomers is found, probably corresponding to a rearrangement of double bonds from the original position, likely to be the 9 position of oleate. These data also illustrate an advantage to the acetonitrile CACI method over DMOX and related methods used for double bond

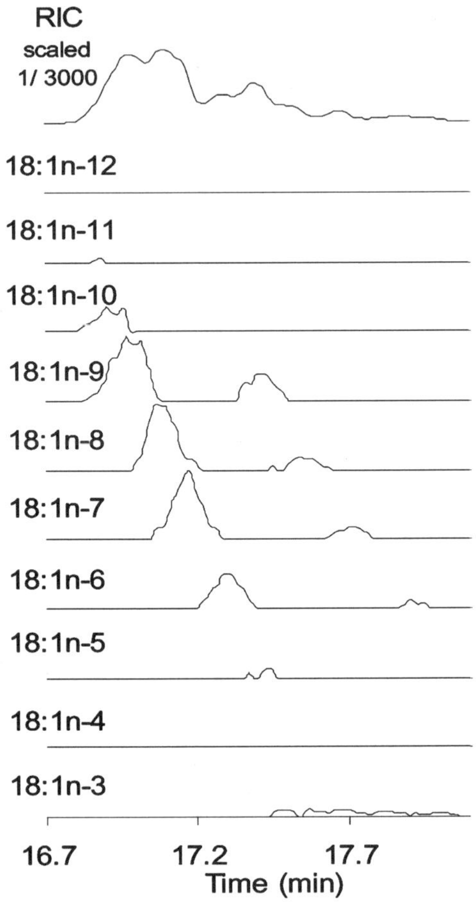

Fig. 7. Region of the chromatogram of esterified partially hydrogenated vegetable oil in which 18:1 methyl esters elute. The upper trace is a reconstructed ion chromatogram from the original chemical ionization mass spectrometry (CI MS) data. The lower traces sum the abundances of the diagnostic ions of methyl octadecaenoates with the specified double bond geometry. The *trans*-isomer elutes first, followed by the *cis* isomers when present.

localization. In those methods, the signal is distributed among many fragments, which effectively reduces sensitivity. Because very little signal is present in any ion, diagnostic fragments are typically very close to baseline and thus similar plots are rarely possible, even when the molecular weight is known. The diagnostic ions in the acetonitrile CACI MS-MS spectra can therefore serve as strong lines for quantitative analysis.

Structures of Homoallylic [M+54]⁺ and Diagnostics Ions. The structure of the homoallylic $[M+54]^+$ ion(s) was established with studies involving isotopically labeled compounds and MS-MS-MS (16). The first step in Figure 8a shows the double bond in methyl oleate (Me18:1n-9) undergoing an electrophilic attack by MIE. MIE adds to the double bond via a [2+2] cycloaddition to yield a 4-membered heterocyclic ring structure at the position of the original double bond. Upon collisional activation, $[M+54]^+$ undergoes a charge-driven rearrangement with transfer of an H ion to the neutral, and fragmentation to yield the α diagnostic ion. Addition of the mirror image (antiparallel) of the Me18:1n-9 ion to MIE produces the ω diagnostic ion. Similar structures and dissociation mechanisms were found for homoallylic PUFAs, with modifications for H transfer and for cleavage at different sites.

From these studies, we have come to understand that peaks appearing at [M+54] are a series of isomers formed from MIE reaction with the neutral FAME, probably at every double bond, and in two different orientations as discussed below. Thus, the number of isomeric ions at [M+54] in any one mass spectrum is twice the number of double bonds in the FAME. Collisional dissociation of this set of isomers gives the MS-2 spectra and the systematics as described. Fragmentation products are likely to be the most energetically favorable dissociation channels among the several isomers present.

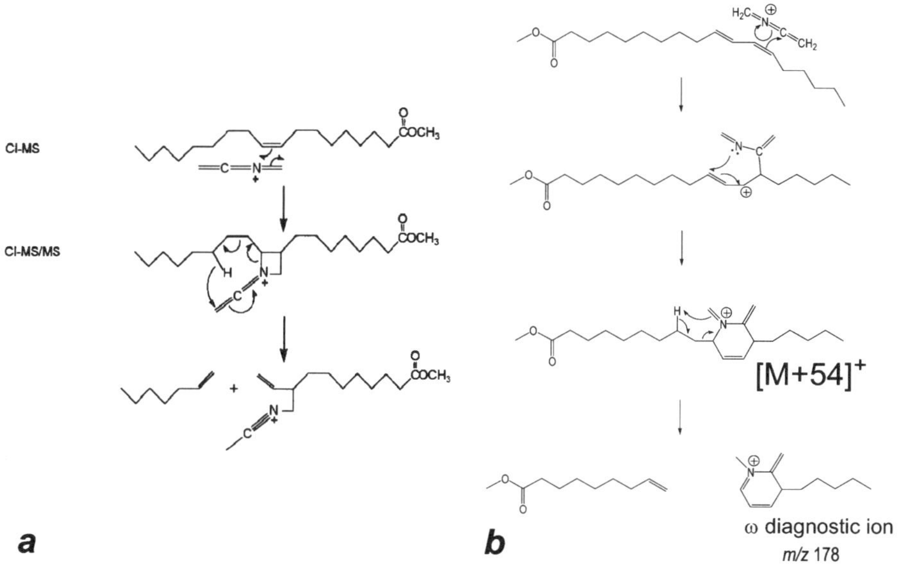

Fig. 8. (a) Structure of homoallylic $[M+54]^+$ ion and the α diagnostic ion for 18:1n-9. The α diagnostic ion corresponds to the structure that would be obtained if 18:1n-9 were antiparallel in the first step. (b) The CLA $[M+54]^+$ ion has a six-membered ring.

NMI-PUFA; CLAs. We found that CLA FAME behave in a qualitatively similar manner to homoallylic FAME. Remarkably, we also found that the MS-1 and MS-2 spectra of CLA FAME yield a richer harvest of information than found for homoallylics, including the rare case of stereochemical information in a mass spectrum. In the MS-1 spectrum, we found that the ratio of the $[M+54]^+/[M+54-32]^+$ is one order of magnitude lower than for homoallylic diene FAME; thus, calculation of this ratio permits assignment as either a homoallylic or a CLA. The structure of the [M+54–32] has not been elucidated, nor is the origin of the difference in its intensity based on double bond conjugation.

Interestingly, the structure of the [M+54] ions for CLA is different than that for homoallylic FAME. Again using labeling studies and MS-MS-MS, we found that the CLA [M+54] are consistent with the development of a six-membered heterocyclic ring as shown in Figure 8b for the case of the [M+54] for *t*10,*c*12-CLA (20). Both mechanisms shown in Figure 8 are consistent with the literature studies of ion-molecule reactions in small hydrocarbons. Fragmentation is again vinylic to the site of the original double bond and, with a few exotic exceptions, double bond positions for most CLAs can be assigned. Armed with the knowledge that a diene is conjugated from MS-1, the diagnostic ions in the MS-2 spectrum fully define the double bond position.

Additional information that defines double bond *geometry* appears in the ratio of the diagnostic ions in MS-2. Figure 9 is a summary slide showing results from the analysis of 7 standards containing 28 CLAs with double bond positions from 6,8 to 12,14 (19). Each sample contained all four possible isomers (*cis-trans*, *trans-cis*, *cis-cis*, *trans-trans*). We calculated the ratio of intensities for the α:ω diagnostic ion abundances. In all cases, it was >4 for *cis*/*trans* geometry and <0.6 for *trans*/*cis* geometry. *Cis*/*cis* and *trans*/*trans* ranged from 0.7 to 3.5. These results indicate that double bond geometry can be determined by abundance ratios of diagnostic ions. Together with GC retention time data, this method provides positive identification for almost all CLA FAME. Figure 10 presents data on one of the first samples containing CLA that we investigated using this method. The sample is milk fat from an experimentally treated cow. Thirteen CLAs were positively identified in this analysis, having the structures presented previously in Figure 2.

Future Work: CLnAs. Detailed studies of the bioactivity of conjugated C18 trienes (CLnA) have not been conducted, although there is indirect evidence of bioactivities. This is in part because of the inability to analyze these compounds, and secondarily because of their relatively low concentration in most systems. At this time, CLnA research is where CLA research was perhaps 10 years ago. We have some preliminary data, presented at an international conference, showing promising results for CACI as an analytical technique for CLnA. Figure 11 illustrates two separate matters regarding CLnA. First, unlike CLA in which two double bonds can be conjugated in only one way, CLnA can be conjugated in two different ways, with all 3 double bonds conjugated together, or with one set of 2 double bonds conjugated

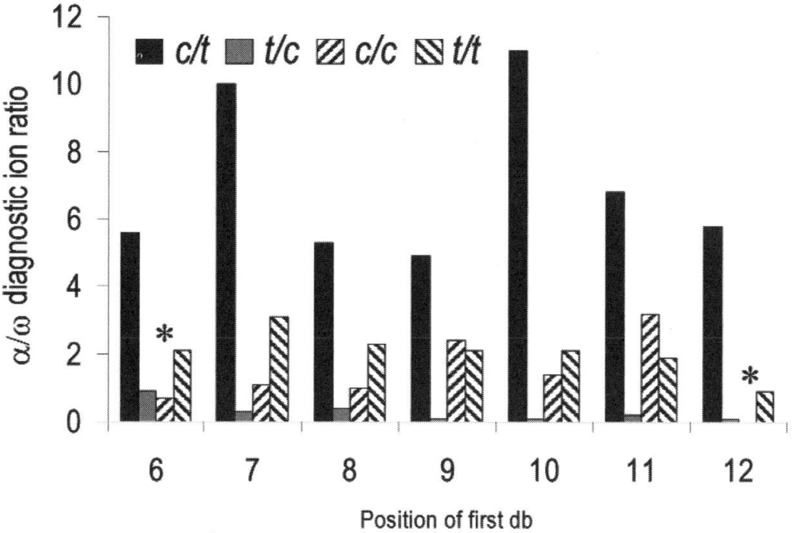

Fig. 9. Diagnostic ion ratios in all conjugated linoleic acid (CLA) standards. Ratios reveal the geometry of the double bond systems. The α:ω ion ratios are >4 for *cis/trans*, <0.6 for *trans/cis*, and for *cis/cis* and *trans/trans* are 0.7–3.5. *Isomers exhibited chromatographic overlap and were excluded.

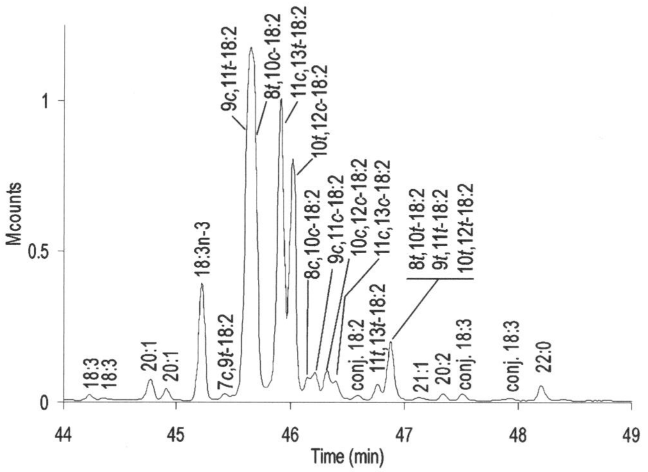

Fig. 10. Total ion chromatogram showing positive identification of 13 conjugated linoleic acids (CLAs) (18:2s) in a sample of milk fat from a cow abomasally infused with a CLA supplement. Other fatty acids (e.g., conj. 18:3s) were labeled in mass spectrometry (MS)1 but abundances did not permit full characterization in MS2. Structures presented in Figure 2.

and one separate double bond. We have termed these "fully" and "partially" conjugated, respectively. The MS-1 spectra of Figure 11 emphasize once again the utility of the $[M+54]^+$:$[M+54-32]^+$ ratio, showing that it differs for homoallylic 18:3 and conjugated 18:3 (CLnA). Figure 12 is a compilation of 18:3 $[M+54]^+$:$[M+54-32]^+$ ratios for several classes of CLnAs as well as 18:3 FAME of unknown double bond position/geometry. The known, fully conjugated 18:3 all have ratios between 0.7 and 0.9, the partially conjugated CLnA have ratios of 1.6–2.4, and the known homoallylics have ratios of 7–15. Importantly, all unknown 18:3 fall into one of these three classes, indicating that the trend is a reliable index of structure.

Conclusions

Among the most important features of acetonitrile CACI is that the resulting spectra are interpretable. That is, the resulting fragmentation does not simply result in a fin-

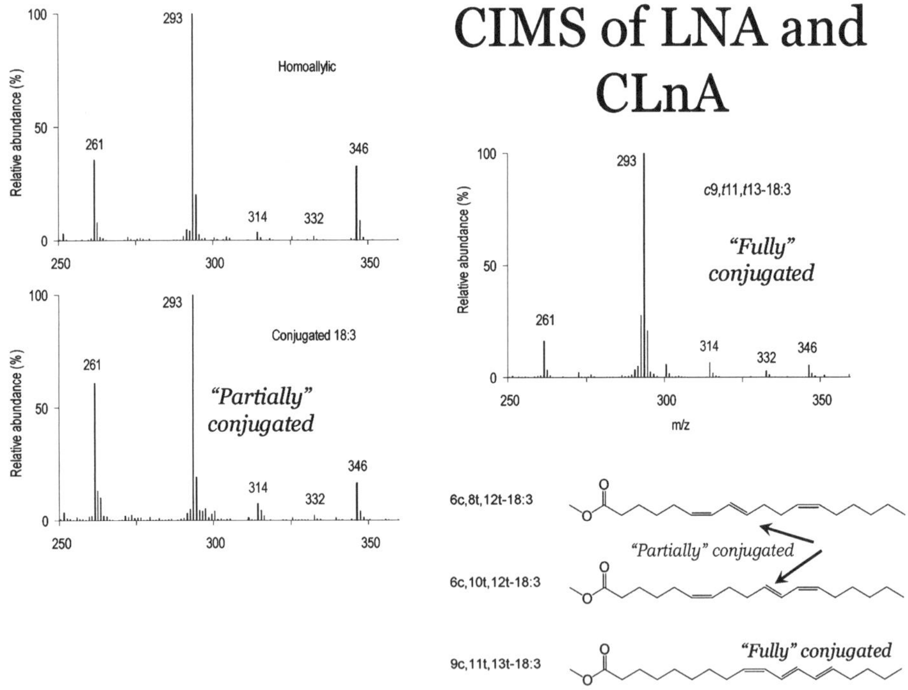

Fig. 11. "Partially" conjugated fatty acids are those with at least 2 conjugated double bonds and one lone double bond; "fully" conjugated fatty acids are those in which all double bonds are conjugated. Mass spectrometry (MS)-1 spectra of partially and fully conjugated C18 trienes (CLnAs). Ratios of *m/z* 346/314 ([M+54]]/[[M+54–32]) depend on the arrangement of double bonds.

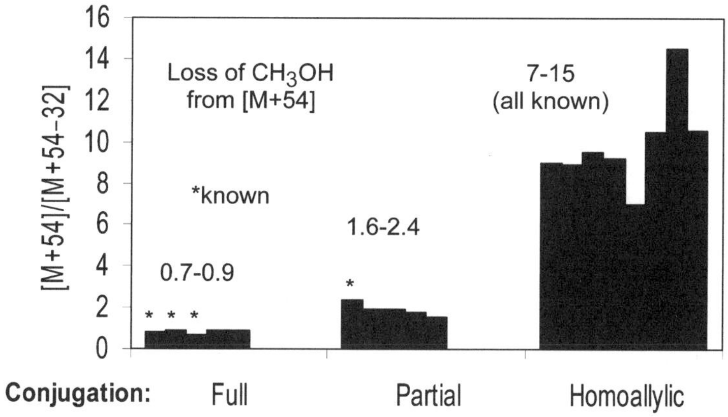

Fig. 12. Compilation of CH_3OH loss ratios for 18:3 isomers, including several unknowns.

gerprint of ions that can be searched only with pattern recognition software; rather, the peaks can be rationalized and, most importantly, predicted for unknown FAs. To emphasize this point, the extraordinary results of Figures 2 and 10 were the very first CLA data that we obtained, other than two CLA standards published some years ago. All of the peaks in that milk fat sample were assigned by interpretation. Subsequent acquisition of CLA standards and publication of the mass spectra demonstrated that every peak in the milk fat sample was properly assigned without standards. This repeats our experience with the independent discovery of the first octaene FA, 28:8n-3 (17). Results of ongoing studies of conjugated PUFA FAME with ≥3 double bonds are promising and may lead to a completely instrument-based method for positive characterization of PUFAs of arbitrary double bond structure, with ions that are useful for quantitative analysis.

Acknowledgments

Original contributions were supported by NIH grants GM49209 and GM071534. Colleen Van Pelt, Anthony Michaud, and Peter Lawrence generated most of the original data. Our collaborators provided samples of milk fat, hydrogenated vegetable oil, and CLnAs (Dale Bauman and co-workers), and synthesized CLA standards (Pierluigi Delmonte, Peter Yurawecz and co-workers).

References

1. Valenzuela, A., and N. Morgado, *Trans* Fatty Acid Isomers in Human Health and in the Food Industry, *Biol. Res.* *32*:273–287 (1999).
2. Salmeron, J., F.B. Hu, J.E. Manson, M.J. Stampfer, G.A. Colditz, E.B. Rimm, and W.C. Willett, Dietary Fat Intake and Risk of Type 2 Diabetes in Women, *Am. J. Clin. Nutr.* *73*:1019–1026 (2001).

3. Pariza, M.W., Y. Park, and M.E. Cook, The Biologically Active Isomers of Conjugated Linoleic Acid, *Prog. Lipid Res. 40*:283–298 (2001).
4. Ip, C., S.F. Chin, J.A. Scimeca, and M.W. Pariza, Mammary Cancer Prevention by Conjugated Dienoic Derivative of Linoleic Acid, *Cancer Res. 51*: 6118–6124 (1991).
5. Belury, M.A., Inhibition of Carcinogenesis by Conjugated Linoleic Acid: Potential Mechanisms of Action, *J. Nutr. 132*:2995–2998 (2002).
6. Gutnikov, G., Fatty Acid Profiles of Lipid Samples, *J. Chromatogr. B Biomed. Appl. 671*:71–89 (1995).
7. Schneider, B., and H. Budzikiewicz, A Facile Method for the Localization of a Double Bond in Aliphatic Compounds, *Rapid Commun. Mass Spectrom. 4*: 550–551 (1990).
8. Gross, M.L., Charge-Remote Fragmentation: An Account of Research on Mechanisms and Applications, *Int. J. Mass Spectrom. 200*:611–624 (2000).
9. Christie, W.W., E.Y. Brechany, and R.T. Holman, Mass Spectra of the Picolinyl Esters of Isomeric Mono- and Dienoic Fatty Acids, *Lipids 22*:224–228 (1987).
10. W.W. Christie, Mass Spectrometry of Fatty Acid Derivatives. The Lipid Library. http://www.lipidlibrary.co.uk/masspec.html.
11. Griffiths, W.J., Tandem Mass Spectrometry in the Study of Fatty Acids, Bile Acids, and Steroids, *Mass Spectrom. Rev. 22*:81–152 (2003).
12. Pulfer, M., and R.C. Murphy, Electrospray Mass Spectrometry of Phospholipids, *Mass Spectrom. Rev. 22*:332–364 (2003).
13. Duffin, K.L., M.G. Obukowicz, W.J. Salsgiver, D.J. Welsch, C. Shieh, A. Raz, and P. Needleman, Lipid Remodeling in Mouse Liver and Plasma Resulting from Delta6 Fatty Acid Desaturase Inhibition, *Lipids 36*:1203–1208 (2001).
14. Jensen, N.J., K.B. Tomer, and M.L. Gross, Fast Atom Bombardment and Tandem Mass Spectrometry of Phosphatidylserine and Phosphatidylcholine, *Lipids 21*:580–588 (1986).
15. Van Pelt, C.K., and J.T. Brenna, Acetonitrile Chemical Ionization Tandem Mass Spectrometry to Locate Double Bonds in Polyunsaturated Fatty Acid Methyl Esters, *Anal. Chem. 71*:1981–1989 (1999).
16. Van Pelt, C.K., B.K. Carpenter, and J.T. Brenna, Studies of Structure and Mechanism in Acetonitrile Chemical Ionization Tandem Mass Spectrometry of Polyunsaturated Fatty Acid Methyl Esters, *J. Am. Soc. Mass Spectrom. 10*:1253–1262 (1999).
17. Van Pelt, C.K., M.C. Huang, C.L. Tschanz, and J.T. Brenna, An Octaene Fatty Acid, 4,7,10,13,16,19,22,25-Octacosaoctaenoic Acid (28:8n-3), Found in Marine Oils, *J. Lipid Res. 40*:1501–1505 (1999).
18. Michaud, A.L., G.Y. Diau, R. Abril, and J.T. Brenna, Double Bond Localization in Minor Homoallylic Fatty Acid Methyl Esters Using Acetonitrile Chemical Ionization Tandem Mass Spectrometry, *Anal. Biochem. 307*:348–360 (2002).
19. Michaud, A.L., M.P. Yurawecz, P. Delmonte, B.A. Corl, D.E. Bauman, and J.T. Brenna, Identification and Characterization of Conjugated Fatty Acid Methyl Esters of Mixed Double Bond Geometry by Acetonitrile Chemical Ionization Tandem Mass Spectrometry, *Anal. Chem. 75*:4925–4930 (2003).
20. Michaud, A.L., P. Lawrence, R.O. Adlof, and J.T. Brenna, On the Formation of Conjugated Linoleic Acid Diagnostic Ions with Acetonitrile Chemical Ionization Tandem Mass Spectrometry, *Rapid Commun. Mass Spectrom. 19*:363–368 (2005).
21. Oldham, N.J., Ion/Molecule Reactions Provide New Evidence for the Structure and Origin of, *Rapid Commun. Mass Spectrom. 13*:1694–1698 (1999).

Part III. Chromatographic Techniques

7 Recent Advances in Silver-Ion HPLC Utilizing Acetonitrile in Hexane as Solvent: Temperature Effects on Lipid Elution Orders/Resolution

Richard O. Adlof

Food and Industrial Oil Research, National Center for Agricultural Utilization Research, USDA, Agricultural Research Service, Peoria, IL 61604 USA

Introduction

Over the last several decades, silver-ion high-performance liquid chromatography (Ag-HPLC), utilizing columns packed with silver ions bonded to silica or a similar substrate and a variety of different solvent systems [1,2-dichloroethane, dichloromethane, or hydrocarbons (hexane, heptane) with acetonitrile or alcohols (methanol, isopropanol) as polar modifier(s)] has proven to be a tremendously powerful technique for the analysis of a wide variety of geometric (*cis* and *trans*) and positional (location of olefinic/acetylenic bonds on the fatty acid backbone) fatty acid (FA), fatty acid methyl ester (FAME), mono-/diacylglycerol (MAG/DAG) and triacylglycerol (TAG) isomers (see Ref. 1–6 for reviews). Specific examples include the separation/quantitation of *cis* and *trans* fatty acids (7,8), FA methyl (9–11)/phenacyl (12,13)/*p*-methoxyphenacyl (10,14,15) esters, positional isomers as FAME (10) or TAG (16,17) from partially hydrogenated vegetable oils, conjugated FA (18)/FA esters (19–21)/TAG (22–24), FAME labeled with deuterium atoms on the double bond carbons (25), 1,2-DAG (26,27), TAG positional isomers (28–30) and to the separation of FAME or TAG mixtures containing FAs of widely differing chain lengths (31) or non-methylene–interrupted double bonds (32). Despite the limited solubility of ACN [~1.5% (vol/vol) at 23°C], the solvent system of acetonitrile (ACN) in hexane has proven to be a tremendously versatile solvent system for many of these analyses.

Although temperature control and programming have become important and integral parts of gas chromatography (GC), the technology has found only limited application(s) in HPLC [reversed-phase (33,34), silica/alumina (35,36), silver nitrate/silica (37), and in Ag-HPLC (26)]. Utilizing an Ag-HPLC column and an eluting solvent gradient composed of 1,2-dichloroethane, dichloromethane, and acetonitrile, Juaneda and co-workers (9) noted that "changing the composition of the solvents and/or the temperature did not improve the separation" of FAME but, "the phenacyl derivatives of the eight isomers were separated according to the number of *trans* double bonds (mono-, di- and tri-*trans*)." The author also noted that *decreasing* the column temperature from 38 to 10°C *increased* the retention time and improved the resolution of the di- and mono-*trans* 18:3 phenacyl esters. Recent studies (38) on the effects of temperature on FAME and TAG retention and resolution, utilizing commercially available Ag-HPLC columns and isocratic solvent systems of ACN in hexane, unexpectedly demonstrated that unsaturated FAME and TAG samples eluted more rapidly at lower column temperatures. These "unusual" results were exactly opposite the temperature effects (in which samples elute more rapidly at higher temperatures) noted in gas and most liquid (reversed-phase or silica gel substrates) chromatography systems. [See Results and Discussion section for an overview of the theories/factors contributing to retention (Ag-HPLC/ACN in hexane solvent system) and how temperature affects this retention.]

To better understand the ACN in hexane solvent system (and whether ACN is even required for sample elution) and the relation(s) between Ag-HPLC column temperature and sample elution order, retention, and resolution, a series of FAME (0–6 double bonds; *cis*/*trans* isomers) and TAG [homogeneous (triacetyl-, tristearoyl-, trioleoyl-, trilinoeyl- and trilinolenoyl-glycerols) and positional (e.g., 1,3-distearoyl,2-monolinolenoyl- and 1,2-distearoyl, 3-monolinolen-oylglycerol) isomers] were analyzed on two ChromSpher© Lipids columns connected in series, with isocratic solvent systems of 0.3–1.5% ACN in hexane, at 10, 20, 30, or 40°C.

Materials

Hexane (ACS grade, Allied Fisher Scientific, Orangeburg, NY), diethyl ether (Fisher Scientific, Fair Lawn, NJ) and ACN (Merck, Darmstadt, Germany) were used as received. TAG isomers POP, PPO, SLnS, SSLn, LOL and LLO [where P = palmitic acid (16:0), S = stearic acid (18:0), O = oleic acid (9*c*-18:1), L = linoleic acid (9*c*,12*c*-18:2) and Ln = linolenic acid (9*c*,12*c*,15*c*-18:3)] were readily synthesized by the method of Kodali (39) with fatty acids obtained from Nu-Chek-Prep, Elysian Fields, MN, and used to prepare TAG mixtures I (POP/PPO), II (SLnS/SSLn) and III (LOL/LLO) [~50/50 each isomer pair (weight %); 1- and 3-positions of TAG assumed equal; ~10 mg/mL isooctane]. TAG mixture IV [~20% each AcAcAc (where Ac = acetate), SSS, OOO, LLL and LnLnLn] and FAME mixtures I [~20% each S, O, L, arachidonate (5*c*,8*c*,11*c*,14*c*-20:4) and docosahexaenoate (4*c*,7*c*,10*c*,13*c*,16*c*,19*c*-22:6)] and II [18:2 isomers (9*t*,12*t*- and 9*c*,12*t*-18:2)] were prepared (excluding triacetin) using TAG and FAME obtained from the same source. Tripalmitin and triacetin were obtained from Sigma-Aldrich, St. Louis, MO.

Methods

HPLC

A Spectra (Thermo Finnigan, San Jose, CA) HPLC system composed of a SCM1000 degasser, a P4000 solvent delivery system, an AS3000 autoinjector (20 μL injection loop), and a UV6000 detector (at absorbance of 206 nm) and/or a Sedex 75 evaporative light scattering detector (ELSD; S.E.D.E.R.E., Alfortville, France) was utilized with data collection via an SS420X converter and ChromQuest 3.0 software (ThermoQuest, San Jose, CA). The HPLC column temperatures were set (manually or programmed) using a Phenomenex ThermaSphere TS-430 HPLC Column Chiller/Heater (5–40°C). The ChromSpher Lipids columns (Cat. No. 28313; 4.6 mm i.d. × 250 mm stainless steel; 5 μm particle size; silver ion impregnated) were purchased from Varian-Chrompack International, Middelburg, The Netherlands, and used as received. Two of the Lipids columns were connected in series (the void volume of each column was 2.1 mL).

The solvent (0.3–1.5% ACN in hexane; magnetic stirring) was freshly prepared each morning and used during that day for the temperature (10, 20, 30, or 40°C)/retention studies. Because the ACN dissolved very slowly into hexane, the ACN in hexane solvent was thoroughly stirred for 15 min before use. A single solvent reservoir was used to minimize possible changes in TAG retention(s) and resolution(s) due to batch-to-batch variations in mixed solvent composition or differences in instrument configurations (number of solvent pumps, reservoirs, mixing chambers, valves) between different HPLC systems; to simplify retention time comparisons, solvent flow was standardized at 1.0, 1.5, or 2.0 mL/min. The sequence began with the column temperature set at the starting temperature (usually 10°C) and, after a 0.5-h stabilization period, the TAG or FAME standard was injected. After elution of the last component (~1 h), the column temperature was set/programmed to the next temperature, and allowed to stabilize at that temperature (15 min) before the next injection, and so on. The total time required for retention studies at four temperatures was ~6 h. An identical study, but utilizing a solvent system of 1.0% ACN and 0.5% ethyl ether (EE) in hexane, was also completed to determine whether the presence of EE as co-solvent had any effects on TAG retention(s). Finally, the pattern was repeated at 20°C, then at 40°C, then again at 20°C (dual injections at each temperature) to confirm that the observed changes in retention times were due to changes in column temperature, not solvent composition, over the 6-h duration of the test. Reproducible patterns and times were achieved even when equilibration times were shortened to 15 min at each temperature.

Gas Chromatography

Elution orders were confirmed by injection of known FAME or TAG standards, or by collection of FAME or TAG fractions [and conversion of the TAG to FAME (34)], analysis on a Varian 3400 Gas Chromatograph (Varian Instruments, Palo Alto, CA) equipped with a 30 m × 0.32 mm SP2380 (Supelco, Bellefonte, PA) capillary column, flame ionization detector (FID) and utilizing He as carrier gas. GC conditions were as follows: 155°C (10 min) programmed to 200°C at 3°C/min (20-min hold).

Specific Studies

ACN Requirement Study. The hypothesis concerning the function of the ACN in the ACN in hexane solvent system, i.e., competition with π-electrons of unsaturated samples/unpaired electrons of carboxyl group of carboxylic acids for the Ag^+ ions, was tested (Dual-column Ag-HPLC; solvent flow: 1.0 mL/min; 20°C; UV at 206 nm) by comparing the elution of Me 17:0 and Me 9*c*-18:1 FAME standards (10 mg/mL hexane each) with solvent systems of 100% hexane and 0.3% ACN in hexane. In all tests, the solvent was collected in a single round bottom flask (changed at each solvent change). The solvent(s) were removed by rotary evaporator; the residue was dissolved in a minimum volume of isooctane, and then analyzed by GC. TEST 1: With solvent 100% hexane, injection of Me 17:0 FAME sample resulted in a peak at 19 min. The eluting solvent was collected; analysis by GC confirmed the presence of Me 17:0. TEST 2: Using the same solvent supply (100% hexane) as in TEST 1, a Me 9*c*-18:1 sample was injected. No eluting peaks were observed, even after 60 min. When the eluting solvent (60 mL) was collected, concentrated, and analyzed by GC, no Me 9*c*-18:1 was noted. But when 100% hexane solvent was replaced with 0.3% ACN in hexane, and one fraction (1 mL/min for 60 min flow rate) was collected, concentrated, and analyzed by GC, the presence of Me 9*c*-18:1 was confirmed. TEST 3: Using the same solvent supply (0.3% ACN in hexane) as in the second part of TEST 2 and Me 9*c*-18:1 (same sample as used in TEST 2) was injected, a peak was noted at 13.6 min. Collection, concentration, and analysis by GC again indicated the presence of Me 9*c*-18:1.

Reproducibility Study. TAG II (not shown) had very good reproducibility. The TAG standard was injected at 20°C, then at 40°C, and then again at 20°C, with 1 h equilibration and dual injections at each temperature, to (i) confirm that the observed changes in retention times over the ~5-h duration of the test were due to changes in temperature, not solvent composition, and (ii) to determine whether 1 h equilibration time at each temperature was sufficient for reproducibility. Neither the number of samples injected nor changes in column temperatures affected column stability/sample retention times or sample-to-sample reproducibility.

Solvent Volume/Temperature Study. Eluting solvent volumes (1 mL/min, 1.0% ACN in hexane) were collected in cooled, graduated cylinders over a timed 2-h period at 10°C and then at 40°C, and compared to determine whether solvent flow rates were influenced by solvent temperature. Total elution volumes after 2 h (10 vs. 40°C) were identical at 119.0 mL. Similar results were obtained when the study was repeated with 0.3 and 1.4% ACN in hexane. [Pump head pressures (dual-column system) were found to increase slightly at lower temperatures (800 psi at 40°C vs. 900 psi at 10°C)].

Solubility of ACN in Hexane Study. Solvent samples (10 mL) of 1.0 and 1.5% ACN in hexane (vol/vol) in scintillation vials) were stored at 20, 10, 0, and –15°C overnight, then examined for phase separation(s). Both samples remained the homogeneous at 10°C, but phase separation was noted in the 1.5% ACN in hexane sample at 0°C, and for 1.0% at –15°C.

Temperature Effects: TAG and FAME Separations

TAG I, II, and III. The column temperature was set at 10°C and, after 1 h of stabilization, 2–5 μL of TAG I was injected. After 30–40 min at 10°C [to allow enough time for elution of the TAG isomer(s)], the column temperature was changed to 20°C and again allowed to stabilize (1 h total) before re-injection of TAG I. Column temperatures can be set manually or programmed; the total time required for TAG retention studies at the four temperatures was 6–7 h. The effects of column temperatures on the elution of the SLnS/SSLn TAG isomer pair are illustrated in Figure 2 and tabulated in Table 1. (Chromatograms of the SLnS/SSLn elution patterns and resolutions at 10 and 30°C are illustrated in Fig. 1.) The increase in column temperature from 10 to 40°C significantly decreased the elution rate (increased retention time) for the SLnS/SSLn isomer pair by ~50% (from 12.9 min at 10°C to 18.6 min at 40°C). A study using the LOL/LLO TAG isomer pair (all other conditions the same) resulted in an even more dramatic increase in elution times, from 16.8 min at 10°C to 34.8 min at 40°C.

TABLE 1
Ag-HPLC Column Temperature (CT) vs. TAG Elution Times[a,b]

TAG Double bonds (n)		PPP (0)	POP/PPO (1)	SLnS/SSLn (3)	LLO/LOL (5)
	CT (°C)	Elution times (min)			
Peak 1	10	9.9	10.0	12.2	16.8
Peak 2			10.3	12.9	16.9
Peak 1	20	10.1	10.5	14.0	
Peak 2			10.8	14.9	
Peak 1	30	10.2	11.1	15.7	
Peak 2			11.5	16.8	
Peak 1	40	10.2	11.6	17.3	34.1
Peak 2			11.8	18.6	34.8
R (%)[c]		3	16	42	103
% Value/Db bonds (n)		0	16	14	21

[a]System: Dual-column Ag-HPLC; 1.0 mL/min 1.0% acetonitrile (ACN) in hexane. Sample: TAG Standard, 10 mg/mL, 2 μL injection.
[b]Abbreviations: HPLC, high-performance liquid chromatography; TAG, triacylglycerol; P, palmitic acid; O, oleic acid; S, stearic acid; Ln, linolenic acid; L, linoleic acid.
[c][(Peak 1/40°C - Peak 1/10°C)/Peak 1/10°C] × 100%. Retention time increase of Peak 1 between 10 and 40°C.

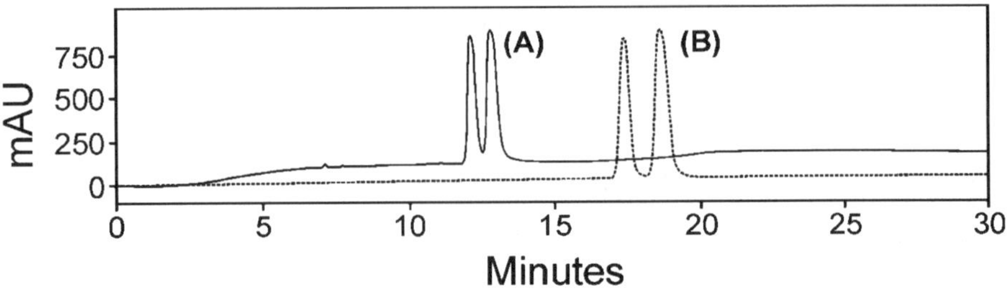

Fig. 1. Chromatograms of the SLnS/SSLn elution patterns and resolutions at 10 and 30°C.

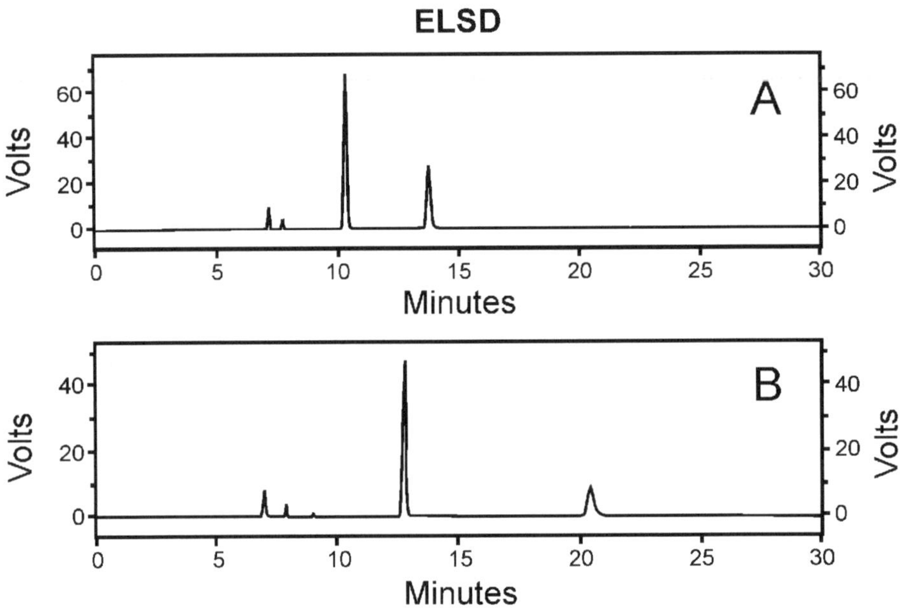

Fig. 2. The effects of column temperatures on the elution of the SLnS/SSLn TAG isomer pair.

TAG II: EE as Co-Solvent. The purpose was to determine whether the presence of EE as co-solvent had any effect on TAG retention times. Conditions were identical to those used in the above study (Table 1), but a solvent system of 1.0% ACN and 0.5% EE in hexane (by vol) was used. The addition of 0.5% EE as co-solvent to the isocratic 1.0% ACN in the hexane solvent system at the four temperature levels (Table 1) resulted in only a slight (<1%) increase (at 10 and 40°C) or decrease (at 20 and 30°C) in retention time(s).

Homogeneous TAG mixture. Conditions were similar to those in the first of these studies, but with equilibration times of 30 min, solvent composition of 1.5% ACN in hexane, and 60-min elution time for all TAG samples but AcAcAc (2.0 mL/min; 30 min equilibration; 90-min run time) (see Table 2). Stabilization times of 0.5–1.0 h were found sufficient (retention times ± 0.6%) and may perhaps be reduced even further by use of temperature-controlled solvent reservoirs, inlet lines, and/or injectors. At temperatures >40°C, mobile phase preheating may be required to prevent a temperature gradient from forming along the column (40).

FAME I. Conditions were similar to those in the first of these studies. Unsaturated FAME retention times increased as column temperature was increased from 10 to 40°C (Figure 2; Table 3); the magnitude of the effect was related to the total number of double bonds in the FAME (a 2.6-fold increase for 18:1 vs. an 8-fold increase for 22:6).

FAME II. Conditions were similar to those in the first of these studies. Improved resolution of the 9*t*,12*t*- vs. 9*t*,12*c*-18:2 FAME was noted as column temperature

TABLE 2
Ag-HPLC Column Temperature (CT) vs. Homogeneous TAG (Including Triacetin) Elution Times[a,b]

TAG Double bonds (n)		SSS (0)	OOO (3)	LnLnLn (9)	AcAcAc (0)
Run #[c]	CT (°C)	Elution times (min)			
Run 1	10	6.7	7.5	12.6	66.5
Run 2		6.6	7.3	11.8	
Run 1	20	7.0	8.1	15.9	35.0
Run 2		7.2	8.4	16.3	70.6[d]
Run 1	30	7.5	9.4	21.6	26.0
Run 2		7.5	9.4	21.6	
Run 1	40	7.7	10.3	28.2	20.2
Run 2		7.7	10.4	28.4	
R40°C–R10°C (%)[e]		15	37	124	–70
% Value/Db bonds (n)		—	12	14	—

[a]System: Dual-column Ag-HPLC; 1.0 mL/min 1.5% acetonitrile (ACN) in hexane, evaporative light scattering detector (ELSD) detection. Sample: 18:0, 18:1, and 18:3 TAG Standard, 10 mg/mL, 2 μL injection.
[b]Abbreviations: as in Table 1. Ac, acetate.
[c]Run #1 vs Run #2 retention times result from stabilization times of 30 and 60 min, respectively.
[d]2.0 mL/min. Solvent composition and other conditions were not changed.
[e][(Run 1 at 40°C – Run 1 at 10°C)/Run 1 at 10°C] × 100%. Retention time increase of Peak 1 between 10 and 40°C.

TABLE 3
Ag-HPLC Column Temperature (CT) vs. Fatty Acid Methyl Esters (FAME) Elution Times[a,b]

	CT (°C)	Elution times (min)				
		18:0	9*c*-18:1	9*c*,12*c*-18:2	5,8,11,14-20:4	4,7,10,13,16,19-22:6
1	10	7.2	7.7	8.4	10.3	13.8
2	20	7.1	7.8	8.6	11.1	15.8
3	30	7.1	7.9	8.8	11.9	18.1
4	40	7.0	7.9	9.0	12.8	20.4
R40°C–R10°C (%)[c]		—	2.6	7.1	24.3	47.8
% Value/Db bonds (n)			2.6	3.5	6.1	8

[a]System: Dual-column Ag-HPLC; 1.0 mL/min 1.0% acetonitrile (ACN) in hexane, ultraviolet at 206 nm and evaporative light scattering detector (ELSD) detection. Sample: 18:1, 18:2 (9*c*,12*c*), 20:4, 22:6 FAME Standard, 5.0 mg/mL, 2 μL injection.
[b]Abbreviations: as in Table 1.
[c][(Peak at 40°C – Peak at 10°C)/Peak at 10°C] × 100%. Retention time increase of Peak 1 between 10 and 40°C.

increased, a result similar to the improved resolution between SOS/SSO TAG isomers (silver nitrate impregnated silica gel and toluene as solvent) noted by Hammond and co-workers (37) at lower column temperature(s). Retention times of the *t,t*-isomer (10 vs. 40°C) changed little, whereas that of the *t,c*-isomer increased (see Table 4).

Temperature Effects on TAG Elution Order. The system conditions were as follows: Dual-column Ag-HPLC; solvent composition/flow, 1.0 mL/min 0.7%

TABLE 4
Ag-HPLC Column Temperature (CT) vs. Fatty Acid Methyl Esters (FAME) Elution Times[a,b]

	CT (°C)	Elution times (min) 9*t*,12*t*-18:2	9*t*,12*c*-18:2	T2 – T1	Resolution[c]	Half-height resolution[d]
1	10	7.73	7.95	0.22	1.10	1.10
2	20	7.76	8.09	0.33	1.48	1.62
3	30	7.79	8.23	0.44	1.91	
4	40	7.76	8.29	0.55	2.14	

[a]System: Dual-column Ag-HPLC; 1.0 mL/min 1.0% acetonitrile (ACN) in hexane. Sample: FAME, 18:2 (9*t*,12*t*- and 9*c*,12*t*-18:2) Standard, 0.0025 mg/mL, 5 μL injection.
[b]Abbreviations: as in Table 1.
[c]R = [(T2 – T1)(*X* cm/*Y* min)/[(W2 + W1)/2] where T2 is retention time (min) of Peak 2 and (*X* cm/*Y* min) is the ratio to convert min to cm; W2 is width (cm) of Peak 2 at baseline by extending sides at peak half-height.
[d]Resolution at peak half-height: 1.18 [(T2 – T1)(*X* cm/*Y* min)/[(W1/2 pk2 + W1/2 pk1)].

ACN in hexane; injection volume, 1.0 μL PPL/OOP TAG standard (25/75; 10 mg/mL isooctane) (Fig. 3) ELSD. As the column temperature increased from 10 to 40°C, the retention of both TAG increased, but for PPL, the TAG containing the FA with the largest number of double bonds in the 1-/3-position on the TAG glycerol backbone retention, retention increased more rapidly (eluted more slowly) than OOP. The result was a reversal of elution orders with PPL, well-separated and eluting before OOP at 10°C, merging with the OOP peak at 30°C, then becoming a well-defined shoulder after OOP at 40°C. [Peak compositions were determined by replacement of the ELSD by the UV detector (206 nm), collection of the fractions and, after removal of the solvents, analysis by GC (see Methods: Gas Chromatography)]. But when OOP/PPL were analyzed by Ag-TLC (Ref. 41; silica gel; AgNO3 impregnated; chloroform/methanol solvent), the TAG with double bonds concentrated in one acyl group (PPL; vs. in 2 groups for OOP) was retained more firmly (OOP eluted before PPL). In Ag-HPLC, competing factors (41) would seem to affect OOP and PPL retention: the total number of olefinic bonds, the FA in the 1- or 3-position of the TAG (considered to have the greatest effect on retention) with the exception when a 3-centered bond can be formed by two neighboring fatty acids. This may be noted in Figure 3, where the two neighboring oleate FA molecules of OOP resulted in increased bond stability/TAG retention, and OOP eluted after PPL (10°C). The retention orders of OOP and PPL were reversed as column temperatures increased, a situation somewhat analogous to the temperature rate-dependent reversal of Me 9*c*,12*c*,15*c*-18:3 and Me 11*c*-20:1 elution orders in GC when cyanoalkylpolysiloxane capillary columns are utilized (42).

Discussion

In silver-ion chromatography, the most highly unsaturated FAME/TAG are usually the most strongly retained [due to the interaction of the silver ions (bonded to the silica or similar substrate of the column) and the olefinic π-electrons of the sample molecule(s)] and exhibit the greatest temperature effect(s); the stability of the silver ion-double bond complex is influenced by the spatial arrangement of the overlapping orbitals, the basicity of electronic pairs in the olefinic molecule, and by solvent effects (43). This effect (10 vs. 40°C) is demonstrated in both TAG [see %R, POP/PPO vs. LLO/LOL (Table 1) and OOO vs. LnLnLn (Table 2)] and FAME [18:1 vs. 18:2 vs. 20:4 vs. 22:6 (Table 3)], and is more noticeable with *cis* than with *trans* double bonds (see 18:2, Tables 3 and 4). The lack of this effect is demonstrated for saturated TAG (SSS, Table 2), whereas the opposite effect ("normal phase," more rapid sample elution at higher temperatures) is noted to a very slight degree in saturated FAME (18:0, Table 3) and significantly in the TAG AcAcAc (Table 2). A similar result was noted by Fevrier et al. (6) during the separation of DAG isomers (solvent program: hexane/2-propanol/ethyl acetate and hexane/2-propanol/ACN), in which "...normal-phase effects tended [to] override the silver ions contributions resulting in a shorter retention time of triolein com-

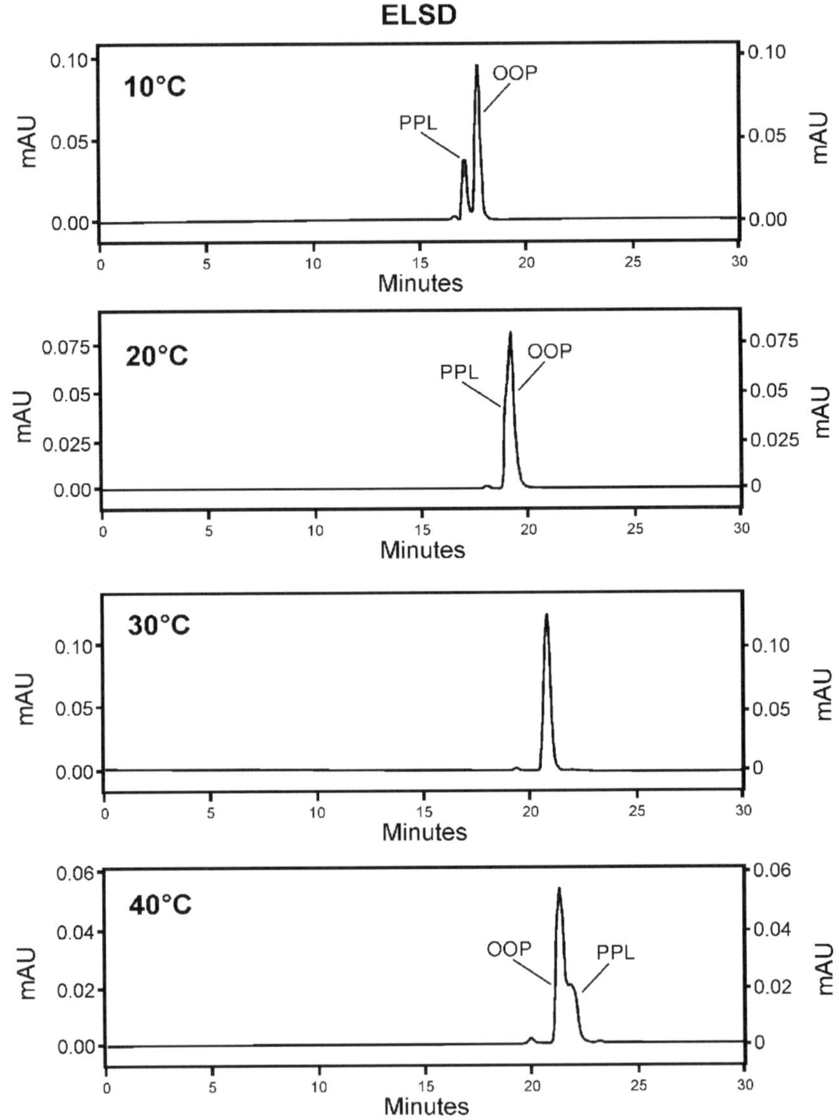

Fig. 3. two neighboring oleate FA molecules of OOP resulted in increased bond stability/TAG retention, and OOP eluted after PPL (10°C).

pared with CLC and CCL" (where C = caprylic acid; 8:0). Again, this effect (the more rapid sample elution at lower temperatures) was not observed with the chlorinated hydrocarbon solvent-based systems utilized by Christie (private communication, 44), and may be limited (38) to Ag-HPLC with ACN/hexane solvent systems.

One contributing factor for this effect may be a temperature-induced change in the partition coefficient between ACN and hexane. In GC, the more-rapid exchange between the sample molecule(s) and the liquid stationary phase in the GC column was used to explain (3) in part the more rapid elution of samples with increasing column temperatures. In Ag-HPLC, solvent modifiers such as isopropanol and acetone may reduce the interaction of the sample with silanol and other polar groups of the substrate (45). ACN complexes strongly with silver ions and is often used as a phase modifier in Ag-HPLC (3) and, as demonstrated, is required in the ACN/hexane solvent systems for elution of unsaturated substrates. Unsaturated solvents, such as benzene, toluene, and ACN, are also assumed to complex with the silver ions, thus reducing their interaction with the double bond(s)/unpaired electrons of the eluting sample(s) (46,47). ACN, at lower temperatures, is less soluble in the hexane and therefore available for this competition [a situation analogous to a higher percentage of ACN in the eluting solvent].

Substrate retention on Ag-HPLC systems (ACN in hexane as solvent) may therefore be influenced by a variety of temperature-dependent factors. These factors include, but may not be limited to (i) the equilibrium between Ag ion(s) and sample FA double-bonds/unpaired carbonyl oxygen electrons, (ii) the solubility of ACN in hexane, (iii) changes in sample flexibility/structure(s) at higher temperatures (steric argument: sample), (iv) changes in flexibility of phenyl/butylsulfonic acid or residual silanol groups on the silica backbone of the Ag-HPLC column, (v) solubility of TAG or FAME sample(s) in ACN/hexane solvent, (vi) the solvent viscosity (27), and/or (vii) the dielectric constant of the solvent(s). Cross et al. (48) noted that only 50% of the available silanol groups of the silica particles in ChromSphere Lipids columns are derivatized with alkyl benzenesulfonic acid moieties. [No evidence of ACN insolubility in the hexane was noted at the ACN concentrations (1.0 and 1.5% ACN) at 10–40°C, although phase separation was noted for 1.5% ACN at 0°C and both 1.0 and 1.5% ACN in hexane at −15°C.] The influence of temperature on relative retention orders is demonstrated by the reversal of the elution order of PPL and OOP [see the Temperature Effects on TAG Elution Orders section, and Figure 3] and illustrates one of the problems encountered when attempting to predict relative elution orders on Ag-HPLC. This is similar to the reversal in elution order observed for 9*c*-20:1 and 9*c*,12*c*,15*c*-18:3 FAME on capillary GC [cyanoalkyl polysiloxane stationary phases (42)] and related to the temperature program rate.

The isocratic ACN in hexane solvent system and the commercially available ChromSpher Lipids column(s) were shown to be a useful and versatile combination for the analysis of a variety of FAME and TAG isomers, a methodology potentially further enhanced by the utilization of other hexane-based solvent systems and, unlike the silver nitrate-impregnated silica system and potential loss of silver nitrate, capable of reproducible retention times even after injection of 70+ samples over a 4-wk period. These results are in contrast to the "...loss of column performance, which is probably caused by the acetonitrile in the mobile phase...

affected the resolution..." of the reverse isomers of 1,2-diacyl-rac-glycerols (as 3,5-dinitrophenylurethanes) noted by Itabashi and co-workers (27) on an ODS RP-HPLC system. The authors noted improved resolution/longer retention times at lower temperatures and that the elution orders of the 1,2-diacylglycerol isomers were reversed as temperatures decreased, with the 18:3/16:0 (1-/2-position) DAG isomer eluting at 182/188 min (10°C) vs. the 16:0/18:3 isomer at 113/116 min (18°C).

Within the last 1–2 years, a notable amount of research has been published relating HPLC column temperatures with column capacity and sample retention/resolution. Substrates include silica (34), alumina (36), reverse-phase (40,49–53), chiral/enantiomeric (54,55), and diol (56). In general, the silica component may contribute to the retention/elution effects (stronger retention/slower sample elution at higher temperatures) noted for some silica- and alumina (binary solvent system of hexane/THF and isopropanol)-based HPLC systems (34 and 36, respectively); the CN and diol columns were unaffected, whereas some, but not all (51) of the C18 RP (40) and chiral (54) columns yielded the opposite (more rapid sample elution at higher temperatures) effect. Recent studies include ultrahigh-speed chromatography at temperatures >150°C (52), but none of these studies related these variables to the separation of TAG, FAME, (or any other FA esters), or other lipids. For Ag-HPLC, predicting elution orders is not limited to the interaction of the Ag+ ions of the column with unsaturated samples, but also the configuration/location of (unpaired or double/triple bond π-) electrons and their ability to form 3-D complexes with the Ag+ ions, and the effect of unreacted silanol groups, all related to the eluting solvent(s) and column temperature.

Future studies will perhaps address the low solubility of ACN in the ACN/hexane solvent system by using co-solvents such as EE or alcohol/ACN/chlorinated solvent–based systems, which could be utilized at lower temperatures (0 to −40°C) to overcome the "eight double bonds per TAG molecule" (Ref. 27; TAG isomer separation limited to LnLnL vs. LnLLn), the optimization of temperature and eluent composition (51,57), or the application of temperature programming to yield improved sample elution times/resolution(s). Given the more rapid elution of FAME and TAG at below room temperatures (10 vs. 20°C), the application of Ag-HPLC with solvents of 0.03–0.3% ACN in hexane and below zero (−5 to −40°C) temperatures provides another useful area for future studies. [A possible limiting factor: pump head pressures (0.5% ACN in hexane) increase from ~800 psi to >1200 psi at temperatures below −20°C, with the solubility of the sample in the eluting solvent a possible contributing factor.]

References

1. Christie, W.W., Silver-Ion High-Performance Liquid Chromatography, in *New Trends in Lipid and Lipoprotein Analyses*, AOCS Press, Champaign, IL, 1995, p. 59.
2. Dobson, G.W., W.W. Christie, and B. Nikolova-Damyanova, Silver Ion Chromatography of Lipids and Fatty Acids, *J. Chromatogr. B 671*:197–222 (1995).

3. Nikolova-Damyanova, B., Silver Ion Chromatography and Lipids, in *Advances in Lipid Methodology—One*, edited by W.W. Christie, The Oily Press, Ayr, Scotland, 1992, p. 181.
4. Nikolova-Damyanova, B., Lipid Analysis by Silver Ion Chromatography, in *Advances in Lipid Methodology—Five*, edited by R.O. Adlof , P.J. Barnes & Associates, Bridgewater, UK, 2003, p. 43.
5. Ratnayake, W.M.N., Overview of Methods for the Determination of *trans* Fatty Acids by Gas Chromatography, Silver-Ion Thin-Layer Chromatography, Silver-Ion Liquid Chromatography, and Gas Chromatography/Mass Spectrometry, *J. Assoc. Off. Anal. Chem Int. 87*:523–539 (2004).
6. Nikolova-Damyanova, B., S. Momtchilova, S. Bakalova, H. Zuilhof, W.W. Christie, and J. Kaneti, Computational Probes into the Conceptual Basis of Silver Ion Chromatography: I. Silver(I) Ion Complexes of Unsaturated Fatty Acids and Esters, *J. Mol. Struct. (Theochem)* 589–590, 239–249 (2002).
7. Cross, R.F., and H. J. Zackari, AG^+-HPLC of Conjugated Linoleic Acids on a Silica-Based Stationary Phase. III; Model Compounds, *Sep. Sci. 25*:897–903 (2002).
8. Cross, R.T., and H. J. Zackari, AG^+-HPLC of Conjugated Linoleic Acids on a Silica-Based Stationary Phase. Part IV: A Reference Stationary Phase and Retention Mechanisms, *Sep. Sci. 26*:480–488 (2003).
9. Juaneda, P., J.L. Sébédio, and W.W. Christie, Complete Separation of the Geometrical Isomers of Linolenic Acid by High Performance Liquid Chromatography with a Silver Ion Column, *J. High Res. Chromatogr. Commun. 17*:321–324 (1994).
10. Adlof, R.O., Separation of *cis* and *trans* Unsaturated Fatty Acid Methyl Esters by Silver Ion High-Performance Liquid Chromatography, *J. Chromatogr. A 659*:95–99 (1994).
11. Adlof, R.O., L.C. Copes, and E.A. Emken, Analysis of the Monoenoic Fatty Acid Distribution in Hydrogenated Vegetable Oils by Silver-Ion High Performance Liquid Chromatography, *J. Am. Oil Chem. Soc. 72*:571–574 (1995).
12. Nikolova-Damyanova, B., B.G. Herslof, and W.W.J. Christie, Silver Ion High-Performance Liquid Chromatography of Derivatives of Isometric Fatty Acids, *J. Chromatogr. 609*:133–140 (1992).
13. Nikolova-Damyanova, B., and S.J. Momchilova, Silver Ion HPLC for the Analysis of Positionally Isomeric Fatty Acids, *J. Liq. Chromatogr. Relat. Technol. 25*:1947–1965 (2002).
14. Momchilova, S., and B. Nikolova-Damyanova, Silver Ion HPLC of p-Methoxyphenacyl Derivatives of Unsaturated Fatty Acids. III. Mobile Phase Effects on *trans* 6-, 9-, and 11-18:1, *J. Liq. Chromatogr. Relat. Technol. 25*: 615–625 (2002).
15. Sandra, P., A. Medvedovici, Y. Zhao, and F. David, Characterization of Triglycerides in Vegetable Oils by Silver-Ion Packed-Column Supercritical Fluid Chromatography Couple to Mass Spectroscopy with Atmospheric Pressure Chemical Ionization and Coordination Ion Spray, *J. Chromatogr. A 974*:231–234 (2002).
16. Macher, M-B., and A. Holmquist, Triacylglycerol Analysis of Partially Hydrogenated Vegetable Oils by Silver Ion HPLC, *J. Sep. Sci. 24*:179–185 (2001).
17. Fevrier, P., A. Binet, L. Dufosse, R. Gree, and F.J. Yvergnaux, Separation of Glycericed Positional Isomers by Silver Ion Chromatography, *J. Chromatogr. A 923*:53–57 (2001).
18. Cross, R.F., E. Ostrowska, M. Muralitharan, and F.R. Dunshea, Mixed Mode Retention and the Use of Competing Acid for the Ag^+-HPLC Analysis of Underivatized Conjugated Linoleic Acids, *J. High Res. Chromatogr. 23*:317–323 (2000).

19. Sehat, N., R. Rickert, M. Mossoba, J.K.G. Kramer, M.P. Yurawecz, J.A.G. Roach, R.O. Adlof, K. Morehouse, J. Fritsche, K.D. Eulitz, H. Steinhart, and Y. Yuoh, Improved Separation of Conjugated Fatty Acid Methyl Esters by Silver Ion-High-Performance Liquid Chromatography, *Lipids 34*:407–413 (1999).
20. Delmonte, P., M.P. Yurawecz, M.M. Mossoba, C. Cruz-Hernandez, and J.K.G. Kramer, Improved Identification of Conjugated Linoleic Acid Isomers Using Silver-Ion HPLC Separations, *J. Assoc. Off. Anal. Chem. Int. 87*:563–568 (2004).
21. Kramer, J.K.G., C. Cruz-Hernandez, Z.Y. Deng, J.Q. Zhou, G. Jahreis, and M.E.R. Dugan, Analysis of Conjugated Linoleic Acid and *trans* 18:1 Isomers in Synthetic and Animal Products, *Am. J. Clin. Nutr. 79*:1137–1145 (2004).
22. Adlof, R.O., Separation of Conjugated Linoleic Acid-Containing Mono-, Di-, and Triacylglycerols by Silver-Ion High-Performance Liquid Chromatography, *Eur. J. Lipid Sci. Technol. 103*:594–632 (2001).
23. Cruz-Hernandez, C., Z. Dena, J. Zhou, A.R. Hill, M.P. Yurawecz, P. Delmonte, M.M. Mossoba, M.E.R. Dugan, and J.K.G. Kramer, Methods for Analysis of Conjugated Linoleic Acids and *trans*-18:1 Isomers in Dairy Fats by Using a Combination of Gas Chromatography, Silver-Ion Thin-Layer Chromatography/Gas Chromatography, and Silver-Ion Liquid Chromatography, *J. Assoc. Off. Anal. Chem. Int. 87*:545–560 (2004).
24. Adlof, R.O., A. Menzel, and V. Dorovska-Taran, Analysis of CLA-enriched Triacylglycerol Mixtures by Silver-Ion HPLC, *J. Chromatogr. A 953*:293–297 (2002).
25. Adlof, R.O., Silver Ion HPLC Separation of Fatty Acid Methyl Esters with Deuterium Atoms on the Double Bonds, *J. Chromatogr. A 685*:178–181 (1994).
26. Adlof, R.O., Analysis of Fatty Acid Mono- and Diacylglycerol Positional Isomers by Silver Ion High-Performance Liquid Chromatography, *J. Chromatogr. A 741*:135–138 (1996).
27. Itabashi, Y., J. Myher, and A. Kuksis, High-Performance Liquid Chromatographic Resolution of Reverse Isomers of 1,2-diacyl-rac-glycerols as 3,5-dinitrophenylurethanes, *J. Chromatogr. A 893*:261–279 (2000).
28. Adlof, R.O., Analysis of Triacylglycerol Positional Isomers by Silver Ion High Performance Liquid Chromatography, *J. High Res. Chromatogr. 18*:105–107 (1995).
29. Blomberg, L.G., M. Demirbuker, and P.E Andersson, Argentation Supercritical Fluid Chromatography for Quantitative Analysis of Triacylglycerols, *J. Am. Oil Chem. Soc. 70*:939–946 (1993).
30. Adlof, R.O., and G.R. List, Synthesis and Analysis of Symmetrical and Non-Symmetrical Disaturated/Monounsaturated Triacylglycerols for Use in the Study of Food Oil Systems, *J. Agric. Chem. 51*:2096–2099 (2003).
31. Adlof, R.O., Normal-Phase Separation Effects with Lipids on a Silver Ion High-Performance Liquid Chromatography Column, *J. Chromatogr. A 764*:337–340 (1997).
32. Joh, Y.-G., and S.-J. Kim, Analysis of Molecular Species of Triacylglycerols from Vegetable Oils Containing Fatty Acids with Non-Methylene-Interrupted Double Bonds, by HPLC in the Silver-Ion Mode, *J. Jpn. Oil Chem. Soc. 47*:927–936 (1998).
33. Chen, S.-H., K.-C. Chen, and H.-M. Lien, Determination of Fatty Acids in Vegetable Oil by Reversed-Phase Liquid Chromatography with Fluorescence Detection, *J. Chromatogr. A 849*:357–369 (1999).
34. Murphy, E.J., T.A. Rosenberger, L.A. Horrocks, Separation of Neutral Lipids by High-Performance Liquid Chromatography: Quantification by Ultraviolet, Light Scattering and Fluorescence Detection, *J. Chromatogr. B 685*:9–14 (1996).

35. Hirano, T., and K. Takahashi, Improvement in Resolution of Low Melting Point Triglyceride Molecular Species on Chilled Column Isocratic Liquid Chromatography, *J. Chromatogr. Sci. 34*:341–348 (1996).
36. Moreau, R., K. Kohout, and V. Singh, Temperature-Enhanced Alumina HPLC Methods for the Analysis of Wax Esters, Sterol Esters, and Methyl Esters, *Lipids 37*:1201–1204 (2002).
37. Smith, E.C., A.D. Jones, and E.W. Hammond, Investigation of the Use of Argentation High-Performance Liquid Chromatography for the Analysis of Triglycerides, *J. Chromatogr. 188*:205–212 (1980).
38. Adlof, R.O., and G. List, Analysis of Triglyceride Isomers by Silver-Ion High-Performance Liquid Chromatography. Effect of Column Temperature on Retention Times, *J. Chromatogr. A 1052*:109–113 (2004).
39. Kodali, D., R.D. Atkinson, T.G. Redgrave, and S.M. Small, Structure and Polymorphism of 18-Carbon Fatty Acyl Triacylglycerols: Effect of Unsaturation and Substitution in the 2-Position, *J. Lipid Res. 28*:403–413 (1987).
40. Wolcott, R.G., J.W. Dolan, L.R. Snyder, S.R. Bakalyar, M.A. Arnold, and J.A. Nichols, Control of Column Temperature in Reversed-Phase Liquid Chromatography, *J. Chromatogr. A 869*:211–230 (2000).
41. Christie, W.W., The Analysis of Fatty Acids, in *Lipid Analysis*, Pergamon Press Ltd., NY, 1973, p. 85.
42. Wolff, R.L., Analysis of Alpha-Linolenic Acid Geometrical Isomers in Deoderized Oils by Capillary Gas-Liquid Chromatography on Cyanoalkyl Polysiloxane Stationary Phases: A Note of Caution, *J. Am. Oil Chem. Soc. 71*:907–909 (1994).
43. Nonadek, L, Complexation in Chromatography, in *Complexation Chromatography*, edited by D. Cagniant, Marcel Dekker, New York, 1991, pp. 1–31.
44. Christie, W.W., A Stable Silver-Loaded Column for the Separation of Lipids by High-Performance Liquid Chromatography, *J. High Res. Chromatogr. Chromatogr. Commun. 10*:148–149 (1987).
45. Powell, W.S., Separation of Icosenoic Acids, Monohydroxyiconsenoic Acids, and Prostaglandins by High-Pressure Liquid Chromatography on a Silver Ion-Loaded Cation-Exchange Column, *Anal. Biochem. 115*:267–277 (1981).
46. Heath, R.R., J.H.Tumlinson, R.E. Doolittle, and A.T. Proveaux, Silver Nitrate High Pressure Liquid Chromatography of Geometrical Isomers, *J. Chromatogr. Sci. 13*:380 (1975).
47. Powell, W.S., Argentation-High-Pressure Liquid Chromatography of Prostaglandins and Monohydroxyeicosenoic Acids, *Methods Enzymol. 86*:530–543 (1982).
48. Cross, R.F., E. Ostrowka, M. Muralitharan, and F.R. Dunshea, Mixed Mode Retention and the Use of Competing Acid for the Ag^+-HPLC Analysis of Underivatized Conjugated Linoleic Acids, *J. High Resolut. Chromatogr.*:317–323 (2000).
49. Bidlingmeyer, B.A., and J. Henderson, Investigation of Retention on Bare Silica Using Reversed-Phase Mobile Phases at Elevated Temperatures, *J. Chromatogr. A 1060*:187–193 (2004).
50. Wolcott, R.G., J.W. Dolan, and, L.R. Snyder, Computer Simulation for the Convenient Optimization of Isocratic Reversed-Phase Liquid Chromatographic Separations by Varying Temperature and Mobile Phase Strength, *J. Chromatogr. A 869*:3–25 (2000).
51. Kiridena, W., C.F. Poole, and W.W.J. Koziol, Effect of Solvent Strength and Temperature on Retention for a Polar-Endcapped, Octadecylsiloxane-Bonded Silica Stationary Phase with Methanol-Water Mobile Phases, *Chromatogr. A 1060*:177–185 (2004).

52. Guillarme, D., S. Heinisch, and J.L. Rocca, Effect of Temperature in Reversed Phase Liquid Chromatography, *J. Chromatogr. A 1052*:39–51 (2004).
53. Buckenmaier, S.M.C., D.V. McCalley, and M.R. Euerby, Rationalisation of Unusual Changes in Efficiency and Retention with Temperature Shown for Bases in Reversed-Phase High-Performance Liquid Chromatography at Intermediate pH, *J. Chromatogr. A 1060*:117–126 (2004).
54. Cirilli, R., R. Ferretti, B. Gallinella, L. Zanitti, and F.J. La Torre, A New Application of Stopped-Flow Chiral HPLC: Inversion of Enantiomer Elution Order, *J. Chromatogr. A 1061*:27–34 (2004).
55. Iwasaki, Y., M. Yasui, T. Ishikawa, R. Irimescu, K. Hata, and T. Yamane, Optical Resolution of Asymmetric Triacylglycerols by Chiral-Phase High-Performance Liquid Chromatography, *J. Chromatogr. A 905*:111–118 (2001).
56. Silversand, C., and C. Haux, Improved High-Performance Liquid Chromatographic Method for the Separation and Quantification of Lipid Classes: Application to Fish Lipids, *J. Chromatogr. B 703*:7–14 (1997).
57. Thompson, J.D., and P.W. Carr, High-Speed Liquid Chromatography by Simultaneous Optimization of Temperature and Eluent Composition, *Anal. Chem. 74*:4150–4159 (2002).

8 Analysis of *trans*-18:1 Fatty Acids by Silver Ion HPLC

Pierluigi Delmonte[a], John K.G. Kramer[b] and Martin P. Yurawecz[a]

[a]*U.S. Food and Drug Administration, College Park, MD, USA;* [b]*Food Research Program, Agriculture and Agri-Food Canada, Guelph, ON, Canada*

Introduction

The interest in the quantitation of *trans*-fatty acids (TFA) in food increased recently after a new regulation regarding the *trans*-fat content was introduced in the North American food market. The analysis of TFA in foods and oils is generally achieved by gas chromatography (GC), using long polar capillary columns (1–3). The total TFA content of fat or oils (regardless of the composition in fatty acids) can also be measured by Fourier transform infrared (FTIR) spectroscopy (4). In partially hydrogenated vegetable oils (PHVO), TFA may account for as much as 50% of the total fat (5). *Trans* octadecenoic acids (*trans*-18:1) generally represent more than 80% of the total TFA in PHVO, and 90% of TFA in ruminant fat (6). The level of polyunsaturated TFA can be as high as 3% of total fatty acid methyl esters (FAMEs) and they are produced mainly during the deodorization step of vegetable oil refining that isomerizes linoleic (*cis* 9, *cis* 12-18:2) and α-linolenic acids (*cis* 9, *cis* 12, *cis* 15-18:3) (5). The current method for the determination of TFA by GC (3) is limited by a consistent overlap of several *cis*-18:1, *trans*-18:1 and 18:2 *cis/trans* isomers (1). Silver ion thin-layer chromatography (Ag^+-TLC) is often used for the earlier separation of *trans*-18:1 from *cis*-18:1 before GC analysis, making the analysis very labor intensive (1). Based on the common principles of argentation chromatography (7), silver ion HPLC (Ag^+-HPLC) should be equally applicable to the fractionation of geometric isomers of FAMEs before GC analysis, to obtain direct quantitative results for all *trans*-18:1 isomers in foods and oils.

Brief History

Since its discovery, during the early 1960s (8,9), silver ion chromatography has been considered a valuable technique for the separation of complex mixtures of geometric and positional isomers of FAMEs in several branches of lipid research. The basic principle is that silver ion chromatography involves transition metals such as silver that act as electron acceptors. In the complex formed, the charge is transferred from the double bonds of unsaturated organic compounds (donors) to the transition metal

(10). By applying different experimental conditions, silver ion chromatography can be used to separate different lipid classes, triglycerides, and fatty acids. The major improvements of this technique were dependent on new developments in liquid chromatography. The first applications involved adsorbing silver nitrate onto a silica base that was used in packed columns or spread on plates (11). However, because the silver ions were not bonded to the silica, they were slowly released, resulting in poor reproducibility. A major improvement in the development of silver ion chromatography was achieved in the late 1980s when Christie reported the preparation of a stable silver ion–loaded high-performance liquid chromatography (HPLC) column (12). The silica used for the HPLC column (Nucleosil 5SA; HPLC tech., Macclesfield, UK) was derivatized with phenylsulfonic acid moieties. The sulfonic acid groups were then loaded by passing an aqueous silver nitrate solution through the column. A few years later the first commercial silver ion–loaded HPLC columns (Chromspher 5 Lipids) became available from Chrompack (Middelburg, Netherlands), currently a part of Varian Inc. Adsorbing the silver ions on a strong cation exchange stationary phase, instead of on silica, increased the life expectancy of the column, and increased the stability of the column by decreasing the silver ion bleed.

The first separation of *trans*-18:1 isomers using these new HPLC columns was achieved by preparing the phenacyl derivatives of the fatty acids (13,14). Later improvements in the separation of the *trans*-18:1 positional isomers were reported by preparing the *p*-methoxy-phenacyl derivatives (15). All of these separations were achieved using chlorinated solvents as the mobile phase; this limited the use of ultraviolet (UV) detection. Fatty acid derivatives were therefore selected with strongly absorbing chromophores, such as phenacyl- and *p*-methoxy-phenacyl-esters that absorb at 245–255 nm. A few years later the separation of *trans*-18:1 positional isomers was reported using derivatives of several aromatic and nonaromatic esters (7). The analysis of fatty acids as aromatic esters required an additional derivatization step that was not only time consuming but could result in the modification of the fatty acid profile (16).

Adlof and co-workers proposed the separation of unsaturated FAME, using acetonitrile (MeCN) in hexane solution as the mobile phase (17,18). MeCN and hexane both present a UV cut-off wavelength < 200 nm, allowing for detection at 205–210 nm. The MeCN/hexane mixture was used to separate the *cis*/*trans* geometric isomers of linoleic (*cis* 9, *cis* 12-18:2), α-linolenic (*cis* 9, *cis* 12, *cis* 15-18:3), and arachidonic (*cis* 5, *cis* 8, *cis* 11, *cis* 14-20:4) acids (17). The method was also applied to the separation of the *trans*-18:1 positional isomers present in PHVO (18). All of the chromatograms were acquired using UV detection at 206 nm.

Separation of trans-18:1 as FAMEs

The analysis of *trans*-18:1 as FAME has several advantages over the aromatic derivatives, including ease of sample preparation, quantitative derivatization, and the flexibility of using the same derivatives for both GC and Ag^+-HPLC analyses.

Several official procedures for FAME preparation are currently available (3). The separation of *trans*-18:1 as FAME was achieved using a commercial Ag^+-HPLC column and 0.15% MeCN in hexane as the mobile phase (18). *Trans*-18:1 FAME can be separated under conditions similar to those used for the separation of conjugated linoleic acid (CLA) (19). Figure 1 shows the separation of a mixture containing *trans*-18:1, *cis*-18:1, and conjugated 18:2 positional isomer reference materials, as methyl esters. Although the UV wavelength of 201 nm detects all of the fatty acids containing one or more double bonds, the wavelength of 233 nm is specific for conjugated systems. Three Chromspher 5 Lipids HPLC columns (Varian Inc.) were used in series, and the mobile phase was a 0.1 % MeCN:0.5% diethyl ether in hexane solution at 1.0 mL/min (20). The use of three columns in series was considered a compromise between the quality of the FAME separation and time of analysis (20). Figure 2 shows a typical separation of total milk fat converted to FAMEs using the same chromatographic conditions.

The identification of the *trans*- and *cis*-18:1 isomers requires special attention because some isomers overlap and the trend of double bond position and relative

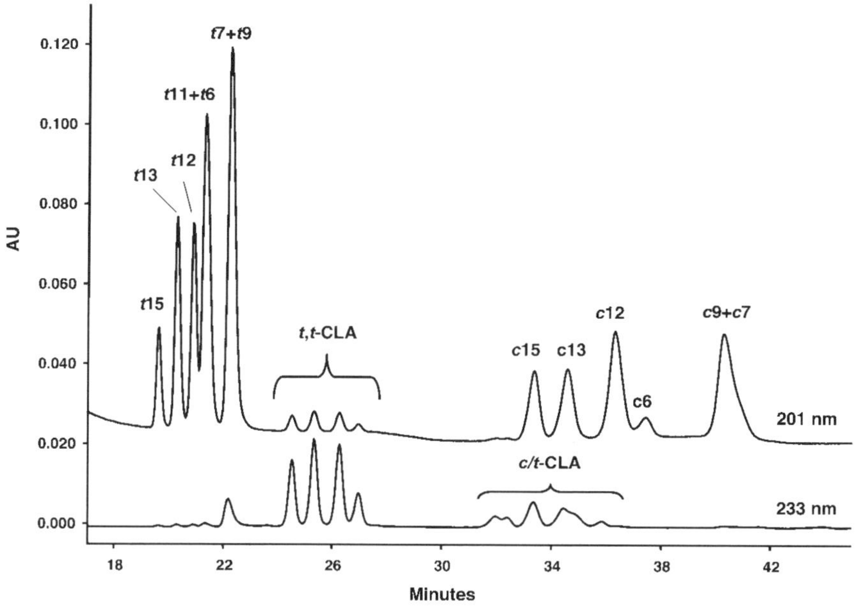

Fig. 1. Ag^+-HPLC separation of a prepared mixture of *trans*-18:1, *cis*-18:1, and CLA positional isomers using standard reference materials. Conditions: three Chromspher 5 Lipids (250 × 4.6 mm, 5 μm particle size, Varian Inc.) columns in series; mobile phase 99.4:0:5:0.1 hexane/diethyl ether/MeCN (by vol) at 1.0 mL/min; UV detection at 201 and 233 nm; operating at room temperature.

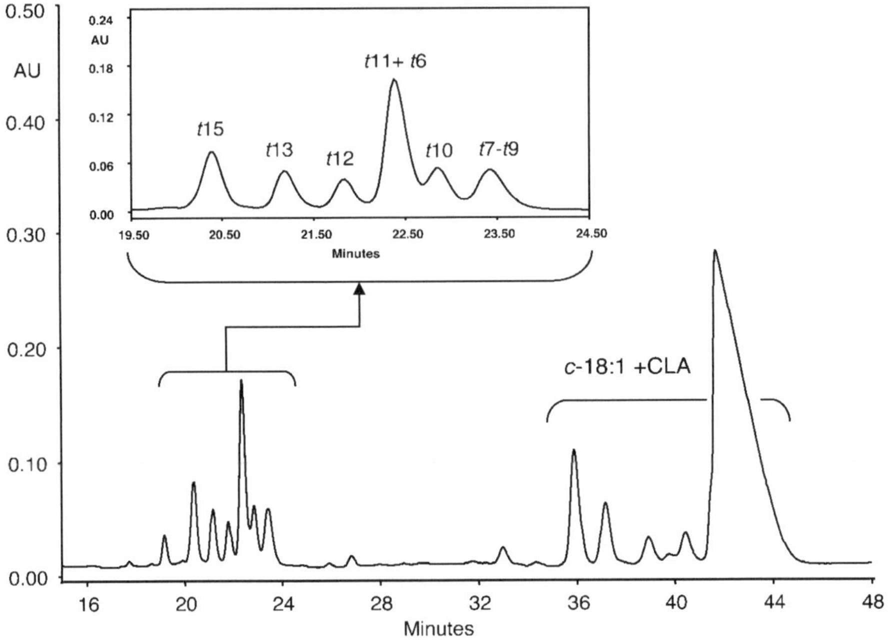

Fig. 2. Ag^+-HPLC separation of total cow's milk fat converted to methyl esters. Column and operating conditions as in Fig. 1, except that only the UV tracing at 201 nm is shown.

retention time (RRT) is not linear. In addition, several factors such as sample size injected, instability of the mobile phase, column age, and elution temperature consistently affect the reproducibility of Ag^+-HPLC separations. The age of the columns, elution temperature, and sample load can be standardized, but the effects of the mobile phase are more complex. On standing, the amount of MeCN dissolved in hexane will progressively decrease, resulting in increased retention time of the FAMEs with time. More producible retention times can be obtained by continuous agitation of the mobile phase (21,22), by adding diethyl ether to the mobile phase (1,21), and by maintaining the mobile phase in a sealed bottle (22). However, despite these efforts, the elution of the FAMEs cannot be identified solely on the basis of retention time comparisons.

Selection of an Internal Standard for Ag^+-HPLC Separations

An internal standard is needed for Ag^+-HPLC separations not only for quantitation but also for the purpose of identification. The comparison of the separations presented here was obtained by introducing an internal reference into every sample and reference material solution, and redrawing the chromatograms based on the use of a RRT scale. The *n*-butyl ester of *trans* 11–18:1 (FABE) was chosen as the internal reference due to the low cost of the fatty acid and because the retention time of this fatty acid decreases with an increase in the chain length of the esterified alcohol. Figure 3 shows the separation of the *trans* 9-18:1 as methyl-, ethyl- (FAEE), *n*-propyl- (FAPE) and *n*-butyl-ester (FABE), using 3 Ag^+-HPLC columns in series and 0.1% MeCN in hexane as the mobile phase at 1 mL/min. Chromatograms were acquired at a different elution temperature, from −10°C to +30°C, and transformed into RRT scale using *trans* 11-18:1 FABE as a reference. The RRT of the *trans* 9-18:1 fatty acid decreased progressively, whereas the chain length of the esterified alcohol increased from 1 to 4 carbons. In all of the chromatograms presented in Figure 3 the *trans* 11-18:1 FABE eluted at ~30 ± 2 min, and the variability in RT due to the mobile phase composition was higher than that due to the temperature. Although no appreciable differences were observed between the separations at 20 and 30°C, RRT values of *trans* 9-18:1 FAME and *trans* 9-18:1 FAEE increased progressively at 0 and −10°C.

Elution Order of trans*-18:1 FAME Isomers*

Figure 4 shows the elution order of the *trans*-18:1 isomers from Δ6 to Δ15, except *trans* 14–18:1, which was not available. The separations of the individual *trans*-18:1 isomers were performed at −10°C. A sample of PHVO is included in Figure 4 because all of the *trans*-18:1 isomers are present in these products. *Trans* 15-18:1 is the first eluting positional isomer, whereas *trans* 7-18:1 and *trans* 8-18:1, which are not separated, elute last in this series. The *trans*-18:1 positional isomers in samples can be more easily identified after converting the chromatograms to a RRT scale, and overlaying the chromatograms relative to the internal reference standard, FABE of *trans* 11-18:1. Figure 5 shows the separation at different elution temperatures of all the *trans*-18:1 isomers present in PHVO from Δ6 to Δ15. Again, no appreciable differences in separation were observed between the results acquired at 20 and 30°C, but *trans* 15-18:1 co-eluted with the reference. The separation between the *trans* 9- and *trans* 7-/*trans* 8-18:1 pair was increased progressively by lowering the elution temperature; at 0°C, the *trans* 15-18:1 was well resolved from the internal reference standard. Figure 6 shows the effect of the fatty acid chain length on the elution of the *trans*-9 monounsaturated FAMEs at 0°C. *Trans* 9-16:1 and *trans* 9-18:1 exhibited the same RRT, whereas *trans* 9-14:1 was only partially resolved from the other two isomers at 0°C.

Ag^+-HPLC Fractionation of trans*-18:1*

Thin-layer chromatography is often used to separate *trans*-18:1 from *cis*-18:1 FAME before GC analysis. TLC is a manual procedure and is labor intensive. Fractionation of the FAMEs by argentation chromatography can be automated by

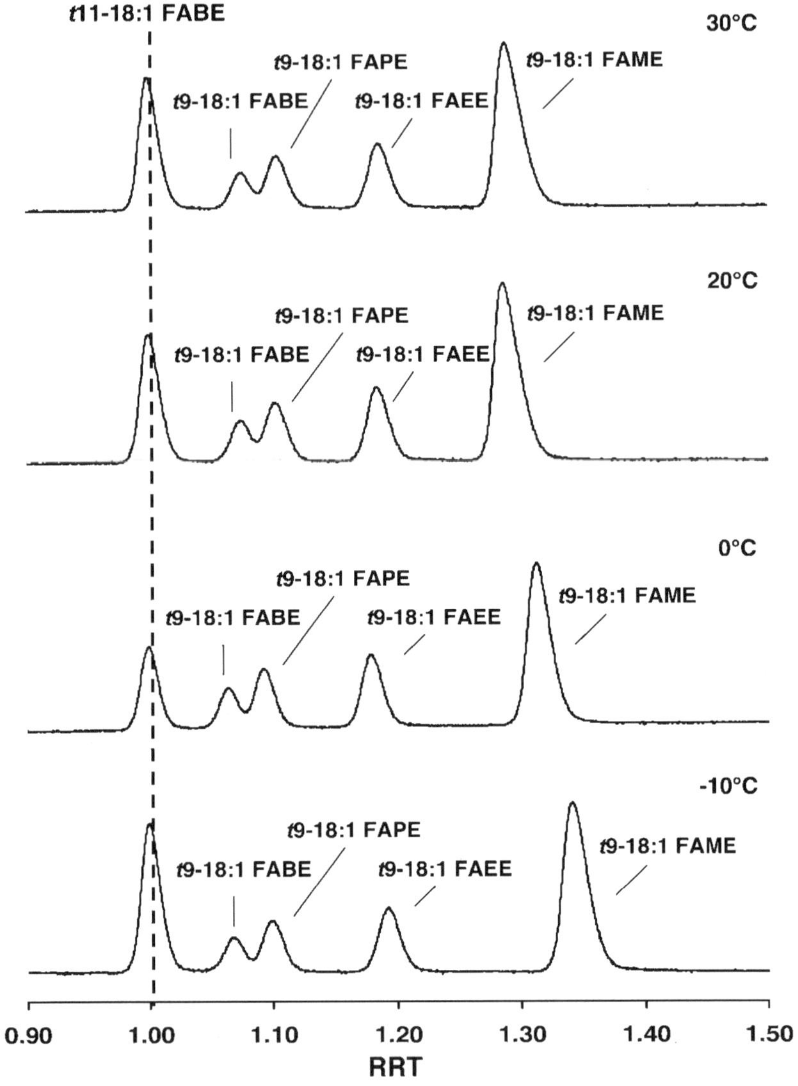

Fig. 3. Ag^+-HPLC separations of a mixture containing *trans* 9–18:1 fatty acid as methyl ester (FAME), ethyl ester (FAEE), *n*-propyl ester (FAPE), and *n*-butyl ester (FABE), plus *trans* 11-18:1 FABE as an internal reference. Chromatograms were transformed into a relative retention time (RRT) scale using *trans* 11-18:1 FABE as reference. Conditions: three Chromspher 5 Lipids (250 × 4.6 mm, 5 μm particle size, Varian Inc.) columns in series; mobile phase 99.9:0.1 hexane/MeCN (vol/vol) at 1.0 mL/min; and UV detection at 200 nm. Separations were performed at different operating temperatures from 30°C to −10°C.

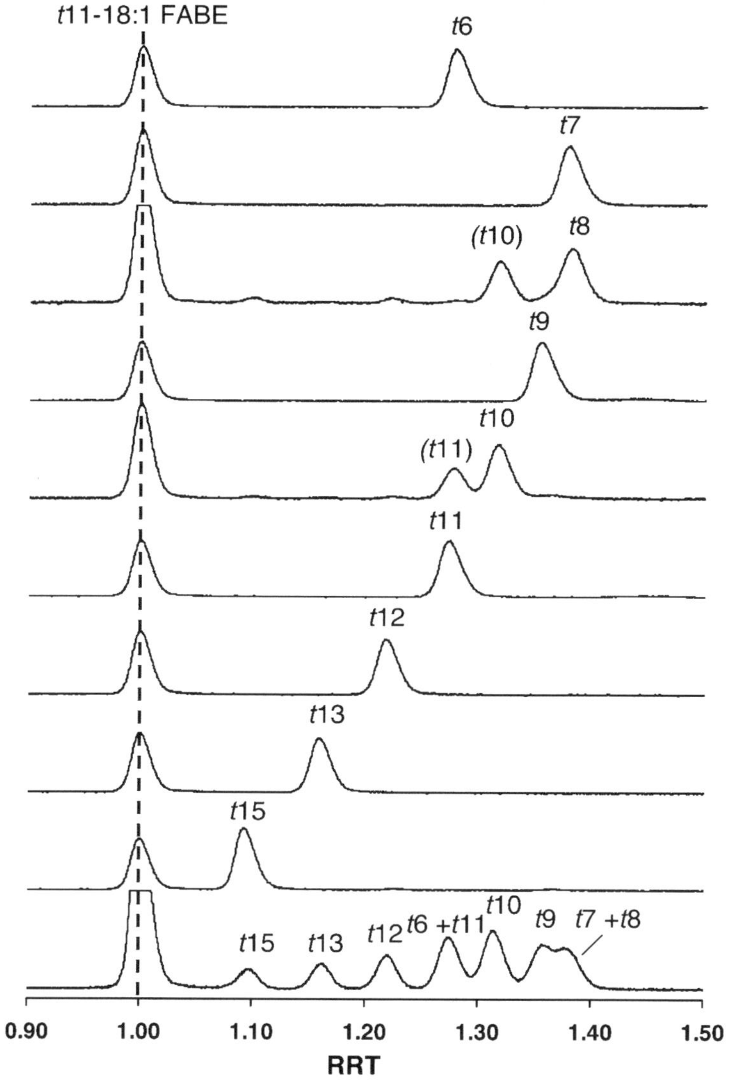

Fig. 4. Ag^+-HPLC separations of *trans*-18:1 positional isomers reference materials, each spiked with *trans* 11-18:1 *n*-butyl ester (FABE) as internal reference. A partially hydrogenated vegetable oil (PHVO) sample converted to fatty acid methyl esters that included *trans* 11-18:1 FABE as reference was added for comparison. Chromatograms were transformed into RRT scales using *trans* 11-18:1 FABE as a reference. Conditions: three Chromspher 5 Lipids (250 × 4.6 mm, 5 μm particle size, Varian Inc.) columns in series; mobile phase 99.9:0.1 hexane/MeCN (vol/vol) at 1.0 mL/min; and UV detection at 200 nm. Chromatograms were acquired at −10°C.

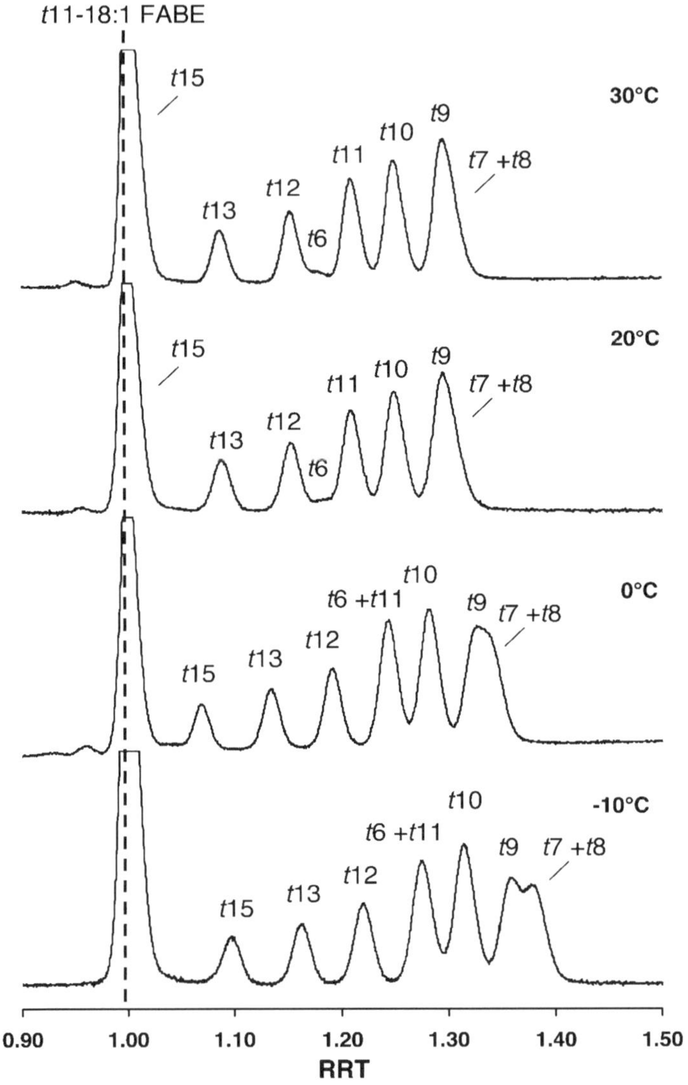

Fig. 5. Ag^+-HPLC separations of a mixture of *trans*-18:1 positional isomer reference materials, and a partially hydrogenated vegetable oil (PHVO) sample converted to fatty acid methyl esters. All samples contained *trans* 11-18:1 *n*-butyl ester (FABE) as an internal reference. Chromatograms were transformed into a relative retention time (RRT) scale using *trans* 11-18:1 FABE as a reference. Conditions: three Chromspher 5 Lipids (250 × 4.6 mm, 5 μm particle size, Varian Inc.) columns in series; mobile phase 99.9:0.1 hexane/MeCN (vol/vol) at 1.0 mL/min; and UV detection at 200 nm. Chromatograms were acquired at different operating temperatures from 30°C to −10°C.

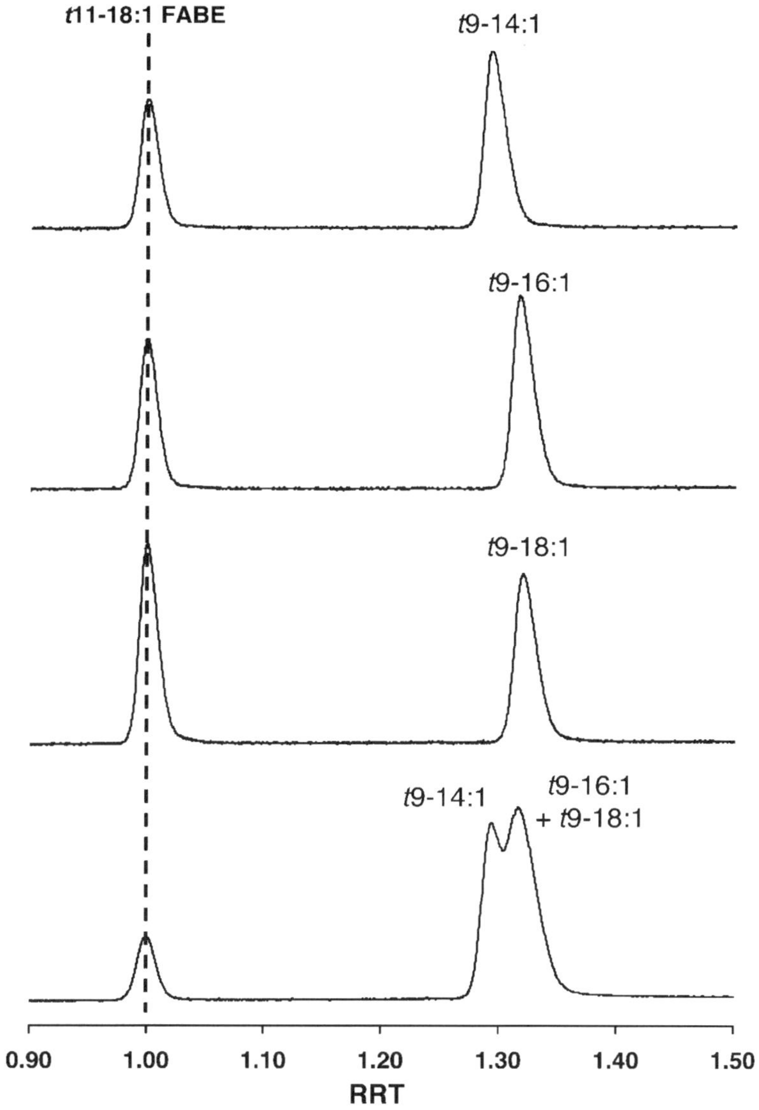

Fig. 6. Ag^{+}-HPLC separations of *trans* 9-14:1, *trans* 9-16:1, *trans* 9-18:1 as fatty acid methyl esters, and the combination of all three. *Trans* 11-18:1 *n*-butyl ester (FABE) was added as a reference. Chromatograms were transformed into a relative retention time (RRT) scale using *trans* 11-8:1 FABE as a reference. Conditions: three Chromspher 5 Lipids (250 × 4.6 mm, 5 μm particle size, Varian Inc.) columns in series; mobile phase 99.9:0.1 hexane/MeCN (vol/vol) at 1.0 mL/min; and UV detection at 200 nm. Chromatograms were acquired at an operating temperature of –10°C.

using HPLC, thus reducing the time of analysis. The Recommended Practice Ce 1g-96 published by the AOCS describes a Ag^+-HPLC fractionation of FAME based on numbers of double bonds and geometry (23). FAMEs were fractionated using one analytical Chromspher 5 Lipids column (250 × 4.6 mm) and a gradient of dichloromethane, MeCN, and acetone. The high cut-off wavelength of the eluent used in that procedure does not allow the application of UV detection; therefore an evaporative light-scattering detector (ELSD) is recommended. Because this detector destroys the sample portion analyzed, a flow splitter is required for collection of the fractions, making reproducible and quantitative recovery of the FAMEs fractions by this technique questionable.

Figure 7A shows an alternative approach to Ag^+-HPLC fractionation of *trans*-18:1 FAME, based on the chromatographic conditions discussed in the previous section. *Trans*-18:1 FAMEs were separated using one semipreparative Chromspher 5 Lipids column (250 × 10 mm, 5 μm particle size, Varian Inc.), a 0.1% MeCN in hexane mobile phase at the flow rate of 3 mL/min, and a UV detector at 206 nm. Sample solutions were prepared by mixing 20 mg of FAME (neat), 0.5 mg of *trans*-11 18:1 FABE (0.5 mL of a 1 mg/mL solution), and sufficient isooctane to fill a 1.8-mL glass vial. A 100-μL aliquot of the FAME solution (~1 mg) was loaded into the column. The appropriate *trans* fraction was collected based on the chromatogram monitored at 206 nm (Fig. 7A). The solvent of the collected fraction was removed under a stream of nitrogen. The *trans* FAME mixture was dissolved in an appropriate volume of isooctane and analyzed by GC. Figure 7B shows the GC chromatogram of the *trans* fraction collected, as identified in Figure 7A. *Trans*-11 18:1 FABE was used as the internal standard to quantitate the *trans*-18:1 isomers by GC. Several different experimental conditions were successfully applied to the fractionation of *trans*-18:1 by Ag^+-HPLC.

To date, none of the reported methods for the quantitative analyses of *trans*-18:1 fatty acids are sufficiently reliable or applicable for routine analysis. A comprehensive HPLC-GC methodology for *trans*-18:1 analysis in food, oils, or fats should include a suitable internal standard added before sample preparation (extraction and derivatization). Meanwhile, the application of Ag^+-HPLC fractionation followed by GC analysis of the isolated *trans*-18:1 FAMEs appears to show great potential, but it is presently limited to research laboratories and skilled analysts.

Conclusions

Ag^+-HPLC as a stand-alone technique, or coupled with GC, is currently being evaluated as a direct quantitative method for the analysis of *trans*-18:1 or in combination with GC after isolation of the *trans*-18:1 fraction. Silver ion chromatography is considered by most analysts to be only a sample preparation procedure used before GC determination, or they simply do not use it. However, as demonstrated here, Ag^+-HPLC can be applied successfully to the direct quantitation of *trans*-18:1 FAME after separating most the positional isomers from Δ6 to Δ15. A major

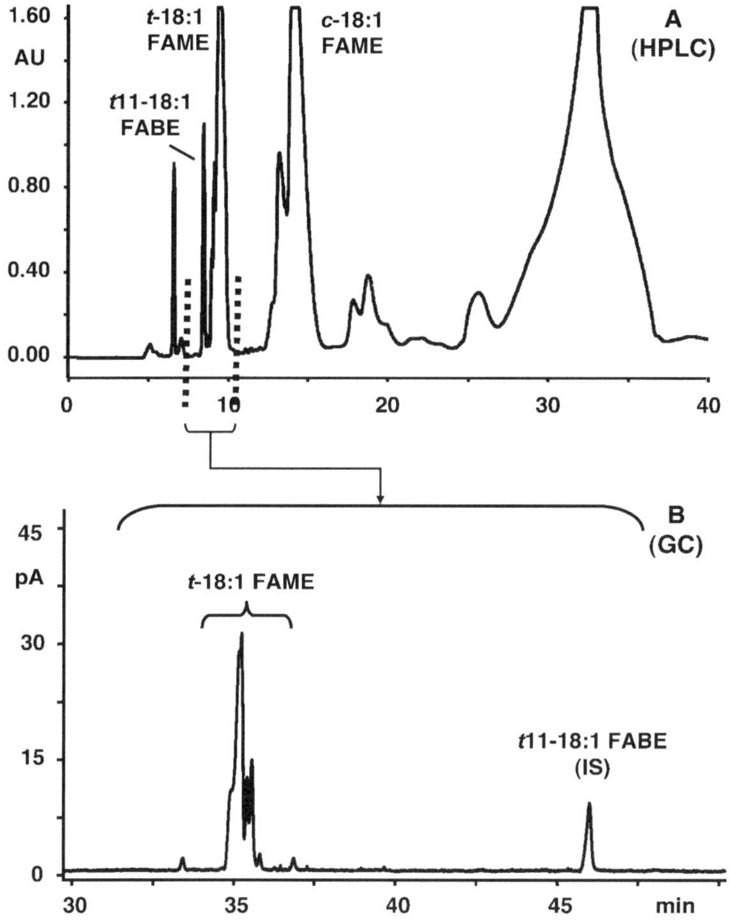

Fig. 7. Semipreparative Ag^+-HPLC separation of a total milk fat sample converted to fatty acid methyl esters (FAMEs) and spiked with the internal standard *trans* 11-18:1 *n*-butyl ester (FABE) (A). The region containing the *trans*-18:1 FAMEs plus *trans* 11-18:1 FABE was isolated and analyzed by gas chromatography (GC). The Ag^+-HPLC conditions in (A) were as follows: one semipreparative Chromspher 5 Lipids (250 × 10 mm, 5 μm particle size, Varian Inc.) column operated at room temperature; mobile phase 99.9:0.1 hexane/MeCN (vol/vol) at 3.0 mL/min; and UV detection at 206 nm. The GC separation was performed using a 100-m CP-Sil 88 capillary column (Varian Inc.) with hydrogen as the carrier gas at 1.0 mL/min. The temperature program was as follows: 2 min at 75°C, programmed at 5°C/min to 175°C and maintained for 33 min, then programmed again at 4°C/min up to 225°C, and maintained for an additional 33 min.

limitation for Ag^+-HPLC is the difficulty in standardizing the separation of FAMEs. The age of the columns, the differences between columns even from the same supplier, and the stability of the solvent mixtures used in the elution appear to be the main factors affecting the reproducibility of Ag^+-HPLC separations. This requires skill in setting up the initial separation, and continued monitoring of subsequent analyses. Fine tuning of this method will be required before this technique can be recommended for routine analysis of *trans* FAMEs.

References

1. Cruz-Hernandez, C., Z. Deng, J. Zhou, A.R. Hill, M.P. Yurawecz, P. Delmonte, M.M. Mossoba, M.E.R. Dugan, and J.K.G. Kramer, Methods for Analysis of Conjugated Linoleic Acids and *trans*-18:1 Isomers in Dairy Fats Using a Combination of Gas Chromatography, Silver-Ion Thin-Layer Chromatography/Gas Chromatography, and Silver-Ion Liquid Chromatography, *J. Assoc. Off. Anal. Chem. Int. 87*:545–562 (2004).
2. Ratnayake, W.M.N., Overview of Methods for the Determination of *trans* Fatty Acids by Gas Chromatography, Silver-Ion Thin-Layer Chromatography, Silver-Ion Liquid Chromatography, and Gas Chromatography/Mass Spectrometry. *J. Assoc. Off. Anal. Chem. Int. 87*:523–539 (2004).
3. American Oil Chemists' Society, AOCS Official Method Ce 1g-04, Determination of *cis*-, *trans*-, Saturated, Monounsaturated and Polyunsaturated Fatty Acids in Oils and Fats by Capillary GLC, in *Official Methods and Recommended Practices*, AOCS, Champaign, IL, 2005.
4. Mossoba, M.M., M.P. Yurawecz, P. Delmonte, and J.K.G. Kramer, Overview of Infrared Methodologies for *trans* Fat Determination. *J. Assoc. Off. Anal. Chem. Int. 87*: 540–544 (2004).
5. Fritsche, J., and H. Steinhart, Analysis of *trans* Fatty Acids, in *New Techniques and Applications in Lipid Analysis*, edited by R.E. McDonald, and M.M. Mossoba, AOCS Press, Champaign, IL, 1997, pp. 234–265.
6. Wolff, R.L., and D. Precht, A Critique of 50-m CP-Sil 88 Capillary Columns Used Alone to Assess *trans*-Unsaturated FA in Foods: The Case of TRANSFAIR Study, *Lipids 37*:627–629 (2002).
7. Nikolova-Damyanova, B., and S. Momchilova, Silver Ion HPLC for the Analysis of Positionally Isomeric Fatty Acids, *J. Liquid Chromatogr. Relat. Technol. 25*:1947–1965 (2002).
8. Morris, L.J., Separation of Higher Fatty Acid Isomers and Vinylogues by Thin-Layer Chromatography, *Chem. Ind.*:1238–1240 (1962).
9. De Vries, B., Quantitative Separation of Lipid Materials by Column Chromatography on Silica Impregnated with Silver Nitrate, *Chem. Ind.*:1049–1050 (1962).
10. Nikolova-Damyanova, B., Silver Ion Chromatography of Lipids, in *Advances in Lipid Methodology—One*, edited by W.W. Christie, The Oily Press Ltd., Dundee, Scotland, 1992, pp. 181–188.
11. Gunstone, F.D., I.A. Ismail, and M. Lie Ken Jie, Fatty Acids, Part 16. Thin Layer and Gas-Liquid Chromatographic Properties of the *cis* and *trans* Methyl Octadecenoates and of Some Acetylenic Esters, *Chem. Phys. Lipids 1*:376–385 (1967).
12. Christie, W.W., A Stable Silver-Loaded Column for the Separation of Lipids by High Performance Liquid Chromatography, *J. High Res. Chromatogr. 10*:148–150 (1987).

13. Nikolova-Damyanova, B., B.G. Herslof, and W.W. Christie, Silver Ion High-Performance Liquid Chromatography of Derivatives of Isomeric Fatty Acids, *J. Chromatogr. A 609*:133–140 (1992).
14. Christie, W.W., and G.H.McG. Breckenridge, Separation of *cis* and *trans* Isomers of Unsaturated Fatty Acids by High-Performance Liquid Chromatography in the Silver Ion Mode, *J. Chromatogr. A 469*:261–269 (1989).
15. Momchilova, S., B. Nikolova-Damyanova, and W.W. Christie, Silver Ion High-Performance Liquid Chromatography of Isomeric *cis-* and *trans-*Octadecenoic Acids: Effect of the Ester Moiety and Mobile Phase Composition, *J. Chromatogr. A 793*:275–282 (1998).
16. Wood, R., and T. Lee, High-Performance Liquid Chromatography of Fatty Acids: Quantitative Analysis of Saturated, Monoenoic, Polyenoic and Geometrical Isomers, *J. Chromatogr. A 254*:237–246 (1983).
17. Adlof, R., Separation of *cis* and *trans* Unsaturated Fatty Acid Methyl Esters by Silver Ion High-Performance Liquid Chromatography, *J. Chromatogr. A 659*:95–99 (1994).
18. Adlof, R.O., L.C. Copes, and E.A. Emkin, Analysis of the Monoenoic Fatty Acid Distribution in Hydrogenated Vegetable Oils by Silver-Ion High-Performance Liquid Chromatography, *J. Am. Oil Chem. Soc. 72*:571–574 (1995).
19. Sehat, N., M.P. Yurawecz, J.A.G. Roach, M.M. Mossoba, J.K.G. Kramer, and Y. Ku, Silver-Ion High-Performance Liquid Chromatographic Separation and Identification of Conjugated Linoleic Acid Isomers, *Lipids 33*:217–221 (1998).
20. Sehat, N., R. Rickert, M.M. Mossoba, J.K.G. Kramer, M.P. Yurawecz, J.A.G. Roach, R.O. Adlof, K.M. Morehouse, J. Fritsche, K. Eulitz, H. Steinhart, and Y. Ku, Improved Separation of Conjugated Linoleic Acid Methyl Esters by Silver-Ion High-Performance Liquid Chromatography, *Lipids 34*:407–413 (1999).
21. Corl, B.A., L.H. Baumgard, J.M. Griinari, P. Delmonte, K.M. Morehouse, M.P. Yurawecz, and D.E. Bauman, *Trans*-7,*cis*-9 CLA Is Synthesized Endogenously by Δ9-Desaturase in Dairy Cows, *Lipids 37*:681–688 (2002).
22. Delmonte, P., J.A.G. Roach, M.M. Mossoba, G. Losi, and M.P. Yurawecz, Synthesis, Silver Ion HPLC Clean-Up, and GC Relative Retention Times for All *cis/trans* Isomers of CLA FAME from the 6,8- to 13,15-Position, *Lipids 39*:185–191 (2004).
23. American Oil Chemists' Society, AOCS Official Method Ce 1g-96, *Trans* Fatty Acids by Silver-Ion Exchange HPLC, in *Official Methods and Recommended Practices*, AOCS, Champaign, IL, 2002.

9 High-Performance Size-Exclusion Chromatography for Lipid Analysis in Organic Media

G. Márquez-Ruiz and M.C. Dobarganes

Instituto de la Grasa (CSIC), 41012 Sevilla, Spain

Introduction

Among chromatographic techniques, high-performance size-exclusion chromatography (HPSEC) is unique in that the separation of compounds is based on their molecular size, normally related to their molecular weight (MW), provided that the compounds have a similar shape. Originally, size-exclusion chromatography was used mainly for characterization of high-MW molecules, either synthetic polymers or biopolymers, but applications have expanded to other areas in the last decades, supported by important technical advances. Fundamental developments and general applications of HPSEC in recent years were reported in various reviews (1–8).

Because of the basis of HPSEC, the technique is appropriate for separating groups of compounds that differ in MW by at least 10–15%; this is necessary to achieve well-resolved peaks. Hence it is not applicable for the separation of compounds that differ only slightly in MW as a result of variable fatty acid composition or degree of unsaturation. In this respect, gas chromatography (GC) or high-performance liquid chromatography (HPLC) are the most convenient techniques for the separation of lipid molecules. However, the limitation of size-exclusion chromatography constitutes, at the same time, the basis of its most specific applications in lipid analysis, which include powerful analytical techniques in both organic and aqueous media (9–15).

Even though application of HPSEC in aqueous media is not intended to be the subject of this review, which is otherwise focused on the analysis of lipids in organic media, it is important to mention that HPSEC constitutes a useful technique for biological samples, thus contributing to the knowledge of the role of lipids in their natural biological environment. Specifically, applications in lipoprotein analysis and size estimation of liposomes, micelles, and vesicles stand out (11). In these applications, conventional and high-performance stationary phases based on silica or cross-linked polymer-based phases compete to give solutions to diverse problems. Distinctive features of these applications are that different aqueous buffers are used to preserve the structural integrity of the solute or to mimic physiological conditions, previous equilibration of the column with salt solutions or lipids may be necessary to reduce interactions between the stationary phase and the solute, and detection is normally carried out by absorbance measurements. In the lipoprotein field, differences in size between the different classes make size-exclusion chromatography a good alternative for lipoprotein separation, offering several advantages, i.e., the procedure is gentle and nondestructive, the size distribution of particles can be visualized easily during separation, and the recovery is high. For low density lipoprotein (LDL) size measurement, good reproducibility and high precision were found using HPSEC with ultraviolet (UV) detection (16–18). A number of systematic studies of both genetic and dietary effects on lipoprotein phenotypes were recently undertaken with the aid of size-exclusion chromatography (19–24). In addition, size-exclusion chromatography has been used for fractionation of liposomes, proteoliposomes, and biomembrane vesicles of up to ~500 nm in size (25). The suitability of the technique for the analysis of liposomes was reviewed recently (26), including a description of methodologies based on HPSEC coupled with multidetection modes and examples of applications that concern polydispersity, size and encapsulation stability, bilayer permeabilization, liposome formation, and reconstitution and incorporation of amphiphilic molecules.

This chapter deals with the use of HPSEC in lipid analysis and, specifically, in organic media. After a brief summary of the main characteristics of the technique in organic media, the most interesting applications developed, which are listed and described briefly in Table 1, are thoroughly discussed. For ease of discussion, applications of HPSEC have been classified into three groups according to increasing complexity of the samples, namely, intact samples, fatty acid methyl esters (FAME), and concentrated fractions.

Characteristics of HPSEC in Organic Media

HPSEC analyses in organic media are characterized by their simplicity, regardless of the application. Once it is decided that the technique is useful for a given purpose, selection of the conditions involves a limited number of possibilities; these are related mainly to the stationary phase, solvent used as the mobile phase, and detection system. The most common conditions are summarized in Table 2.

Migration of molecules between the stationary phase and the mobile phase occurs essentially by diffusion, and the elution order is inversely related to molecular size or weight. Thus, larger molecules are excluded and emerge first, whereas smaller molecules can diffuse into the pores of the gel, partially or completely, and elute later. It is possible to estimate the MW of unknown molecules by plotting retention volume vs. the logarithm of MW for a series of known standards such as polystyrene standards. These plots provide accurate determinations of MW for molecules that adopt a conformation in solution similar to that of the standard (27,28). However, it is thought that complex analyte species should be much better represented by their absolute MW determined by a universal calibration method utilizing plots of log [$\eta \times$ MW] vs. elution volume, where η is intrinsic viscosity (29).

TABLE 1
Summary of the Main Applications of High-Performance Size-Exclusion Chromatography to Lipid Analysis in Organic Media[a]

	Sample	Application	Remarks
Intact lipids, fats or oils:	Used frying oils	Determination of TG polymers	Quality control parameter.
	Fish oils	Determination of TG polymers	Quality evaluation.
	Mixtures of partial glycerides	Separation and quantitation of TG, DG, MG, FFA and other lipids	Control of chemical and enzymatic reactions of hydrolysis or esterification.
	Fats and oils replacers	Separation and quantitation of high-MW replacers	Analysis of mixtures with fats and oils. Evaluation of absorption and fate.
	Minor lipid compounds	Determination of specific lipid compounds in combination with other techniques	Applications in foods, biomedical and biological samples.
	Food lipid extracts	Removal of TG	Clean-up method before pesticide analysis.
	Industrial oils	Determination of polymers and their MW distribution	Control of residues from oil refineries. Control of drying process in paints.
FAME derivatives:	Discarded used frying oils	Determination of FAME polymers	Indication for the suitability of biodiesel.
	Oleochemicals	Determination of FAME monomers, dimers and polymers	Characterization of technical mixtures.
	Mixtures of FAME and partial glycerides	Separation and quantitation of TG, FAME and partial glycerides	Control of transesterification and hydrolysis reactions.
Concentrated fractions:	Used frying and thermoxidized oils	Quantitation of oxidized TG monomers, TG dimers, TG polymers, DG and FFA	Quality parameters for oxidative, thermal and hydrolytic degradations.

(continued)

TABLE 1 *(continued)*

	Sample	Application	Remarks
Concentrated fractions:	Crude and refined oils	Quantitation of oxidized TG monomers, TG dimers, TG polymers, DG and FFA	Control of refining process. Characterization of virgin and refined olive oils.
	Oxidized fats, oils and food lipids	Quantitation of oxidized TG monomers, TG dimers and TG polymers	Concomitant evaluation of primary and secondary oxidation products.
	Used frying and thermoxidized oils (FAME derivatives)	Quantitation of oxidized FAME monomers, dimers and polymers	Specific analysis of oxidized and polymerized fatty acyls included in TG.
	Fecal lipids (FAME derivatives)	Quantitation of oxidized FAME monomers, dimers and polymers	Determination of digestibility coefficients.

[a]*Abbreviations:* TG, triacylglycerols; DG, diacylglycerols; MG, monoacylglycerols; FFA, free fatty acids; FAME, fatty acid methyl esters; MW, molecular weight.

TABLE 2
High-Performance Size-Exclusion Chromatography Conditions for Lipid Analyses in Organic Media

Stationary phase	Copolymers of styrene-divinyl benzene
	Particle size: 3, 5[a], 10 μm
	Pore size: 50, 100[a], 500[a], 1000 Å
Mobile phase	Tetrahydrofuran[a], toluene, dichloromethane
Detection system	Refractive index detector[a]
	Evaporative light-scattering detector
	Viscometry detector

[a]Most commonly used.

Figure 1 shows a typical calibration graph. As can be observed, a species is eluted at a volume (V_e) exactly equal to the volume available to it in the column, following the general equation:

$$V_e = V_o + K_d \times V_i$$

where V_o is the volume external to the beads or void volume, V_i is the internal (pore) volume and K_d is the distribution or partition coefficient. V_o represents the elution volume for completely excluded molecules and thereby $K_d = 0$, whereas V_t

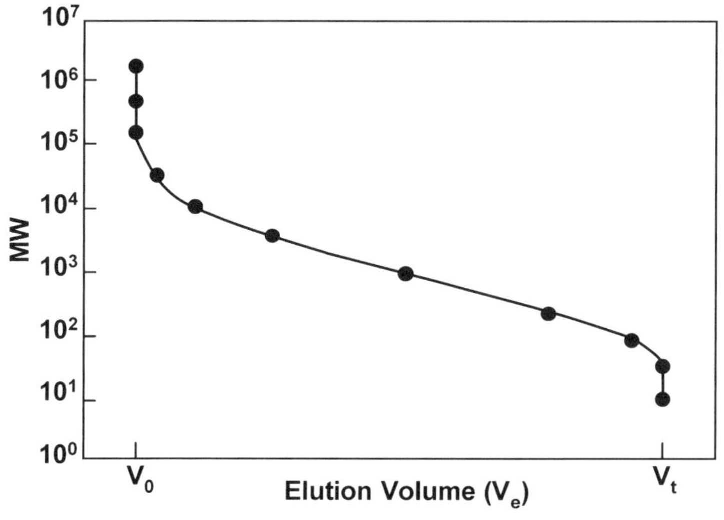

Fig. 1. Representation of a typical calibration graph for high-performance size-exclusion chromatography.

is the elution volume of the material experiencing total penetration of the gel pores, and it is equal to the sum of V_o and V_i ($K_d = 1$). When $K_d > 1$, the separation by size is modified by some interaction such as adsorption to the stationary phase.

The latest developments in calibration methodologies, including direct calibration by standards and various instrumental methods [nuclear magnetic resonance (NMR), mass spectrometry (MS), light scattering] as well as universal calibration with viscometry detectors, were recently reported in detail (30).

Stationary Phase

Typically, stationary phases consist of macromolecules cross-linked to form a three-dimensional network characterized by a specific pore size. The most important parameters influencing resolution are the pore volume, pore-size distribution, and particle size; improvements in the control of these parameters have contributed to obtaining columns of high efficiency and separation capacity (31).

The stationary phases based on copolymers of styrene divinyl benzene have been the most promising to date and are widely used for applications in organic media. Copolymers of styrene divinyl benzene are available over a wide range of pore sizes, but 50, 100, and 500 Å are essential porosities for low-MW separations (100–20,000 MW). They are sold under trade designations such as Ultrastyragel from Waters Associates and PL-Gel from Agilent Technologies, among others. Size-exclusion columns are generally larger than those used in other chromatographic modes, so that the amount of stationary phase and thus the effective pore volume available is increased. Packed columns are normally ~30 cm × 0.8 cm i.d. and are often used in series. Thus, effective selection within a broad range is accomplished by the first column and fractionation within a more defined range is achieved on the second or third column. Optimization of stationary phase particles is essential to enhance mass transfer given that resolution in HPSEC is dependent on diffusion. In this respect, the development of spherically shaped monosized particles contributed considerably to improving particle-size and pore-size distribution and hence efficiency and separation capacity (31,32). Particle sizes of 5 and 10 μm are the most commonly used.

Mobile Phase

For the mobile phase, a single solvent is used, and HPSEC columns are normally filled with the same solvent used to dissolve the sample. Supports of copolymer styrene divinyl benzene were designed to operate across a wide spectrum of solvents. Columns can even be transferred easily and rapidly between solvents of differing polarity without damage to the packed bed. Tetrahydrofuran (THF) is the most commonly used solvent, although toluene and dichloromethane are used for certain applications. Flow rates between 0.5 and 1.5 mL/min are the most usual, thus allowing a complete analysis in <30 min.

Detection System

For applications in organic media, detection with nonselective, mass-sensitive detectors, such as a refractive index detector (RID) and an evaporative light scattering detector (ELSD), are most common. RID is limited to isocratic separations but provides good linearity for mass-sensitive measurements. ELSD, on the other hand, has excellent gradient capabilities and a slightly superior sensitivity but provides nonlinear (exponential or slightly sigmoid) response curves because of underlying light scattering mechanisms (33–36). For quantitative purposes, a refractometer is simpler in that a single solvent is used and linear responses are normally obtained in the ranges of interest. Recently, the application of viscometric detection for accurate determination of MW was reported (29).

In recent years, HPSEC development in detection systems has advanced considerably. In particular, special emphasis was paid to new multidetector combinations that improve elucidation of molecular properties of polymers, including detection with Fourier transform infrared (FTIR) spectrometry with solvent-evaporation interfaces (37,38). Some examples of the combinations used are the coupling of multiangle light scattering and differential refractometry detectors (39), the combination of RID and ELSD with fluorescence spectroscopy (40), or the coupling of UV detection and on-line NMR spectroscopy and MS combined with on-line collection of the chromatographic eluent for subsequent FTIR (41). In summary, a combination of different detection systems is directed to gain complementary information on the sample components, i.e., identification of specific structures, exact determination of MW, and quantitation of the different fractions. Although these complex combinations of detectors are not normally applied in the applications discussed in this review, the future seems to be promising, particularly for the elucidation of the main structures present in unknown and complex fractions, i.e., polymerization compounds.

Applications on Intact Lipid Samples

An excellent application of HPSEC that is of great utility for certain fields of lipid analysis consists in the direct use of the technique in entire, intact lipid samples, thus constituting a simple and rapid analytical approach. It is necessary only to dilute the lipids, fat, or oil in the appropriate solvent; thereafter, chromatographic analysis is performed in only 10–30 min, depending on the flow rate and number of columns.

Used Frying Oils

Evaluation of polymerized triacylglycerols (TG) by HPSEC is particularly valuable in the area of heated and used frying oils because they are the major degradation compounds among those formed. Perrin and co-workers (42) were the first to propose high-performance columns of polystyrene divinyl benzene for the analysis of TG polymers, and the potential possibilities of the technique were later confirmed (43,44), even though exact quantitative data could not be provided. Good correlations were found between amounts of TG polymers and polar compounds separated by column chromatography (45–47); the latter was the method proposed for quality control in used frying fats and oils (48,49). HPSEC quickly proved to be the most useful technique for the analysis of polymerized TG because of its simplicity; it is necessary only to dilute the oil or fat in the appropriate solvent, and the chromatographic determination is short and performed with a single solvent. In consequence, after two interlaboratory tests carried out in 1986–87, the International Union of Pure and Applied Chemistry (IUPAC) Commission on Oils, Fats and Derivatives adopted a method for determination of polymerized TG in used frying fats and oils for samples containing not <3% polymers (50,51). The method proposes a single column of 30 cm × 0.77 cm i.d. packed with a high-performance spherical gel made of copolystyrene divinyl benzene of 5 μm, with THF as the mobile phase, and a RID with a sensitivity at full scale at least 1×10^{-4} of the refractive index. The sample concentration suggested is 50 mg/mL for an injection valve with a 10-μL loop. The analysis time is ~10 min at a flow rate of 1 mL/min. Figure 2 shows a typical chromatogram of a used frying oil injected directly into HPSEC, which clearly reflects the simplicity of the determination. The sample selected for this figure was a used frying oil with a polar compound level close to the limit for rejection (23.1%) that contained 13.7% TG polymers.

Analysis of TG polymers as a quality control parameter for used frying fats is used extensively at present (52–62). In this context, a new index called OSET (oxidative stability at elevated temperature) was recently described for estimating the stabilizing activity of additives at simulated frying temperature. The test consists of accelerating polymerization with acid-catalyzed silica gel; oils or fats are heated for 2 h at 170°C in the presence of the additive tested, and HPSEC is used to evaluate dimeric and polymeric TG contents (56).

Fish Oils

In the case of highly unsaturated oils, such as fish oils, polymerization is very rapid even at low temperatures, because of the high instability of unsaturated hydroperoxides. Hence the simple determination of TG polymers by direct application of HPSEC was also used satisfactorily for quality evaluation of commercial fish oil capsules (63,64) and routine assessment of marine oils (65). For the chromatographic conditions applied to the analysis of fish oils, not only was a single copolystyrene divinyl benzene column used but also three columns in series, to improve separation of polymers, and both THF and dichloromethane were selected as the mobile phase.

In comparison to the methods most commonly used to evaluate oxidation in highly unsaturated lipids, determination of TG polymers has proved to be a simple and valuable alternative. Thus, determination of polymers was suggested to be a

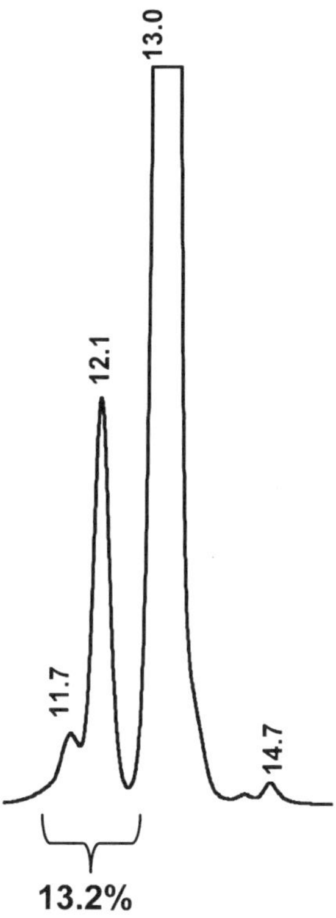

Fig. 2. High-performance size-exclusion chromatogram of a used frying oil. Retention times (min): 11.7, triacylglycerol polymers; 12.1, triacylglycerol dimers; 13.0, triacylglycerol monomers plus diacylglycerols; and 14.7, free fatty acids.

good indication of oxidation at low temperature for model polyenoic fatty acids, which were also analyzed for oxygen uptake, peroxide value, and conjugated dienes (66); the advantages of TG polymer analysis vs. determinations of polyene index and thiobarbituric acid reactive substances to monitor oxidation of fish oils during storage were also reported (67).

Partial Glycerides

Partial glyceride evaluation is an important field for direct applications of HPSEC in lipid analysis because differences in MW between the main lipid classes are sufficient for successful separation. Hence HPSEC is convenient to use as an alternative technique to TLC, HPLC and GLC. Control of food emulsifiers, technical oleochemicals, chemical and enzymatic reactions of hydrolysis, or esterification and analysis of special seed oils are typical examples of applications.

The most complete study among HPSEC applications on the separation and quantitation of mixtures of TG, diacylglycerols (DG), monoacylglycerols (MG), and fatty acids (FA) in oils was accomplished by Christopoulou and Perkins in 1986

(68). Two columns packed with 5 μm styrene-divinylbenzene copolymer were connected in series, toluene was employed as the eluant, and lipid components were monitored by refractometry. Discrimination among TG, FA, and other lipid compounds, such as steryl esters or free sterols, was also achieved successfully in more complex samples such as lipophilic extracts (69).

The technique is useful to verify the yield reached in transesterification reactions because TG, unreactive partial glycerides, glycerol, and FA methyl or ethyl esters can be determined easily in a single analysis (70,71). In the same line, direct HPSEC was used to analyze products of glycerolysis and rate of reaction in MG- and DG-based emulsifier production (72).

In nutritional studies directed at the study of the absorbability and fate of dietary thermoxidized oils, more complex mixtures, also containing dimeric and polymeric TG, were satisfactorily evaluated by direct HPSEC, including the nonabsorbed nonhydrolyzed lipids found in feces (73), the hydrolysates resulting from *in vitro* enzymatic lipolysis of thermoxidized oils (74,75), and the products of depolymerization reactions under *in vitro* acidic conditions simulating those occurring in the stomach (76).

Figure 3 shows HPSEC chromatograms of a heated oil before and after lipolysis, showing the increase in complexity once DG, FA monomers, dimers, and polymers were released. For analysis, two columns (100 and 500 Å), packed with highly cross-linked styrene divinylbenzene copolymer (particle size: 5 μm), were connected in series, THF served as the mobile phase, at a flow rate of 1 mL/min, and a RID was used. Even though some hydrolytic products co-eluted with other compounds, useful information was obtained from the distribution of the compounds quantitated into three groups according to their MW. First, at retention times >14 min are the compounds produced exclusively by hydrolysis, MG and FA monomers; their global level indicates the minimum of hydrolytic products originated. On the other hand, the amount of glycerides of MW higher than that of TG monomers (TGM) (retention time <12 min) measures the minimum of nonhydrolyzed compounds remaining after hydrolysis. Finally, the 12–14 min region includes both partially hydrolyzed compounds and complex fatty acids (FA dimers and trimers). Results showed that the extent of hydrolysis was reduced because total oil alteration was greater and, more specifically, the increase in TG dimers (TGD) and TG polymers (TGP) markedly affected the hydrolysis rate of TGM.

In recent years, Sánchez-Muniz and co-workers have continued to investigate these aspects, and have contributed greatly to the knowledge of the *in vitro* enzymatic hydrolysis kinetics in thermally oxidized and used frying oils and fats (77–79).

Fats and Oils Replacers

HPSEC evaluation has been extended to lipid samples other than classical fats and oils, such as certain fat replacers or substitutes of low digestibility. Among the fat substitutes proposed, sucrose polyesters, i.e., olestra, approved for use in foods by the

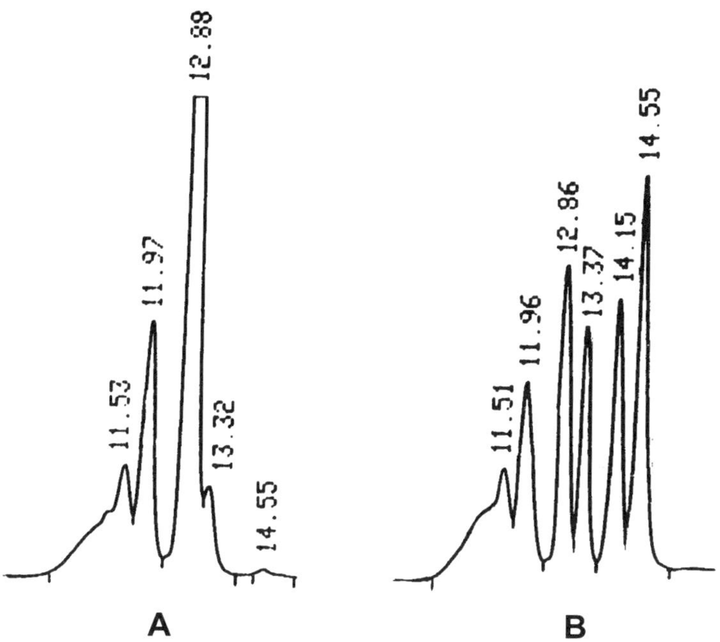

Fig. 3. High-performance size-exclusion chromatograms of a heated oil before (A) and after (B) lipolysis. Retention times (min): 11.5, triacylglycerol polymers; 11.9, triacylglycerol dimers; 12.8, triacylglycerol monomers plus fatty acid polymers; 13.3, diacylglycerols plus fatty acid dimers; 14.1, monoacylglycerols and 14.5, fatty acid monomers.

FDA in 1996, stand out. In this context, HPSEC has served to measure absorption of sucrose polyesters through determinations in diets and human feces (80) as well as in animal tissues (81); in the last-mentioned case, this was in conjunction with radiolabeling techniques, with results showing that intact olestra is virtually not absorbed.

HPSEC seems to be the simplest, most general and reproducible method for analysis of sucrose polyesters/fats or oils blends (82,83), which is of great utility for production and quality control, and nutritional labeling of commercialized products. For example, peaks corresponding to sucrose polyesters, TG, and FAME can be separated neatly, thus making it possible to verify the absence of residual FAME after sucrose polyester synthesis or to quantitate sucrose polyesters and TG in mixtures. Figure 4 illustrates the efficacy of the HPSEC separation for sucrose polyesters prepared from olive oil mixed with sunflower oil in a 1:4 ratio. The chromatographic conditions used were the same as those for the sample in Figure 3. The validity of the method relies on the elution of sucrose polyesters as a single peak and the excellent separation achieved between that and the TG peak, whereas those compounds present in minor amounts were not detected. HPSEC served not only to quantitate the relative proportions of both parts in the mixture, which is essential to define the energy value of the product, but also to know the fatty acid composition of each component after recovery of the separately eluted fractions; this is of additional nutritional interest given that only a fraction of natural fat or oil can be digested and absorbed.

Similarly, HPSEC applications can be extended to the analysis of mixtures of natural oils with other fat substitutes provided that they differ significantly in MW, as occurs for sorbitol polyesters, trehalose polyesters, and raffinose polyesters. Polymer formation at high temperature was also determined in sucrose polyesters (84) and propoxylated glycerol esters (85) by HPSEC.

Minor Lipid Compounds

Even though applications of size-exclusion chromatography to minor lipid compounds, such as neutral hydroxyl lipids, phospholipids, unsaponifiable fractions, antioxidants, waxes and hydrocarbons, started in the 1970s, their acceptability was nominal due to simultaneous development of other chromatographic techniques

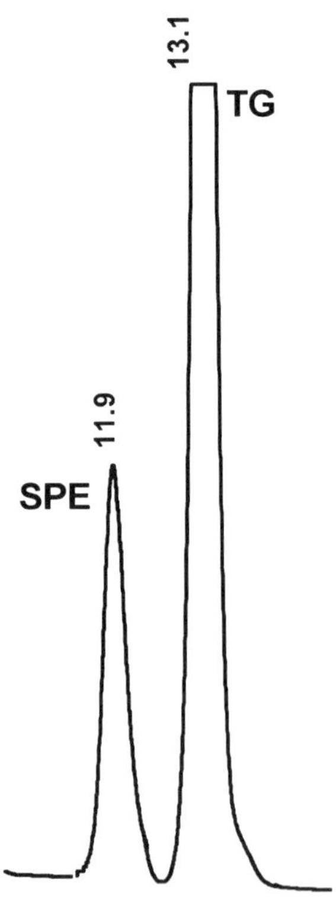

Fig. 4. High-performance size-exclusion chromatogram of a 1:4 mixture of sucrose polyesters:sunflower oil. *Abbreviations:* SPE, sucrose polyesters and TG, triacylglycerols.

based on much more convenient differentiating principles than that of molecular size-exclusion (11).

However, there have been some promising applications to biomedical, biological, and food samples in recent years that are worth noting. A size-exclusion chromatographic method was described for measuring the absorption of the steroid-based lipids cholesterol and cortisone into a polyurethane used in biomedical implants, using dual (UV diode-array and refractive) detection (86). High-MW ether lipids were determined in algae by a combination of methods, including HPSEC (87). Interestingly, HPSEC was used to test oxidative stress in rat membrane hepatocytes in cultures, by measuring extracellular free malondialdehyde (88). Studies on lipid metabolism can also benefit from the possibilities of this technique, as reported recently by Zarini and Murphy (89), who assessed the factor responsible for the conversion of the precursor of a metabolite of arachidonic acid with important biological activities.

In foods, a novel method for prefractionation of aroma extracts of goat cheese was developed to obtain a purified fraction of volatile compounds devoid of TG for direct analysis by GLC (90). Both diethyl ether and dichloromethane were tested as mobile phases, the effect of column load on separation between TG and a large variety of aroma compounds was studied, and the method was satisfactorily applied to the analysis of goat cheese volatiles. Another application of HPSEC for analysis of minor lipids in foods consisted in the detection of hydroperoxy TG and hydroperoxy cholesterol esters after reduction in emulsion samples, using the postcolumn fluorometric diphenyl-1-pyrenylphosphine oxidation principle (91).

Others

One of the former applications of size-exclusion chromatography that has endured to date is its use as a lipid clean-up method before pesticide analysis by GLC, directed primarily toward the removal of high-MW material with TG present in the largest concentrations. Most applications correspond to fruit or vegetable samples (92–96), and specifically olive oil (97–99), although other fat extracts were analyzed, i.e., milk samples (100), chicken and fish samples (92), as well as soil and dust samples (101,102). The technique saves considerable time in preparative work and is highly reproducible; the main drawbacks are related to the requirement of miniaturized columns for on-line coupling with GLC. Improvements were achieved by inserting a liquid chromatography step on silica gel before GLC analysis (94). Other alternatives proposed are the combination of size-exclusion chromatography with HPLC-MS (95,96,99,103), and the introduction of a solid-matrix dispersion partition step before HPSEC on a minicolumn (100,104).

In the area of petrochemicals, there is a growing interest in the use of size-exclusion chromatography as the only existing methods that can easily and routinely provide information on relative MW distributions of residues produced in oil refineries (105–109). Another application worth mentioning is the use of size-exclusion chromatography as a complementary technique to follow the oxidative degradation of linseed oils, used in paints and varnishes, during the drying process (110,111).

Applications on FAME

Size-exclusion chromatography has also been applied for the evaluation of FAME. In fact, the first application of low-performance size-exclusion chromatography in lipid analysis was the evaluation of polymerized FA in the area of oleochemicals, to characterize technical mixtures of monomeric, dimeric, and trimeric FA derivatives (112–114). Other studies widened the application to the analysis of both TG polymers and polymerized FA after transesterification or hydrolysis (115–117). Later, the application of high-performance supports considerably improved the results obtained for characterization of polymerized FA, as reported for raw materials used as commercial feed fats (118). Comparisons of HPSEC with other chromatographic techniques available, such as GLC and TLC with flame ionization detection, were also carried out, and good correlations were found (119,120).

FAME are well established as an alternative fuel called biodiesel and, for economic reasons, used frying oils represent an interesting feedstock for biodiesel production. The utility of HPSEC in this context is supported by the need to control the amount of polymers present, which constitutes a good indicator of the suitability for biodiesel. The amount of polymeric FAME has a negative influence on fuel characteristics because viscosity must not exceed a certain level according to the existing specifications for biodiesel (121). On the other hand, as was previously stated, determination of FAME by HPSEC is an adequate approach to verify the yield reached in transesterification reactions (70,71).

Applications on Concentrated Fractions

The possibilities of application of HPSEC were widened considerably by the preliminary separation of fractions concentrated in the compounds of interest. Thus, methodologies based on a combination of adsorption chromatography and HPSEC have numerous applications. In this section, the main procedures for obtaining concentrated fractions, either from intact oil samples or from FAME, and subsequent analysis by HPSEC, are detailed first; then, the main applications of such procedures are described.

Methodologies for Concentrated Fractions from Intact Oil Samples

Combination Silica Column-HPSEC. By virtue of a previous separation in a silica column of the most abundant fraction in oils, i.e., nonpolar or nonoxidized TG, determination by HPSEC in the concentrated fraction of polar compounds allowed quantitation of different groups of oxidized and hydrolytic compounds which differ substantially in MW: TGP, TGD, oxidized TGM (oxTGM), DG, MG, and free fatty acids (FFA), in that order of elution (122). Briefly, starting from 1 g of fat or oil, nonpolar and polar fractions are eluted with 150 mL of a mixture of hexane:diethyl ether 90:10 and 150 mL diethyl ether, respectively. Nonpolar and polar fractions are evaporated under reduced pressure, and separation is checked by TLC. The polar compound fraction is determined gravimetrically and further analyzed by HPSEC, using two columns (100 and 500 Å), packed with a porous, highly cross-linked styrene divinylbenzene copolymer (particle size: 5 μm), connected in series. THF serves as the mobile phase, at a flow rate of 1 mL/min, and a RID is used. HPSEC analysis requires a total run time of only 15 min. Concentrations are calculated from peak areas and gravimetric determination of the polar fraction. As discussed below, this analytical scheme was proposed as a useful procedure to obtain complete information on different types of degradation, i.e., thermal, oxidative, and hydrolytic, all of which are involved in the frying process. The methodology was standardized recently by the IUPAC, with slight modifications (123), for analysis of used frying oils, and also for virgin and refined oils.

Resolution in the range of lowest or highest MW can be improved by adding one styrene/divinylbenzene copolymer column of 50 Å (33) or 500 Å (124), respectively, or using 3-μm mixed-bed styrene/divinylbenzene copolymer columns (35).

Combination Solid Phase Extraction (SPE)-HPSEC. A modification of the former methodology was introduced to reduce the quantity of sample and solvents, and shorten analysis time (125). Briefly, 2 mL of the sample solution, containing 50 mg of oil and 1 mg of monostearin, used as internal standard, is placed on the silica cartridge for SPE. Monostearin is used as internal standard because MG are normally present in negligible, undetectable amounts in fats and oils. The nonpolar fraction is eluted with 15 mL of hexane:diethyl ether 90:10. A second fraction containing polar compounds and the internal standard is eluted with 15 mL of diethyl ether. Nonpolar and polar fractions are evaporated under reduced pressure and redissolved in 1 mL of THF for further analyses, i.e., TLC to verify the efficiency of the separation and HPSEC. Fractions of polar compounds are analyzed by HPSEC under theconditions described previously. Precision, accuracy and recovery data were determined. Samples containing levels of polar compounds ranging from 3.7 to 24.3% were analyzed by both this procedure and that based on gravimetric determination of the polar fraction; the mean values obtained did not differ. However, the lower the polar compound concentration, the lower the SD for the internal standard method. In view of these results, it seems that the SPE-HPSEC method is useful for a wide range of samples, and is especially adequate for samples with a low oxidation level.

Methodology for Concentrated Fractions from FAME

After transesterification, through separation by adsorption chromatography, a fraction of concentrated polar FAME can be obtained, which allows specific analysis of different groups of oxidized and polymerized fatty acyls included in TG molecules by HPSEC (126). Briefly, FAME are obtained by transesterification of 1 g of sample with sodium methoxide and hydrochloric acid:methanol. FAME are recovered quantitatively and separated by silica column chromatography into two fractions, which are analyzed by HPSEC, under the chromatographic conditions described above. Depending on the objective of the application, two possibilities for the analytical scheme were proposed; these are represented in Figure 5. When using 150 mL hexane:diethyl ether 95:5, the first fraction includes exclusively the most abundant nonoxidized FAME. Then, elution with 150 mL diethyl ether yields a minor polar fraction including three groups of FAME (polymers, dimers, and oxidized monomers). Alternatively, when using hexane:diethyl ether 88:12 for elution of the first fraction, the combined chromatographic analysis permits discrimination between nonpolar and oxidized FAME dimers. Thus, both nonoxidized FAME and the nonpolar or nonoxidized FAME dimers, linked by C-C bonds and lacking extra-oxygenated functions in their structure, are quantitated in the first fraction. FAME polymers, oxidized FAME dimers, and oxidized FAME monomers are determined in turn in the polar fraction. Therefore, global quantitation of the compounds eluted in the second fraction provides a measurement of the total oxidized fatty acyl groups included in TG molecules.

Used Frying Fats and Oils

The methodology based on silica column-HPSEC has proved to be an excellent alternative for the evaluation of used frying oils. In addition to providing both the determination of total polar compounds and polymerized TG, broader knowledge of the different groups of compounds formed is gained (122,127–129). Advantages in comparison with the analysis of the total oil sample are clearly reflected in a comparison of Figure 1 (used frying oil directly injected in HPSEC) with Figure 6 (HPSEC profiles of polar fractions of used frying oils). First, a substantial increase in the possibilities for quantitation of all of the groups of compounds that are affected is achieved because of the effect of concentration. Second, oxTGM, a measurement of oxidative degradation, can be determined independently because the co-eluting major peak of nonoxidized TG is separated in the nonpolar fraction. Third, concomitant evaluation of DG as a marker of hydrolytic degradation is possible; otherwise, it overlaps with the abundant TG in direct HPSEC analysis of the whole sample. Finally, a substantial increase in sensitivity is achieved in quantitation of polymerization compounds due to the effect of concentration, overcoming the limitation of the IUPAC method to a minimum of 3% content for analyses of total samples by HPSEC (51). The groups of compounds quantitated can be differentiated between thermally oxidized compounds (oxTGM, TGD and TGP), associated with negative physiological effects, and hydrolytic products (DG and FFA), released naturally from lipolysis in the gut before absorption. The contribution of both types of compounds helps determine the real nutritional importance of the limitation estab-

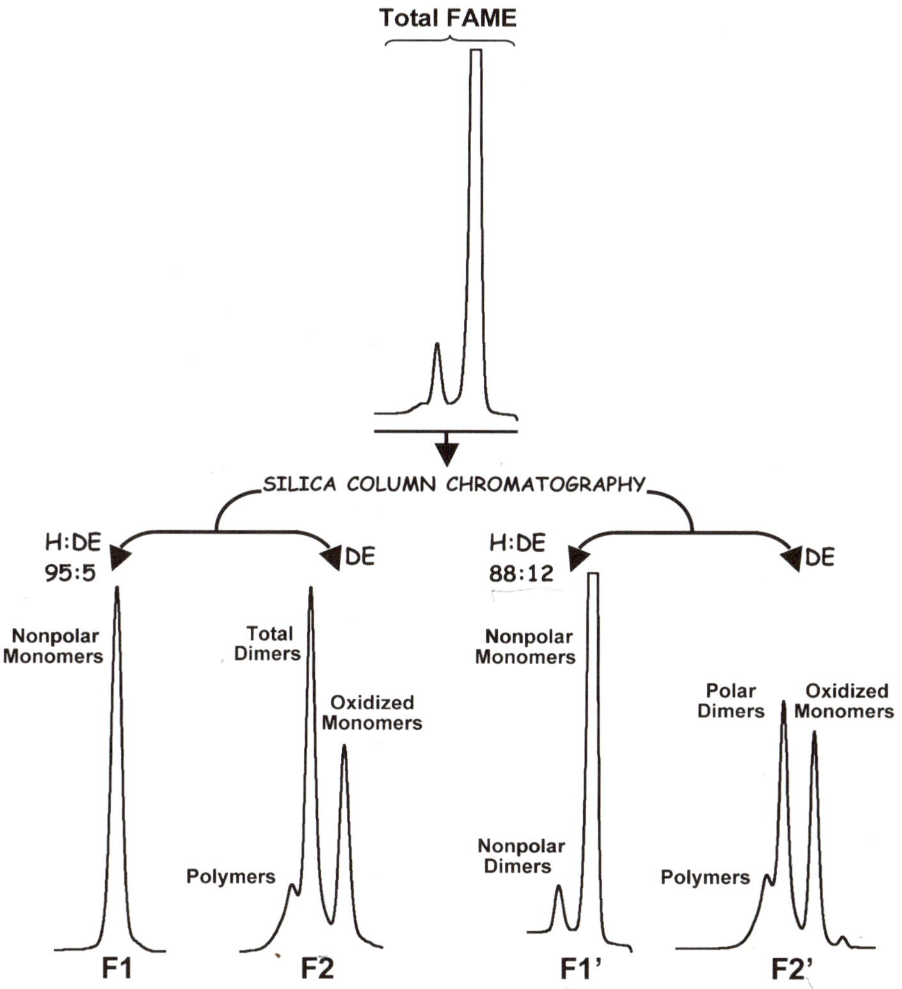

Fig. 5. Schematic representation of methodologies based on separation of fatty acid methyl esters (FAME) by adsorption chromatography and high-performance size-exclusion chromatography. Abbreviations: H, hexane; DE, diethyl ether; and F, fraction.

lished to discard used frying oils (25% polar compounds). Figure 6 reflects the very different patterns of polar compound distribution that can be obtained for used frying oils with practically identical content of total polar compounds (122). Although thermally oxidized compounds predominate in sunflower oil (A), the contribution of hydrolytic compounds is greater in olive oil (B).

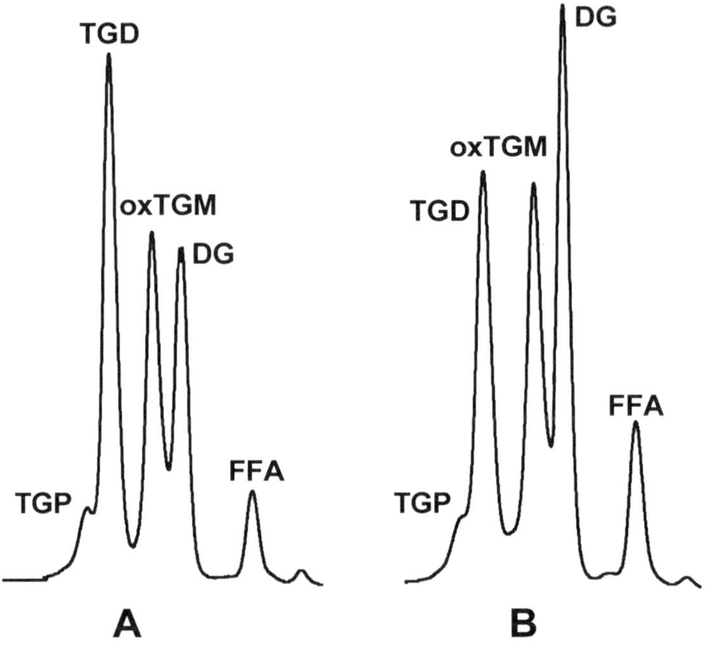

Fig. 6. High-performance size-exclusion chromatograms of polar fractions from used frying sunflower oil (A) and used frying olive oil (B). Abbreviations: TGP, triacylglycerol polymers; TGD, triacylglycerol dimers; oxTGM, oxidized triacylglycerol monomers; DG, diacylglycerols; and FFA, free fatty acids.

In recent years, a plethora of studies on frying fats and oils based on application of this methodology was published, and the results obtained have contributed to improve our knowledge on important issues of the frying process, such as frying performance of different oils (29,52,130–139), the composition of the oils absorbed by the fried food, the lipid interchange between frying oil and food (132,140–142), the relevance of hydrolysis among the reactions occurring during the frying process (52,130,131,140), the action of the main variables involved in continuous and discontinuous frying processes (142,143), and the effect of oil replenishment on oil quality during frying (144–148). In addition to applications for used frying oils, others reported concern thermally oxidized fats and oils (35,149–153), and food lipids and oils subjected to microwave heating (154–158).

The application of HPSEC to the concentrated fractions obtained from both intact samples and FAME, through the silica column-HPSEC and transesterification-silica column-HPSEC procedures, respectively, has contributed greatly to the intensive study of the composition of the polar fraction of used frying oils

(159,160). Analyses of used frying fats collected by Food Inspection Services in Spain showed that samples with polar compounds levels around the limit for rejection (21.1–27.6% polar compounds) gave values of total altered FAME from 8.1 to 11.3%; among these, the major group was constituted by oxidized FAME monomers (~30 mg/g oil). Also, results gave some insight into the complexity of the TGP structure, by comparing TG and FAME dimer and polymer values. In general, the low FAME polymers:TG polymers ratios in contrast to the FAME dimers:TG dimers ratios gave evidence of the considerable contribution of dimeric linkages to the structures of trimeric and higher oligomeric TG (159).

Oxidized fatty acyl groups whose formation does not involve the breakdown of hydroperoxides include epoxy, keto, and hydroxy fatty acyl groups. Interest on the levels and nature of such oxidized compounds in the diet is based on their reported association with promotion and development of cardiovascular disease and cancer (161). As we pointed out, their abundance as a global group of oxidized fatty acyl groups was demonstrated in used frying fats and oils. More recently, a step forward in this line involved the determination of the contribution of epoxides, through analysis of FAME by GLC-MS, to total oxidized FAME monomers, as determined by the transesterification-silica column-HPSEC procedure. Results showed that monoepoxy compounds were one of the major oxidized groups formed at frying temperatures (162–164). Specifically, monoepoxy compounds accounted for ~25% of total oxidized fatty acyl groups in real used frying oils at the limit of rejection. Among future applications, we foresee its use as a complementary preparative technique to concentrate oxidation compounds. In this respect, HPSEC might constitute an excellent analytical tool to facilitate the analysis of specific oxidation compounds by other chromatographic techniques.

Crude and Refined Oils

One of the most interesting applications of HPSEC applied to concentrated fractions is the evaluation of the quality of crude and refined oils; this adds valuable information to the quality specifications introduced in legislation, such as FFA, peroxide value, flavor, color, stability, smoke point, and fatty acid composition. Although such indices are useful to verify whether a certain oil is well refined, no information is provided on its oxidation status, which depends on that of the original crude oil.

Even if the fraction of polar compounds in refined fats and oils is low (unchanged TG normally constitute >95% of the oil), four peaks are well resolved in the polar fraction. This is clearly illustrated in Figure 7, corresponding to polar fractions of virgin olive oil (A) and refined olive oil (B). TGD, formed during the deodorization step, are present in refined oils (B), but are absent or detected in very low amounts in crude oils (A). The oxTGM and DG levels remain practically unchanged after refining because they are not volatile under refining conditions and are not eliminated at any other processing stage; and FFA decrease due to the neutralization step. Thus, quantitation of TGD, oxTGM, and DG provides information

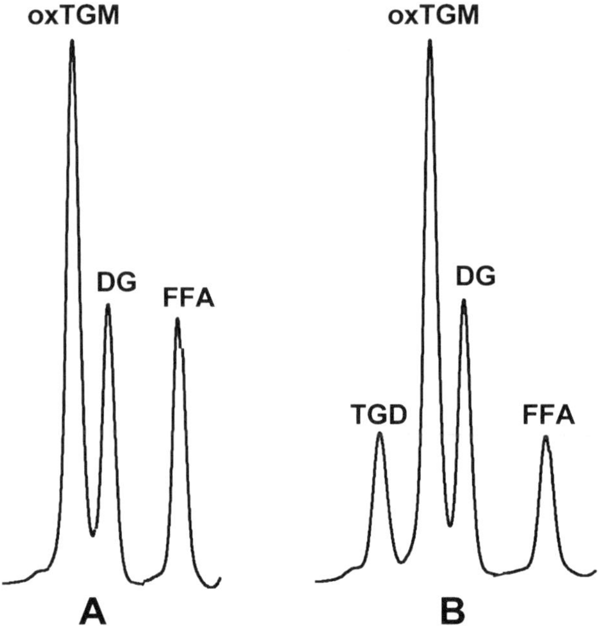

Fig. 7. High-performance size-exclusion chromatograms of polar fractions from virgin olive oil (A) and refined olive oil (B). Abbreviations: TGD, triacylglycerol dimers; oxTGM, oxidized triacylglycerol monomers; DG, diacylglycerols; and FFA, free fatty acids.

on the initial polymerization, oxidation, and hydrolysis, which could affect the subsequent performance of the oil to different extents during storage and utilization for food preparation. This analytical approach was adopted in recent years to study the influence of the refining conditions on fats and oils quality (165–171) and for characterization of virgin vs. refined olive oils because the presence of TGD and a high ratio of DG/FFA indicates that the oil has been refined (172,173).

Oxidized Fats and Oils

For applications to oxidized lipids, in which the polar fraction normally accounts for a relatively small amount, the best-suited methodology is a combination of SPE-HPSEC using an internal standard, which gave lower SD for samples with low concentrations of polar compounds (125). Among the groups of compounds quantitated, oxTGM are of great relevance because this group of compounds comprises all monomeric TG containing at least one oxidized fatty acyl group. Oxygenated functions may be peroxide groups, the primary oxidation products formed, or those characteristic of decomposition and/or further reactions of hydroperoxides, i.e., epoxy compounds, alcohols, or ketones. Therefore, quantitation of oxTGM constitutes an excellent global measurement of the nonvolatile oxidation compounds formed from the initiation reactions through the advanced stages of oxidation. Quantitation of TGD and TGP completes the information obtained on the oxidation state because their increase indicates the onset of the accelerated stage of oxidation. Accordingly, and in contrast to the evaluation of oxidation status through various analytical indices, HPSEC in combination with adsorption chromatography permits accurate quantitation of the total primary and secondary oxidation compounds in a single analysis.

A representative example of a polar fraction obtained using the combination SPE-HPSEC is shown in Figure 8, which corresponds to a sample of sunflower oil stored at room temperature, taken right after the end of the induction period, when tocopherols are exhausted. At that point, oxTGM had accumulated up to 17.7% of the oil (from only 0.9% in the starting sample) and a significant increase in TGD was found (from 0.5% initially to 1.4%), thus indicating that the sample was entering the period of advanced oxidation.

In a recent review, the potential of HPSEC for oxidation studies was described in detail (15); briefly, the main applications addressed were those directed toward clarifying the oxidation profile and progress in oils and food lipids (174–179), gaining information on kinetic parameters (180,181), and detecting the main differences between oxidation in continuous lipid phase, i.e., monophasic lipid systems, and noncontinuous lipid phase (67,182–184).

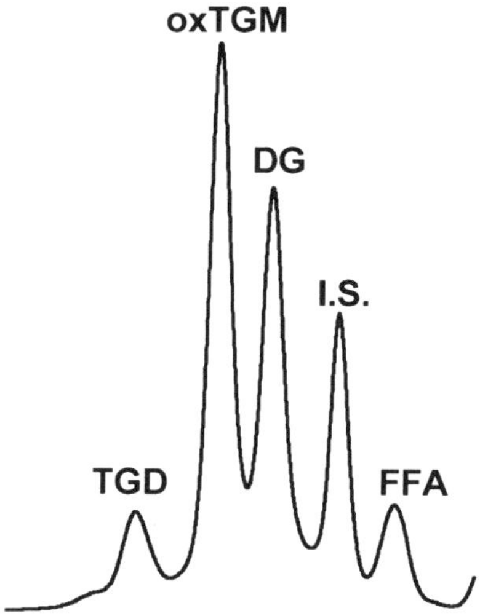

Fig. 8. High-performance size-exclusion chromatogram of polar fractions from sunflower oil oxidized at 25°C. Abbreviations: TGD, triacylglycerol dimers; oxTGM, oxidized triacylglycerol monomers; DG, diacylglycerols; I.S., internal standard (monostearin); and FFA, free fatty acids.

Others

In animal studies, the transesterification-silica column-HPSEC procedure has offered an excellent tool with which to examine digestibility coefficients of various groups of compounds (76,185), given that the products of lipid digestion ultimately absorbed are largely the fatty acids released by pancreatic lipase. After analysis of dietary thermoxidized oils and fecal lipids, high digestibility coefficients, averaging 80%, were found for oxidized fatty acid monomers, thus indicating that such oxidized compounds are of utmost importance from the nutritional standpoint, supported also by their quantitative relevance in the diet. Among polymeric fatty acids, the lowest digestibilities were found for nonpolar dimers, whereas oxidized dimers and polymers possessed higher apparent absorbability than expected. This could be due in part to depolymerization reactions occurring under the strongly acidic conditions in the stomach (76).

Interestingly, digestibility of nonoxidized fatty acids was negatively correlated with the global alteration level of the dietary oil. This finding was attributed to impaired hydrolysis of TGP and TGD which include in part nonoxidized fatty acids, due to the difficulties involved in the pancreatic lipase action on complex glyceridic molecules, as reflected in in vitro experiments (74,75). Also, analyses of fecal lipids confirmed that a considerable fraction of the high-MW compounds ingested remained unhydrolyzed (73).

A different approach was later used by Sánchez-Muniz and co-workers (186–188) to study these nutritional aspects, based on true digestibility measurements by means of esophageal probes. After 4-h experiments in Wistar rats, true digestibility coefficients of thermoxidized palm oleins, measured by lipid disappearance from the intestinal lumen, were significantly lower than those of the unused palm olein. Application of the methodology based on combination silica column-HPSEC to the remaining luminal lipids showed that true digestibility values of TG polymers and TG dimers decreased as the changes in the oil increased and that hydrolysis of nonoxidized TG was negatively affected by the presence of large amounts of thermally oxidized compounds.

Acknowledgment

This work was supported in part by Ministerio de Educación y Ciencia (Project AGL 2004–00148).

References

1. Barth, H.G., and B.E. Boyes, Size Exclusion Chromatography, *Anal. Chem. 62*:381R–394R (1990).
2. Barth, H.G., and B.E. Boyes, Size Exclusion Chromatography, *Anal. Chem. 64:*428R–442R (1992).
3. Barth, H.G., B.E. Boyes, and C. Jackson, Size Exclusion Chromatography, *Anal. Chem. 66:*595R–620R (1994).

4. Barth, H.G., B.E. Boyes, and C. Jackson, Size Exclusion Chromatography, *Anal. Chem. 68:*445R–466R (1996).
5. Barth, H.G., B.E. Boyes, and C. Jackson, Size Exclusion Chromatography and Related Separation Techniques, *Anal. Chem. 70*, 251R–278R (1998).
6. Balke, S.T., T.H. Mourey, and T.C. Schunk, Size Exclusion Chromatography: Practical Methods for Quantitative Results, *Polymer React. Eng. 7*:429–452 (1999).
7. Balke, S.T., T.H. Mourey, and A. Karami, Evaluating Size Exclusion Chromatography Fractionation, *Int. J. Polymer Anal. Charact. 6*:13–33 (2000).
8. Stulik, K., V. Pacakova, and M. Ticha, Some Potentialities and Drawbacks of Contemporary Size-Exclusion Chromatography, *J. Biochem. Biophys. Methods 56:*1–13 (2003).
9. Christie, W.W., *High-Performance Liquid Chromatography and Lipids. A Practical Guide*, Pergamon Press, Oxford, 1987.
10. Christie, W.W., Size Exclusion Chromatography of Lipids, *Lipid Technol. 7*:7–18 (1995).
11. Dobarganes, M.C, and G. Márquez-Ruiz, Size Exclusion Chromatography in the Analysis of Lipids, in *Advances in Lipid Methodology-Two,* edited by W.W. Christie, The Oily Press, Dundee, Scotland, 1993, pp. 113–137.
12. Dobarganes, M.C., and G. Márquez-Ruiz, High-Performance Size-Exclusion Chromatography Applied to the Analysis of Edible Fats, in *New Trends in Lipid and Lipoprotein Analysis*, edited by J.L. Sébédio and E.G. Perkins, AOCS Press, Champaign, IL, 1995, pp. 81–92.
13. Márquez-Ruiz, G., and M.C. Dobarganes, Analysis of Lipid Oxidation Products by Combination of Chromatographic Techniques, in *New Techniques and Applications in Lipid Analysis*, edited by R.E. McDonald and M.M. Mossoba, Technomic Publishing Company, Lancaster, PA, 1997, pp. 216–233.
14. Dobarganes, M.C., and G. Márquez-Ruiz, Analytical Evaluation of Fats and Oils by Size-Exclusion Chromatography, *Analusis 26:*M61–M65 (1998).
15. Márquez-Ruiz, G., and M.C. Dobarganes, Analysis of Non-Volatile Lipid Oxidation Compounds by High-Performance Size-Exclusion Chromatography, in *Analysis of Lipid Oxidation*, edited by A. Kamal-Eldin and J. Pokorny, AOCS Press, Champaign, IL, 2005.
16. Scheffer, P.G., S.J.L. Bakker, R.J. Heine, and T. Teerlink, Measurement of Low-Density Lipoprotein Particle Size by High-Performance Gel-Filtration Chromatography, *Clin. Chem. 43:*1904–1912 (1997).
17. Scheffer, P.G., S.J.L. Bakker, R.J. Heine, and T. Teerlink, Measurement of LDL Particle Size in Whole Plasma and Serum by High-Performance Gel-Filtration Chromatography Using a Fluorescent Lipid Probe, *Clin. Chem. 44:*2148–2151 (1998).
18. Scheffer, P.G., G. Bos, H.G.F.M. Volwater, J.M. Dekker, R.J. Heine, and T. Teerlink, Associations of LDL Size with In Vitro Oxidizability and Plasma Levels of In Vivo Oxidized LDL in Type 2 Diabetic Patients, *Diabetic Med. 20:*563–567 (2003).
19. Nanjee, M.N., J.R. Crouse, J.M. King, R. Hovorka, S.E. Rees, E.R. Carson, J.J. Morgenthaler, P. Lerch, and N.E. Miller, Effects of Intravenous Infusion of Lipid-Free Apo A-I in Humans, *Arterioscler. Thromb. Vasc. Biol. 16*:1203–1214 (1996).
20. Orth, M., M. Hanisch, G. Kroning, M. Porsch-Ozcurumez, H. Wieland, and C. Luley, Fluorometric Determination of Total Retinyl Esters in Triglyceride-Rich Lipoproteins, *Clin. Chem. 44*:1459–1465 (1998).
21. Nanjee, M.N., C.J. Cooke, W.L. Olszewski, and N.E. Miller, Concentrations of Electrophoretic and Size Subclasses of Apolipoprotein A-I-Containing Particles in Human Peripheral Lymph, *Arterioscler. Thromb. Vasc. Biol. 20*:2148–2155 (2000).

22. Westphal, S., M. Orth, A. Ambrosch, K. Osmundsen, and C. Luley, Postprandial Chylomicrons and VLDLs in Severe Hypertriacylglycerolemia Are Lowered More Effectively than Are Chylomicron Remnants After Treatment with n-3 Fatty Acids, *Am. J. Clin. Nutr. 71:*914–920 (2000).
23. Nanjee, M.N., C.J. Cooke, J.S. Wong, R.L. Hamilton, W.L. Olszewski, and N.E. Miller, Composition and Ultrastructure of Size Subclasses of Normal Human Peripheral Lymph Lipoproteins: Quantification of Cholesterol Uptake by HDL in Tissue Fluids, *J. Lipid Res. 42*:639–648 (2001).
24. Yamaguchi, Y., M. Kunitomo, and J. Haginaka, Assay Methods of Modified Lipoproteins in Plasma, *J. Chromatogr. B 781*:313–330 (2002).
25. Lundahl, P., C.M. Zeng, C.L. Hagglund, I. Gottschalk, and E. Greijer, Chromatographic Approaches to Liposomes, Proteoliposomes and Biomembrane Vesicles, *J. Chromatogr. B 722*:103–120 (1999).
26. Grabielle-Madelmont, C., S. Lesieur, and M. Ollivon, Characterization of Loaded Liposomes by Size Exclusion Chromatography, *J. Biochem. Biophys. Methods 56:*189–217 (2003).
27. Stellwagen, E., Gel-Filtration, *Methods Enzymol. 182*:317–328 (1990).
28. Stogiou, M., C. Kapetanaki, and H. Iatrou, Examination of the Universality of the Calibration Curve of Size Exclusion Chromatography by Using Polymers Having Complex Macromolecular Architectures, *Int. J. Polymer Anal. Charact. 7*:273–283 (2002).
29. Abidi, S.L., and K. Warner, Molecular-Weight Distributions of Degradation Products in Selected Frying Oils, *J. Am. Oil Chem. Soc. 78*:763–769 (2001).
30. Kostanski, L.K., D.M. Keller, and A.E. Hamielec, Size-Exclusion Chromatography—A Review of Calibration Methodologies, *J. Biochem. Biophys. Methods. 58*:159–186 (2004).
31. Kulin, L.I., P. Flodin, T. Elligsen, and J. Ugelstad, Monosized Polymer Particles in Size-Exclusion Chromatography. 1. Toluene as Solvent, *J. Chromatogr. 514*:1–9 (1990).
32. Ugelstad, J., P.C. Moerk, K. Herder Kaggerud, T. Ellingsen, and A. Berge, Swelling of Oligomer-Polymer Particles—New Methods of Preparation of Emulsions and Polymer Dispersions, *Adv. Colloid Interface Sci. 13*:101–140 (1980).
33. Hopia, A., V.I. Piironen, P.E. Koivistoinen, and L.E.-T. Hyvoenen, Analysis of Lipid Classes by Solid-Phase Extraction and High-Performance Size-Exclusion Chromatography, *J. Am. Oil Chem. Soc. 69*:772–776 (1992).
34. Hansen, S.L., and W.E. Artz, The Evaporative Light-Scattering Detector, *INFORM 6*:170–176 (1995).
35. Abidi, S.L., I.H. Kim, and K.A. Rennick, Determination of Nonvolatile Components of Heated Soybean Oils Separated with High-Efficiency Mixed-Bed Polystyrene/Divinylbenzene Columns, *J. Am. Oil Chem. Soc. 76*:939–944 (1999).
36. Kaufmann A., B. Ryser, and B. Suter, HPLC with Evaporative Light Scattering Detection for the Determination of Polar Compounds in Used Frying Oils, *Eur. Food Res. Technol. 213*:372–376 (2001).
37. Karami, A., S.T. Balke, and T.C. Schunk, Quantitative Interpretation of Fourier-Transform Infrared Spectroscopic Data from a Size-Exclusion Chromatography Solvent-Evaporation Interface, *J. Chromatogr. 911*:27–37 (2001).
38. Torabi, K., A. Karami, S.T. Balke, and T.C. Schunk, Quantitative FTIR Detection in Size-Exclusion Chromatography, *J. Chromatogr. 910*:19–30 (2001).

39. Laguna, M.T.R., R. Medrano, M.P. Plana, and M.P. Tarazona, Polymer Characterization by Size-Exclusion Chromatography with Multiple Detection, *J. Chromatogr. 919*:13–19 (2001).
40. Mrkvickova, L., V. Pokorna, and J. Pecka, Study of Polymer Complexes by Size Exclusion Chromatography Coupled with Light Scattering Combination with Fluorescence Spectroscopy, *Macromol. Symp. 162*:227–233 (2000).
41. Ludlow, M., D. Louden, A. Handley, S. Taylor, B. Wright, and I.D. Wilson, Size-Exclusion Chromatography with On-Line Ultraviolet, Proton Nuclear Magnetic Resonance and Mass Spectrometric Detection and On-Line Collection for Off-Line Fourier Transform Infrared Spectroscopy, *J. Chromatogr. 857*:89–96 (1999).
42. Perrin, J.L., F. Redero, and A. Prevot, Dosage Rapide des Polymères de Triglycérides par Chromatographie d'Exclusion, *Rev. Fr. Corps Gras 31*:131–133 (1984).
43. Kupranycz, D.B., M.A. Amer, and B.E. Baker, Effects of Thermal-Oxidation on the Constitution of Butterfat, Butterfat Fractions and Certain Vegetable Oils, *J. Am. Oil Chem. Soc. 63*:332–337 (1986).
44. White, P.J., and Y.C. Wang, A High-Performance Size-Exclusion Chromatographic Method for Evaluating Heated Oils, *J. Am. Oil Chem. Soc. 63*:914–920 (1986).
45. Schulte, V.E., Gelchromatographische Bestimmung Polymerisierter Triglyceride, *Fette Seifen Anstrichm. 84*:178–180 (1982).
46. Gere, A., Studies of the Changes in Edible Fats During Heating and Frying, *Nahrung 26*:923–932 (1982).
47. Perrin, J.L., P. Perfetti, C. Dimitriades, and M. Naudet, Etude Analytique Approfondie d'Huiles Chauffées I. Techniques Analytiques et Essais Préliminaires, *Rev. Fr. Corps Gras 32*:151–158 (1985).
48. Waltking, A.E., and M. Wessels, Chromatographic Separation of Polar and Non-Polar Components of Frying Fats, *J. Assoc. Off. Anal. Chem. 64*:1329–1330 (1981).
49. IUPAC Standard Method 2.507: Determination of Polar Compounds in Frying Fats, in *Standard Methods for the Analysis of Oils, Fats and Derivatives*, 7th ed., International Union of Pure and Applied Chemistry, Blackwell, Oxford, England, 1987.
50. Wolff, I.P., F.X. Mordret, and A. Dieffenbacher, Determination of Polymerized Triglycerides in Oils and Fats by High-Performance Liquid Chromatography, *Pure Appl. Chem. 63*:1163–1171 (1991).
51. IUPAC Standard Method 2.508: Determination of Polymerized Triglycerides in Oils and Fats by High Performance Liquid Chromatography, in *Standard Methods for the Analysis of Oils, Fats and Derivatives*, 7th ed., International Union of Pure and Applied Chemistry, Blackwell, Oxford, England, 1992.
52. Masson, L., P. Robert, N. Romero, M. Izaurieta, S. Valenzuela, J. Ortiz, and M.C. Dobarganes, Comportamiento de Aceites Poliinsaturados en la Preparación de Patatas Fritas para Consumo Inmediato: Formación de Nuevos Compuestos y Comparación de Métodos Analíticos, *Grasas Aceites 48*:273–281 (1997).
53. Lampi, A.-M., L.H. Dimberg, and A. Kamal-Eldin, A Study on the Influence of Fucosterol on Thermal Polymerization of Purified High Oleic Sunflower Triacylglycerols, *J. Sci. Food Agric. 79*:573–579 (1999).
54. Gertz, C., Chemical and Physical Parameters as Quality Indicators of Used Frying Fats, *Eur. J. Lipid Sci. Technol. 102*:566–572 (2000).
55. Gertz, C., S. Klostermann, and S.P. Kochhar, Testing and Comparing Oxidative Stability of Vegetable Oils and Fats at Frying Temperature, *Eur. J. Lipid Sci. Technol. 102*:543–551 (2000).

56. Gertz, C., and S.P. Kochhar, New Practical Aspects About Deep Frying Process, *INFORM 13*:386–389 (2002).
57. Soheili, K.C., W.E. Artz, and Preeyanooch-Tippayawat, Pan-Heating of Low-Linolenic Acid and Partially Hydrogenated Soybean Oils, *J. Am. Oil Chem. Soc. 79*:287–290 (2002).
58. Soheili, K.C., Preeyanooch-Tippayawat, and W.E. Artz, Comparison of a Low-Linolenic and a Partially Hydrogenated Soybean Oil Using Pan-Fried Hash Browns, *J. Am. Oil Chem. Soc. 79*:1197–1200 (2002).
59. Neff, W.E., K. Warner, and F. Eller, Effect of γ-Tocopherol on Formation of Non-volatile Lipid Degradation Products During Frying of Potato Chips in Triolein, *J. Am. Oil Chem. Soc. 80:*801–806 (2003).
60. Kiatsrichart, S., M.S. Brewer, K.R. Cadwallader, and W.E. Artz, Pan-Frying Stability of Nusun Oil, a Mid-Oleic Sunflower Oil, *J. Am. Oil Chem. Soc. 80:*479–483 (2003).
61. El-Sayed, F.E., and S.S.M. Allam, Thermoxidative and Hydrolytic Changes in Oils Used for Frying of Frozen Prefried Foods, *J. Food Lipids 10*:285–300 (2003).
62. Coscione, A.R., P.C. Osidacz, S. Kiatsrichart, and W.E. Artz, The Pan-Heating Stability of an Expeller Expressed Soybean Oil, *J. Food Lipids 11*:57–64 (2004).
63. Shukla, V.K.S., and E.G. Perkins, The Presence of Oxidative Polymeric Materials in Encapsulated Fish Oils, *Lipids 26:*23–26 (1991).
64. Sagredos, A.N., On the Quality and Contamination of Fish Oil Capsules, *Fett Wiss. Technol. 94*:101–111 (1992).
65. Burkow, I.C., and R.J. Henderson, Isolation and Quantification of Polymers from Autoxidized Fish Oils by High-Performance Size-Exclusion Chromatography with an Evaporative Mass Detector, *J. Chromatogr. 552:*501–506 (1991).
66. Cho, S.Y., K. Miyashita, T. Miyazawa, K. Fujimoto, and T. Kaneda, Autoxidation of Ethyl Eicosapentaenoate and Docosahexaenoate, *J. Am. Oil Chem. Soc. 64:*876–879 (1987).
67. Márquez-Ruiz, G., J. Velasco, and M.C. Dobarganes, Evaluation of Oxidation in Dried Microencapsulated Fish Oils by Combination of Adsorption and Size Exclusion Chromatography, *Eur. Food Res. Technol. 211*:13–18 (2000).
68. Christopoulou, C.N., and E.G. Perkins, High-Performance Size Exclusion Chromatography of Fatty-Acids, Monoglyceride, Diglyceride and Triglyceride Mixtures, *J. Am. Oil Chem. Soc. 63:*679–684 (1986).
69. Wallis, A.F.A., and R.H. Wearne, Characterization of Resin in Radiata Pine Woods, Bisulfite and Mill Pitch Samples, *Appita J. 50*:409–414 (1997).
70. Fillieres, R., B. Benjellounmlayah, and M. Delmas, Ethanolysis of Rapeseed Oil—Quantitation of Ethyl Esters, Monoglycerides, Diglycerides and Triglycerides and Glycerol by High-Performance Size-Exclusion Chromatography, *J. Am. Oil Chem. Soc. 72*:427–432 (1995).
71. Darnoko, D., M. Cheryan, and E.G. Perkins, Analysis of Vegetable Oil Transesterification Products by Gel Permeation Chromatography, *J. Liquid Chromatogr. Relat. Technol. 23*:2327–2335 (2000).
72. Campbell-Timperman, K., J.H. Choi, and R. Jiménez-Flores, Mono- and Diglycerides Prepared by Chemical Glycerolysis from a Butterfat Fraction, *J. Food Sci. 61*, 44–47 (1996).
73. Márquez-Ruiz, G., M.C. Pérez-Camino, and M.C. Dobarganes, Evaluation of Hydrolysis and Absorption of Thermally Oxidized Olive Oil in Non-Absorbed Lipids in the Rat, *Ann. Nutr. Metab. 37*:121–128 (1993).

74. Márquez-Ruiz, G., M.C. Pérez-Camino, and M.C. Dobarganes, In Vitro Action of Pancreatic Lipase on Complex Glycerides from Thermally Oxidized Oils, *Fat Sci. Technol. 94*:307–312 (1992).
75. Márquez-Ruiz, G., G. Guevel, and M.C. Dobarganes, Application of Chromatographic Techniques to Evaluate Enzymatic Hydrolysis of Oxidized and Polymeric Triglycerides by Pancreatic Lipase *In Vitro*, *J. Am. Oil Chem. Soc. 75:*119–126 (1998).
76. Márquez-Ruiz, G., M.C. Pérez-Camino, and M.C. Dobarganes, Assessments on the Digestibility of Oxidized Compounds from [1-^{14}C]-Linoleic Acid Using a Combination of Chromatographic Techniques, *J. Chromatogr. B 675*:1–8 (1995).
77. Arroyo, R., F.J. Sánchez-Muniz, C. Cuesta, F.J. Burguillo, and J.M. Sánchez-Montero, Hydrolysis of Used Frying Palm Olein and Sunflower Oil Catalyzed by Porcine Pancreatic Lipase, *Lipids 31*:1133–1139 (1996).
78. Arroyo, R., F.J. Sánchez-Muniz, C. Cuesta, J.V. Sinisterra, and J.M. Sánchez-Montero, Thermoxidation of Substrate Models and Their Behaviour During Hydrolysis by Porcine Pancreatic Lipase, *J. Am. Oil Chem. Soc. 74*:1509–1516 (1997).
79. Sánchez-Muniz, F.J., R. Arroyo, J.M. Sánchez-Montero, and C. Cuesta, In Vitro Digestibility Study of Thermal Oxidized Palm Oleins, *Food Sci. Technol. Int. 6*:449–456 (2000).
80. Birch, C.G., and F.E. Crow, Determination of Sucrose Polyesters by High Performance Gel Permeation Chromatography, *J. Am. Oil Chem. Soc. 53:*581–583 (1976).
81. Daher, G.C., K.D. Lawson, D.H. Tallmadge, P. Vanderploeg, and K.W. Miller, Disposition of Ingested Olestra in Weanling Mini-Pigs, *Food Chem. Toxicol. 34*:693–699 (1996).
82. Ríos, J.J., M.C. Pérez-Camino, G. Márquez-Ruiz, and M.C. Dobarganes, Isolation and Characterization of Sucrose Polyesters, *J. Am. Oil Chem. Soc. 71*:385–390 (1994).
83. Márquez-Ruíz, G., J.J. Ríos, M.C. Pérez-Camino, and M.C. Dobarganes, Characterization of Sucrose Polyesters/Triacylglycerols Mixtures, *J. Am. Oil Chem. Soc. 71:*1017–1020 (1994).
84. Gardner, D.R., and R.S. Sanders, Isolation and Characterization of Polymers in Heated Olestra and Olestra/Triglyceride Blend, *J. Am. Oil Chem. Soc. 67:*788–796 (1990).
85. Artz , W.E., S.L. Hansen, and M.R. Myers, Heated Fat-Based Oil Substitutes, Oleic and Linoleic Acid-Esterified Propoxylated Glycerol, *J. Am. Oil Chem. Soc. 74*:367–374 (1997).
86. Dillon, J.G., and M.K. Hughes, Determination of Cholesterol and Cortisone Absorption in Polyurethane. 1. Methodology Using Size-Exclusion Chromatography and Dual Detection, *J. Chromatogr. B 572*:41–49 (1991).
87. Metzger, P., and E. Casadevall, Ether Lipids from *Botrycoccus braunii* and Their Biosynthesis, *Phytochemistry 31:*2341–2349 (1992).
88. Sergent, O., I. Morel, P. Cogrel, M. Chevanne, N. Pasdeloup, P. Brissot, G. Lescoat, P. Cillard, and J. Cillard, Simultaneous Measurements of Conjugated Dienes and Free Malondialdehyde, Used as a Micromethod for the Evaluation of Lipid-Peroxidation in Rat Hepatocyte Cultures, *Chem. Phys. Lipids 65:*133–139 (1993).
89. Zarini, S., and R.C. Murphy, Biosynthesis of 5-Oxo-6,8,11,14-Eicosatetranoic Acid from 5-Hydroperoxyeicosatetranoic Acid in the Murine Macrophage, *J. Biol. Chem. 278:*11190–11196 (2003).
90. Lubke, M., J.L. LeQuere, and D. Barron, Prefractionation of Aroma Extracts from Fat-Containing Foods by High-Performance Size-Exclusion Chromatography, *J. Chromatogr. 729*:371–379 (1996).

91. Hartvigsen, K., L.F. Hansen, P. Lund, K. Bukhave, and G. Holmer, Determination of Neutral Lipid Hydroperoxides by Size Exclusion HPLC with Fluorometric Detection. Application to Fish Oil Enriched Mayonnaises During Storage, *J. Agric. Food Chem. 48*:5842–5849 (2000).
92. Grob, K., and I. Kalin, Attempt for an Online Size Exclusion Chromatography Gas-Chromatography Method for Analyzing Pesticide-Residues in Foods, *J. Agric. Food Chem. 39:*1950–1953 (1991).
93. Vanrhijn, J.A., and L.G.M.T. Tuinstra, Miniaturization of Size-Exclusion Chromatography as a Powerful Cleanup Tool in Residue Analysis, *J. Chromatogr. 552*:517–526 (1991).
94. Depaoli, M., M.T. Barbina, R. Mondini, A. Pezzoni, A.Valentino, and K. Grob, Determination of Organophosphorus Pesticides in Fruits by Online Size-Exclusion Chromatography Liquid-Chromatography Gas-Chromatography Flame Photometric Detection, *J. Chromatogr. 626:*145–150 (1992).
95. Barnes, K.A., J.R. Startin, S.A. Thorpe, S.I. Reynolds, and R.J. Fussell, Determination of the Pesticide Diflubenzuron in Mushrooms by High-Performance Liquid-Chromatography Atmospheric-Pressure Chemical-Ionization Mass-Spectrometry, *J. Chromatogr. A 712*:85–93 (1995).
96. Bester, K., G. Bordin, A. Rodríguez, H. Schimmel, J. Pauwels, and G. Van Vyncht, How to Overcome Matrix Effects in the Determination of Pesticides in Fruit by HPLC-ESI-MS-MS. *Fresenius J. Anal. Chem. 371:*550–555 (2001).
97. Vreuls, J.J., R.J.J. Swen, V.P. Goudriaan, M.A.T. Kerkhoff, G.A. Jongenotter, and U.A.T. Brinkman, Automated On-Line Gel Permeation Chromatography-Gas Chromatography for the Determination of Organophosphorus Pesticides in Olive Oil, *J. Chromatogr. 750*:275–286 (1996).
98. Jongenotter, G.A., M.A.T. Kerkhoff, H.C.M. van der Knaap, and B.G.M. Vandeginste, Automated On-Line GPC-GC-FPD Involving Co-Solvent Trapping and the On-Column Interface for the Determination of Organophosphorus Pesticides in Olive Oils, *HRC-J. High Resolut. Chromatogr. 22*:17–23 (1999).
99. Barrek, S., O. Paisse, and M.F. Grenier-Loustalot, Determination of Residual Pesticides in Olive Oil by GC-MS and HPLC-MS After Extraction by Size-Exclusion Chromatography, *Anal. Bioanal. Chem. 376:*355–359 (2003).
100. DiMuccio, A., P. Pelosi, D.A. Barbini, T. Generali, A. Ausili, and F. Vergori, Selective Extraction of Pyrethroid Pesticide Residues from Milk by Solid-Matrix Dispersion, *J. Chromatogr. 765*:51–60 (1997).
101. Puig, D., and D. Barcelo, Comparative-Study of Various Size-Exclusion Chromatographic Columns for the Cleanup of Selected Pesticides in Soil Samples, *J. Chromatogr. 673:*55–64 (1994).
102. Moate, T.F., M. Furia, C. Curl, J.F. Muniz, J.B. Yu, and R.A. Fenske, Size Exclusion Chromatographic Cleanup for GC/MS Determination of Organophosphorus Pesticide Residues in Household and Vehicle Dust, *J. Assoc. Off. Aanal. Chem. Int. 85:*36–43 (2002).
103. Bester, K, and H. Huhnerfuss, Improvements of a Combined Size Exclusion Chromatography and Solid Phase Extraction Approach for the Clean-Up of Marine Sediment Samples for Trace Analysis of Pesticides. *Fresenius J. Anal. Chem. 358*:630–634 (1997).
104. DiMuccio, A., P. Pelosi, D.A. Barbini, T. Generali, S. Girolimetti, P. Stefanelli, A. Leonelli, G. Amendola, L Vergori, and E.V. Fresquet, Determination of Pyrethroid Pesticide Residues in Fatty Materials by Solid-Matrix Dispersion Partition, Followed by Mini-Column Size-Exclusion Chromatography, *J. Chromatogr. 833:*19–34 (1999).

105. Barman, B.N., V.L Cebolla, and L. Membrado, Chromatographic Techniques for Petroleum and Related Products, *Crit. Rev. Anal. Chem. 30:*75–120 (2000).
106. Bermejo, J., R. Menendez, A.L. Fernández, M. Granda, I. Suelves, A.A. Herod, and R. Kandiyoti, A Comparative Study of the Composition of Anthracene Oil Polymerized by Different Treatments, *Fuel 80*:2155–2162 (2001).
107. Carbognani, L., Preparative Subfractionation of Petroleum Resins and Asphaltenes. I. Development of Size Exclusion Separation Methodology, *Petrol. Sci. Technol. 21:*1685–1703 (2003).
108. Merdrignac, I., C. Truchy, E. Robert, I. Guibard, and S.P. Kressmann, Size Exclusion Chromatography: Characterization of Heavy Petroleum Residues. Application to Residue Desulfurization Process, *Petrol. Sci. Technol. 22*:1003–1022 (2004).
109. Müller, H., and J.T. Andersson, Gel Permeation Chromatography of Sulfur Containing Aromatics in Vacuum Residues, *Polycyclic Aromatic Compounds 24*:299–308 (2004).
110. Lazzari, M., and O. Chiantore, Drying and Oxidative Degradation of Linseed Oil, *Polymer Degrad. Stab. 65:*303–313 (1999).
111. Van den Berg, J.D.J., N.D. Vermist, L. Carlyle, M. Holcapek, and J.J. Boon, Effects of Traditional Processing Methods of Linseed Oil on the Composition of Its Triacylglycerols, *J. Sep. Sci. 27:*181–199 (2004).
112. Chang, T.L., Analysis of Tall Oil by Gel Permeation Chromatography, *Anal. Chem. 40*:989–992 (1968).
113. Inoue, H., K. Konishi, and N. Taniguchi, Analysis of Thermal Polymeric Fatty Acids, Methyl Esters and Alcohols on Sephadex LH-20, *J. Chromatogr. 47*:348–354 (1970).
114. Harris, W.C., E.P. Crowell, and B.B. Burnett, Quantitative Analysis of Polymerized Fatty-Acids Using Gel-Permeation Chromatography, *J. Am. Oil Chem. Soc. 50*:537–539 (1973).
115. Aitzetmüller, K., Gel Filtration Liquid Chromatography of Fatty Acids and Other Fat Samples Using a Pye Liquid Chromatography Detector, *J. Chromatogr. 71:*355–360 (1972).
116. Perkins, E.G., R. Taubold, and A. Hsieh, Gel Permeation Chromatography of Heated Fats, *J. Am. Oil Chem. Soc. 50:*223–225 (1973).
117. Unbehend, V.M., H. Scharmann, H.-J. Strauss, and G. Billek, Anwendung der Gelpermeationschromatographie auf die Untersuchung Thermisch-Oxydativ Belasteter Fette, *Fette Seifen Anstrichm. 75:*689–696 (1973).
118. Jensen, O.N., and J. Moller, Dimer and Trimer Fatty Acids in Feed Fats Characterized by HPLC-GPC and MS, *Fette Seifen Anstrichm. 88*:352–357 (1986).
119. Christopoulou, C.N., and E.G. Perkins, High-Performance Size Exclusion Chromatography of Monomer, Dimer and Trimer Mixtures, *J. Am. Oil Chem. Soc. 66*:1338–1343 (1989).
120. Rao, T.C., V. Kale, P. Vijayalakshmi, A. Gangadhar, R. Subbararo, and G. Lakshminarayama, Chromatographic Methods for the Determination of Monomer, Dimer and Trimer Fractions in Dimer Fatty Acids, *J. Chromatogr. 466:*403–406 (1989).
121. Mittelbach, M., and H. Enzelsberger, Transesterification of Heated Rapeseed Oil for Extending Diesel Fuel, *J. Am. Oil Chem. Soc. 76:*545–550 (1999).
122. Dobarganes, M.C., M.C. Pérez-Camino, and G. Márquez-Ruiz, High Performance Size Exclusion Chromatography of Polar Compounds in Heated and Non-Heated Fats, *Fat Sci. Technol. 90*:308–311 (1988).
123. Dobarganes, M.C., J. Velasco, and Dieffenbacher, The Determination of Polar Compounds, Polymerized Triacylglycerols, Oxidized Triacylglycerols and Diacylglycerols in Fats and Oils, *Pure Appl. Chem. 72:*1563–1575 (2000).

124. Gomes, D., Oligopolymer, Diglyceride and Oxidized Triglyceride Contents as Measures of Olive Oil Quality, *J. Am. Oil Chem. Soc. 69*:1219–1223 (1992).
125. Márquez-Ruiz G., N. Jorge, M. Martín-Polvillo, and M.C. Dobarganes, Rapid, Quantitative Determination of Polar Compounds in Fats and Oils by Solid-Phase Extraction and Exclusion Chromatography Using Monostearin as Internal Standard, *J. Chromatogr. 749*:55–60 (1996).
126. Márquez-Ruiz, G., M.C. Pérez-Camino, and M.C. Dobarganes, Combination of Adsorption and Size-Exclusion Chromatography for the Determination of Fatty Acid Monomers, Dimers and Polymers, *J. Chromatogr. 514*:37–44 (1990).
127. Dobarganes, M.C., and G. Márquez-Ruiz, Dimeric and Higher Oligomeric Triglycerides, in *Deep Frying: Chemistry, Nutrition and Practical Applications*, edited by E.G. Perkins and M.D. Erickson, AOCS Press, Champaign, IL, 1996, pp. 89–111.
128. Dobarganes, M.C., Formation and Analysis of High-Molecular Weight Compounds in Frying Fats and Oils, *Oléagineux Corps Gras Lipides 5*:41–47 (1998).
129. Dobarganes, M.C., G. Márquez-Ruiz, O. Berdeaux, and J. Velasco, Determination of Oxidized Compounds and Oligomers by Chromatographic Techniques, in *Frying of Foods*, edited by D. Boskou and I. Elmadfa, Technomic Publishing Company, Lancaster, PA, 1999, pp. 143–161.
130. Dobarganes, M.C., G. Márquez-Ruiz, and M.C. Pérez-Camino, Thermal Stability and Frying Performance of Genetically Modified Sunflower Oils, *J. Agric. Food Chem. 41:*678–681 (1993).
131. Arroyo, R., C. Cuesta, J.M. Sánchez-Montero, and F. Sanchez-Muniz, High Performance Size Exclusion Chromatography of Palm Olein Used for Frying, *Fat Sci. Technol. 95*:292–296 (1995).
132. Sébédio, J.L., M.C. Dobarganes, G. Márquez, I. Wester, W.W. Christie, G. Dobson, F. Zwobada, J.M. Chardigny, Th. Mairot, and R. Lahtinen, Industrial Production of Crisps and Prefried French Fries Using Sunflower Oils, *Grasas Aceites 47*:5–13 (1996).
133. Márquez-Ruiz, G., R. Garcés, M. León-Camacho, and M. Mancha, Thermoxidative Stability of Triacylglycerols from Mutant Sunflower Seeds, *J. Am. Oil Chem. Soc. 76*:1169–1174 (1999).
134. Masson, L., P. Robert, M. Izaurieta, N. Romero, and J. Ortiz, Fat Deterioration in Deep Fat Frying of French Fries Potatoes at Restaurant and Food Shop Sector, *Grasas Aceites 50*:460–468 (1999).
135. Bastida, S., and F.J. Sanchez-Muniz, Thermal Oxidation of Olive Oil, Sunflower Oil and a Mix of Both Oils During Forty Discontinuous Domestic Fryings of Different Foods, *Food Sci. Technol. Int. 7:*15–21 (2001).
136. Bastida, S., and F.J. Sanchez-Muniz, Polar Content *vs.* TAG Oligomer Content in the Frying-Life Assessment of Monounsaturated and Polyunsaturated Oils Used in Deep-Frying, *J. Am. Oil Chem. Soc. 79*:447–451 (2001).
137. Masson, L., P. Robert, M.C. Dobarganes, C. Urra, N. Romero, J. Ortiz, E. Goicoechea, P. Pérez, M. Salame, and R. Torres, Stability of Potato Chips Fried in Vegetable Oils with Different Degrees of Unsaturation. Effect of Ascorbyl Palmitate During Storage, *Grasas Aceites 53*:190–198 (2002).
138. Abidi, S.L., and K.A. Rennick, Determination of Nonvolatile Components in Polar Fractions of Rice Bran Oils, *J. Am. Oil Chem. Soc. 80*:1057–1062 (2003).
139. Houhoula, D.P., V. Oreopoulou, and C. Tzia, The Effect of Process Time and Temperature on the Accumulation of Polar Compounds in Cottonseed Oil During Deep-Fat Frying, *J. Sci. Food Agric. 83*:314–319 (2003).

140. Pérez-Camino, M.C., G. Márquez-Ruiz, M.V. Ruiz-Méndez, and M.C. Dobarganes, Lipid Changes During Frying of Frozen Prefried Foods, *J. Food Sci. 56*:1644–1648 (1992).
141. Pozo-Díaz, R.M., T.A. Masoud-Musa, M.C. Pérez-Camino, and M.C. Dobarganes, Intercambio Lipídico durante la Fritura de Patatas Prefritas Congeladas en Aceite de Girasol Alto Oleico, *Grasas Aceites 46*:85–91 (1995).
142. Jorge, N., G. Márquez-Ruiz, M. Martín-Polvillo, M.V. Ruiz-Méndez, and M.C. Dobarganes, Influence of Dimethylpolysiloxane Addition to Edible Oils: Performance of Sunflower Oil in Discontinuous and Continuous Laboratory Frying, *Grasas Aceites 47*:20–25 (1996).
143. Jorge, N., G. Márquez-Ruiz, M. Martín-Polvillo, M.V. Ruiz-Méndez, and M.C. Dobarganes, Influence of Dimethylpolysiloxane Addition to Edible Oils: Dependence on the Main Variables of the Frying Process, *Grasas Aceites 47*:14–19 (1996).
144. Cuesta, C., F.J. Sánchez-Muñiz, C. Garrido-Polonio, S. López-Varela, and R. Arroyo, Thermoxidative and Hydrolytic Changes in Sunflower Oil Used in Fryings with a Fast Turnover of Fresh Oil, *J. Am. Oil Chem. Soc. 70*:1069–1073 (1993).
145. Romero, A., C. Cuesta, and F.J. Sánchez-Muniz, Quantitation and Distribution of Polar Compounds in an Extra Virgin Olive Oil Used in Fryings with Turnover of Fresh Oil, *Fett Wiss. Technol. 97*:403–407 (1995).
146. Romero, A., C. Cuesta, and F.J. Sánchez-Muniz, Effect of Oil Replenishment During Deep-Fat Frying of Frozen Foods in Sunflower Oil and High-Oleic Acid Sunflower Oil, *J. Am. Oil Chem. Soc. 75*:161–167 (1998).
147. Cuesta, C., and F.J. Sánchez-Muniz, Quality Control During Repeated Fryings, *Grasas Aceites 49*:310–318 (1998).
148. Romero, A., C. Cuesta, and F.J. Sánchez-Muniz, Does Frequent Replenishment with Fresh Monoenoic Oils Permit the Frying of Potatoes Indefinitely?, *J. Agric. Food Chem. 47*:1168–1173 (1999).
149. Barrera-Arellano, D., M.V. Ruiz-Méndez, G. Márquez-Ruiz, and M.C. Dobarganes, Loss of Tocopherols and Formation of Degradation Compounds in Triacylglycerol Model Systems Heated at High Temperature, *J. Sci. Food Agric. 79*:1923–1928 (1999).
150. Barrera-Arellano, D., M.V. Ruiz-Méndez, J. Velasco, G. Márquez-Ruiz, and M.C. Dobarganes, Loss of Tocopherols and Formation of Degradation Compounds at Frying Temperatures in Oils Differing in Degree of Unsaturation and Natural Antioxidant Content, *J. Sci. Food Agric. 82*:1696–1702 (2002).
151. Verleyen, T., A. Kamal-Eldin, M.C. Dobarganes, R. Verhe, K. Dewettinck, and A. Huyghebaert, Modeling of α-Tocopherol Loss and Oxidation Products Formed During Thermoxidation in Triolein and Tripalmitin Mixtures, *Lipids 36*:719–726 (2001).
152. Verleyen, T., A. Kamal-Eldin, R. Mozuraityte, R. Verhe, K. Dewettinck, A. Huyghebaert, and W. de Grey, Oxidation at Elevated Temperatures: Competition Between Alpha-Tocopherol and Unsaturated Triacylglycerols, *Eur. J. Lipid Sci. Technol.10*:228–233 (2002).
153. Kamal-Eldin, A., J. Velasco, and M.C. Dobarganes, Oxidation of Mixtures of Triolein and Trilinolein at Elevated Temperatures, *Eur. J. Lipid Sci. Technol. 105*:165–170 (2003).
154. García-Ayuso, L. E., J. Velasco, M.C. Dobarganes, and M. D. Luque de Castro, Determination of the Oil Content in Seeds by Focused Microwave-Assisted Söxhlet Extraction, *Chromatographia 52*:103–108 (2000).

155. Caponio, F., A. Pasqualone, M.T. Bilancia, M.T., D. Sacco, D. Delcuratolo, and T. Gomes, Measure of Thermo-Oxidation of Vegetable Oils Microwave-Heated in Model Systems, *Ind. Alimentar. 40*:628–632 (2001).
156. Caponio, F., A. Pasqualone, and T. Gomes, Effects of Conventional and Microwave Heating on the Degradation of Olive Oil, *Eur. Food Res. Technol. 215*:114–117 (2002).
157. Luque-García, J.L., J. Velasco, M.C. Dobarganes and M.D. Luque de Castro, Fast Quality Monitoring of Oil from Prefried and Fried Foods by Focused Microwave-Assisted Söxhlet Extraction, *Food Chem. 76:*241–248 (2002).
158. Priego-López, E., J. Velasco, M.C. Dobarganes, G. Ramis-Ramos, and M.D. Luque de Castro, Focused Microwave-Assisted Söxhlet Extraction: An Expeditive Approach for the Isolation of Lipids from Sausage Products. *Food Chem. 83:*143–149 (2003).
159. Márquez-Ruiz, G., M. Tasioula-Margari, and M.C. Dobarganes, Quantitation and Distribution of Altered Fatty Acids in Frying Fats, *J. Am. Oil Chem. Soc. 72*:1171–1176 (1995).
160. Jorge, N., L.A. Guaraldo-Goncalves, and M.C. Dobarganes, Influence of Fatty Acid Composition on the Formation of Polar Glycerides and Polar Fatty Acids in Sunflower Oils Heated at Frying Temperatures, *Grasas Aceites 48*:17–24 (1997).
161. Dobarganes, M.C., and G. Márquez-Ruiz, Oxidized Fats in Foods, *Curr. Opin. Clin. Nutr. Metab. Care 6*:157–163 (2003).
162. Berdeaux, O., G. Márquez-Ruiz, and M.C. Dobarganes, Characterization, Quantitation and Evolution of Monoepoxy Compounds Formed in Model Systems of Fatty Acid Methyl Esters and Monoacid Triglycerides Heated at High Temperature, *Grasas Aceites 50*:53–59 (1999).
163. Velasco, J., S. Marmesat, G. Márquez-Ruiz, and M.C. Dobarganes, Formation and Occurrence of Short-Chain Fatty Acids and Oxidized Monomeric Triglycerides During Deep-Frying, *Eur. J. Lipid Sci. Technol 106*:728–735 (2004).
164. Velasco, J., S. Marmesat, O. Berdeaux, G. Márquez-Ruiz, and M.C. Dobarganes, Formation and Evolution of Monoepoxy Fatty Acids in Thermoxidized Olive and Sunflower Oils and Quantitation in Used Frying Oils from Restaurants and Fried Food Outlets, *J. Agric. Food Chem, 52:*4438–4443 (2004).
165. Dobarganes, M.C., M.C. Pérez-Camino, G. Márquez-Ruíz, and M.V. Ruiz-Méndez, New Analytical Possibilities for Quality Evaluation of Refined Oils, in *Edible Fats and Oils: Basic Principles and Modern Practices*, edited by D.R. Erickson, AOCS Press, Champaign, IL, 1990, pp. 427–429.
166. Hopia, A., Analysis of High Molecular Weight Autoxidation Products Using High Performance Size Exclusion Chromatography: I. Changes During Processing, *Food Sci. Technol. 26*, 568–571 (1993).
167. Ruiz-Méndez, M.V., G. Márquez-Ruiz, and M.C. Dobarganes, Relationships Between Quality of Crude and Refined Edible Oils Based on Quantitation of Minor Glyceridic Compounds, *Food Chem. 60*:549–554 (1997).
168. De Greyt, W.F., M.J. Kellens, and A.D. Huyghebaert, Polymeric and Oxidized Triglyceride Content of Crude and Refined Vegetable Oils, *Fett/Lipid 99*:287–290 (1997).
169. Gomes, T., and F. Caponio, A Study of Oxidation and Polymerization Compounds During Vegetable Oil Refining, *Riv. Ital. Sost. Grasse 75*:97–100 (1997).
170. Gomes, T., and F. Caponio, Evaluation of the State of Oxidation of Olive-Pomace Oils. Influence of the Refining Process, *J. Agric. Food Chem. 46*:1137–1142 (1998).
171. Gomes, T., F. Caponio, and D. Delcuratolo, Fate of Oxidized Triglycerides During Refining of Seed Oils, *J. Agric. Food Chem. 51*:4647–4651 (2003).

172. Dobarganes, M.C., M.C. Pérez-Camino, and G. Márquez Ruiz, Application of Minor Glyceridic Component Determination to the Evaluation of Olive Oil, in *Actes du Congres International Chevreul pour l'Etude des Corps Gras. Premier Congrès Eurolipid*, Angers, France, 1989, Vol. 2, pp. 578–584.
173. Pérez-Camino, M.C., M.V. Ruiz-Méndez, G. Márquez-Ruiz, and M.C. Dobarganes, Aceites de Oliva Vírgenes y Refinados: Diferencias en Componentes Menores Glicerídicos, *Grasas Aceites 44*:91–96 (1993).
174. Pérez-Camino, M.C., G. Márquez-Ruiz, M.V. Ruiz-Méndez, and M.C. Dobarganes, Determinación Cuantitativa de Triglicéridos Oxidados para la Evaluación Global del Grado de Oxidación en Aceites y Grasas Comestibles, *Grasas Aceites 41*:366–370 (1990).
175. Pérez-Camino, M.C., G. Márquez-Ruiz, M.V. Ruiz-Méndez, and M.C. Dobarganes, Lipid Oxidation in Fats and Fatty Foods. Quantitative Determination of Oxidized Triglycerides, in *Proceedings of Euro. Food Chem. VI,* edited by W. Baltes, T. Eklund, R. Fenwick, W. Pfannhauser, A. Ruiter, and H.P. Thier, Lebensmittelchemische Gesellschaft, Frankfurt, 1991, Vol. 2, pp. 569–574.
176. Hopia, A., Analysis of High Molecular Weight Autoxidation Products Using High Performance Size Exclusion Chromatography: I. Changes During Autoxidation, *Food Sci. Technol. 26*:563–567 (1993).
177. Martín-Polvillo, M., G. Márquez-Ruiz, N. Jorge, M.V. Ruiz-Méndez, and M.C. Dobarganes, Evolution of Oxidation During Storage of Crisps and French Fries Prepared with Sunflower Oil and High Oleic Sunflower Oil, *Grasas Aceites 47*:54–58 (1996).
178. Márquez-Ruiz, G., M. Martín-Polvillo, N. Jorge, M.V. Ruiz-Méndez, and M.C. Dobarganes, Influence of Used Frying Oil Quality and Natural Tocopherol Content on Oxidative Stability of Fried Potatoes, *J. Am. Oil Chem. Soc. 76*:421–425 (1999).
179. Martín-Polvillo, M., G. Márquez-Ruiz, and M.C. Dobarganes, Oxidative Stability of Sunflower Oils Differing in Unsaturation Degree During Long-Term Storage at Room Temperature, *J. Am. Oil Chem. Soc, 81*:577–583 (2004).
180. Márquez-Ruiz, G., M. Martín-Polvillo, and M.C. Dobarganes, Quantitation of Oxidized Triglyceride Monomers and Dimers as a Useful Measurement for Early and Advanced Stages of Oxidation, *Grasas Aceites 47*:48–53 (1996).
181. Márquez-Ruiz, G., M. Martín-Polvillo and M.C. Dobarganes, Effect of Temperature and Addition of α-Tocopherol on the Oxidation of Trilinolein Model Systems, *Lipids 38*:233–240 (2003).
182. Márquez-Ruiz, G., J. Velasco, and M.C. Dobarganes, Oxidation in Dried Microencapsulated Oils, in *Lipid Oxidation Pathways*, edited by A. Kamal-Eldin, AOCS Press, Champaign, IL, 2003, pp. 245–264.
183. Velasco, J., M.C. Dobarganes, and G. Márquez-Ruiz, Oxidation of Free and Encapsulated Oil Fractions in Dried Microencapsulated Oils, *Grasas Aceites 51*:441–448 (2000).
184. Velasco, J., M.C. Dobarganes, and G. Márquez-Ruiz, Antioxidant Activity of Phenolic Compounds in Model Oil-In-Water Emulsions Containing Sunflower Oil, Sodium Caseinate and Lactose, *Eur. J. Lipid Sci. Technol. 106*:325–333 (2004).
185. Márquez-Ruiz, G., M.C. Pérez-Camino, and M.C. Dobarganes, Digestibility of Fatty Acid Monomers, Dimers and Polymers in the Rat, *J. Am. Oil Chem. Soc. 69*:930–934 (1992).
186. González-Muñoz, M.J., C. Tulasne, R. Arroyo, and F.J. Sánchez-Muniz, Digestibility and Absorption Coefficients of Palm Olein—Relationships with Thermal Oxidation Induced by Potato Frying, *Fett/Lipid 98*:104–108 (1996).

187. González-Muñoz, M.J., S. Bastida, and F.J. Sánchez-Muniz, Short-Term In Vivo Digestibility of Triglyceride Polymers, Dimers and Monomers of Thermoxidized Palm Olein Used in Deep-Frying, *J. Agric. Food Chem. 46*:5188–5193 (1998).
188. Sánchez-Muniz, F.J., S. Bastida, and M.J. Gónzalez-Muñoz, Column and High-Performance Size Exclusion Chromatography Applications to the In Vivo Digestibility Study of a Thermoxidized and Polymerized Olive Oil, *Lipids 34*:1187–1192 (1999).

10 Lipid Separations Using Packed-Column Supercritical Fluid Chromatography

Douglas G. Hayes

Department of Biosystems Engineering and Soil Science, University of Tennessee, Knoxville, TN 37996–4531

Introduction

Supercritical fluid chromatography (SFC), which first received major attention in the 1980s, is now an established method for several niche applications for which it is slightly better suited than gas or high performance liquid chromatography, GC and HPLC, respectively. [See Table 1 for a list of review articles (1–10).] After 20+ years of use of commercially available SFC systems, it is now apparent that SFC offers a mixture of advantages and disadvantages compared with GC and HPLC (Table 2). The largest area of growth for SFC has been in the area of pharmaceuticals and for the scale-up of chromatographic separations, due to the environmental friendliness of supercritical CO_2, or SC-CO_2, the most commonly employed supercritical fluid mobile phase (and the focus of this chapter) (11). In addition, CO_2 is inexpensive, and possesses a relatively low critical pressure and temperature, P_c and T_c, 31.1°C and 73.9 bar, respectively. In recent years, SFC has been categorized within the broader term "Unified Chromatography," which most frequently refers to compressible fluids as mobile phases at high pressures and often high temperatures. Common examples of Unified Chromatography mobile phases are substances that exist as gases at standard pressure and temperature, but are liquified at high pressure, and liquids at standard pressure and temperature that are pressurized at temperatures above the normal boiling points to remain in the liquid state. The concept of "unified chromatography" is also related to the common theoretical aspects of column-based chromatographic methods, independent of the state of the mobile phase—gas, liquid, or supercritical fluid [reviewed in (12,13)].

For analytical-scale lipid separations, SFC has found utility for the following analyte matrices as a favorable alternative to HPLC or GC: (i) TAG and other high-molecular-weight lipids, particularly those enriched with multiple double bonds and oxygenated side-chain moieties; (ii) detection of nonsaponifiable minor seed oil components, such as sterols, tocopherols, and carotenes; and (iii) rapid analyses of free fatty acids (FFA) or their methyl esters (FAME).

For the first two applications, GC analysis requires high temperature for elution, which can lead to degradation of double bonds or oxygenated (hydroxyl or epoxy) side-chain groups, and HPLC analysis requires long run times due to the

TABLE 1
Review Articles for Supercritical Fluid Chromatography and Its Applications for Lipid Analyses

Subject	Year published	Comments	Reference
SFC analysis of TAG, DAG, MAG, FFA, and steroids	1992	Emphasis on capillary-SFC	(1)
SFC analysis of lipids	1992	Good early review	(2)
SFC analysis of saccharides and their derivatives	1996	Introduction to SFC, apparatus, and columns; coverage of capillary and packed analyses	(3)
SFC: State of the Art	1996	Good discussion of SFC equipment	(4)
Capillary-SFC analysis of TAG, DAG, MAG, FFA, sterol esters, and PAG	1996	Focus is on comparison of capillary-SFC to capillary GC	(5)
Comparison of SFC with GC and HPLC; overview of applications in lipid separations	1996	Emphasis is on capillary-SFC	(4)
SFC analysis of carotenoids, steroids, and cholesterol	1996	Emphasis is on capillary-SFC	(6)
SFC Analysis of TAG that contain PUFA and poly(hydroxy acids)	1997	Emphasis on capillary-SFC	(7)
SFC analysis sterols, steryl esters, and carotenoids	2001	Includes overview of separation with capillary and packed columns; effect of modifier on retention	(8)
SFC analysis of FFA and FAME	2002	Micro-packed capillary, capillary, and packed-column separations reviewed; emphasis on food products	(9)
SFC: Current State of the Art	2004	Overview of recent developments in instrumentation and applications	(10)

Abbreviations: DAG, diacylglycerol; FAME, fatty acid methyl ester; FFA, free fatty acid; GC, gas chromatography; HPLC, high-performance liquid chromatography; MAG, monoacylglycerol; PAG, phosphoacylglycerol; PUFA, polyunsaturated fatty acids; SFC, supercritical fluid chromatography; TAG, triacylglycerol.

lower diffusivity of analytes in organic solvents compared with SFC. GC analysis of FFA and nonsaponifiable solutes often requires derivatization.

The majority of SFC-related research that has occurred for the three lipid analysis categories listed above has employed systems based on capillary columns. As discussed below, the majority of commercially available systems today are based on packed columns. Since the earlier work emphasizing capillary-SFC is well documented (Table 1), the emphasis of this review will be on packed-column-SFC of lipids.

TABLE 2
Advantages and Disadvantages of Supercritical Fluid Chromatography for Lipid Separations

Advantages compared with HPLC	
	Higher efficiency (longer columns are allowable due to lower pressure drop per unit length)
	Quicker run times by factors of 3–5 (due to increased diffusion and lower pressure drop per unit length, the latter in turn due to decreased viscosity)
	Much lower solvent and solvent disposal costs
	Lower solvent-induced background signal improves ultraviolet-visible and mass spectroscopic detection
	Easier modification of solvent strength: via changing density of CO_2 through pressure and temperature programming rather than gradient elution used in HPLC
	Reduction in equilibration time (e.g., for a change of solvent systems)
Disadvantages compared with HPLC	
	The precision of peak area from run to run is often worse
	Higher capital cost
Advantages compared with GC	
	Nonvolatile and thermally unstable analytes can be separated and detected at moderately low temperatures
	No need for derivatization of FFA, saccharides, or sterols
	Gentler on column wear and tear
Disadvantages compared with GC	
	Lower efficiency

Abbreviations given in Table 1.

Equipment

Packed vs. Capillary Columns

During the 1980s and early 1990s, SFC systems featuring capillary (or "open tubular") columns were the most frequently employed systems to perform the three categories of lipid separations discussed above. Such systems more closely mimic GC behavior, separating analytes based on their relative volatility, hence by their molecular weight. (See Table 1 for reviews.) A common capillary-SFC system in the 1980s and early 1990s was the Lee Scientific Model 600 system from Dionex (Sunnyvale, CA), featuring a syringe pump to transport pure CO_2 through the capillary column and timed-split injection for sample introduction. The column effluent was depressurized using a capillary tube known as a frit restrictor before entrance into a Flame Ionization Detector, or FID, a commonly employed GC detector for analysis of neutral lipids due to its sensitivity and the linearity of its signal. Although Dionex no longer manufactures SFC systems, Selerity, Inc. (Salt Lake City, UT) provides systems using technology similar to that of the Dionex/Lee Scientific models. An alternative to capillary "open tubular" columns are "packed microcolumns," in which capillary tubing is packed with HPLC-like packing beads. Cited as disadvantages for packed microcolumns are low flow rate and reproducibility (14).

The majority of the SFC systems available today feature packed columns (Table 3). The underlying cause of the current shift in direction of the SFC market appears to be the application of SFC for pharmaceuticals on both an analytical and preparative scale, due to the environmental friendliness of CO_2 as a mobile phase, the savings in solvent and solvent disposal costs, and the more effective concentration of solutes from fractions collected merely via depressurization (10,11). In contrast, capillary-SFC, similar to its "cousin," capillary-GC, cannot be effectively scaled up.

Packed-column-SFC systems are HPLC-like in their design and mode of separation [reviewed in (15)]. One such system is depicted in Figure 1. The typical packed-column-SFC system consists of an HPLC-like diaphragm or reciprocating pump used to transport CO_2. This stream is mixed with an organic solvent modifier solution at a T-junction downfield of the CO_2 pump, with the modifier also transported by an additional HPLC-like pump. Effective mixing characteristics are readily obtained, as long as pressure and temperature coincide with the 1-phase region of the phase diagram. The combined stream is passed through an HPLC-like multiposition injection valve with sample loop, and then transported through the packed column. In general terms, the latter is usually very similar to an HPLC column except that it is packed at a higher pressure. The pressurized stream then passes through a detector, of which the most common is an ultraviolet-visible (UV-vis) detector equipped with a high-pressure sample cell. Unlike most HPLC lipid separations, UV detection at low wavelengths (190–230 nm) is possible because SC-

TABLE 3
Manufacturers of Supercritical Fluid Chromatography Equipment and Columns

Manufacturer	Product
Chiral Technologies, France	Analytical and preparative chiral packed columns
Jasco, Easton, MD	Analytical systems with ultraviolet (UV)-visible, circular dichroism, and evaporative light scattering detectors
Mettler-Toledo Auto-Chem, Columbia, MD	Berger analytical and preparative models; UV, diode array, polarimeter, and mass spectrometry detectors; analytical and preparative packed columns
NovaSep, Boothwyn, PA	Preparative scale and simulated moving bed systems
Princeton Chromatography, Cranbury, NJ	Analytical and preparative packed columns
Selerity, Salt Lake City, UT	Model 400 Series, designed for capillary columns; capillary and silica packed columns
Thar Technologies, Pittsburgh, PA	Analytical and preparative systems
Western Analytical Products, Murrieta, CA	Analytical columns, including cyanopropyl and diol stationary phases

CO_2 does not promote a high background signal. FID detectors are less common because the presence of most modifiers interferes with the detector signal, with water and formic acid at low levels serving as exceptions (16). Perhaps the second most abundant packed-SFC detector is the evaporative light scattering detector, or ELSD. Compared with the UV-vis detector, it provides greater sensitivity, lower baseline noise, and can be used to detect a wider variety of analytes. A disadvantage compared with the UV-vis detector is the nonlinearity of the relation between detector signal and mass of analyte; typically, a log-log relation exists. In addition, MS detectors compatible with SFC systems comprise a continuing area of research and development, with SFC-compatible MS detectors commercially available (Table 3) (10,17).

A major complaint about traditional capillary-SFC systems is the variability of mobile phase flow rate, hence to peak retention times, due to the ever-changing resistance to flow occurring in the frit restrictor (e.g., clogging). Most packed-SFC systems employ back-pressure regulators downstream of the detectors to control

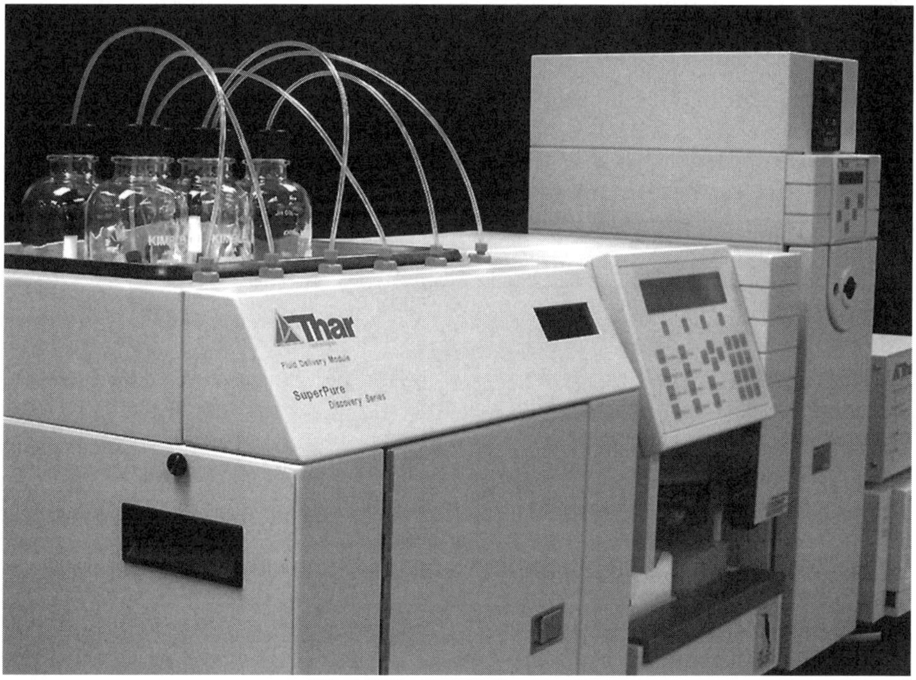

Fig. 1. Analytical packed-column SFC system from Thar Technologies. (A) SFC pump, with modifier solvent reservoirs; (B) fraction collector; (C) column oven; (D) ultraviolet-visible detector; (E) back-pressure regulator. Not pictured: a personal computer for system control and data acquisition. Reproduced with permission from Thar Technologies, Pittsburgh, PA.

column pressure, hence flow rate. Back-pressure regulators are interfaced to pressure transducers to allow for continuous control for maintaining constant pressure. In capillary-SFC systems, frit restrictors control both the mobile phase pressure and velocity. In contrast, pump reciprocation can be used for packed-SFC to control velocity independently of the back-pressure regulator setting.

Modifiers

For capillary-SFC, the most important operational parameter employed for optimizing the degree of separation is the density of the mobile phase, most frequently controlled by the pump's discharge pressure and the column oven temperature (as well as the condition of the frit restrictor). In contrast, for packed SFC, mobile phase pressure and column temperature are secondary in importance to type, concentration, and programming of co-solvent, or "modifier" (15). Common modifiers and their physical properties are presented in Table 4 (16,18,19). The most common modifier is methanol, due to its strong polarity and relatively large miscibility with SC-CO_2. The main role of the modifier is to increase the solvent strength and deactivate stationary phase desorption sites, particularly unmodified silanol sites of silica-based packing materials in the presence of strongly adsorbing polar analytes. It is common to include a basic or acidic additive in the mobile phase and to

TABLE 4
Properties of Common Modifiers

Modifier	T_c^a (°C)	P_c^a (bar)	Dielectric constant at 20°C[a]	Log P
CO_2	31.1	73.9		
Trifluoroacetic acid	218	32.58	8.55[b]	−2.1[c]
Formic acid	307	315.0	58.5[b]	−1.55[d]
Dimethyl sulfoxide	465	229.9	46.7	−1.35
Methanol	239	81.0	32.7	−0.65
Acetonitrile	275	48.3	37.5	−0.33
Ethanol	243	63.8	24.3	−0.30
1,4-Dioxane	314	52.1	2.25	−0.27[e]
Acetone	235	47.0	20.7[b]	−0.23
Water	374	220.5	80.1	0
1-Propanol	264	51.7	20.3	0.02
2-Propanol	235	47.6	18.3	0.08
Tetrahydrofuran	267	51.9	7.58	0.46
Dichloromethane	237	60.8	8.93	0.60
Triethylamine	263	30.3	2.42[f]	1.45

[a] *Source:* Reference 16.
[b] At 25°C. *Source:* http://www.cem.msu.edu/~reusch/OrgPage/solvent.htm.
[c] *Source:* http://www.afeas.org/environ.html.
[d] *Source:* Reference 19.
[e] *Source:* Reference 18.
[f] *Source:* http://www.daylight.com/dayhtml/doc/clogp/clogp_examples.html.

increase the column oven temperature to further decrease the strong adsorption (20). As little as 1–5% of modifier frequently fulfills the desired role. An optimal modifier type, amount, or program is difficult to predict a priori because the modifier's behavior strongly depends on the combination of stationary phase and analytes employed. (21). Density and phase behavior of binary systems can be estimated using the approaches discussed in standard textbooks (22), with the density of SC-CO_2 as a function of temperature and pressure readily calculated using software kindly provided to the general public by Prof. D. Bush of Georgia Institute of Technology on his website (23).

An important consideration is the lowering of the phase boundaries of the 1-phase supercritical region due to the elevation of P_c and T_c relative to pure CO_2 . For P≤90 bar and T≤29°C, at most 12 mol% methanol can be solubilized; the solubility decreases to 1.5% for $CHCl_3$ and acetonitrile as modifiers (16). Phase behavior data for the SC-CO_2-methanol binary system are contained in the literature (24,25).

Subcritical Fluid Chromatography

The use of near- or subcritical solvent systems, e.g., with temperatures just below the T_c for CO_2, 31.1°C, has increased in popularity in recent years for packed-column separations. Subcritical fluid chromatography allows for enhanced control of the selectivity by the stationary phase relative to SFC. Although solute diffusivity and solvent viscosity are not as favorable as those that occur in the supercritical fluid state, values are highly favorable compared with typical HPLC solvent systems. The solubility and retention behavior of analytes can differ greatly using subcritical fluid chromatography compared with HPLC (26). Faster run times are reported for analytical-scale subcritical fluid chromatography compared with HPLC due to the former's higher mobile phase flow rate (3–5 mL/min, compared with ~1.0 mL/min for HPLC) (27). An analysis of modifier type and concentration on subcritical fluid chromatography was made recently, yielding relations between retention time and modifier amount that are difficult to express mathematically and cannot be applied universally (28,29).

Applications of Packed-column-SFC of Lipids

Triacylglycerols (TAG), FFA, and Glyceride Mixtures

Packed-column-SFC separations of TAG and FFA mixtures contained in the literature are given in Table 5 (26,29–59). In general, the columns employed, hence the mode for separation, closely resemble that encountered for HPLC. Nonpolar or reversed-phase stationary phases (e.g., RP-C_{18}) provide separation more by means of molecular weight difference, and polar stationary phases (e.g., silica, diol, cyanopropyl, and amino) separate according to the degree of unsaturation. Equiva-

TABLE 5
Example of Packed-SFC-Based Separations of Triacylglycerols (TAG), Free Fatty Acids (FFA), and Glyceride Mixtures Using CO_2-Based Mobile Phases

Analyte matrix	Stationary phase	SFC conditions	Important results	References
TAG containing saturated acyl groups	Cyanopropyl, RP-C_{18}	Methanol modifier (isocratic, 2.4–10 vol %), pressure programming, isothermal (40–150°C), ELSD	Effective separation based on carbon number	(30,31)
Common TAG mixtures separated based on degree of unsaturation	Silica tubing packed with cation exchange stationary phase and impregnated with $AgNO_3$ or K_2MnO_4	Acetonitrile and 2-propanol modifiers, pressure and temperature programming, ELSD	Comparable separation effectiveness as HPLC	(32–36)
Common TAG mixtures separated based on degree of unsaturation	Cation exchange column, impregnated with $AgNO_3$	Acetonitrile/2-propanol modifiers, positive pressure and gradient programming, isothermal (65°C), UV-vis (210 nm), MS-APCI and -CIS detectors	Allowed for identification of acyl species on the glycerol backbone	(37)
Various seed oil TAG (e.g., perilla, soybean); cocoa butter; TAG mixtures; separation based on MW and degree of saturation	RP-C_{18}	No modifiers, isobaric (150 or 200 bar), isothermal (0–25°C)[a], UV-vis (210 nm)	Separations improved in subcritical (<30°C) compared with supercritical region; selectivity significantly affected by temperature; separation well characterized using equivalent chain lengths	(38,39)

(Continued)

TABLE 5 *(continued)*

Analyte matrix	Stationary phase	SFC conditions	Important results	References
TAG from Argan oil; mixtures based on degree of unsaturation	RP-C_{18}	Methanol and acetonitrile modifiers; various pressure temperature[a], and modifier programming investigated; UV-vis (210 nm) and ELSD	Improved separation vs. RP-HPLC; optimization is a complex function of modifier type, concentration, pressure, and temperature; retention detectors times are linearly related to carbon number, and to degree of unsaturation per molecule for analytes of the same molecular weight	(26,29,40)
Separation of TAG from hydroperoxides	Silica	Ethanol modifier (7%), isothermal (40°C), isobaric (120 bar), UV-vis (230 nm)	A photodiode array detector is useful to differentiate TAG from hydroperoxides due to differences in UV spectra	(41)
Separation of MAG and DAG	Silica and RP-C_{18}	Methanol modifier (isocratic, 2.5–5%), isobaric (229 bar), isothermal (40°C), UV-vis and IR detector	SFC provides similar degree of separation as HPLC and GC	(42)
Separation of TAG from DAG, MAG, FFA, carotenes, sterols, and steryl esters	Aminopropyl silica gel (43); RP-C_{18} (44)	Methanol or ethanol modifier (10%), isobaric (150–180 bar), isothermal (70°C), UV-vis (215 nm)	Near-baseline separation of lipid classes; semi-preparative scale	(43,44)
Partial separation of 1,3 and 1,2-DAG	Benzoylamidopropyl	No modifier, pressure programming, isothermal (70°C) MS detector	Partial separation	(45)

(Continued)

TABLE 5 *(continued)*

Analyte matrix	Stationary phase	SFC conditions	Important results	References
FFA mixtures	RP-C_{18}	No modifier, pressure programming, isothermal (45°C), UV-vis (190 nm)	HPLC-like separations	(46)
Mixture of octadecyl-FAME; components differ in number and type of double bonds	Silica	No modifier, CO_2 density held constant at 841 kg/m^3 (200 bar, 45°C), ELSD	Generally, baseline separations; partial separation of 14:0 and 16:0 and 18:1-*cis* and 18:1-*trans*	(47–49)
Fish oil FAME	Silica capillary tubing micro-packed with cation exchange stationary phase and impregnated with $AgNO_3$ or K_2MnO_4	Acetonitrile/2-propanol modifier system (isocratic, 2.6–6.0% and 0.3–0.6%, respectively); pressure and temperature programming, UV-vis (210 nm)	Excellent separations of 18:2-9,12 and 18:3-9,12,15 *cis*/*trans* isomers	(50)
Saturated FAME	Silica, deactivated with polymethylhydrosiloxane and crosslinked 5% phenyl/95% dimethylpolysiloxane	No modifier; pressure programming, isothermal (85°C), FID	Deactivation of silica reduced run time and improved peak shape	(51)
FAME that differ in degree, position, and type of saturation; fish oil-derived FAME	Cyanopropyl micro-packed capillary column deactivated with polymethylhydrosiloxane and treated with Ag^{2+}	No modifier; several different temperatures and pressure programs, FID	Good separations	(52)
FAME that differ in degree of saturation and chain length	Cyanobiphenyl micro-packed capillary column deactivated with polymethylhydrosiloxane and treated with Ag^{2+}	No modifier, pressure programming, 85°C, FID	Good separations	(53)

(Continued)

TABLE 5 *(continued)*

Analyte matrix	Stationary phase	SFC conditions	Important results	References
Saturated FFA	RP-C_8 and cyanopropyl columns	CO_2 saturated with water; 70°C, pressure programming, FID	Addition of water as modifier improved resolution and sensitivity	(54)
Fish oil FAEE (separation of DHA and EPA)	RP-C_{18}	Pure CO_2 mixed with significant amount of ethanol from sample injection, isobaric (145 bar), isothermal (65°C), UV-vis	Preparative scale	(55)
FAME from perilla, soybean, and fish oil; butter fat	2-dimensional: silica, followed by RP-C_{18}	No modifier; isobaric and isothermal; UV-vis and FID	Good separation	(56,57)
FAME mixtures with varying carbon number and degree of saturation	Silica column; packed beads were deactivated and cross-linked	No modifier, isobaric (238–380 bar), isothermal (50°C), ELSD	Good separation; partial separation of 18:1-9 *cis/trans* isomers; degree of saturation was the most influential factor controlling the degree of separation	(58)
Fish oil FFA; other FA mixtures	Aminopropyl	No modifier, isobaric (150 bar), isothermal (40°C) , UV-vis (200 nm)	As stationary phase density increased, resolution based on degree of saturation increased, but resolution based on chain length decreased	(59)

Abbreviations: APCI, atmospheric pressure chemical ionization; CIS, coordination ion spray; ELSD, evaporative light scattering detector; FID, flame ionization detector; MS, mass spectroscopy; and RP, reversed phase; and UV-vis, ultraviolet-visible spectrophotometry detector. For other abbreviations, see Table 1.

[a]Subcritical fluid chromatography ($T < 30°C$).

lent carbon numbers, a technique commonly employed for GC, is an effective predictor of retention for TAG in SFC. Although most cited works employed UV-vis or ELSD detection, a recent investigation demonstrated the potentially efficient detection of MS (37). In that investigation, the chemical identification of chromatographically separated TAG isomers that differed in acyl chain positions occurred via MS (37). FFA analytes were particularly vulnerable to strong adsorption to underivatized silanol stationary phase sites, requiring the use of a modifier. However, good separations were achieved for FFA mixtures without derivatization of the analytes [reviewed in (9)]. Packed-SFC fractionation was successfully scaled up for the isolation of docosahexaenoic acid and eicosapentaenoic acid from fish oils for numerous applications in foods and pharmaceuticals (55,60).

Phosphoacylglycerols (PAG), Sterols, Tocopherols, and Other Seed Oil Minor Components

Examples of packed-SFC separations of minor components are listed in Table 6 (27,61–70) and reviewed elsewhere (8). The majority of separations occurred using a RP-C_{18} column (Table 6), although a recent investigation achieved excellent separation of a tocopherol/tocotrienol mixture using a diol packed column (70). Taylor et al. (71) and Taylor and King (72) successfully coupled supercritical fluid extraction and SFC to purify tocopherol- and sterol-rich products. Choo et al. (44) purified tocopherols and tocotrienols from palm oil using normal-phase packed-SFC.

Packed-column-SFC Separation of Saccharides

Although not classified as lipids, saccharides are common building blocks for many lipid products such as fatty acid-saccharide esters and glucosides. Saccharides are readily separated using polar stationary phases such as diol and cyanopropyl [Table 7 and reviewed in (3)] (3,66,73–76). SFC provides enhanced resolution of saccharides for many saccharide separations compared to HPLC. Unlike GC analysis, derivatization of saccharides is not required. The papers cited employed an ELSD.

Conclusions

Packed-SFC remains an important method for the analytical and preparative scale separations of TAG, FFA, and minor components of seed oils. Although packed-SFC-generated resolution of chromatographic peaks provides a modest improvement at best compared with that achieved using GC or HPLC for many of the lipid separations cited in the literature, interest will continue due to the environmental friendliness of CO_2 as a mobile phase and the low operating temperatures, particularly for preparative-scale separations. The escalating production and waste

TABLE 6
Example of Packed-SFC-Based Separations of Phosphoacylglycerols (PAG), Sterols, Tocopherols, Carotenes, and other Minor Seed Oil Components[a]

Analyte matrix	Stationary phase	SFC conditions	Important results	References
PAG separations based on head group (from soybean)	Silica	Methanol/H_2O/triethylamine ternary system modifier (20.52, 1.01, 0.01%, respectively), isobaric (278 bar),isothermal (45°C),[a] ELSD	Effective separation; several trials required to optimize gradient rate and other conditions	(61)
PAG separations based on head group and acyl chains	RP-C_8	Ethanol/methanol/trifluoro-acetic acid ternary system modifier (4.5, 4.5, and 0.9%, respectively) undergoing a positive gradient elution, isobaric (125 bar), isothermal[a] (70°C), ELSD	Effective separation; several trials to optimize gradient rate and other conditions	(62)
Tocopherols	Silica	Dichloromethane modifier (4.7%),isobaric (120–240 bar), isothermal (37–50°C), UV (290 nm)	Baseline separation of α, β, γ, and δ-tocopherols	(42)
Cholesterol and its esters	RP-C_{18}	No modifier, isobaric (200 bar), isothermal (45°C), UV-vis (190 nm) and FID detectors	Response factors for UV were dependent upon degree of unsaturation for cholesterol esters; baseline separation	(63)

(Continued)

TABLE 6 *(continued)*

Analyte matrix	Stationary phase	SFC conditions	Important results	References
Tocopherols	RP-C_{18}	Methanol modifier (0.5–15%), isobaric (120–300 bar), isothermal (40°C), UV-Vis (290 nm)	Retention was a linear function of fluid density, baseline separations of α-, β-, γ, δ-tocopherols within 25 min; optimal modifier concentration for maximizing resolution	(64)
β-Carotene: retention behavior	RP-C_{18}, NH_2	No modifier, isobaric (100–300 bar), isothermal (308–323°C), UV-vis (450 nm)	Retention for C_{18} columns depended on solubility in fluid phase, retention behavior similar to RP-HPLC	(65)
Steroids	Silica	Methanol modifier (10%), isobaric (210 bar), isothermal (40–70°C), ELSD	Broad peaks, but baseline separation achieved in most instances	(66)
Sterols/cholesterols	Cyanopropyl	Methanol modifier (isocratic, 5%,and positive gradient), either isobaric (200 bar) or positive pressure gradient, isothermal (50°C), ELSD	Baseline separation; run times near 10 min	(67)
β-Carotenes	RP-C_{18}	Methanol and acetonitrile modifiers (5–15%) or methanol/acetonitrile system (0.4 and 5.6%, respectively), isobaric (100–150 bar), isothermal (25–50°C)[a], UV-Vis (450 nm)	Two different types of C^{18} columns coupled together to achieve separation of *cis/trans* isomers	(68)

(Continued)

TABLE 6 *(continued)*

Analyte matrix	Stationary phase	SFC conditions	Important results	References
Tocopherols	RP-C_{18}	Ethanol modifier (isocratic, 0–6.5%),isothermal, isobaric (120–200 bar),UV-vis (205 nm)	Resolution of α, γ-, and δ-tocopherols maximized with optimal proportion of ethanol, and was increased with a decrease in pressure and an increase in temperature	(69)
Sphingolipids and glycolipids	Silica, diol	Methanol (positive gradient), isobaric (100 bar), isothermal (40°C)[a], ELSD	Best separation occurred when diol and silica columns coupled	(27)
Tocopherols and tocotrienols	Diol	Methyl-*tert*-butyl ether modifier (5.7%), isobaric (190 bar), isothermal (40°C), UV-vis	Excellent, near-baseline separation of α-, β-, γ- and δ-tocopherols and -tocotrienols	(70)

[a]Subcritical fluid chromatography (T < 30°C).
Abbreviations given in Table 5.

TABLE 7
Example of Packed-SFC-Based Separations of Saccharides and Their Derivatives[a]

Analyte matrix	Stationary phase	SFC conditions	Important results	References
O-Alkylglucosides	Cyanopropyl and diol	Methanol modifier (6.25%), isobaric, isothermal (40°C), ELSD	Good separation	(73)
Mono- and di- and trisaccharides	Cyanopropyl, diol and NO_2	Methanol modifier (isocratic, 4–16%), isothermal, isobaric, isocratic and positive ramp of methanol, ELSD	Baseline separations; greater range of selectivity than HPLC but same retention mechanism	(3,74,75)
Mono and di-saccharides	Cyanopropyl, β-cyclodextrin	Methanol modifier, isobaric (225–250 bar), isothermal (85–95°C), positive modifier gradient, ELSD	Near-baseline separations within 12 min	(66)
Saccharides and sugar alcohols	Silica, Zorbax TMS	Methanol, methanol/water, and methanol/water/triethylamine modifier systems, isobaric, isothermal[a], ELSD	Temperature increase led to increased retention; co-modifiers of methanol, water, and triethylamine perform best	(76)

[a]Subcritical fluid chromatography (T < 30°C).
Abbreviations given in Table 5.

removal cost of organic solvents and improved equipment and column designs may perhaps increase further interest in SFC as a viable alternative to HPLC.

References

1. Laakso, P., Supercritical Fluid Chromatography of Lipids, in *Advances in Lipid Methodology—One*, edited by W.W. Christie, Oily Press, Ayr, UK, 1992, pp. 81–119.
2. Bartle, K.D., and T.A. Clifford, Supercritical Fluid Extraction and Chromatography of Lipids and Related Compounds, in *Advances in Applied Lipid Research*, edited by F. B. Padley, JAI Press, London, 1992, pp. 217–263.
3. Lafosse, M., B. Herbreteau, and L. Morin-Allory, Supercritical Fluid Chromatography of Carbohydrates, *J. Chromatogr. A 720*:61–73 (1996).
4. Chester, T.L., J.D. Pinkston, and D.E. Raynie, Supercritical Fluid Chromatography and Extraction, *Anal. Chem. 68*:487R–514R (1996).
5. Sandra, P., and F. David, Basic Principles and the Role of Supercritical Fluid Chromatography in Lipid Analysis, in *Supercritical Fluid Technology in Oil and Lipid Chemistry*, edited by J.W. King and G.R. List, AOCS Press, Champaign, IL, 1996, pp. 321–347.
6. Artz, W.E., Supercritical Fluid Chromatography of Trace Components in Oils and Fats, in *Supercritical Fluid Technology in Oil and Lipid Chemistry*, edited by J.W. King and G.R. List, AOCS Press, Champaign, IL, 1996, pp. 376–386.
7. Hayes, D.G., Analysis of Unusual Triglycerides and Lipids Using Supercritical Fluid Chromatography, in *New Techniques and Applications in Lipid Analysis*, edited by R.E. McDonald and M.M. Mossoba, AOCS Press, Champaign, IL, 1997, pp. 163–182.
8. Lesellier, E., Analysis of Non-Saponifiable Lipids by Super-/Subcritical-Fluid Chromatography, *J. Chromatogr. A 936*:201–214 (2001).
9. Señorans, F.J., and E. Ibañez, Analysis of Fatty Acids in Foods by Supercritical Fluid Chromatography, *Anal. Chim. Acta 465*:131–144 (2002).
10. Chester, T.L., and J.D. Pinkston, Supercritical Fluid and Unified Chromatography, *Anal. Chem. 76*:4606–4613 (2004).
11. Harris, C.M., The SFC Comeback, *Anal. Chem. 74*:87A–91A (2002).
12. Parcher, J.F., and T.L. Chester, *Unified Chromatography (ACS Symposium Series No. 748)*, American Chemical Society, Washington, DC, 2000.
13. Tong, D., K.D. Bartle, and T.A. Clifford, Principles and Applications of Unified Chromatography, *J. Chromatogr. A 703*:17–35 (1995).
14. Blomberg, L.G., M. Demirbüker, I.Hågglund, and P.E. Andersson, Supercritical Fluid Chromatography: Open Tubular vs. Packed Columns, *Trends Anal. Chem. 13*:126–137 (1994).
15. Berger, T.A., Chemical Basis for the Instrumentation Used in Packed Column Supercritical Fluid Chromatography, in *Supercritical Fluid Chromatography with Packed Columns: Techniques and Applications (Chromatographic Science Series Vol. 75)*, edited by K. Anton and C. Berger, Marcel Dekker, New York, 1997, pp. 19–58.
16. Lee, M.L., and K.E. Markides, *Analytical Supercritical Fluid Chromatography and Extraction*, Chromatography Conferences, Inc., Provo, UT, 1990.
17. Zhao, Y., P. Sandra, G. Woo, S. Thomas, K. Gahm, and D. Semin, Packed Column Supercritical Fluid Chromatography-Mass Spectrometry for Drug Discovery Applications, *LC-GC Eur. 17*:224–238 (2004).

18. Howard, P.H., *Handbook of Environmental Fate and Exposure Data for Organic Chemicals. Vol II. Solvents*, Lewis Publishers, Ann Arbor, MI, 1990.
19. Verschueren, K., *Handbook of Environmental Data on Organic Chemicals,* 2nd ed., Van Nostrand Reinhold Co., New York, 1983.
20. Kot, A., P. Sandra, and A. Venema, Sub- and Supercritical Fluid Chromatography on Packed Columns: A Versatile Tool for the Enantioselective Separation of Basic and Acidic Drugs, *J. Chromatogr. Sci. 32*:439–448 (1994).
21. Klesper, E., and F.P. Schmitz, Gradient Methods in Supercritical Fluid Chromatography, *J. Supercrit. Fluids 1*:45–69(1988).
22. Reid, R.C., J.M. Prausnitz, and B.E. Polling, *The Properties of Gases and Liquids*, McGraw-Hill, New York, 1987.
23. Bush, D., *EoS Program to Calculate Thermophysical Data for Commonly Used Supercritical Fluids*, http://www.che.gatech.edu/ssc/bush, 2005.
24. Chester, T.L., Unified Chromatography from the Mobile Phase Perspective, in *Unified Chromatography (ACS Symposium Series Vol. 748)*, J.F. Parcher, and T.L. Chester, American Chemical Society, Washington, DC, 2000, pp. 6–29.
25. Wells, P.S., S. Zhou, and J.F. Parcher, Unified Chromatography with CO_2-Based Binary Mobile Phases, *Anal. Chem. 75*:18A–24A (2003).
26. Lesellier, E., J. Bleton, and A. Tchapla, Use and Relationships Between Retention Behaviors and Chemical Structures in Subcritical Fluid Chromatography with CO_2 /Modifier Mixtures for the Identification of Triglycerides, *Anal. Chem. 72*:2573–2580 (2000).
27. Deschamps, F.S., E. Lesellier, J. Bleton, A. Baillet, A. Tchapla, and P. Chaminade, Glycolipid Class Profiling by Packed-Column Subcritical Fluid Chromatography, *J. Chromatogr. A 1040*:115–121 (2004).
28. Lesellier, E., K. Gurdale, and A. Tchapla, Interaction Mechanisms on Octadecyl Packed Columns in Subcritical Fluid Chromatography with CO_2-Modifier Mobile Phases, *J. Chromatogr. A 975*:335–347 (2002).
29. Lesellier, E., and A. Tchapla, Separation of Vegetable Oil Triglycerides by Subcritical Fluid Chromatography with Octadecyl Packed Columns and CO_2 / Modifier Mobile Phases, *Chromatographia 51*:687–694 (2000).
30. Anton, K., M. Bach, C. Berger, F. Walch, G. Jaccard, and Y. Carlier, From Potential to Practice: Relevant Industrial Applications of Packed-Column Supercritical Fluid Chromatography, *J. Chromatogr. Sci. 32*:430–438 (1994).
31. Carraud, P., D. Thiebaut, M. Caude, R. Rosset, M. Lafosse, and M. Dreux, Supercritical Fluid Chromatography / Light-Scattering Detector: A Promising Coupling for Polar Compounds Analysis with Packed Columns, *J. Chromatogr. Sci. 25*:395–398 (1987).
32. Demirbüker, M., and L.G. Blomberg, Group Separation of Triacylglycerols on Micropacked Argentation Columns Using Supercritical Media as Mobile Phases, *J. Chromatogr. Sci. 28*:67–72 (1990).
33. Demirbüker, M., and L.G. Blomberg, Separation of Triacylglycerols by Supercritical-Fluid Argentation Chromatography, *J. Chromatogr. 550*:765–774 (1991).
34. Blomberg, L.G., M. Demirbüker, and P.E. Andersson, Argentation Supercritical Fluid Chromatography for Quantitative Analysis of Triacylglycerols, *J. Am. Oil Chem. Soc. 70*:939–946 (1993).
35. Demirbüker, M., L.G. Blomberg, N. Urban Olsson, M. Bergqvist, B.G. Hersloef, and F. Alvarado Jacobs, Characterization of Triacylglycerols in the Seeds of *Aquilegia vulgaris* by Chromatographic and Mass Spectrometric Methods, *Lipids 27*:436–441 (1992).

36. Demirbüker, M., and L.G. Blomberg, Permanganate-Impregnated Packed Capillary Columns for Group Separation of Triacylglycerols Using Supercritical Media as Mobile Phases, *J. Chromatogr. 600*:358–363 (1992).
37. Sandra, P., A. Medvedovici, Y. Zhao, and F. David, Characterization of Triglycerides in Vegetable Oils by Silver-Ion Packed-Column Supercritical Fluid Chromatography Coupled to Mass Spectroscopy with Atmospheric Pressure Chemical Ionization and Coordination Ion Spray, *J. Chromatogr. A 974*:231–241 (2002).
38. Funada, Y., and Y. Hirata, Retention Behavior of Triglycerides in Subcritical Fluid Chromatography with Carbon Dioxide Mobile Phase, *J. Chromatogr. A 764*:301–307 (1997).
39. Funada, Y., and Y. Hirata, Analysis of Plant Oils by Subcritical Fluid Chromatography Using Pattern Fitting, *J. Chromatogr. A 800*:317–325 (1998).
40. Lesellier, E., and A. Tchapla, Retention Behavior of Triglycerides in Octadecyl Packed Subcritical Fluid Chromatography with CO_2/Modifier Mobile Phases, *Anal. Chem. 71*:5372–5378 (1999).
41. Sugiyama, K., T. Shiokawa, and T. Moriya, Application of Supercritical Fluid Chromatography and Supercritical Fluid Extraction to the Measurement of Hydroperoxides in Foods, *J. Chromatogr. 515*:555–562 (1990).
42. Perrin, J.L., and A. Prevot, La Chromatographie en Phase Supercritique: Impacts dans le Domaine des Corps Gras, *Rev. Fr. Corps. Gras. 35*:485–494 (1988).
43. Medvedovici, A., F. David, and P. Sandra, Analysis of Sterols in Vegetable Oils Using Off-Line SFC/Capillary GC-MS, *Chromatographia 44*:37–42 (1997).
44. Choo, Y.M., A.N. Ma, H. Yahawa, Y. Yamaguchi, M. Bounoshita, and M. Saito, Separation of Crude Palm Oil Components by Semipreparative Supercritical Fluid Chromatography, *J. Am. Oil Chem. Soc. 73*:523–525 (1996).
45. Schmeer, K., G. Nicholson, S. Zhang, E. Bayer, and K. Bohning-Gaese, Identification of the Lipids and the Ant Attractant 1,2-Dioleoylglycerol in the Arils of *Commiphora guillaumini* Perr (Burseraceae) by Supercritical Fluid Chromatorgraphy-Atmospheric Pressure Chemical Ionisation Mass Spectrometry, *J. Chromatogr. A 727*:139–146 (1996).
46. Nomura, A., J. Yamada, K.I. Tsunoda, K. Sakaki, and T. Yockochi, Supercritical Fluid Chromatographic Determination of Fatty Acids and Their Esters on an ODS-Silica Gel Column, *Anal. Chem. 61*:2076–2078 (1989).
47. Cocks, S., and R.M. Smith, Analysis of Fatty Acid Methyl Esters by Using Supercritical Fluid Chromatography with Mass Evaporative Light Scattering Detection, *Anal. Proc. 28*:11–12 (1991).
48. Smith, R.M., and S. Cocks, Separation of Saturated and Unsaturated Fatty Acid Methyl Esters by Supercritical Fluid Chromatography on a Silica Column, *Analyst 119*:921–924 (1994).
49. Cocks, S., and R.M. Smith, Analysis of Fatty Acid Methyl Esters by Using Supercritical Fluid Chromatography with Mass Evaporative Light Scattering Detection, *Anal. Proc. 28*:11–12 (1991).
50. Demirbüker, M., I. Hägglund, and L.G. Blomberg, Separation of Unsaturated Fatty Acid Methyl Esters by Packed Capillary Supercritical Fluid Chromatography: Comparison of Different Column Packings, *J. Chromatogr. 605*:263–267 (1992).
51. Shen, Y., A. Malik, W. Li, and M.L. Lee, Packed Capillary Column Supercritical Fluid Chromatography Using SE-54 Polymer Encapsulated Silica, *J. Chromatogr. A 707*:303–310 (1995).

52. Shen, Y., S.L. Reese, B.E. Rossiter, and M.L. Lee, Silver-Complexed Dicyanobiphenyl-Substituted Polymethylsiloxane Encapsulated Particles for Packed Capillary Column Supercritical Fluid Chromatography, *J. Microcol. Sep. 7*:279–287 (1995).
53. Shen, Y., W. Li, A. Malik, S.L. Reese, B.E. Rossiter, and M.L. Lee, Cyanobiphenyl-Substituted Polymethylsiloxane Encapsulated Particles for Packed Capillary Column Supercritical Fluid Chromatography, *J. Microcol. Sep. 7*:411–419 (1995).
54. Geiser, F.O., S.G. Yocklovich, S.M. Lurcott, J.W. Guthrie, and E.J. Levy, Water as a Stationary Phase Modifier in Packed-Column Supercritical Fluid Chromatography. 1. Separation of Free Fatty Acids, *J. Chromatogr. 459*:173–181 (1988).
55. Alkio, M., C. Gonzalez, M. Jäntti, and O. Aaltonen, Purification of Polyunsaturated Fatty Acid Esters from Tuna Oil with Supercritical Fluid Chromatography, *J. Am. Oil Chem. Soc. 77*:315–321 (2000).
56. Hirata, Y., T. Hashiguchi, and E. Kawata, Development of Comprehensive Two-Dimensional Packed Column Supercritical Fluid Chromatography, *J. Sep. Sci. 26*:531–535 (2003).
57. Hirata, Y., and I. Sogabe, Separation of Fatty Acid Methyl Esters by Comprehensive Two-Dimensional Supercritical Fluid Chromatography with Packed Columns and Programming of Sampling Duration, *Anal. Bioanal. Chem. 378*:1999–2003 (2004).
58. Smith, R.M., M. Hyytiäinen, A.V. Felipe, and P.M. Morris, High Resolution Packed Column Supercritical Fluid Chromatography of Fatty Acid Methyl Esters, *J. Sep. Sci. 24*:208–212 (2001).
59. Sakaki, K., Supercritical Fluid Chromatographic Separation of Fatty Acid Methyl Esters on Aminopropyl-Bonded Silica Stationary Phase, *J. Chromatogr. 648*:451–457 (1993).
60. Lemke, P., Production of High Purity n-3 Fatty Acid-Ethyl Esters by Process Scale Supercritical Fluid Chromatography, in *Supercritical Fluid Chromatography with Packed Columns: Techniques and Applications (Chromatographic Science Series Vol. 75)*, edited by K. Anton and C. Berger, Marcel Dekker, New York, 1997, pp. 429–443.
61. Lafosse, M., C. Elfakir, L. Morin-Allory, and M. Dreux, The Advantages of Evaporative Light Scattering Detection in Pharmaceutical Analysis by High Performance Liquid Chromatography and Supercritical Fluid Chromatography, *J. High Resolut. Chromatogr. 15*:312–318 (1992).
62. Eckard, P.R., L.T. Taylor, and G.C. Slack, Method Development for the Separation of Phospholipids by Subcritical Fluid Chromatography, *J. Chromatogr. A 826*:241–247 (1998).
63. Nomura, A., J. Yamada, A. Takatsu, Y. Horimoto, and T. Yarita, Supercritical Fluid Chromatographic Determination of Cholesterol and Cholesteryl Esters in Serum on ODS-Silica Gel Column, *Anal. Chem. 65*:1994–1997 (1993).
64. Yarita, T., A. Nomura, K. Abe, and Y. Takeshita, Supercritical Fluid Chromatographic Determination of Tocopherols on an ODS-Silica Gel Column, *J. Chromatogr. A 679*:329–334 (1994).
65. Sakaki, K., T. Shinbo, and M. Kawamura, Retention Behavior of β-Carotene on Polar and Nonpolar Stationary Phases in Supercritical Fluid Chromatography, *J. Chromatogr. Sci. 32*:172–178 (1994).
66. Berry, A., E.D. Ramsey, M. Newby, and D.E. Games, Applications of Packed-Column SFC Using Light-Scattering Detection, *J. Chromatogr. Sci. 34*:245–253 (1996).
67. Thompson, J., B. Strode, III, and L.T. Taylor, Evaporative Light Scattering Detection for Supercritical Fluid Chromatography, *J. Chromatogr. Sci. 34*:261–271 (1996).

68. Lesellier, E., K. Gurdale, and A. Tchapla, Separation of *cis/trans* Isomers of β-Carotene by Supercritical Fluid Chromatography, *J. Chromatogr. A 844*:307–320 (1999).
69. Jiang, C., Q. Ren, and P. Wu, Study on Retention Factor and Resolution of Tocopherols by Supercritical Fluid Chromatography, *J. Chromatogr. A 1005*:155–164 (2003).
70. Han, N.M., C.Y. May, M.A. Ngan, C.C. Hock, and M.A. Hasham, Isolation of Palm Tocols Using Supercritical Fluid Chromatography, *J. Chromatogr. Sci. 42*:536–539 (2004).
71. King, J.W., F. Favati, and S.L. Taylor, Production of Tocopherol Concentrates by Supercritical Fluid Extraction and Chromatography, *Sep. Sci. Technol. 31*:1843–1857 (1996).
72. Taylor, S.L., and J.W. King, Preparative-Scale Supercritical Fluid Extraction/Supercritical Fluid Chromatography of Corn Oil, *J. Am. Oil Chem. Soc. 79*:1133–1136 (2002).
73. Lafosse, M., P. Rollin, C. Elfakir, L. Morin-Allory, M. Martens, and M. Dreux, Supercritical Fluid Chromatography with Light-Scattering Detection. I. Preliminary Results of the Analysis of Polar Compounds with Packed Columns, *J. Chromatogr. 505*:190–197 (1990).
74. Herbreteau, B., M. Lafosse, L. Morin-Allory, and M. Dreux, Analysis of Sugars by Supercritical Fluid Chromatography Using Polar Packed Columns, and Light-Scattering Detection, *J. Chromatogr. 505*:299–305 (1990).
75. Morin-Allory, L., and B. Herbreteau, High-Performance Liquid Chromatography and Supercritical Fluid Chromatography of Monosaccharides and Polyols Using Light-Scattering Detection. Chemometric Studies of the Retentions, *J. Chromatogr. 590*:203–213 (1992).
76. Salvador, A., B. Herbreteau, M. Lafosse, and M. Dreux, Subcritical Fluid Chromatography of Monosaccharides and Polyols Using Silica and Trimethylsilyl Columns, *J. Chromatogr. A 785*:195–204 (1997).

11 TLC-FID with Special Reference to Marine Lipids and Other High-Molecular-Weight Organic Compounds

R.G. Ackman and A. Timmins

Canadian Institute of Fisheries Technology, Dalhousie University, Halifax, NS, B3J 2X4, Canada

Introduction

In a world flooded with various chromatographic technologies, it is not easy to determine at what point the sensitivity of the flame ionization detector (FID), popular in gas-liquid chromatography (GLC), was effectively and successfully joined with thin-layer chromatography (TLC). However, Ranny (1) traced it to a U.S. patent issued to T. Okumura and T. Kadano in 1974 (2). It is phenomenally useful for analyzing mixtures of lipid classes.

Ackman's first encounter with TLC-FID was at a conference equipment display in Marseille in 1977; he discovered that the "Iatroscan" was already in use in Northern Europe, although mostly for blood lipids (3). Because Ackman had already explored the responses of the FID in GLC (4,5), this technology appealed to him and promised to solve several lipid class problems in food fish research. Soon, what was probably the first Iatroscan on the North American continent, a Mark I TH-10, was installed in the Halifax Laboratory of the Fisheries Research Board of Canada.

This model had a basic simplicity shown in Figure 1, and originated from the Iatron Laboratories of Japan. The Chromarods themselves were quartz rods with a thin coating of high-quality silica gel held in place with a soft glass frit, illustrated elsewhere (6). Significant advances included stainless steel frames for carrying, spotting, developing, and scanning the 10 Chromarods; later the initial Chromarod-S (S for silica; particles ~10 μm) quality was bettered in Chromarods S-II by reducing the silica particles to ≤5 μm. Alumina-coated rods, the Chromarod-A, are also available. The final Chromarod improvement was achieved by using machine instead of hand manufacture; it was such an improvement in terms of consistent quality that the rods were designated Chromarods-SIII. However, even today, it is wise to purchase more Chromarods than required and test them with the classes of materials to be examined so that they can be sorted into groups of 10 with similar analytical properties. For reasons not now apparent, the Chromarods were scanned on a slope, from the top down.

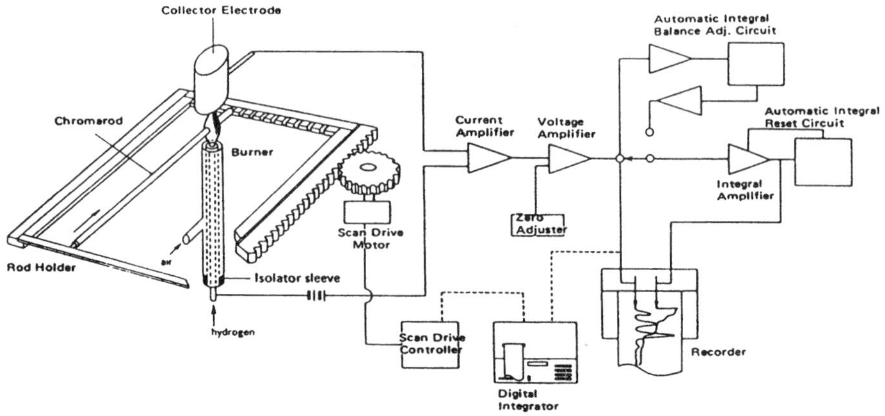

Fig. 1. Schematic diagram of the basic Iatroscan Mark III TH-10 Analyzer. Reproduced courtesy of Iatron Laboratories, Inc.

Review chapters on this topic are numerous. In fact, the most comprehensive review is probably a book (1), but other reviews began to appear from Ackman's laboratory as early as 1981 (6); this was followed a decade later by a chapter in an AOCS (Perkins) book (7) and then by journal articles (8). Among other sources of information, K.D. Mukerjee has been a frequent contributor to the field (9). The 1991 review (8) illustrates one change in basic design, in which the Chromarods are scanned in a horizontal position in the MK V Iatroscan introduced in 1989. This was used to separate paralytic shellfish poisoning toxins (10), with the flame thermionic (nitrogen detection) modification shown in Figure 2. A MK IV had also appeared (8). This unit incorporated flame photometric detection, increasing the sensitivity for heteroatoms such as phosphorus and sulfur; this capability was later included in the MK-6 Iatroscan, which was a true multifunctional system capable of operating with samples well under 1 μg.

Principles

TLC-FID applies to separations of combustible organic materials. Like the FID of GLC, the detector will ignore carbonyl groups, most obviously shown by the successful use of oxalic acid impregnation to modify the silica gel slightly to sharpen up peaks for phospholipids (11). On the other hand, volatile materials such as the hydrocarbon squalene (molecular weight 410 with 6 ethylenic bonds) might give a low response if vaporized by radiant heat before entering the flame. The materials can be made less volatile by exposure to iodine vapor, thus reducing radiant heat loss of such volatiles as the particular band of material approaches the flame (12). However, if sorting operations are carefully standardized, an estimation can be made of the natural squalene content of a large number of marine oil samples rapid-

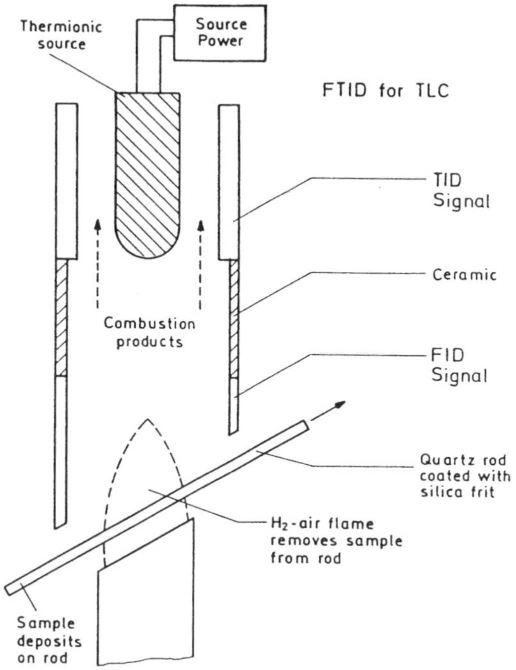

Fig. 2. The Flame Thermionic Ionization Detector (or FTID) introduced in 1984. It added the capability for large signals from nitrogen or halogen compounds in addition to the basic carbon FID signal. Reproduced courtesy of Iatron Laboratories, Inc.

ly and conveniently by TLC-FID without iodine but with some confidence (13). For comparison with GLC, total sterols in the 1–20 μg range could be determined by TLC-FID with the use of cholestane as an internal standard (14). The quantitation results were similar to those from GLC, although the latter supplied individual sterol data. The sterols, of course, are not heavier than squalene (MW 386 for cholesterol) but benefit from a hydroxyl group to interact with the silica gel. Most natural lipid classes have such polar groups to reduce volatility problems.

Silver nitrate has been used to separate *trans* unsaturated fatty acids and at the same time serve to stabilize and separate them on Chromarods (15). Various derivatives of polar groups can add mass to counteract volatility problems, and improved separations of marine lipids have resulted from total hydrogenation (7,16). There is no substitute for patient testing of conditions of operation or samples (17,18), but once achieved, Chromarod life is excellent with little change in properties over hundreds of analyses.

Quantitation and Applications

At times it seems that every natural lipid sample demands a reinvestigation of factors such as hydrogen flow rate, scanning speed, and solvent composition to provide optimal results. This can take many investigations, but it starts with sample preparation, a problem shared with most chromatographic technology (19), and can proceed as far as to a new deconvolution procedure based on a new mathematical function for describing chromatographic peaks and enhancing their resolution. This was applied (20) to neutral, uronic acid and amino sugar subfractions present in marine snow and mucilage samples of Italian seas. It is natural for a review from the Atlantic coast of North America to extend this "quality assurance" concept of TLC-FID results to more readily available marine lipid classes (21), but ocean scientists all over the world appreciate the merits of TLC-FID.

It is interesting to record a series of papers dealing with steady improvements in TLC-FID applied to the mundane lipid field of the separation of mixtures of fatty acid methyl esters and tri-, di-, and monoacylglycerides (22–24). Suddenly this system of lipid class resolution is of keen interest to producers of biodiesel fuels. In turn, Ackman points out that it is not surprising that the petroleum and similar industries (e.g. coal-tar pitch) have benefited from TLC-FID (25–28). Particular attention may have to be paid to purification of relevant standards (8,27). Obviously the solvents used in recovery of hydrocarbons for analysis are critical, even for diesel exhaust particulates (28). The complexity and variety of lipid classes possible in some environmental samples is illustrated in Figure 3, showing that biogenic triacylglycerols and phospholipids may be mixed with fossil fuel materials or residues in such samples. This illustration is taken from a recent review combining planar TLC and the Iatroscan (29), and originally appeared elsewhere. It is wise to keep options open; some samples may possibly need both semipreparative TLC and TLC-FID. To show that freshwater life has lipids as fascinating as marine life, Figure 4 is taken from a chapter in a recent book (30).

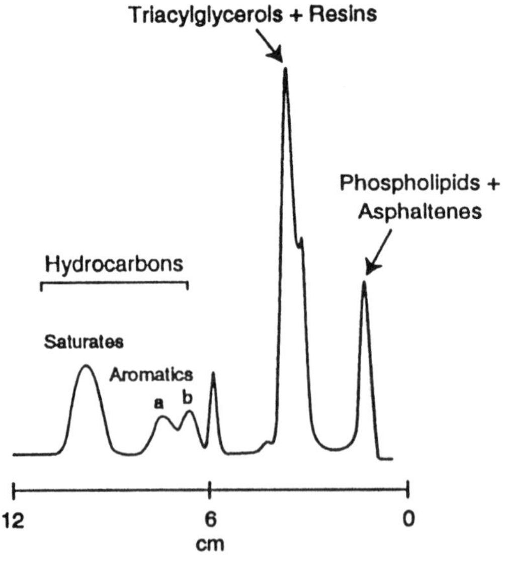

Fig. 3. Thin-layer chromatography-flame ionization detection chromatogram showing the separation of biogenic lipids from hydrocarbons in an extract of a petroleum-contaminated soil. From (29), reproduced courtesy of Blackie and Academic and Professional, London, UK, by permission.

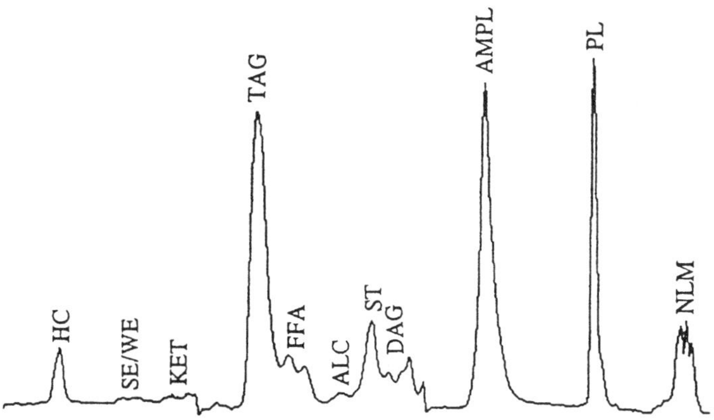

Fig. 4. TLC/FID chromatogram of a net-tow sample (20-μm mesh) taken from a Newfoundland stream in May, 2005. Iatroscan conditions: air flow 2.0 L/min, H_2: 190 mL/min. The chromatogram is a composite of a sequence of three separate scans. Peak identities: HC, hydrocarbon; SE/WE, steryl and wax esters; KET, ketone; TAG, triacylglycerol; FFA, free fatty acid; ALC, alcohol; ST, sterol; DAG, diacylglycerol; AMPL, acetone-mobile polar lipids; PL, phospholipid; NLM, nonlipid material. For sequence details see Figure 5. Reproduced courtesy of Springer-Verlag New York, Inc., New York by permission.

A peculiar and unique advantage of TLC-FID is sequential scanning (31). One can select which type of lipid in a given sample is to be determined and develop it into an upper part of the Chromarod, using solvents suitable for moving hydrocarbons, sterol esters, wax esters, or other lipids of low polarity, and then scanning that part of the rod only. From the safely stored part of the sample at the point of application, another lipid class can be moved out and along the rod for another scanning, and finally the rest can be developed and determined as shown in Figure 4. Details are provided in Figure 5 and the original published paper (30).

A strong feature of the Iatroscan TLC-FID technology has been the continuous development of the basic detector, first into the flame thermionic detector for nitrogen (Figure 2). It was promptly (1988) carefully evaluated for marine lipids where nitrogen can be an important factor for humic and other polar lipid classes (31). The most recent adaptation (Figure 6) is a flame photometric detector (FPD) for detecting phosphorus and sulfur concurrently with the FID carbon ion recording.

Time is money. A scan of a Chromarod may take only 30 s, but the mechanics of passing the flame head back to the starting point of the next rod double that. The 10 rods permit combinations of rods for statistical evaluation, or of as many as 10 samples for comparison. An automated spotting system is available and could assist in the numerous samples generated by quality control work. Development of a set of 10 rods may be no more time consuming (~1 h) than planar TLC of one plate, and can be integrated into busy times in a laboratory when hands are needed elsewhere.

Sequence Leading to the First Chromatogram (HC to KET)

a. Blank scan the rods three times in the Iatroscan.
b. Apply samples and standards with an Autosampler or a Hamilton syringe fitted with a repeating dispenser.
c. Focus twice (three times if sample is very concentrated) in acetone to produce a narrow band of lipid material near the lower end of the rods.
d. Dry and condition in a constant-humidity chamber (over saturated $CaCl_2$) for 5 minutes.
e. Develop twice in hexane:diethyl ether:formic acid (98.95:1:0.05 by vol). The first development is for 25 minutes; the rods are dried in the constant humidity chamber for 5 minutes and redeveloped for 20 minutes.
f. Dry for 5 minutes in the Iatroscan.
g. Scan to the lowest point behind the KET peak (pps 25 on the Iatroscan).

↓

Sequence Leading to the Second Chromatogram (TAG to DAG)

a. Condition for 5 minutes.
b. Develop for 40 minutes in hexane:diethyl ether:formic acid (79:20:1 by vol).
c. Dry and scan to the lowest point behind the DAG peak (pps 11 on the Iatroscan).

↓

Sequence Leading to the Third Chromatogram (AMPL and PL)

a. Condition for 5 minutes.
b. Develop twice for 15 minutes in 100% acetone.
c. Condition for 5 minutes.
d. Develop twice for 10 minutes in chloroform:methanol:chloroform extracted water (5:4:1 by vol).
e. Dry and scan entire length of rods.

Fig. 5. Developing and conditioning sequences used for the separation of aquatic lipid classes by development and sequence scanning as illustrated in Figure 4. See also Figure 6. Lipid class abbreviations are explained in the legend to Figure 4. Reproduced courtesy of Springer-Verlag New York, Inc., New York, by permission.

Once developed, most analytes are nonvolatile and stable enough to be scanned at leisure. Precision and accuracy problems have been considered fully (21,32).

For commercial fish species and other foods, there is usually no shortage of sample, but the Iatroscan can be very revealing (33,34), and is excellent for examining used frying oils in the food industry (35). The adaptation of sequential development reveals fish muscle lipids (Figure 7) in detail, an achievement that historically would require weeks of work. Note the small peak for oxidized lipids. Marine lipids contain highly unsaturated fatty acids and prolonged handling or storage inevitably

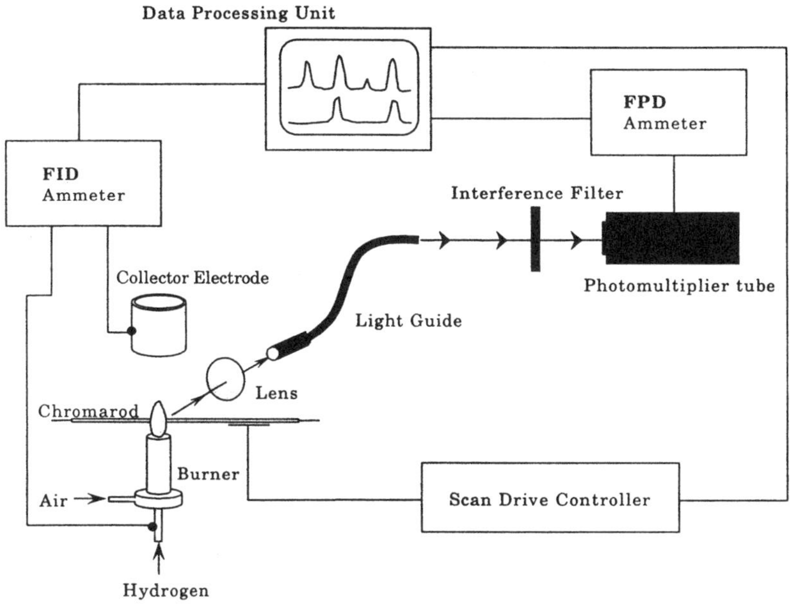

Fig. 6. Schematic of the Flame Ionization-Flame Photometric Detector System for the Iatroscan MK-6 basic TLC-FID instrument. Courtesy of Shell-USA, Fredericksburg, VA, USA.

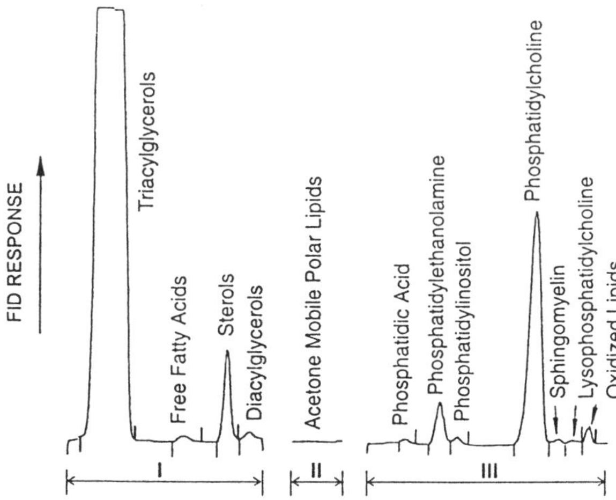

Fig. 7. Sequential Iatroscan thin-layer chromatography-flame ionization detector profiles of the lipid classes extracted from silver hake muscle tissue. I, II, and III represent partial chromatograms from three-stage development sequences of total lipids on Chromarods S-III as described in (34). Reproduced courtesy of AOCS Press, Champaign, IL.

leads to some loss; thus, rapid Iatroscan analysis is much more accurate for lipid classes. The adaptation of the Iatroscan-Chromarod system to direct mass spectroscopy (16) of the isolated classes of lipids is a very promising advance for many lipid biochemists to consider.

However, Ackman emphasizes that the Iatroscan is ideal for the pure research scientist. Thousands of small organisms have nonvolatile components, not necessarily conventional lipids, that can be recovered in often minimal amounts compatible with TLC-FID. The challenges, continuously explored by one of Ackman's former students (16–18,30–32), are to (i) resolve these factors, (ii) identify them if necessary, and (iii) quantitate them. Good hunting and good luck are the philosophies to apply. In a 1980 review (36), to introduce the Iatroscan and other then "modern" developments in lipid analysis Ackman suggested that no journal titles starting with "A Simple Method (or Apparatus)..." should be permitted. Ackman certainly did not foresee the modifications of the basic Iatroscan and wide applications, some of which are mentioned in this review for this durable and well-engineered laboratory tool.

References

1. Ranny, M., ed., *Thin-Layer Chromatography with Flame Ionization Detection A Laboratory Handbook,* Kluwer Academic Publishers Group, Hingham, MA, 1987.
2. Okumura, T., and T. Kadano, U.S. Patent 3,829,205 (1974).
3. Vandamme, D., G., Vankerhoven, R. Vercaemst, F. Soeteway, V. Blaton, H. Peeters, and M. Rosseneu, A Simple Screening Method for Plasma Lipids by Thin-Layer Chromatography with Flame Ionization Detection, *Clin. Chem. Acta 89*:231–238 (1978).
4. Ackman, R.G., and J.C. Sipos, Application of Specific Response Factors in the Gas Chromatographic Analysis of Methyl Esters of Fatty Acids with Flame Ionization Detectors, *J. Am. Oil Chem. Soc. 41*:377–378 (1964).
5. Addison, R.F., and R.G. Ackman, Flame Ionization Detector Molar Responses for Methyl Esters of Some Polyfunctional Metabolic Acids, *J. Gas Chromatogr. 6*:135–138 (1968).
6. Ackman, R.G., Flame Ionization Detection Applied to Thin-Layer Chromatography on Coated Quartz Rods, in *Methods in Enzymology*, edited by J.M. Lowenstein, Academic Press, New York, 1981, Vol. 72, pp. 205–252.
7. Ackman, R.G., Application of Thin-Layer Chromatography to Lipid Separations: Detection Methods, in *Analyses of Fats Oils and Lipoproteins*, edited by E.G. Perkins, American Oil Chemists' Society, Champaign, IL, 1991, pp. 97–121.
8. Ackman, R.G., C.A. McLeod, and A.K. Banerjee, An Overview of Analyses by Chromarod-Iatroscan TLC-FID, *J. Planar Chromatogr. 3*:450–462 (1990).
9. Mukerjee, K.D., Applications of Flame Ionization Detectors in Thin-Layer Chromatography, in *Handbook of Thin-Layer Chromatography*, 2nd ed., edited by J. Sherma and B. Fried, Marcel Dekker, New York, 1996, pp. 361–372.
10. Indrasena, W.M., R.G. Ackman, and T.A. Gill, Separation of Paralytic Shellfish Poisoning Toxins on Chromarods-SIII by Thin Layer Chromatography with Iatroscan (Mark 5) and Flame Thermionic Detection, *J. Chromatogr. A 855*:657–668 (1999).
11. Banerjee, A.K., W.M.N. Ratnayake, and R.G. Ackman, Effect of Oxalic Acid Impregnation of Chromarods on the Separation of Phospholipids for Determination by the Iatroscan TLC/FID, *Lipids 20*:121–125 (1985).

12. Banerjee, A.K., W.M.N. Ratnayake, and R.G. Ackman, Enhanced Response of More Highly Unsaturated Lipids on Thin-Layer Chromatography-Iatroscan Flame Ionization Detection After Exposing the Developed Chromarods to Iodine Vapour, *J. Chromatogr.* *319*:215–217 (1985).
13. Thankappan, T.K., and K. Gopahumer, A Rapid Method of Separation and Estimation of Squalene from Fish Liver Oils Using Iatroscan Analyzer, *Fish. Technol. 28*:63–66 (1991).
14. Walton, C.G., W.M.N. Ratnayake, and R.G. Ackman, Total Sterols in Seafoods: Iatroscan TLC/FID Versus the Kovacs GLC/FID Method, *J. Food Sci. 54*:793–795, 804 (1989).
15. Sébédio, J.L., and R.G. Ackman, Chromarods-S Modified with Silver Nitrate for the Quantitation of Isomeric Unsaturated Fatty Acids, *J. Chromatogr. Sci. 19*:552–557 (1981).
16. Hudson, E.D., R.J. Helleur, and C.C. Parrish, Thin-Layer Chromatography-Pyrolysis-Gas Chromatography-Mass Spectrometry: A Multidimensional Approach to Marine Lipid Class and Molecular Species Analysis, *J. Chromatogr. Sci. 39*:146–152 (2001).
17. Parrish, C.C., and R.G. Ackman, The Effect of Developing Solvents on Lipid Class Quantification in Chromarod Thin-Layer/Flame Ionization Detection, *Lipids 18*:563–565 (1985).
18. Parrish, C.C., and R.G. Ackman, Calibration of the Iatroscan-Chromarod System for Marine Lipid Class Analysis, *Lipids 20*:521–530 (1985).
19. Moldoveanu, S.C., Solutions and Challenges in Sample Preparation for Chromatography, *J. Chromatogr. Sci. 42*:1–14 (2004).
20. Mecozzi, M., and Z. Pápai, Application of Curve Fitting in Thin-Layer Chromatography-Flame Ionization Detection Analysis of the Carbohydrate Fraction in Marine Mucilage and Marine Snow Samples from Italian Seas, *J. Chromatogr. Sci. 42*:268–274 (2004).
21. Bergen, B.J., J.G. Quinn, and C.C. Parrish, Quality-Assurance Study of Marine Lipid-Class Determination Using Chromarod-Iatroscan Thin-Layer Chromatography-Flame Ionization Detector, *Environ. Tox. Chem. 19*:2189–2197 (2000).
22. Freedman, B., E.H. Pryde, and W.F. Kwolek, Thin Layer Chromatography/Flame Ionization Analysis of Transesterified Vegetable Oils, *J. Am. Oil Chem. Soc. 61*:1215–1220 (1984).
23. Peyrou, G., V. Rakotondrazafy, Z. Mouloungui, and A. Gaset, Separation and Quantitation of Mono-, Di-, and Triglycerides and Free Oleic Acid Using Thin-Layer Chromatography with Flame-Ionization Detection, *Lipids 31*:27–32 (1996).
24. Striby, L., R. Lafont, and M. Goutx, Improvement in the Thin-Layer Chromatographic-Flame Ionization Detection Analysis of Marine Lipids. Separation and Quantitation of Monoacylglycerols and Diacylglycerols in Standards and Natural Samples, *J. Chromatogr. A. 849*:371–380 (1999).
25. Poirier, M.A., P. Rahimi, and S.M. Ahmed, Quantitative Analysis of Coal-Derived Liquids Residues by TLC with Flame Ionization Detection, *J. Chromatogr. Sci. 22*:116–119 (1984).
26. Cebolla, V.L., J. Vela, L. Membrado, and A.C. Ferrando, Coal-Tar Pitch Characterization by Thin-Layer Chromatography with Flame Ionization Detection, *Chromatographia 42*:295–299 (1996).
27. Bharati, S., R. Patience, N. Mills, and T. Hanesand, A New North Sea Oil-Based Standard for Iatroscan Analysis, *Org. Geochem. 26*:49–57 (1997).

28. Obuchi, A., H. Aoyama, A. Ohi, and H. Ohuchi, Application of Thin-Layer Chromatography with Flame Ionization Detection to the Characterization of Organic Extracts from Diesel Exhaust Particulates, *J. Chromatogr. 288*:187–194 (1984).
29. Shantha, N.C., and G.E. Napolitano, Lipid Analysis Using Thin-Layer Chromatography and the Iatroscan, in *Lipid Analysis in Oils and Fats*, edited by R.J. Hamilton, Blackie Academic and Professional, London, UK, 1998, pp. 1–33.
30. Parrish, C.C., Determination of Total Lipid, Lipid Classes, and Fatty Acids in Aquatic Samples, in *Lipids in Freshwater Ecosystems*, edited by M.T. Arts and B.C. Wainman, Springer-Verlag New York, Inc., New York, 1998, pp. 4–20.
31. Parrish, C.C., X. Zhou, and L.R. Herche, Flame Ionization and Flame Thermionic Detection of Carbon and Nitrogen in Aquatic Lipid and Humic-type Classes with an Iatroscan Mark IV, *J. Chromatogr. 435*:350–356 (1988).
32. Parrish, C.C., Separation of Aquatic Lipid Classes by Chromarod Thin-Layer Chromatography with Measurement by Iatroscan Flame Ionization Detection, *Can. J. Fish. Aquat. Sci. 44*:722–731 (1987).
33. Whitsett, J.F., J.M. Kennish, D.E. Kramer, and J.S. French, Fish Oil Analysis Using Combined Thin-Layer Chromatography and Flame Ionization Detection (TLC-FID), in *Seafood Quality Determination*, edited by D.E. Kramer and J. Liston, Elsevier Science Publishers B.V., Amsterdam, 1987, pp. 161–176.
34. Ackman, R.G., and H. Heras, Recent Advances of Iatroscan TLC-FID Technology, in *Applications in Lipid Analysis*, edited by E. McDonald and M.M. Mossoba, AOCS Press, Champaign, IL, 1997, pp. 325–340.
35. Sébédio, J.L., P.O. Astorg, C. Septier, and A. Grandgirard, Quantitative Analysis of Polar Components in Frying Oils by the Iatroscan Thin-Layer Chromatography-Flame Ionization Detection Technique, *J. Chromatogr. 405*:371–378 (1987).
36. Ackman, R.G., Potential for More Efficient Methods for Lipid Analysis, *J. Am. Oil Chem. Soc. 57*:821A–829A (1980).

12 Fast GC for Cellular FAME Analysis of Bacteria

Jeffrey S. Buyer

Agricultural Research Service, United States Department of Agriculture, Beltsville, MD

Introduction

A wide variety of fatty acids are found throughout the bacterial kingdom. They appear in cell membranes and cell walls, primarily attached to other molecules through ester linkages, but also may occur (although rarely) as free fatty acids. One to three fatty acids may be linked to glycerol via ester bonds, forming monoacylglycerols, diacylglycerols, and triacylglycerols. Fatty acids may be esterified to various derivatives of phosphatidic acid, leading to a wide variety of phospholipids. Although cellular fatty acid analysis may be limited to a determination of the total fatty acid composition, it is also possible to determine the fatty acids that occur in various classes of lipids. Although lipid and fatty acid analyses are important in studies of biochemistry, physiology, and metabolism, the focus of this article will be on fatty acid analysis for taxonomic purposes.

Fatty acids vary widely in carbon chain length. Fatty acids from 14 to 20 carbons long are most common, but chain lengths from 2 to 80 are known (1). Long-chain fatty acids are particularly noted in *Mycobacteria*. Saturated fatty acids are linear, branched, or contain rings. Unsaturated fatty acids may be *cis* or *trans*, and may contain one or several double bonds. Other functional groups include hydroxyl, oxo, and epoxy moieties.

Fatty acid length and degree of unsaturation are indicated by 2 numbers separated by a colon. For example, 10:0 indicates decanoic acid, 10 carbons long with no double bonds, whereas 18:1 indicates a fatty acid with 18 carbons and 1 double bond. A complete description of an unsaturated fatty acid includes the *cis* or *trans* designation and indicates where the double bond or bonds occur in the molecule. Unfortunately, there are several systems for designating the location of the double bonds (2). Branching in fatty acids most commonly consists of a methyl group on the alkyl chain, and is indicated by the prefix "br." If the methyl group occurs at the end of the chain the prefix "i" for iso is used, whereas if the methyl group occurs at the penultimate carbon, the prefix "a" for anteiso is used (2).

Microbial Identification by Fatty Acid Analysis

The cellular fatty acid composition of bacteria is a useful taxonomic determinant. The great variety of fatty acids found in bacteria provide a large number of taxonomic features. Although the qualitative and quantitative fatty acid composition varies with growth conditions, including temperature and substrates, under carefully controlled conditions, the fatty acid composition is quite reproducible. The fatty acid composition of many species of bacteria has been reported, and correlations between some groups of bacteria and certain types of fatty acids have been observed (3).

A variety of methods are available for determining the fatty acid composition of bacteria. Most commonly, the fatty acids are cleaved from the lipids, converted to fatty acid methyl esters (FAMEs), and analyzed by gas chromatography (GC). The cleavage and methylation may be accomplished by saponification followed by methylation, acid hydrolysis followed by methylation (4), or direct transesterification (5). Other methods, including supercritical fluids to derivatize and extract FAMEs (6), pyrolytic methylation of intact bacteria followed by GC and mass spectrometry (GC-MS) (7,8), and the analysis of fatty acids from the surface of cells by secondary ion MS (9), have not been widely adopted.

The only commercially available system for microbial identification by fatty acid analysis that we are aware of is the Microbial Identification System (MIS) manufactured by MIDI, Inc., Newark, DE, USA. The system consists of an Agilent GC, Agilent Chemstation software to operate the GC, and Sherlock software provided by MIDI. The Sherlock software consists of peak naming tables to identify the fatty acids, databases of the fatty acid composition of various microorganisms, and pattern-matching algorithms used to compare the fatty acid composition of unknown samples to those in the databases. To use the databases, microorganisms must be grown with the same media and temperatures used by MIDI when building the databases. Furthermore, the samples must be prepared by the same method used by MIDI, which is a four-step procedure (saponification, methylation, extraction, and washing). To identify peaks correctly the GC must be calibrated correctly, and MIDI provides a calibration mix specifically for this purpose. In our experience, it is not difficult to obtain reliable results with the MIS as long as one is careful to follow the protocols provided.

Gas Chromatography

Fatty acids are not very volatile; thus, they are typically derivatized to form the methyl esters before GC. FAMEs have been resolved on a variety of nonpolar and polar columns, depending on the specific application and which FAMEs require resolution. Nonpolar columns separate primarily by the number of carbons and then by degree of unsaturation and branching, whereas polar columns are better for separating *cis*/*trans* isomers. Absolute retention times vary among instruments and columns;

thus, a relative measure of retention time called the equivalent chain length (ECL) is frequently used. The ECL is defined as follows:

$$ECL_x = \frac{R_x - R_n}{R_{n+1} - R_n} + n$$

where ECL_x and R_x are the ECL and retention time, respectively, for the peak of interest, R_n is the retention time for the straight-chain saturated fatty acid n carbons long eluting immediately before the peak of interest, and R_{n+1} is the straight-chain saturated fatty acid n+1 carbons long eluting immediately after the peak of interest (10).

In chromatography, there is generally a trade-off between speed and resolution. Faster GC separations, while maintaining resolution, can be obtained by using shorter columns with smaller internal diameters (i.d.) and thinner liquid phase coatings (11). Retention time is proportional to capillary column i.d. (12); therefore, reducing the column diameter is an obvious way to achieve faster separations. Higher carrier gas flow rates and faster column oven heating rates lead to faster separations (11), but excessively high flow rates and heating rates reduce resolution. It is not a trivial task to balance all of these parameters when developing a new method, particularly if one wants to maintain relative retention times. Mathematical equations exist to describe the relations among these factors (13), and computer software is available to facilitate the process (14).

Materials and Methods

Bacteria

The following standard strains of bacteria were obtained from the American Type Culture Collection (ATCC, Manassas, VA, USA): *Acinetobacter baumanii*, *Pseudomonas aeruginosa*, *Stenotrophomonas maltophilia*, *Escherichia coli*, *Bacillus cereus*, *Staphylococcus aureus*, and *Enterococcus faecalis*. MIDI provided the following strains: *Burkholderia cepacia*, *Methylobacterium mesophilicum*, *Nocardioides simplex*, *Chryseobacterium balustinum*, *Hydrogenophaga palleronii*, *Pseudoalteromonas haloplanktis*, *Pseudoalteromonas denitrificans*, *Flavobacterium johnsoniae*, *Vibrio mediterranei*, *Acetobacter aceti*, *Shewanella putrefaciens*, *Pseudomonas chlororaphis*, and *Variovorax paradoxus*.

Sample Preparation

The standard MIS procedure and a modified rapid procedure were used. For both procedures, bacteria were quadrant streaked onto trypticase soy broth agar (TSBA) and incubated at 28°C for 24 h for identification with the TSBA40 library or onto trypticase soy agar with 5% defibrinated sheep blood and incubated at 35°C for 24 h for identification with the CLIN40 library. In the standard procedure, biomass from the third quadrant was harvested with a 10-µL loop and transferred to a screw-cap test tube.

Methanolic base (1 mL; 45 g NaOH, 150 mL methanol, 150 mL H_2O) was added and the tube mixed on a vortex mixer. After 5 min in a boiling water bath, the tube was vortexed again and then returned to the boiling water bath for another 25 min. Acidic methylation reagent (2.0 mL; 325 mL 6.00 N HCl, 275 mL methanol) was added and the tube vortexed. After 10 min in an 80°C water bath, the FAMEs were extracted with 1.25 mL of 1:1 hexane:methyl *tert*-butyl ether. The aqueous (bottom) phase was removed and the organic phase was washed with 3.0 mL of dilute base (10.8 g NaOH, 900 mL H_2O). The organic phase was then transferred to chromatography vials.

The modified rapid procedure used exactly the same reagents, but each step was conducted on a smaller scale so that a multichannel pipettor could be used for all liquid transfers (15). Biomass from the third quandrant was harvested with a disposable 1-µL inoculating loop and transferred to 1000-µL glass vials held in a 96-well aluminum plate (Kombi-Screen QL deep-well plate, Kimble/Kontes, Vineland, NJ, USA). Methanolic base (0.15 mL) was added to each tube using a 6-channel pipettor (Matrix Impact EXP, 1250 µL) and disposable pipette tips (Matrix 1250 µL TallTip) (Matrix, Hudson, NH, USA). The tubes were covered using a 96-position Teflon-silicon mat (Kombi-Screen) and clamped between 2 aluminum plates held together with bolts. After shaking, the tubes were placed in a boiling water bath for 5 min, shaken again, and returned to the bath for another 25 min. After cooling in a water bath and removal of the clamp and mat, 0.3 mL of acidic methylation reagent was added by multichannel pipettor, the assembly was sealed and clamped, and the tubes were vigorously shaken and heated at 80°C for 10 min. The assembly was cooled in a water bath and the clamp and mat removed. Then 0.3 mL 1:1 hexane:methyl *tert*-butyl ether was added by multichannel pipettor and the tubes clamped and vigorously shaken. The bottom phases were removed with the multichannel pipettor (0.5 mL) and discarded. Dilute base (0.6 mL) was added by multichannel pipettor and the tubes clamped and vigorously shaken. Emulsions were eliminated by adding 0.05 mL of saturated NaCl. The top phases were transferred to chromatography vials using the multichannel pipettor.

Chromatography

An Agilent 6890 GC equipped with electronic pressure control, autoinjector, split-splitless inlet, and flame ionization detector was used (Agilent Technologies, Palo Alto, CA, USA). The system was controlled with Chemstation (Agilent) and Sherlock (MIDI) software. The standard MIS method, TSBA40, and the rapid MIS method, RTSB50, used a 25-m long × 0.20-mm i.d. × 0.33-µm film thickness Ultra 2 column (Agilent, Newark, DE, USA). For fast GC, we used a 10-m long × 0.10-mm i.d. × 0.17-µm film thickness DB-5 column (Agilent). Both columns contain a 5% phenyl-methylpolysiloxane bonded phase, and both columns have the same phase ratio (film thickness divided by column i.d.). All methods used hydrogen as the carrier gas. The GC methods are summarized in Table 1. Methods FAST2 and FAST3 required the use of an oven insert (Agilent) to achieve the high ramp rates employed (16). Chromatographic parameters for the FAST methods were developed with the Agilent Method Translation software (http://www.chem.agilent.com/cag/servsup/usersoft/main.html#mxlator).

Results And Discussion

Chromatographic Methods

The FAST method (15) was a direct translation of the original TSBA40 method developed for the MIS, which used a 0.2-mm i.d. column, to a 0.1-mm i.d. column. The smaller column permitted the use of faster temperature programming and higher carrier gas flow rate while preserving relative retention times and efficiency of the separation. However, it was necessary to reduce injection size and increase the split ratio to achieve sufficiently sharp peaks for adequate chromatographic resolution.

The FAST2 method (16) was designed to optimize efficiency rather than serve as a direct translation of TSBA40. The computer software employed calculated the temperature program necessary to preserve relative resolution times while using the theoretically optimal gas flow rate. The faster oven temperature programming rate necessitated the use of an oven liner, which increased the maximum oven heating rate by reducing the effective volume of the oven. We found it necessary to reduce the inlet liner diameter to obtain sharper peaks. We also switched from constant pressure to constant flow, which provided a slight additional decrease in run time. The smaller column diameter, thinner phase, and higher gas flow rates compared with TSBA40 resulted in peaks eluting at lower oven temperatures; thus, we were able to reduce the maximum oven temperature in the method.

The FAST3 method (16) was a translation of TSBA40 for faster analysis. The software used a gas flow rate twice that of the theoretical optimum, and then calculated the necessary temperature program to maintain relative resolution times. As in FAST2, the oven liner and smaller diameter inlet liner were needed, and constant flow was used.

Chromatograms had a similar appearance with all of the methods (Fig. 1), and all methods calibrated successfully after corrected retention times were entered into the peak naming tables. Various bacteria were run by the TSBA40 and FAST methods to check for shifts in ECL values. A number of small shifts were found in the original FAST method, particularly with hydroxy fatty acids. These shifts were corrected in the FAST peak naming table, although, in retrospect, the corrections were likely unnecessary. No ECL corrections were made to the FAST2 and FAST3 naming tables.

After we developed our FAST methods, MIDI introduced the rapid method RTSB50. This method used the standard MIS column, unlike our FAST methods,

TABLE 1
Methods for Gas Chromatography

Item	TSBA40	RTSB50	FAST	FAST2	FAST3
Inlet	Constant pressure 62.1 kPa	Constant flow 1.3 mL/min	Constant pressure 1493 kPa	Constant flow 0.5 mL/min	Constant flow 1.0 mL/min
Inlet liner	4 mm i.d.	4 mm i.d.	4 mm i.d.	2.3 mm i.d.	2.3 mm i.d.
Oven program (°C)	170 to 260°C at 5°/min 260 to 310°C at 40°C/min 310 for 1.5 min	170 to 288°C at 28°/min 288 to 310°C at 60°C/min 310 for 1.25 min	170 to 260°C at 18°C/min 260 to 310°C at 40°C/min 310 for 0.4 min	170 to 260°C at 30°C/min 260°C for 0.3 min	170 to 260°C at 45°C/min 260°C for 0.1 min
Oven equilibration time (min)	1.5	0.1	1.5	0.25	0.25
Split ratio	100:1	40:1	200:1	200:1	200:1
Injection volume (μL)	2	2	1	1	1
Cycle time (min)	25.1	9.0	10.75	6	4.75

and the standard inlet liner. In RTSB50, constant flow, at a rate slightly greater than the theoretically optimal gas flow rate and much higher than that achieved in TSBA40, was used. The oven temperature program was designed to maintain relative resolution times and minimize overall run time.

There were some substantial differences in sensitivity among these methods. The FAST methods all injected one-quarter as much volume as the TSBA40 method. Integrated peak areas were reduced proportionally, but the peaks were much sharper, which compensated somewhat. We diluted the calibration mix and ran it with the various methods to compare actual sensitivity. The FAST methods were ~50% as sensitive as the TSBA40 method. The RTSB50 method, on the other hand, injects 2.5 times as much sample as in the TSBA40 method, and is correspondingly more sensitive.

Bacterial Identification

Results obtained with TSBA40 and the various FAST methods are summarized in Table 2. The FAST methods generally gave results similar to TSBA40 in a fraction of the time. However, the match for *Stenotrophomonas maltophilia* was consistently higher with TSBA40 and FAST2 than with FAST or FAST3. The fatty acid 16:1n-9 *cis*, which resolved well from summed feature 3 (16:1n-7 *cis* and 15:0 iso 2OH) in TSBA40 and FAST2, resolved poorly in FAST and FAST3, resulting in a lower match (Fig. 2). FAST2 therefore appears to be the most generally useful of the FAST methods.

We have not conducted any systematic investigations of RTSB50, but initial experiments indicate that bacteria are identified at least as well as with TSBA40; as shown in Figure 2, the resolution of 16:1n-9 *cis* and summed feature 3 is quite good. The reduced sensitivity of the FAST methods made it difficult to identify slow-growing organisms for which it was a challenge to obtain sufficient biomass. We found that RTSB50, with its increased sensitivity, was the best method for samples with inadequate biomass, clearly superior to TSBA40 and the FAST methods (data not shown).

Sample Preparation

With the standard MIS sample processing method, reagents are dispensed using bottle-top pipettors into screw-cap test tubes. Because each dispensing operation requires a cap to be unscrewed, reagent dispensed, and the cap rescrewed, the method is inefficient for large numbers of samples. In addition, each tube must be vortexed three times, and it is difficult to vortex several tubes at a time. During the extraction step the bottom phase is removed; this is done one tube at a time, and the final transfer to the GC vial is done individually for each sample. With all of these limitations, our laboratory is able to process only 48 samples, plus a blank and a positive control, in 4 h.

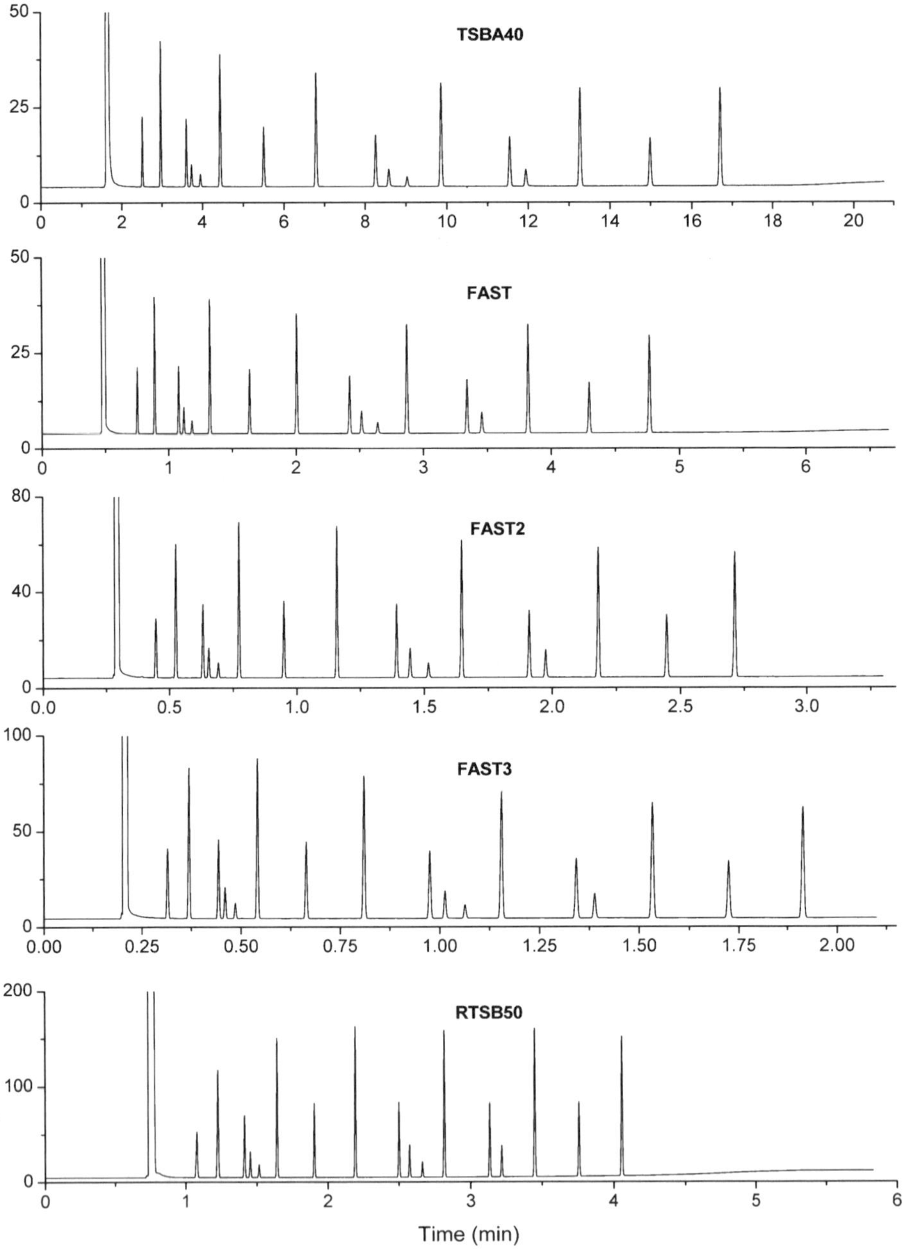

Fig. 1. Chromatograms of Microbial Identification System Calibration Mix using various methods.

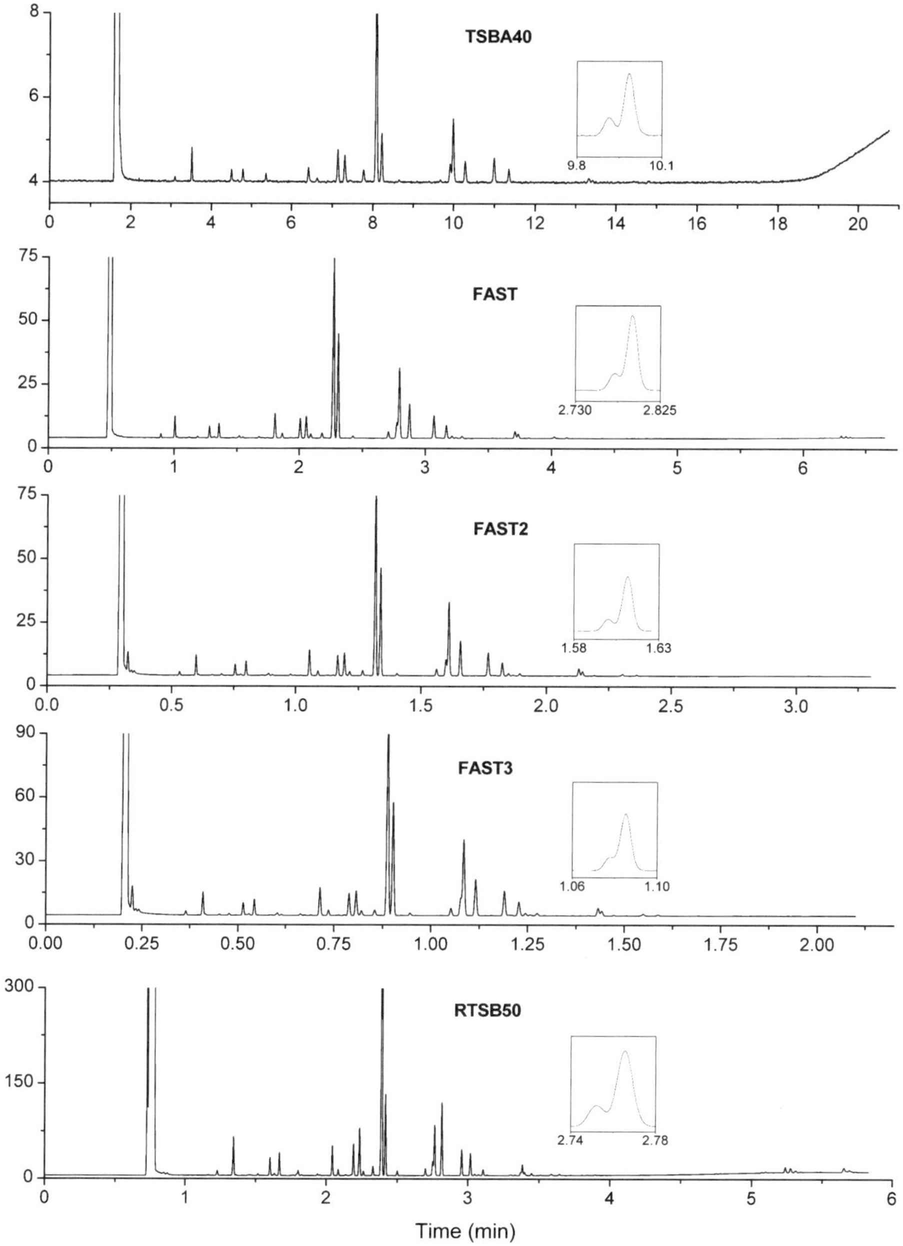

Fig. 2. Chromatograms of *Stenotrophomonas maltophilia* fatty acid methyl esters using various methods. The inset shows 16:1n-9 *cis* and summed feature 3.

TABLE 2
Identification of Bacteria Using the TSBA40 and FAST Methods[a]

	TSBA				Blood agar			
Organism	TSBA40	FAST	FAST2	FAST3	TSBA40	FAST	FAST2	FAST3
Acinetobacter baumanii	0.900	0.901	0.898	0.896	0.701	0.749	0.711	0.725
Bacillus cereus	0.395	0.431	0.430	0.436	0.457	0.494	0.437	0.449
Burkholderia cepacia	0.375	0.475	0.481	0.492	0.877	0.900	0.847	0.905
Chryseobacterium balustinum	0.915	0.861	0.871	0.871	ND	ND	ND	ND
Enterococcus faecalis	ND	ND	ND	ND	0.569	0.304	0.537	0.314
Escherichia coli	ND	ND	ND	ND	0.394	0.447	0.394	0.398
Flavobacterium johnsoniae	0.700	0.734	0.706	0.713	ND	ND	ND	ND
Methylobacterium mesophilicum	0.593	0.328	ND	ND	ND	ND	ND	ND
Pseudomonas aeruginosa	0.733	0.782	0.753	0.764	0.485	0.535	0.509	0.530
Pseudomonas chlororaphis	0.831	0.801	ND	ND	ND	ND	ND	ND
Shewanella putrefaciens	0.380	0.399	0.380	0.390	0.636	0.647	0.628	0.643
Staphylococcus aureus	0.483	0.512	0.512	0.519	0.717	0.680	0.725	0.727
Stenotrophomonas maltophilia	0.945	0.808	0.943	0.817	0.822	0.790	0.809	0.774
Variovorax paradoxus	0.724	0.650	ND	ND	ND	ND	ND	ND

[a]ND, not detected.

To speed up sample preparation, we modified the method to use multichannel pipettors. We used 1-mL glass test tubes held in an aluminum block and sealed with a teflon-silicon mat. With this method, we could add and remove reagents to six tubes at a time, and unclamping and clamping the block was much faster than unscrewing and screwing individual caps. Using this technique, we were able to process 94 samples, plus a blank and a positive control, in 3 h. Bacteria were identified as accurately with this method as with the conventional MIS method. Although we used nuts and bolts to clamp the assembly, the use of a quick release clamp would be faster and more convenient.

It does take some practice to master multichannel pipetting. In particular, the final step requires transferring the top phase to chromatography vials. It is critical that only the top phase be transferred, because otherwise the aqueous bottom phase will be injected into the GC, which will almost certainly destroy the GC column. To accomplish this, the multichannel pipettor must be held level with the tips far enough into the top phase to suck up sufficient sample, but not at or below the interface with the bottom phase. Also, a typical multichannel pipettor will not be usable for this final transfer because the GC vials are wider than the spacing between the pipette tips. Therefore, a multichannel pipettor with variable spacing, such as the one we used in this work, is useful.

Future Opportunities for Research

We performed some experiments using 50-μm i.d. columns. Although these gave adequate resolution, the oven on our GC did not heat rapidly enough to take advantage of these smaller columns. Also, we found that 50-μm columns required a split ratio of 1000:1, which gave very poor sensitivity. We were able to obtain reasonable sensitivity by injecting at a split ratio of 100:1 with the column oven at 110°C and then heating rapidly to 170°C. Again, there was no advantage to this method with a conventional GC oven. Metal-clad columns that are resistively heated allow heating rates of up to 1200°C/min (11), and would allow much faster separations than we have achieved.

We used the increased sensitivity available through splitless injection to identify single colonies of bacteria (17). This method was, however, an adaptation of the TSBA40 method, and consequently relatively slow. If this method were combined with 50- or 100-μm resistively heated columns, one might be able to achieve both high sensitivity and high speed. An alternative to conventional splitless injections would be to use a cryofocusing injection loop (18).

There may be instances in which higher speed would be so advantageous that lower resolution could be tolerated. For example, in a suspected bioterrorism situation, one might wish to screen a large number of samples as rapidly as possible, using FAST3 or an even faster method. Potentially positive samples could be reanalyzed using a slower method with higher resolution such as FAST2. Ideally, the software controlling the system would identify the potentially positive sample, switch methods, reanalyze the sample, and alert the operator to the results.

Although our modification of the four-step sample processing procedure is much faster than the original method, it may be possible to develop a direct transesterification procedure that would combine the first two steps, resulting in further time savings. Ultimately, the slowest step in the entire procedure is the culturing and incubation of pure strains. If strains could be identified by fatty acid analysis of the original sample, without subculturing and incubating the final culture, up to several days would be saved. Whether this would be possible would depend on the complexity and population density of the original sample. Some research has been done on identifying bacteria in mixed cultures by fatty acid analysis (19).

Summary

Cellular fatty acid analysis is a useful method for identification of bacteria. The commercially available MIDI provides the equipment and software needed. The technique requires growing pure isolates of bacteria, forming and purifying FAME in four steps, and GC. Modifications described here include a streamlined method for preparing FAME and fast GC. Together, these modifications increase throughput severalfold.

References

1. Kerwin, J.L., Evolution of Structure and Function of Fatty Acids and Their Metabolites, in *Isopentenoids and Other Natural Products: Evolution and Function*, edited by W.D. Nes, American Chemical Society, Washington, D.C., 1994, pp. 163–201.
2. Ratledge, C., and S.G. Wilkinson, Fatty Acids, Related and Derived Lipids, in *Microbial Lipids*, edited by C. Ratledge and S.G. Wilkinson, Academic Press, London, Vol. 1, pp. 23–53.
3. Lechevalier, H. and M.P. Lechevalier, Chemotaxonomic Use of Lipids—An Overview, in *Microbial Lipids*, edited by C. Ratledge and S.G. Wilkinson, Academic Press, London, Vol. 1, pp. 869–902.
4. Lambert, M.A., and C.W. Moss, Comparison of the Effects of Acid and Bass Hydrolyses on Hydroxy and Cyclopropane Fatty Acids in Bacteria, *J. Clin. Microbiol. 18*:1370–1377 (1983).
5. Carrapiso, A.I., and C. Garcia, Development in Lipid Analysis: Some New Extraction Techniques and In Situ Transesterification, *Lipids 35*:1167–1177 (2000).
6. Gharaibeh, A.A., and K.J. Voorhees, Characterization of Lipid Fatty Acids in Whole-Cell Microorganisms Using In Situ Supercritical Fluid Derivatization/Extraction and Gas Chromatography/Mass Spectrometry, *Anal. Chem. 68*:2805–2810 (1996).
7. Holzer, G., T.F. Bourne, and W. Bertsch, Analysis of In Situ Methylated Microbial Fatty Acid Constitutents by Curie-Point Pyrolysis-Gas Chromatography-Mass Spectrometry, *J. Chromatogr. 468*:181–190 (1989).
8. Basile, F., M.B. Beverly, K.J. Voorhees, and T.L. Hadfield, Pathogenic Bacteria: Their Detection and Differentiation by Rapid Lipid Profiling with Pyrolysis Mass Spectrometry, *Trends in Anal. Chem. 17*:95–109 (1998).
9. Ingram, J.C., W.F. Bauer, R.M. Lehman, S.P. O'Connell, and A.D. Shaw, Detection of Fatty Acids from Intact Microorganisms by Molecular Beam Static Secondary Ion Mass Spectrometry, *J. Microbiol. Methods 53*:295–307 (2003).

10. Sasser, M., Identification of Bacteria by Gas Chromatography of Cellular Fatty Acids, Technical Note #101, MIDI Inc., Newark, DE, 2001.
11. Reed, G.L., K. Clark-Baker, and H.M. McNair, Fast Gas Chromatography of Various Sample Types Using Fast Oven Temperature Programming, *J. Chromatogr. Sci. 17*:300–305 (1999).
12. Schutjes, C.P.M., E.A. Vermeer, J.A. Rijks, and C.A. Cramers, Increased Speed of Analysis in Isothermal and Temperature-Programmed Capillary Gas Chromatography by Reduction of the Column Inner Diameter, *J. Chromatogr. 253*:1–6 (1982).
13. Blumberg, L.M. and M.S. Klee, Method Translation and Retention Time Locking in Partition GC, *Anal. Chem. 70*:3828–3839 (1998).
14. Klee, M.S., and V. Giarrocco, Predictable Translation of Capillary Gas Chromatography Methods for Fast GC, Application Note 5965-7673E, Agilent Technologies, Wilmington, DE, USA, 1997.
15. Buyer, J.S., Rapid Sample Processing and Fast Gas Chromatography for Identification of Bacteria by Fatty Acid Analysis, *J. Microbiol. Methods 51*:209–215 (2002).
16. Buyer, J.S., Improved Fast Gas Chromatography for FAME Analysis of Bacteria, *J. Microbiol. Methods 54*:117–120 (2003).
17. Buyer, J.S., Identification of Bacteria from Single Colonies by Fatty Acid Analysis, *J. Microbiol. Methods 48*:259–265 (2002).
18. Borgerding, A.J., and C.W. Wilkerson, Cryogenically Cooled Microloop System for Sampling and Injection in Fast GC, *Anal. Chem. 68*:701–707 (1996).
19. Lehtonen, L., R. Peltonen, and E. Eerola, Computerised Gas-Liquid Chromatography of Bacterial Cellular Fatty Acids in Analysis of Bacterial Mixtures, *J. Microbiol. Methods 25*:317–327 (1996).

Part IV. Vibrational Spectroscopic Techniques

13 Use of Cellular Fatty Acids to Identify Food-Borne Pathogens by Infrared Spectroscopy and Capillary Gas Chromatography

M.M. Mossoba[a] and S.F. Al-Khaldi[b]

Food and Drug Administration (FDA), Center for Food Safety and Applied Nutrition (CFSAN), [a]Division of General Scientific Support, OSAS and [b]Division of Microbiological Studies, College Park, MD 20740–3835

Introduction

According to Mead et al. (1), it is estimated that millions of Americans contract food-borne illnesses each year. There has been active surveillance for laboratory-diagnosed cases of 10 food-borne diseases resulting from infection with *Campylobacter* spp., *Escherichia coli* O157, *Listeria monocytogenes*, *Salmonella* spp., *Shigella* spp., *Vibrio* spp., *Yersinia enterocolitica, Cryptosporidium parvum, Cyclospora cayetanensis*, and hemolytic uremic syndrome (HUS) (2). Since September 11, 2001, there has been an additional requirement to identify select agents. These include toxins such as botulinum toxin and staphylococcal enterotoxin, chemical agents (e.g., sarin nerve gas and mustard gas), viruses, as well as pathogenic bacteria including *Bacillus anthracis*, *Variolla* (smallpox), *Clostridium botulinum*, and *C. perfringens* (3). This need has led to a sudden surge in the demand for accurate and rapid methods of identification that include routine rapid tests as well as fast microbiological assays and analytical instrumentation methodologies.

The identification of bacteria using traditional microbiological methods can take at least a day and generally much longer; it requires that microorganisms be streaked on selective media, and toxin identification requires purification of the toxin before testing (4,5). Methods in molecular biology have also been used to confirm the identities of genes and bacteria. Bacterial contaminants were identified using PCR, testing with surrogate biochemical and immunological markers (4,5), as well as DNA and protein microarray hybridization, and phenotype microarray methodologies. However, genotypic methods are generally not always regarded as ideally suitable for purposes of routine bacteria identification due to their time-consuming nature and expense.

In research laboratories, chemical analytical instrumentation methods have also been used increasingly for bacterial speciation. These techniques include capillary gas chromatography (GC) (6–21), Fourier transform infrared (FTIR) spectroscopy (22–30), and mass spectrometry (MS) (9,27). Whole-cell bacteria were identified by vibrational spectroscopic methods including ultraviolet (UV) resonance Raman spectroscopy (25), and FTIR spectroscopy in the mid-infrared and recently near-infrared (31) ranges. FTIR spectroscopy was also used to measure bacterial cellular fatty acids (32). This chapter will be limited to a review of the application of IR spectroscopy as well as capillary GC to the analysis of methyl ester derivatives of fatty acids extracted from food-borne pathogens.

Gas Chromatography and Cellular Fatty Acid Methyl Esters (FAMEs)

Capillary gas chromatographs equipped with flame ionization detection (FID) (6–8,18–21,26,33) or interfaced to mass spectrometers (21,34) have been used for the identification of both endospore and vegetative cellular FAMEs. For over a decade, there has been a commercial database and software, the Microbial Identification System (MIS) (MIDI Incorporated, Newark, DE) (7,17), that can identify microorganisms based on unique FAME GC-FID patterns of known species and strains. This automated system compares observed cellular FAME profiles for test samples with those in its proprietary database, and identifies them by using its pattern recognition software. This commercial database consists of >2000 entries, including aerobic and anaerobic bacteria and yeasts. This chromatographic system applies a 20-min capillary GC-FID procedure. With the advent of high-speed GC technology (35), it has become possible to dramatically reduce the chromatographic run times, and rapid (~5 min) GC separations have also been applied to the analysis of bacterial FAMEs (36) (see Chapter 12 by Buyer).

Fatty Acid Structures in Cellular Lipids

Cellular fatty acid components exhibit a wide variety of chemical structures and include saturated, unsaturated, branched fatty acids, and molecules with hydroxy groups (e.g. 3-OH-$C_{14:0}$) and rings (13). Branched fatty acids consist of the iso series, $(CH_3)_2CH(CH_2)_nCOOH$, and the antiso series, $CH_3CH_2CH(CH_3)(CH_2)_mCOOH$ or CH_3 $(CH_2)_xCH(CH_3)(CH_2)_yCOOH$ with branching at a carbon in the middle of a fatty acid chain, such as iso-$C_{15:0}$, antiso-$C_{15:0}$, and 10-Me-$C_{19:0}$. *Cis*-11,12-methylene-octadecanoic acid ($C_{19:0}$cyc11,12) is an example of a saturated 3-membered cyclopropane ring.

The lipid component of the cell membranes (including phospholipids) or the lipid A component of lipopolysaccharides in gram-negative bacteria and lipoteichoic acid in gram-positive bacteria are the primary source of cellular fatty acids (15). Many bacterial cellular fatty acid structures have chain lengths ranging from 10 to 19 carbon atoms, and most are fatty acids with 16–18 carbons. Observed differences in fatty acid composition and relative amounts of individual fatty acid components in GC profiles allow identification of various microorganisms. For instance, branched structures are predominant in some gram-positive bacteria, whereas fatty acids with cyclopropane rings and hydroxy fatty acids are characteristic of lipopolysaccharides of gram-negative bacteria. Gram-negative bacteria appear to have a larger proportion of saturated and monounsaturated fatty acids with an even number of carbon atoms than gram-positive bacteria. The latter generally have a larger proportion of saturated branched-chain fatty acids with an odd number of carbon atoms and low levels of saturated straight-chain fatty acids.

Profiling Cellular FAMEs by Gas Chromatography

Since the early 1960s, GC has been used as an instrumental tool to analyze cellular fatty acids and for the speciation of bacteria (11,14). Intact bacterial cells are usually treated with sodium hydroxide and alcohol to cause hydrolysis and release of cellular membrane fatty acids. Sodium salts are formed and subsequently converted to the corresponding esters. The identification and determination of the mixture of FAMEs is then carried out by GC. Esterification increases volatility and improves the GC resolution. In recent years, the application of multivariate analysis, such as principal component analysis, to GC FAME profiles reportedly facilitated the identification of microorganisms (12,14).

Since the advent of long-fused silica capillary columns in the 1980s, the application of GC to the determination of cellular FAMEs has become more widely used. The automated MIDI system (17), which applies a GC procedure with FID and multivariate analysis to the profiling of fatty acids ranging in length from 9 to 20 carbon atoms, has been used increasingly by research institutions. The generally recommended GC-FID procedure (17) is outlined below. GC procedures were used to identify bacterial species and strains mostly from clinical, environmental, plant, and soil, and only to a limited extent from food matrices (13).

Cultivation of Cells

The analytical chemistry of bacterial identification depends on the comparison of characteristic differences in the chemical composition of either intact cells or their constituents, e.g., lipids, proteins, or carbohydrates. The various microorganisms must be grown under identical conditions to minimize variability and improve precision. The microorganisms must be cultured to reach a biomass of ~40 mg reportedly required for GC-FID analysis (17).

The accurate identification of unknown microorganisms using a particular FAME database requires the use of microbiological media and growth and incubation conditions identical to those used to create that database. These conditions include the temperature used in the cultivation of bacteria, the age of the culture, and the nature of the growth medium (17). It was established (37) that differences in these factors would lead to significant variability in the lipid content and composition of bacteria. Aerobic bacteria are usually grown on trypticase soy broth agar (TSBA) medium, consisting of 30 g trypticase soy broth and 15 g agar. For aerobic bacteria, other media that depend on the nature of the organism investigated are commonly used. A temperature of 28°C is used to grow a wide range of microorganisms on TSBA. For anaerobic bacteria, agar cultures are incubated at 35°C on brain heart infusion (BHI) with supplements. Broth cultures are harvested at a given turbidity, and plate cultures are grown for 24 (aerobes) or 48 h (anaerobes) to minimize incubation time variability. Longer incubation times are used for bacteria that grow slowly. For plate cultures, the physiological age is commonly standardized by selecting a particular sector from a plate that is streaked into quadrants. Often, all quadrants must be harvested to reach a biomass of 40 mg. The weighed cells are placed in a culture tube with a teflon-lined screw cap for further analysis by GC.

Preparation of Cellular FAMEs for Analysis

The saponification, methylation, and extraction steps (13,17) are carried out to prepare FAMEs for subsequent separation by GC. Initially, ~40 mg of bacterial cells are saponified with 1 mL of a sodium hydroxide (NaOH) solution prepared from 45 mg NaOH, 150 mL H_2O, and 150 mL methanol. Tubes containing the cells and NaOH solution are securely capped, vortexed for 5–10 s, heated in a boiling water bath for ~5 min, again vortexed for 5–10 s, and returned to the water bath for 25 min. The tubes are allowed to cool down slowly to room temperature, and the fatty acids are methylated with 2 mL of an acidic methanol stock solution prepared from 6N HCl (325 mL) and methanol (275 mL). The tubes are vigorously vortexed for 5–10 s, recapped, carefully heated at 80 ± 1°C for 10 ± 1 min, and rapidly cooled. The FAME products are then extracted with 3 mL of a 1:1 solution of hexane:*tert*-butyl ether by gently shaking the tubes for 10 min. The lower aqueous phase is removed with a pipette and discarded. To clean the sample, the remaining organic phase is washed with 3 mL NaOH solution prepared from 10.8 g NaOH in 900 mL H_2O. This is carried out by recapping and gently shaking the tubes for 5 min. Two-thirds of the top organic phase is removed with a pipette and placed in a GC vial that is capped and saved for subsequent GC analysis.

GC-FID Methodology

The remaining GC-FID procedure includes automated steps (17). An autosampler allows FAME test samples to be injected and separated while the experiment is unattended by an analyst. The chromatographic step is usually carried out on a 25-m and 0.2-mm i.d. phenyl methyl silicone fused silica capillary column. The temperature program used consists of ramping the temperature from 170 to 270°C at 5°C/min. GC peaks are automatically integrated and fatty acid composition data are stored. GC-FID profiles of FAME test samples are automatically compared with those in the MIDI database and analyzed by the MIS pattern recognition algorithms that apply statistical techniques to reduce the dimensionality of multivariate data while preserving most of the variance (17).

Limitations of GC-FID and Alternative Methodologies

A major disadvantage of using GC for fatty acid profiling of bacteria is the requirement of labor-intensive fatty acid derivatization procedures. Attempts to overcome this drawback include the use of supercritical fluid derivatization/extraction and GC-MS (34), in situ thermal hydrolysis methylation (THM)-GC-MS (38), and in situ THM-MS (9) for the determination of cellular FAMEs. With in situ THM-MS, the time-consuming extraction/methylation (60 min) and the GC separation (20 min) steps are eliminated, and this procedure may potentially be automated (9). Trimethylsulfonium hydroxide (TMSH) pyrolysis was used by Muller et al. (20) to complement the MIS procedure for the identification of FAMEs. This rapid single-step TMSH pyrolysis was used to transesterify bound fatty acids in phospholipids and triacylglycerol molecules to FAMEs at room temperature. This procedure also released secondary alcohols and, for *Mycobacterium* spp., mycolic acid cleavage products with chain lengths from C22 to C26 that were also amenable to GC-FID analysis. Pyrolysis-GC-atomic emission (39) was also used for identification of microorganisms. The carbohydrate composition of bacterial cell hydrolysates was reportedly analyzed using an automated alditol acetate derivatization procedure, GC-MS, and GC-MS-MS (16). Recently, Han and Gross (40,41) reported the analysis of mixtures of cellular lipidomes by electrospray ionization MS (see Chapter 3 by Han and Gross).

Application of GC-FID Methodology to Food-Borne Pathogens

Although many of the methodologies discussed above can provide high specificity and sensitivity in detecting trace amounts of chemical markers from bacterial samples, the profiling of cellular fatty acids from vegetative cells and spores by GC-FID is a more widely used analytical procedure (18,42). This is probably due to the availability of the MIS software and, in particular, the MIDI commercial database for a large number of microorganisms, as well as the availability of the GC-FID AOAC Official Method for *B. anthracis,* namely, AOAC 2004.11 (7,8). Although the great majority of GC studies involve the identification of clinical, environmental, and other microorganisms (20,43–45), there is a paucity of publications on the analysis of FAME GC profiles of food-borne pathogens (6,42,46,47). The last-mentioned studies are discussed next.

In a recent study (6), a capillary GC-FID system equipped with the MIDI database and MIS software was applied to the identification of the whole-cell fatty acid profiles of microbial pathogens that cause food-borne illnesses. These microbes were compiled by the American Medical Association, Centers for Disease Control and Prevention, United States Food and Drug Administration, and Food Safety and Inspection Service of the United States Department of Agriculture. In that study, the FAME profiles of spores and vegetative cells of the same endospore-forming bacilli were compared. Fifteen bacteria representing the eight genera *Staphylococcus, Liste-*

ria, Bacillus, Yersinia, Salmonella, Shigella, Escherichia, and *Vibrio,* and 11 species were used. Endospore-forming bacilli were processed to obtain pure spores and vegetative cell FAMEs for analysis by GC-FID. A data set for each bacterial agent was prepared using FAME profiles from five replicate batches prepared on different days. The observed GC-FID chromatograms exhibited unique FAME intensity profiles for each of the 11 species and were used to discriminate among these organisms. The cellular FAME GC data for *Bacillus anthracis* and *Bacillus cereus* indicated that there are two unique fatty acids with branched chains, iso 17:1 ω10*c* and 17:1 antiso. The former fatty acid is present in *B. cereus* vegetative cells and spore, but not in *B. anthracis*. The latter fatty acid is present in *B. anthracis* cells but not in *B. cereus* cells. The 16:0 2OH and 17:0 iso 3OH FAMEs are present in *B. anthracis* and *B. cereus* spores but not in the corresponding vegetative cells.

In a validation study (46), the performance of the GC-FID-based MIS was compared with that of four different commercially available automated microbial identification systems, namely, the MicroScan WalkAway 40 system (Dade Diagnostics Corp., Mississauga, ON, Canada), the MicroLog system (Biolog, Inc., Hayward, CA), the VITEK system (bioMerieux Vitek, Hazelwood, MO), and the Replianalyzer system (Oxoid Inc., Nepean, ON, Canada). These four microbial identification systems depend on substrate utilization and bacterial growth to induce changes in pH, which in turn trigger changes in the color of indicators. The five systems were evaluated for their ability to identify six common food-borne pathogens. The sensitivities, specificities, and repeatabilities of the different systems were tested by identifying isolates of *B. cereus, Campylobacter jejuni, Listeria monocetogenes, Staphylococcus aureus, Salmonella* spp., and *verotoxigenic Escherichia coli* (VTEC).

In this GC validation study by Odumeru et al. (46), 40 reference positive isolates (RPIs) and 40 reference negative isolates (RNIs) of these six microorganisms were investigated. Of the RPIs, 35 were obtained from food products and five from the American Type Culture Collection (ATCC). Of the RNIs, five were ATCC strains, and 35 were cultured from food, clinical, or environmental samples and consisted of laboratory isolates that were related to but not identical to the bacteria of interest. The RNIs were similar in their biochemical reactions and gram-staining results to those of the microorganisms of interest. Sensitivity was defined in this study as the proportion of the RPIs that were correctly identified with an acceptable identification confidence level defined by the system's manufacturer. Specificity was measured by the proportion of RNIs that were not identified as the pathogen of interest with an acceptable confidence rating. Repeatability tests were based on replicate analyses for 20 randomly selected ATCC strains and laboratory isolates from the RPIs and RNIs. Repeatability of identification was defined as the proportion of replicate analyses that yielded the same result at similar confidence levels.

According to Odumeru et al. (46), the sensitivities of the five commercial systems used for the identification of microorganisms fell in the wide range of 42.5–100%. In particular, the sensitivity of the MIS was 90% for *Listeria* spp., 47.5% for *S. aureus*, 55% for *B. cereus*, 72.5% for *C. jejuni*, 85% for *Salmonella* spp., and 52% for *E. coli*. The lower sensitivities found for some of these pathogens with the MIS were attributed to the fact that it is based on fatty acid composition, whereas the reference systems for these species were based on biochemical reactions. The specificities were usually close to 100%; those of the MIS were 100% for *Listeria* spp., 100% for *S. aureus*, 97.5% for *B. cereus*, 32.5% for *C. jejuni*, 100% for *Salmonella* spp., and 97.5% for *E. coli*. The repeatabilities of the MIS for the identification of test organisms were somewhat lower, ranging from 30 to 90%, than those of the remaining four systems (60–100%). The repeatability of the MIS for RPIs was 30% for *L. monocytogenes*, 60% for *S. aureus*, 90% for *B. cereus*, 90% for *C. jejuni*, 90% for *Salmonella* spp., and 70% for *E. coli*, whereas the repeatability of the MIS for RNIs was 55% for *L. monocytogenes*, 90% for *S. aureus*, 80% for *B. cereus*, 75% for *C. jejuni*, 65% for *Salmonella* spp., and 65% for *E. coli*. The selection of an automated system for the identification of food-borne bacteria depends on many variables, such as the nature of the available range of organisms in the system's database and the ability of the system to correctly identify the pathogen of interest.

Lin et al. (42) successfully applied the GC-FID procedure and the MIS system to investigate the incidence and distribution of *B. cereus* vegetative cells and spores in raw and pasteurized milk. The presence of *B. cereus* in pasteurized milk is a concern for the dairy industry because it could lead to off-flavors, sweet curdling, and even outbreaks of food poisoning. *B. cereus* is a gram-positive, spore-forming microorganism that can produce toxins in pasteurized milk even at refrigeration temperatures. *B. cereus* found in pasteurized milk may originate from spores that are present in the raw milk or from the dairy plant environment.

In the study of Lin et al. (42), a total of 232 milk samples from sampling points along milk processing lines and 122 environmental swabs were collected in two dairy plants over several months. The FAME of each *B. cereus* isolate was determined by GC-FID. Using MIS, a database of *B. cereus* FAME profiles for 229 *B. cereus* isolates from milk samples and environmental swabs was generated and used to characterize the relations between *B. cereus* isolate test samples.

In raw milk, <10% of test samples were positive for *B. cereus* vegetative cells. The average *B. cereus* count in positive test samples was <50 cfu/mL after enrichment at 8°C for 3 d. The incidence of *B. cereus* spores in raw milk test samples was measured by the presence of *B. cereus* in heat-treated (75°C for 20 min) milk and was excessively high. Of the heat-treated raw milk test samples, >80% contained *B. cereus*, and the average *B. cereus* count in positive samples was >1.1×10^5 cfu/mL after enrichment at 8°C for 14 d.

In pasteurized milk, the incidence of *B. cereus* was significantly high. After enrichment at 8°C for 14 d, 76–94% of these test samples were contaminated with *B. cereus* and the average count reached 3.7×10^5 cfu/mL. Of the pasteurized test samples from milk in cartons or plastic bags, >90% contained *B. cereus*, and the average count reached 5.5×10^6 cfu/mL after enrichment. Most *B. cereus* isolates obtained from the pasteurized milk and final milk products belonged to the same subgroups as the *B. cereus* strains germinated from spores in raw milk. Therefore, Lin et al. (42)

concluded that *B. cereus* spores in raw milk were the major source of *B. cereus* contamination in pasteurized milk, and that postpasteurization contamination occurred. However, analysis of environmental swabs suggested that the dairy plant environment was a potentially minor source of *B. cereus* in pasteurized milk.

Sundhein et al. (47) also used the MIS-based GC-FID technique as an effective complementary technique to carbon source assimilation (CSA) tests for the identification of pseudomonads from fresh and chill-stored chicken carcasses. The shelf life of fish, red meat, and poultry in air is limited by the presence of psychrotrophic pseudomonads that cause the formation of slime and the production of off-odors. In this study, hundreds of bacterial isolates from 18 chicken carcasses were assigned to one of 17 defined groups of chicken pseudomonads. Isolates that could not be matched to any known species by CSA exhibited FAME profiles that corresponded to *P. fluorescens*, *P. Lundensis*, or *P. fragi.* The *P. fluorescens* biovars had greater levels of *cis* 9-16:1 (21–37%) and *cis* 9-18:1 (10–19%) relative to those of 17:0 cyclo (1–17%) and 19:0 cyclo (0–1%), whereas the opposite was found for *P. Lundensis* and *P. fragi.* Sundhein et al. (47) concluded that none of the species were dominant and that the relative incidence of the various species may vary with flock or even individual birds.

Infrared Spectroscopy and Cellular FAMEs

An attenuated total reflectance (ATR)-FTIR spectroscopic procedure was recently developed at the FDA infrared facility to identify bacteria by analyzing mixtures of FAMEs extracted from cellular lipids (32). Advantages of using an ATR-FTIR procedure for the measurement of mixtures of bacterial FAMEs included the elimination of the GC separation step, the speed (2 min) of the infrared spectroscopic measurement, and the inherent precision of the ATR sampling technique. A description of this technique is outlined below, with the experimental protocol used and results of this ATR-FTIR study.

ATR-FTIR Methodology

ATR spectroscopy has been reviewed extensively (48–50). ATR is an internal reflection infrared sampling technique that can significantly increase the precision of measurements. When a FAME test sample is placed on the surface of a crystal such as diamond, the infrared light penetrates a distance of only a few micrometers into the FAME analyte when the conditions of total internal reflection apply. These conditions occur when light traveling in a transparent medium of high refractive index (η_1) (such as diamond) strikes the interface between this medium and another transparent medium of lower refractive index (η_2) (such as air or a FAME test sample) at an angle of incidence (θ) exceeding the critical angle (θ_c) defined by:

$$\theta_c = \sin^{-1}(\eta_2/\eta_1)$$

Normally, light is partially transmitted and partially reflected. However, under these conditions, it is not transmitted, but is totally reflected inside the diamond crystal. Moreover, as the light bounces (one or more times) inside the crystal, a so-called evanescent wave also propagates away from the surface of the crystal through the FAME test sample, in this example. At the surface of the crystal, the intensity of this wave decays exponentially with distance. It is also attenuated by the absorption of IR light by the FAME test sample. The depth of penetration (dp) of the IR light into the test sample is minuscule. It typically varies between 1 and 4 μm and depends on θ, η_2, η_1, and the wavelength (γ) as given by the relation:

$$dp = \gamma/2\pi\, \eta_1\, [\sin^2(\theta) - (\eta_2/\eta_1)]^{1/2}$$

As a result, the depth of penetration (which is also the effective path length of the light through the test sample) will be higher the greater λ or the smaller the frequency. Therefore, an interferogram (raw IR spectrum) is a measure of the attenuation by a FAME test sample of the totally internally reflected IR light. The interferogram of a reference background (e.g., air) is measured similarly. They are subsequently used to obtain an absorption spectrum. ATR-FTIR measurements are easy, convenient, and require ~2 min/test sample (49).

ATR-FTIR Protocol

The FTIR system (Equinox 55 FTIR spectrometer, Bruker Optics, Billerica, MA) used at the FDA facility was equipped with a heated diamond ATR IR cell (SensIR Technologies, Danbury, CT) that requires ~1 μL of neat (without solvent) FAME. The diamond ATR IR cell was preheated to 65 ± 1°C for 1 h before measurement. Each bacterial FAME solution (~0.8 mL) was slowly evaporated under a gentle stream of argon gas in a fume hood until the volume was ~100 μL. Using a disposable plastic pipette the solution was transferred to a glass conical insert and evaporated to ~1–2 μL. Using a syringe equipped with a fused silica needle, ~1–2 μL of hexane was added to the glass conical insert containing the FAME solution, and 1–2 μL of the hexane-treated FAME solution was then transferred to the heated diamond surface of the ATR cell using the fused silica needled syringe. ATR-FTIR spectra for neat melted FAMEs were collected over the wave number range of 4000–600 cm^{-1} at a resolution of 4 cm^{-1}; to improve the signal-to-noise ratio, 128 scans were co-added and the signal averaged (2 min). The reference background was collected in the absence of any test sample. The acquired ATR-FTIR spectra were imported into the multivariate statistics program PIROUETTE™ 3.0 (InfoMetrix, Inc., Woodinville, WA). To minimize differences among absorbance spectra, such as baseline shifts, all of the measured spectra were mean-centered and transformed to second derivative spectra. Data analysis and investigation of clustering were accomplished using principal component analysis and soft independent modeling of class analogy (SIMCA) algorithms (51,52) in PIROUETTE™ 3.0 software.

Application of ATR-FTIR Methodology to Food-Borne Pathogens

Fourteen bacteria were identified in this proof-of-concept study (32). Gram-positive bacteria included *S. aureus* ATCC 19075, *L. monocytogenes* Scott A, *B. anthracis* Sterne, and *B. cereus* ATCC 700282. Gram-negative bacteria included *Y. enterocolitica* 289, *S. typhimurium* DT104B, *S. sonnei* 20143, *V. cholerae* 2194C, *V. vulnificus* C7684, *V. parahaemolyticus* 4037, *E. coli* O157:H7 933J, *E. coli* 153:H2 393, *E. coli* O-:H11 402, and *E. coli* O157:H7 933W. The cultures were grown aerobically on BHI agar plates at 37°C for 24 h. *V. vulnificus* C7684 was also grown with 1% NaCl incorporated into the medium. Each of these bacteria comprised a separate category. The total number of categories was 15. The maximum number of bacterial samples in the training sets was 168 samples (averaging ~11 samples per category) comprised of same-day (usually 1 or 2) and different-day (between 4 and 7 d) replicates. Using IR spectroscopy to measure bond vibrations of functional groups within FAMEs obtained from bacterial cells, fingerprint-like infrared spectra were obtained and used for identification.

The precision advantage of ATR is highly desired whenever replicate IR spectra observed for bacterial FAMEs on same and/or different days are subsequently used for identification. As a result of this advantage, many of the observed ATR-FTIR spectra (Fig. 1) consistently exhibited maximum absorbance values near 0.2 absorbance units. Major IR bands were correlated to the structure of FAME molecules and utilized in creating models that discriminated among bacterial categories (genera, species, or strains). Long-chain FAME mid-IR spectra (Fig. 1) exhibited bands between 2960 and 2850 cm^{-1} due to the carbon-hydrogen (C-H) stretch absorptions of the hydrocarbon chain. These are attributed to the asymmetric C-H stretching of methyl (2960 cm^{-1}, shoulder) and methylene (2925 cm^{-1}) groups, and symmetric C-H stretching of methyl (2870 cm^{-1}, very weak) and methylene (2850 cm^{-1}) groups. The presence of double bonds within the fatty acid carbon backbone shifts the unsaturated carbon-hydrogen (=C-H or $=CH_2$) stretch absorptions to lower wavelengths, typically between 3125 and 2990 cm^{-1}. These weak, yet highly characteristic unsaturation bands, were found at 3060 cm^{-1} ($=CH_2$) and near 3000 cm^{-1} (=C-H). The ester group exhibited a very intense symmetric C=O stretching vibration at 1747 cm^{-1}. The medium intensity bands near 1459, 1161, and 722 cm^{-1} are due to the CH_2 scissoring deformation, symmetric C-O stretching, and CH_2 rocking vibrations, respectively.

To build the bacteria FAME data sets, a total of 168 spectra were acquired from bacterial samples grown on 4–7 different days. Different-day infrared spectra replicate measurements of the bacterial FAME were necessary to investigate the possible sources of variance in the sampling procedure.

Multivariate Statistical Analysis of Observed Infrared Spectra

To extract and analyze unique information about a particular bacterium from a complex infrared spectrum, sophisticated statistical methods are required. With the super-

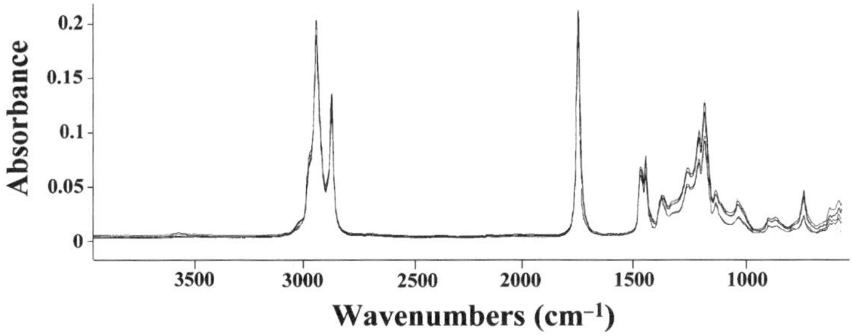

Fig. 1. Examples of observed attenuated total reflectance-Fourier transform infrared spectra in the 4000–600 cm^{-1} range are shown for the fatty acid methyl esters of five bacteria investigated: *Y. enterocolitica, S. typhimurium, S. sonnei, B. cereus,* and *S. aureus.*

vised pattern recognition SIMCA algorithm, prediction models are constructed from training set samples that have been preassigned to a particular category, namely, bacterial genus, species or strain. As an illustration, Figure 2 exhibits the space in which 113 representative samples for 10 bacteria are predicted to cluster as visualized in the SIMCA 3-dimensional analysis plot using the first three principal components. This and other models are then used to predict the identities of test samples whose categories were included in the data sets.

In an experiment that included all of the 15 categories investigated, all but two of the 168 samples (98.8%) in this training set were correctly assigned to their category: one of the *E. coli* strains was predicted as another *E. coli* strain, and one of the *V. cholerae* species was assigned as a *V. vulnificus.* Using this optimized training set model, the identities were correctly predicted for a test set that comprised 15 bacterial test samples (genera, species, or strains) that represented each of the 15 different categories included in the prediction models. No misidentification was indicated for these 15 bacterial test samples, a result that can probably be attributed to the use of an optimal spectral range, 3068–2941, 1780–1695, and 1523–673 cm^{-1}.

In this study, ATR-FTIR spectroscopy and multivariate analysis were successfully used to discriminate among the gram-positive bacteria *S. aureus*, *L. monocytogenes*, *B. anthracis*, and *B. cereus* and the gram-negative bacteria *Y. enterocolitica*, *S. typhimurium*, *S. sonnei*, *E. coli* (four strains of *E. coli*), and *V. cholerae*, *V. vulnificus*, and *V. parahemolyticus*. Also, the Enterobacteriaceae were differentiated from the vibrios. In this set, the identification was at the interspecies (for each of the *Bacillus* and *Vibrio* genera) or intraspecies (for the *E. coli* species) level.

Conclusion

The identification of bacteria by analytical techniques requires expertise in several disciplines including microbiology, chromatography, spectroscopy, and multivariate

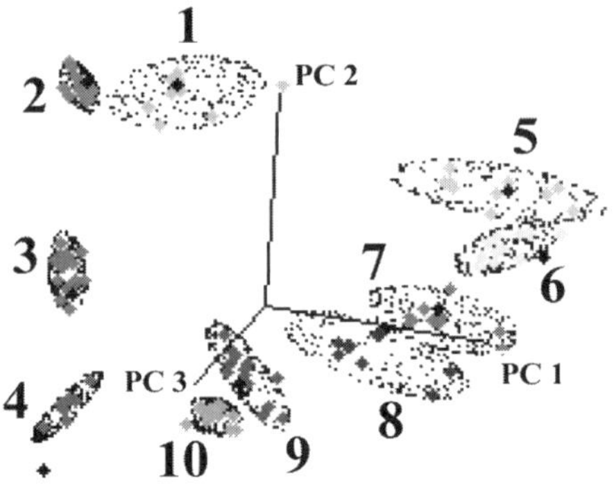

Fig. 2. Soft independent modeling of class analogy (SIMCA) results of an experiment in which the spectral range 3068–2941 cm^{-1} was tested. The space in which 113 samples representing 10 bacterial genera or species (training set) are predicted to cluster was visualized in the SIMCA three-dimensional analysis plot using the first three principal components. The identities of the 10 bacteria used are: gram-positive bacteria (1) *S. aureus* ATCC 19075, (2) *L. monocytogenes* Scott A, (3) *B. anthracis* Sterne, and (4) *B. cereus* ATCC 700282, and gram-negative bacteria (5) *E. coli* O157:H7 933W, (6) *Y. enterocolitica* 289, (7) *S. typhimurium* DT104, (8) *S. sonnei* 20143, (9) *V. cholerae* 2194C, and (10) *V. vulnificus* C7684 grown in the presence of 1% NaCl.

analysis. GC-FID has been a widely used analytical technique for bacteria speciation due to the availability of the MIDI databases and MIS commercial software. Unique FAME profiles from bacteria and spores can provide a powerful tool for the identification of food-borne pathogens. *B. anthracis* from culture can be identified by following the new GC-FID AOAC Official Method 2004.11. Observed ATR-FTIR spectral fingerprints demonstrated the discriminating capability of this newly developed and rapid (2 min) IR procedure. An advantage of using an ATR sampling accessory is its high precision, which can minimize the spectral variability among replicate bacterial FAME test samples, a necessary condition for successful pattern recognition.

The analysis of cellular fatty acids by GC-FID and ATR-FTIR can be complementary analytical tools for biological pathogens together with methods in classical microbiology and molecular biology.

References

1. Mead, P.S., L. Slutsker, V. Dietz, L.F. McCaig, J.S. Bresee, C. Shapiro, P.M. Griffin, and R.V. Tauxe, Food-Related Illness and Death in the United States, *Emerg. Infect. Dis. 5*:607–625 (1999).
2. Centers for Disease Control, Preliminary FoodNet Data on the Incidence of Foodborne Illnesses—Selected Sites, US, 2001, *Morbid. Mortal. Wkly. Rep. 51*:325–329 (2002).

3. Jordan, J.R., Analyzing Bioterrorism, *Inside Laboratory Management Jan/Feb*:19–23 (2002).
4. Feng, P., Development and Impact of Rapid Methods for Detection of Foodborne Pathogens, in *Food Microbiology: Fundamentals and Frontiers*, 2nd ed., edited by M.P. Doyle, ASM Press, Washington, 2001, pp. 755–796.
5. Iqbal, S.S., M.W. Mayo, J.G. Bruno, B.V. Bronk, C.A. Batt, and J.P. Chambers, A Review of Molecular Recognition Technologies for Detection of Biological Threat Agents, *Biosens. Bioelectron. 15*:549–578 (2000).
6. Whittaker, P., F.S. Fry, S.K. Curtis, S.F. Al-Khaldi, M.M. Mossoba, M.P. Yurawecz, and V.C. Dunkel, Use of Fatty Acid Profiles to Identify Foodborne Bacterial Pathogens and Aerobic Endospore-Forming Bacilli, *J. Agric. Food Chem. 53*:3735–3742 (2005).
7. Sasser, M., C. Kunitsky, G. Jackoway, J.W. Ezzell, J.D. Teska, B. Harper, and S. Parker, Identification of *Bacillus anthracis* from Culture Using GC Analysis of FAME, *J. Assoc. Off. Anal. Chem. Int. 48*:178–181 (2005).
8. AOAC International, *Official Methods of Analysis*, 17th ed. Microbiological Method 2004.11, Gaithersburg, MD, 2004.
9. Xu, M., K.J. Voorhees, T.L. Hadfield, Repeatability and Pattern Recognition of Bacterial Fatty Acid Profiles Generated by Direct Mass Spectrometric Analysis of in Situ Thermal Hydrolysis/Methylation of Whole Cells, *Talanta 59*:577–589 (2003).
10. James, A.T., Qualitative and Quantitative Determination of the Fatty Acids by Gas-Liquid Chromatography, *Methods Biochem. Anal. 8*:1–59 (1960).
11. Abel, K., H. de Dchmerting, and J.I. Peterson, Classification of Microorganisms by Analysis of Chemical Composition. I. Feasibility of Utilizing Gas Chromatography, *J. Bacteriol. 85*:1039–1044 (1963).
12. Brondz, I., J. Carlsson, M. Sjostrom, and G. Sundquist, Significance of Cellular Fatty Acids and Sugars in Defining the Genus Porphyromonas, *Int. J. Syst. Bacteriol. 39*:314–318 (1989).
13. Brondz, I., Development of Fatty Acid Analysis by High-Performance Liquid Chromatography, Gas Chromatography, and Related Techniques, *Anal. Chim. Acta 465*:1–37 (2002).
14. Brondz, I., I. Olsen, and M. Sjostrom, Multivariate Analysis of Quantitative Chemical and Enzymatic Characterization Data in Classification of *Actinobacillus*, *Haemophilus* and *Pasteurella* spp., *J. Gen. Microbiol. 136*:507–513 (1990).
15. Welch, D.F., Applications of Cellular Fatty Acid Analysis, *Clin. Microbiol. Rev. 4*:422–438 (1991).
16. Fox, A., Carbohydrate Profiling of Bacteria by Gas Chromatography-Mass Spectrometry and Their Trace Detection in Complex Matrices by Gas Chromatography-Tandem Mass Spectrometry, *J. Chromatogr. A 843*:287–300 (1999).
17. Sasser, M., Identification of Bacteria by Gas Chromatography of Cellular Fatty Acids. MIDI Technical Note #101 (Revised 2001), MIDI, Inc. Newark, DE, 1990.
18. Song , Y., R. Yang, Z. Guo, M. Zhang, X. Wang, and F. Zhou, Distinctness of Spore and Vegetative Cellular Fatty Acid Profiles of Some Aerobic Endospore-Forming Bacilli, *J. Microbiol. Methods 39*:225–241 (2000).
19. Leclercq, A., A. Guiyoule, M. El Lioui, E. Carniel, and J. Decallonne, High Homogeneity of the *Yersinia pestis* Fatty Acid Composition, *J. Clin. Microbiol. 38*:1545–1551 (2000).
20. Muller, K., E.N. Schmid, and R.M. Kroppenstedt, Improved Identification of Mycobacteria by Using the Microbial Identification System in Combination with Additional Trimethylsulfonium Hydroxide Pyrolysis, *J. Clin. Microbiol. 36*:2477–2480 (1998).

21. Basile, F., M.B. Beverly, C. Abbas-Hawks, C.D. Mowry, K.J. Voorhees, and T.L. Hadfield, Direct Mass Spectrometric Analysis of in Situ Thermally Hydrolyzed and Methylated Lipids from Whole Bacterial Cells, *Anal. Chem. 70*:1555–1562 (1998).
22. Naumann, D., D. Helm, and H. Labischinski, Microbiological Characterizations by FT-IR Spectroscopy, *Nature 351*:81–82 (1991).
23. Choo-Smith, L.P., K. Maquelin, T. van Vreeswijk, H.A. Bruining, G.J. Puppels, N.A. Ngo Thi, C. Kirschner, D. Naumann, D. Ami, A.M. Villa, F. Orsini, S.M. Doglia, H. Lamfarraj, G.D. Sockalingum, M. Manfait, P. Allouch, and H.P. Endtz, Investigating Microbial (Micro) Colony Heterogeneity by Vibrational Spectroscopy, *Appl. Environ. Microbiol. 67*, 1461–1469 (2001).
24. Kirschner, C., K. Maquelin, P. Pina, N.A. Ngo Thi, L.P. Choo-Smith, G.D. Sockalingum, C. Sandt, D. Ami, F. Orsini, S.M. Doglia, P. Allouch, M. Mainfait, G.J. Puppels, and D. Naumann, Classification and Identification of Enterococci: A Comparative Phenotypic, Genotypic, and Vibrational Spectroscopic Study, *J. Clin. Microbiol. 39*:1763–1770 (2001).
25. Maquelin, K., C. Kirschner, L. Choo-Smith, N. van den Braak, H.P. Endtz, D. Naumann, and G.J. Puppels, Identification of Medically Relevant Microorganisms by Vibrational Spectroscopy, *J. Microbiol. Methods 51*:255 (2002).
26. Nalik, H.P., K.D. Muller, and R. Ansorg, Rapid Identification of *Legionella* Species from a Single Colony by Gas-Liquid Chromatography with Trimethylsulphonium Hydroxide for Transesterification, *J. Med. Microbiol. 36*:371–376 (1992).
27. Goodacre, R., B. Shann, R.J. Gilbert, E. Timmins, A.C. McGovern, B.K. Alsberg, D.B. Kell, and N.A. Logan, The Detection of the Dipiclonic Acid Biomarker in *Bacillus* Spores Using Curie-Point Pyrolysis Mass Spectrometry and Fourier-Transform Infrared Spectroscopy, *Anal. Chem. 72*:119–127 (2000).
28. Mossoba, M.M., F.M. Khambaty, F.S. Fry, Novel Application of a Disposable Optical Film to the Analysis of Bacterial Strains. A Chemometric Classification of Mid-Infrared Spectra, *Appl. Spectrosc. 56*:732–736 (2001).
29. Mossoba, M.M., S.F. Al-Khaldi, A. Jacobson, L.I. Segarra Crowe, and F.S. Fry, Application of a Disposable Transparent Filtration Membrane to the Infrared Spectroscopic Discrimination Among Bacterial Species, *J. Microbiol. Methods 55*:311–314 (2003).
30. Mossoba, M.M., S.F. Al-Khaldi, J., Kirkwood, F.S. Fry, J. Sedman, and A.A. Ismail, Printing Microarrays of Bacteria for Identification by Infrared Spectroscopy, *Vib. Spectrosc. 38*:229–236 (2005).
31. Rodriguez-Saona, L.E., F.M. Khambaty, F.A. Fry, and E.M. Calvy, Rapid Detection and Identification of Bacterial Strains by Fourier Transform Near-Infrared Spectroscopy, *J. Agric. Food Chem. 49*:574–579 (2001).
32. Whittaker, P., M.M. Mossoba, S.F. Al-Khaldi, F.S. Fry, V.C. Dunkel, B.D. Tall, and M.P. Yurawecz, Identification of Food-Borne Bacteria by Infrared Spectroscopy Using Cellular FAME, *J. Microbiol. Methods 55*:709–716 (2003).
33. Cacciapuoti, B., L. Ciceroni, D. Attard Barbini, Fatty Acid Profiles, a Chemotaxonomic Key for the Classification of Strains of the Family *Leptospiraceae*, *Int. J. Syst. Bacteriol. 41*:295–300 (1991).
34. Gharaibeh, A.A., and K.J. Voorhees, Characterization of Lipid Fatty Acids in Whole-Cell Microorganisms Using in Situ Supercritical Fluid Derivatization/Extraction and Gas Chromatography/Mass Spectrometry, *Anal. Chem. 68*:2805–2810 (1996).
35. Sacks, R., H. Smith, and M. Nowak, High-Speed Gas Chromatography, *Anal. Chem. 70*:29A–37A (1998).

36. Buyer, J.S., Improved Fast Gas Chromatography for FAME Analysis of Bacteria, *J. Microbiol. Methods 54*:117–120 (2003).
37. Asselineau, J., The Bacterial Lipids, in *Chemistry of Natural Products Series*, edited by E Lederer, Holden-Day, San Francisco, 1966.
38. Dworzanski, J.P., L. Berwald, and H.L. Meuzelaar, Pyrolytic Methylation-GC of Whole Bacterial Cells for Rapid Profiling of Cellular Fatty Acids, *Appl. Environ. Microbiol. 56*:1717–1724 (1990).
39. Voisin, S., F.N.R. Renaud, J. Freney, M. de Montclos, R. Boulier, and D. Derauz, Pyrolysis-Gas-Liquid Chromatography with Atomic Emission Detection for the Identification of Corynebacterium Species, *J. Chromatogr. A 863*:243–248 (1999).
40. Han, X., and R.W. Gross, Global Analyses of Cellular Lipodomes Directly from Crude Biological Samples by ESI Mass Spectrometry: A Bridge to Lipidomics, *J. Lipid Res. 44*:1071–1079 (2003).
41. Han, X., and R.W. Gross, (2005) Shotgun Lipidomics: Electrospray Ionization Mass Spectrometric Analysis and Quantitation of the Cellular Lipidomes Directly from Crude Extracts of Biological Samples, *Mass Spectrom. Rev. 24*:367–412.
42. Lin, S., H. Scraft, J.A. Odumeru, M.W. Griffiths, Identification of Contamination Sources of *Bacillus cereus* in Pasteurized Milk, *Int. J. Food Microbiol. 43*:159–171 (1998).
43. Dees, C., D. Ringelberg, T.C. Scott, and T.J. Phelps, Characterization of the Cellulose-Degrading Bacterium NCIMB 10462, *Appl. Biochem. Biotechnol. 51*:263–274 (1995).
44. Vongraevenitz, A., V. Punter, E. Gruner, G.E. Pfyffer, and G. Funke, Identification of Coryneform and Other Gram-Positive Rods with Several Methods, *APMIS 102*:381–389 (1994).
45. Tang, Y.W., N.M. Ellis, M.K. Hopkins, D.H. Smith, D.E. Dodge, and D.H. Persing, Comparison of Phenotypic and Genotypic Techniques for Identification of Unusual Aerobic Pathogenic Gram-Negative *Bacilli*, *J. Clin. Microbiol. 36*:3674–3679 (1998).
46. Odumeru, J.A., M. Steele, L. Fruhner, C. Larkin, J. Jiang, E. Mann, and W.B. McNab, Evaluation of Accuracy and Repeatability of Identification of Foodborne Pathogens by Automated Bacterial Identification Systems, *J. Clin. Microbiol. 37*:944–949 (1999).
47. Sundheim, G., A. Sletten, and R.H. Dainty, Identification of Pseudomonads from Fresh and Chill-Stored Chicken Carcasses, *Int. J. Food Microbiol. 39*:185–194 (1998).
48. Harrick, N.J., *Internal Reflection Spectroscopy*, Wiley-Interscience, New York, 1967.
49. Ismail, A.A., A. Nicodemo, J. Sedman, F.R. van de Voort, and I.E. Holzbauer, Infrared Spectroscopy of Lipids: Principles and Applications, in *Spectral Properties of Lipids*, edited by R.J. Hamilton and J. Cast, Sheffield Academic Press/CRC Press, Boca Raton, FL, 1998, pp. 235–269.
50. Mirabella, F.M., ed., *Internal Reflection Spectroscopy*, Practical Spectroscopy Series, Vol. 15, Marcel Dekker, New York, 1992.
51. Kowalski, B.R., ed., *Chemometrics: Mathematics in Statistics and Chemistry*, D. Reidel Publishing Company, Dordrecht, 1984.
52. Massart, D.V., B. Deming, S. Michotte, and Y. Kaufman, *Chemometrics: A Textbook* (Data Handling in Science and Technology), Elsevier, New York, 1988.

14 Application of FT-NIR for Rapid Determination of the *trans* Fatty Acid Composition in Fats and Oils

Hormoz Azizian[a] and John K.G. Kramer[b]

[a]NIR Technologies Inc. Oakville, Ontario, Canada and [b]Food Research Program, Agriculture and Agri-Food Canada, Guelph, Ontario, Canada

Introduction

The main aim of this paper is not to present an exhaustive presentation of near infrared spectroscopy, but to encourage lipid scientists to consider the potential of using Fourier Transform Near Infrared Spectroscopy (FT-NIR) as a tool for the rapid and direct (neat) analysis of complex lipid mixtures. In the present review the development of FT-NIR models is described based on accurate gas chromatography (GC) data, and the application of chemometric analysis. The FT-NIR model permits the determination of the complete fatty acid (FA) composition of the fat and oil in minutes using the FT-NIR spectral information. The current *trans* fatty acid (TFA) issue is used as an example to demonstrate the capability of FT-NIR to provide both the total TFA content and the complete TFA isomer composition of any fat or oil in minutes. Such an analysis would require hours to be accomplished using highly specific capillary GC columns, and a combination of methods for definitive identification.

Trans *Fatty Acid (TFA) Issue*

The most detailed data on most food labels appears to be the information requested for the type of FAs present in the food product. This information is requested by health conscious consumers, and our general desire to decrease our mortality rate due to coronary heart disease. The information of the type of FA present is currently being expanded to include the content of total isolated TFAs, excluding conjugated linoleic acid (CLA). Legislation is currently being implemented to reduce the intake of TFA in Canada (1–2), United States (3), Argentina, Brazil, Paraguay and Uruguay (4), and Denmark (5). The basis for this legislation is the finding that the consumption of TFAs is associated with higher risk factors of cardiovascular disease (6–8) and inflammation (9,10).

The main sources of TFAs are industrially produced partially hydrogenated vegetable oils (PHVOs) and milk and meat fats from ruminants. TFAs produced during commercial refining of vegetable oils (mainly deodorization) or moderate frying processes (11,12) are minor, as are naturally occurring TFA in plant lipids (13). The TFA isomers present in PHVO, heated vegetable oils and ruminant fats consist mainly of *trans* monounsaturated FAs, and smaller amounts of di- and triunsaturated FAs containing one *trans* double bond. TFAs with two or more *trans* double bonds are rare. Industrial and ruminant fats generally contain the same kind of TFAs but they differ in the relative proportion of the different TFA isomers (14–18). Differences in the TFA isomer distribution among industrially produced fats depend on the catalyst and condition used, while differences in ruminant fats depend on what and how much supplements, if any, are fed, and on feeding practices. The TFA isomers responsible for the negative effects have not been identified, which makes it difficult to assess whether the TFAs from PHVO and ruminant fats present similar risk factors (19,20).

The regulation to declare the total TFA content in a food product excludes CLA. CLA is a mixture of positional and geometric isomers of a C18 diunsaturated FA (18:2) in which one of the two double bonds is generally in the *trans* configuration. CLA appears to be excluded from the total TFA declaration based on evidence showing that certain CLA isomers do not increase LDL (17) and protect against cancer in experimental animals and cell models (21,22). However, by not specifying which CLA isomer is excluded from the total TFA content on the food label, CLA isomers other than 9*cis*,11*trans*-CLA (9*c*11*t*-CLA) could be included with potentially undesirable health effects (23). Furthermore, the present labeling requirement excludes 11*t*-18:1 together with all the other TFA, despite the fact that 11*t*-18:1 is the metabolic precursor of 9*c*11*t*-CLA (24). It is hoped that future modifications to the TFA regulations will exclude both 9*c*11*t*-CLA, 11*t*-18:1 and any other beneficial TFA isomer from mandatory labeling. However, that will depend on the availability of accurate and rapid methods to quantify the different CLA and TFA isomers.

Official GC Methods Available for TFA, but not for CLA Determination

Providing accurate and complete analyses for so many food products will be very demanding and costly, unless more rapid methods are made available that provide accurate information of total TFA and CLA in a more cost effective manner. Searching for a rapid and accurate method that also provides the isomer composition of TFAs and CLA will be an even greater challenge. Official GC methods are presently available to determine the total TFAs content and their isomer composition (25–28). The official GC methods involve acid or base digestion of the test portion, extraction of the lipids with organic solvents, addition of an internal standard, acid- or based-catalyzed methylation to prepare FA methyl esters (FAMEs) for GC analysis, and GC separation using 100 m highly polar fused silica capillary columns [AOAC 994.15 (27); AOCS Ce 1h-05 (28)]. For neat fats and oils prior digestion and lipid extraction is not necessary. Despite the claims that the official GC methods provide a complete analysis of the TFA and its isomers in food products, several *cis*- and *trans*-18:1 isomers still overlap. A definitive analysis requires prior separations of the geometric 18:1 isomers by silver-ion TLC (Ag^+-TLC) (14,16,29,30) or Ag^+-HPLC (see Delmonte et al. Chapter 8 in current volume).

At the present time there is no official method available for the analysis of either total CLA or its isomer composition. The content of CLA in food products will vary depending on the method used for methylation (30–32), the GC column (30,33,34), and whether a complimentary Ag^+-HPLC is used for the separation of the CLA isomers (30,35–38). In many cases the CLA content in foods is represented only as the major GC peak in the CLA region that is reported as 9*c*11*t*-CLA, even though this peak is known to consist of a number of unresolved CLA isomers. For example, the coeluting CLA isomer 7*t*9*c*-CLA can be present in substantial amounts in some dietary conditions in ruminants (30,36–38). Therefore, the value of CLA to be excluded from the declared content of total TFA remains vague and may not be justified in many circumstances.

Official FTIR Methods for Total TFA, but not for Total CLA Determination

Official spectroscopic methods are available for the quantitation of the total TFA content that involves Fourier Transform Infrared Spectroscopy (FTIR) with an attenuated total reflectance (ATR) cell attachment (39–42), i.e., method Cd 14d-99 from the AOCS (41) and method 2000.10 from the AOAC (42). The ATR-FTIR method has the advantage of being rapid and requires no prior derivatization or use of organic solvents, but the information is limited to the measurement of total TFA content greater than 5% in the test samples (40,43), and it provides no information on individual TFAs, or any other FAs in the samples.

The ATR-FTIR method is not suitable for the determination of the CLA content of fat and oils, since the absorption frequencies of CLA near 986 and 950 cm^{-1} cause interferences with the C-H deformation of isolated *trans* bonds at 966 cm^{-1} (31,40,44). Likewise free FA that absorb near 935 cm^{-1} pose a challenge for the analysis of total *trans* by ATR-FTIR (40,43). Christy et al. (45) recently attempted to quantitate isolated TFAs in the presence of CLA using the ATR-FTIR method by preparing calibration models for each of these two FAs and using Chemometrics. Despite obtaining rather reasonable results, the authors concluded that the resolution of TFAs and CLA might be too difficult using this approach. Milosovic et al. (46) reported better resolution of the closely associated IR absorption bands of TFAs by evaluating the second derivatives of the ATR-FTIR spectra.

The limitations of FTIR stand in contrast to the GC methods that can resolve most of the TFA and many of the CLA isomers (26,30), and for this reason the GC method has remained the preferred method for analysis of TFAs for labeling purposes. However, it is becoming evident that the increased number of products that will need to be analyzed urgently requires methods that combine the speed of the

ATR-FTIR method with the greater information available from the GC technique in a cost effective manner.

FT-NIR and its Applications

The near infrared region of light occurs between 800 to 2500 nm or 12,500 to 4000 cm^{-1} wavenumbers (Fig. 1), and the absorption is due to overtones and combination band sequences of functional group vibrations in the mid-infrared region. The strong vibrational bonds include C-H, O-H, C=O, N-H, COOH, and aromatic groups. The resolution of the broad near infrared bands remained a challenge for decades. However, the application of Chemometrics has made it possible to resolve these broad bands using multiple harmonics in the FT-NIR spectra. Identification and quantitation of materials using their FT-NIR spectra requires the application of Partial Least Squares (PLS), principle component analysis (PCA) and Chemometrics. Further improvements in instrumentation were made by manufacturers in combination with enhanced computer power that now performs these calculations in seconds. It is these refinements in FT-NIR measurement techniques, and the availability of Chemometrics, that are making FT-NIR a method to consider for many applications in the food, agricultural, pharmaceutical, medical, and the plastic industries for quality control and assurance measurements. For further details of the theory of FT-NIR the reader is referred to a number of excellent publications on FT-NIR (47–49) as well as shorter reviews (50,51).

The FT-NIR technique has many of the same advantages as ATR-FTIR. It is rapid and nondestructive. In the lipid area it is generally performed directly on the fat or oil (neat), and requires no chemical preparations. FT-NIR has been evaluated for the determination of the total *cis* and *trans* content, iodine value, and saponification number of several fats and oils (52–55), and in 2001 FT-NIR has been approved by the AOCS as an official method for the determination of the iodine value of fats and oil (55,56). Chemometric techniques have been applied to NIR and FT-NIR spectra of products from the food, agriculture, pharmaceutical, medical, and plastics industries for quality control and assurance measurements (47–49). FT-NIR measurements

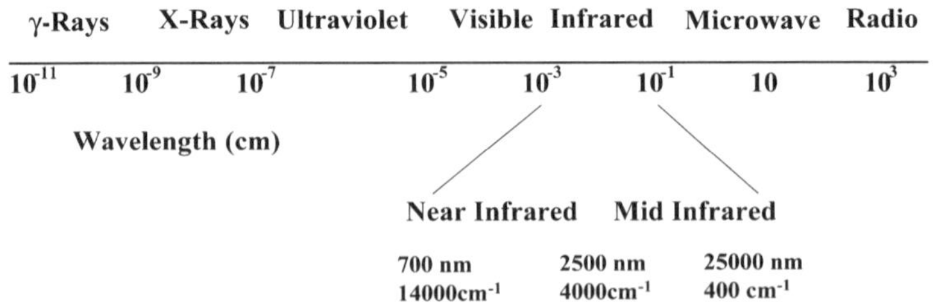

Fig. 1. Electromagnetic spectrum with emphasis on mid infrared and near infrared.

have also been applied to measure oxidative changes and oxidation products in vegetable oils (57–59) and pork products (60). The application of discriminate analysis to the FT-NIR data made it possible to identify adulteration of olive oil with other vegetable oils (61,62). Recent papers have shown limited success in the application of NIR to predict the FA composition of vegetable oils (63–66) and animal tissues (67,68); see review by Garrido-Varo et al. (69). Significant improvements were observed with the application of FT-NIR to the determination of the major FA groups including total TFAs in different edible oil products (52–55).

Recently FT-NIR has been applied to the determination of the FA composition of an oil or fat, including TFAs and CLA (70–73). Such an approach was made possible by using FT-NIR spectral data as a secondary method to develop FT-NIR models and using accurate GC data with 100 m highly polar fused silica capillary columns as the primary source. This new FT-NIR method was shown to be applicable for the rapid determination of individual TFA isomers, comparable to results obtained by using official GC methods. In addition to meeting the requirements for regulatory purposes, the present FT-NIR method described provides quantitative analysis of all the individual TFA and CLA isomers that could be used to exclude specific TFA isomers from the total TFA content on a food label. The FT-NIR method for the determination of individual FA in fats and oils is fairly new and that is why the authors felt it was necessary to review both the GC and FT-NIR experimental procedures again in this chapter.

Experimental Procedures

Materials

For the development of the reference FT-NIR models several commercial vegetable oils were purchased locally which included 3 canola oils, a corn oil, 2 flax oils, an olive oil, 3 soybean oils, 5 sunflower oils, and 2 walnut oils. In addition, 7 margarines, 2 shortenings and lard were acquired. Five fractions from partially hydrogenated canola oil and soybean oil were provided by industrial suppliers. Eighteen mixtures were gravimetrically prepared by combining specific fats and oils in different ratios. Triolein (glycerol tri-9*c*-octadecenoate), trielaidin (glycerol tri-9*t*-octadecenoate), trilinolein (glycerol tri-9*c*12*c*-octadecadienoate), and trilinolenin (glycerol tri-9*c*12*c*15*c*-octadecatrienoate) were purchased from Nu Chek Prep Inc. Elysian, MN. All chemicals and solvents were of analytical grade.

Gas Chromatography (GC)

Accurate GC data is essential for the development of the FT-NIR model because the GC results serve as the primary source of values that are incorporated into the FT-NIR model. This review is limited to the development of FT-NIR models of fats and oils that do not require the prior extraction of total lipids from a given matrix both for GC and FT-NIR analysis. To ensure accurate GC data, both acid- and base-catalyzed methylations were performed on the fats and oils and the resultant FAME were purified by TLC prior to GC analysis. Multiple GC separations were performed using 100 m highly polar fused silica capillary GC columns to resolve as many FAMEs as possible. In addition, complimentary techniques were used to definitively identify and quantitate FAME isomers that were not resolved by GC. The individual parts of the GC method are described in several publications (30–34), and are only briefly summarized here.

Methylation. The fat and oil products and their mixtures (about 20 mg) were separately methylated using anhydrous 5% HCl/methanol (by wt) for 1 h at 80°C, and 0.5% solution of $NaOCH_3$ in methanol (#33080, Supelco Inc., Bellefonte, PA) for 15 min at 50°C. After methylation, water was added (5% by vol) and the resultant FAMEs were extracted with hexane. Hexane was removed and the FAMEs were purified by TLC on silica gel G plates (Fisher Scientific, Ottawa, ON) using the developing solvent hexane/diethyl ether/acetic acid (85:15:1). The FAME bands on TLC were identified after spraying the plates with 2′,7′-dichlorofluorescein in methanol and the bands were visualized under ultraviolet light. Each FAME band on TLC was scraped off, transferred into a Pasteur pipette (5.75 inches) containing a previously cleaned glass wool plug (that had been washed with chloroform/methanol, 1:1), and the FAMEs were eluted with hexane. Purification of the FAMEs prior to GC analysis was performed to remove any non-FA components that might have co-extracted or artifact produced during the methylation procedure, since any organic material would also result in a GC-FID signal. The added benefit of TLC purification of FAMEs was a longer life expectancy of the GC columns. Appropriate concentrations of FAMEs (1–2 μg/μL) in hexane were prepared for GC analysis.

GC Conditions. All the FAME preparations were analyzed by GC (Hewlett-Packard Model 5890 Series II, Palo Alto, CA) equipped with a splitless injection port flushed after 0.3 min, a flame ionization detector, autosampler (Hewlett-Packard, Model 7673), a 100 m CP-Sil 88 fused capillary column (100 m × 0.25 mm i.d. × 0.2 μm film thickness; Varian Inc., Mississauga, ON), and a Hewlett-Packard ChemStation software program (Version A.09). The operating conditions were: injector and detector temperatures both at 250°C; H_2 as carrier gas (1 mL/min) and for the flame ionization detector (30 mL/min), N_2 makeup gas (30 mL/min), and air (300 mL/min). All gases were of highest purity. The following temperature program was used: initial temperature of 45°C and held for 4 min, programmed at 13°C/min to 175°C and held for 27 min, then programmed at 4°C/min to 215°C and held for 35 min (30,33,34). FAMEs were identified by comparison with a GC reference FAME standard (#463) spiked with the four positional CLA isomer mixture (#UC-59M), and long-chain saturated FAMEs 21:0, 23:0, and 26:0 (all samples were obtained from Nu Chek Prep Inc., Elysian, MN). The *trans* isomers of linoleic and linolenic acid were prepared as described previously (34). All GC results are presented as relative percent of total FAME based on the flame ionization response.

There is an absolute requirement for 100 m highly polar fused silica capillary GC columns (i.e., CP Sil 88 by Varian Inc. or SP 2560 by Supelco Inc., or columns having similar properties) to maximize the resolution of most of the TFA and CLA isomers. The resolutions can be further maximized by altering the GC temperature program. For example, lowering the oven temperature from 170 to 150°C will change the elution order of the 20:1 and 18:3 isomers, such that the *cis*-20:1 isomers from 8*c*- to 11*c*-20:1 eluted after linolenic acid, instead of overlapping with the *trans* containing 18:3 isomer (74). Furthermore, the resolution of many *trans*-18:1 isomers can be improved at lower sample load. But not to compromise reliable identification of minor FAMEs, GC analyses were performed at a higher and a lower sample load (30). The quantitation of FAME analysis by GC is not often discussed even though discrimination of FAME based on differences in volatility can often occur when split or splitless injections are used (75). The commonly used injector design appears to be the problem and one may need to switch to using a programmed temperature vaporization (PTV) inlet for improved quantitation. The PTV inlet is now available from a number of GC suppliers. If one operates a split or splitless injection, it is advisable to perform period checks to ensure quantitative GC analysis by using reliably GC FAME mixtures as reference.

Ag^+-TLC Separations. Prior Ag^+-TLC was used when the *trans*- and *cis*-18:1 isomers were impossible to resolve by GC. This generally occurred when the relative distribution of adjacent isomers was drastically different (30). Ag^+-TLC separates FAMEs based on their geometric (*cis* from *trans*) configuration and number of double bonds (14,16,29,30,33,76). These isolated *trans*- and *cis*- TLC bands were then resolved by GC at an oven temperature of 120°C that separated most of the *trans*- and *cis*-18:1 isomers, except for 6*t*- to 8*t*-18:1 (14,16,29,30,33,76). The GC/MS and GC-FTIR techniques were also used to identify the FAMEs (43,77).

The FAMEs prepared by the acid- and base-catalyzed procedure were both analyzed by GC and compared. The results of the two methylation procedures were generally similar except for differences in the CLA isomer composition; acid conditions resulted in a higher content of the *tt*-CLA isomers, because of the acid-catalyzed isomerization (31). The final FA composition was obtained by averaging the results of both methylation procedures, except for the CLA isomer distribution that was taken from the GC results of the base-catalyzed methylation. The CLA isomer composition was confirmed using Ag^+-HPLC separations (30,33,36,38).

FT-NIR Spectral Measurements

Equipment. The FT-NIR spectra of a number of edible oils, shortening, lard, and partially hydrogenated soybean and canola oils were acquired using a Bruker Optics

Matrix F spectrometer equipped with temperature stabilized InGaAs detector and a solid transflectance fiber optic probe with a liquid attachment (Bruker Optics Inc. Milton, ON, Canada). FT-NIR spectra was also obtained using a Vector 22/N spectrometer from the same supplier (Bruker Optics Inc.) equipped with a Ge Diode detector and the same type of solid transflectance fiber optic probe with a liquid attachment over the range of 10,000 to 4,000 cm^{-1} (1000 to 2500 nm). The schematics of the reflection probe are shown in Figure 2. The spectral resolution was set at 8 cm^{-1} and the path length setting was 2 mm. The resolution setting of 4 cm^{-1} gave high background noise while the setting of 16 cm^{-1} provided inadequate resolution. The OPUS software tools from Bruker Optics were used to collect and handle the data using Chemometric analysis.

Spectral Measurements. The FT-NIR spectral measurements consisted of 5 scans obtained of the neat sample in liquid form and using the liquid attachment tool available from the instrument supplier. The actual scan required only 25 seconds. All samples were scanned at 22 ± 2°C except semi-solid fat samples that were melted and scanned at 30 ± 2°C. The complete scanning, cleaning of the probe and analysis was achieved in a few minutes.

Development of FT-NIR Classification Model. The following procedure was used in the development of a classification spectral reference model. An average absorption file for each fat and oil sample was created and used in the development of a classification reference model using Chemometric analysis tools provided with-

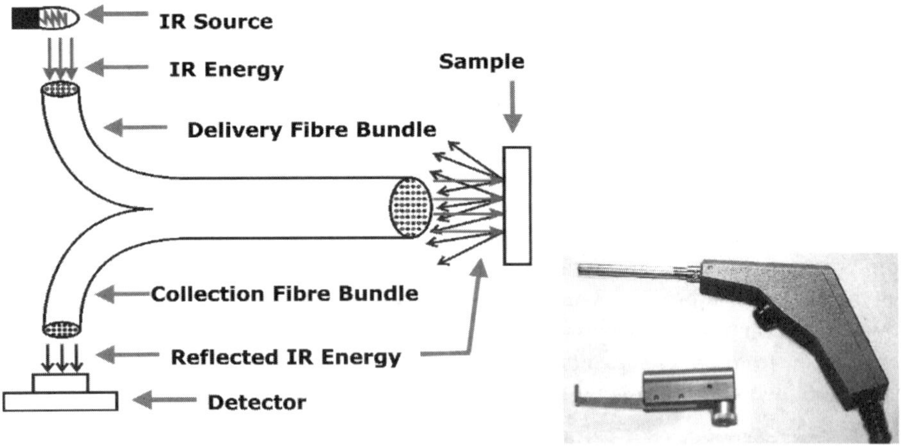

Fig. 2. Schematic of fiber optic probe together with actual probe and liquid attachment tool. (Reproduced with permission from Bruker Optics Inc., Milton, ON, Canada).

in OPUS software. There are two types of classification models, standard and factorized. In the latter, one can deduce additional information based on the factors (representing different components in the matrix) used in the analysis. These models were then used in the classification of fats and oils having a similar FA composition with a 99% confidence interval.

Development of FT-NIR Quantification Model. The quantitative analysis models were developed using the OPUS "Quant 2" program provided with the instrument. The data incorporated into this program included the FA composition obtained by GC (independent variables) and the FT-NIR spectra. A combination of processing parameters was then selected, such as first or second derivatives, vector normalization, spectral regions, and validation type (i.e., cross validation), all of which were then analyzed by PLS1 procedures. Once a reasonable calibration model was established (first derivative plus vector normalization and specific spectral regions 4547–4794, 5423–6013, and 6974–7290 cm^{-1}), it was applied to determine FA concentration in all reference samples (the average spectrum), including TFAs from within the calibration set, as well as external samples that were not part of the model. The final FT-NIR predicted values were then compared to GC data for the same sample.

Chemometric Analysis. The classification process involves a spectral comparison of the test sample to those in the Classification Reference Model. The result of a comparison between two spectra is expressed in terms of the Euclidean distance $D_{\text{test sample}}$:

$$D_{\text{test sample}} = \sqrt{\Sigma[A_{\text{test sample}}(k) - A_{\text{Reference}}(k)]^2}$$

Where $A_{\text{test sample}}(k)$ is the absorption of the test sample spectrum at wavelength k, and $A_{\text{Reference}}(k)$ is the absorption of the reference spectrum at wavelength k. Based on the equation above, the more similar a spectrum is to the reference spectrum the smaller the Euclidean distance $D_{\text{test sample}}$. For any two test samples with identical spectra, $A_{\text{test sample}}(k) = A_{\text{Reference}}(k)$ for every (k), and the value of $D_{\text{test sample}}$ will be zero.

The second parameter used in the classification process was the threshold distance (D_T) that defines the tolerance of the classification model. In this case, a 99% confidence interval was applied in the calculation of the threshold value. The optimal setting for the threshold depended on the reference samples incorporated in the classification model. If the value for D_T was set too high, the model incorrectly identified samples as identical, when in fact they were different. On the other hand, if the D_T was set too low, the model rejected samples that should have been accepted.

The capability of the FT-NIR classification model to identify fats and oils by their vibrational spectra using mathematical treatments is based on the probability that the spectral characteristics of the sample in question are related to the spectral properties of material or materials in the classification reference library. Therefore, FT-NIR comparisons will have one of three possible outcomes:

(i) The hit quality for the test sample was lower than the threshold value of the reference sample, and no other hits were found to meet this criterion. The result was reported as "IDENTICAL," i.e., the test sample was identical to the reference.

(ii) The hit quality of one or more of the test samples was smaller than the threshold of the reference. The identity test was reported as "CAN BE CONFUSED WITH <N> OTHER HITS." This outcome was only possible if the classification reference library contained very similar materials, such as two soybean oils from two different sources with very similar fatty acid compositions, or the same soybean oil analyzed using two separate FT-NIR instruments.

(iii) The hit quality was greater than the threshold, and no match was found of the test sample to any reference in the spectral library. In this case the sample was reported as "NOT IDENTICAL." This situation could also arise if the sample was not scanned properly, contained impurities, the batch-to-batch variations were too large, or if the test sample had a FA composition outside the range of any reference sample in the library.

Results and Discussion

There are fundamental differences between mid-FTIR and FT-NIR both in terms of frequencies of light used (Fig. 1) and the spectral information collected. Figure 3 shows the complete spectral range of FTIR and FT-NIR of a typical partially hydrogenated soybean oil sample. FTIR is well known for its functional group characterization of molecules. For example, the carbonyl group of FAs shows a strong

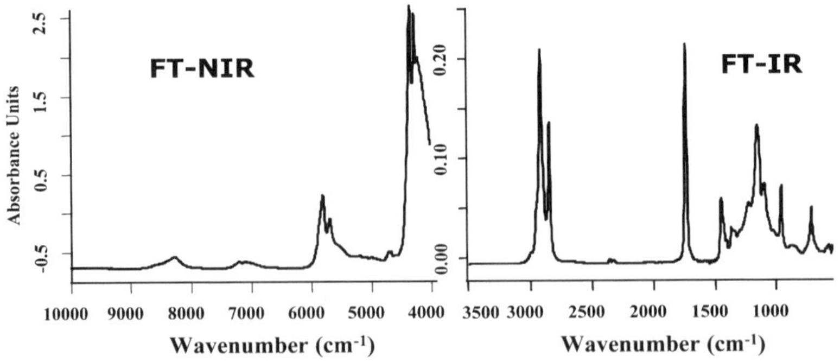

Fig. 3. FTIR and FT-NIR absorption spectra of partially hydrogenated soybean oil. The absorption spectra of FTIR and FT-NIR are from 800 to 3500 cm^{-1} and from 4,000 to 10,000 cm^{-1}, respectively.

absorption at 1743 cm^{-1}, and a *trans* vibration band at 966 cm^{-1} (Fig. 3). As a result, FTIR can easily be used to provide qualitative and quantitative information on oils and fats. In contrast, the broad peaks in the FT-NIR spectra are the result of overtones and combinations of the fundamental vibration bands in the FTIR region. For decades these broad peaks remained unresolved and NIR spectroscopy did not gain the same recognition as FTIR. However, this changed in the last decade with the availability of Chemometric analysis tools, increased computing power, and enhanced instrumentation (47–51). FT-NIR has recently attracted attention in quality control and assurance measurements in conjunction with fiber-optic probes for testing raw material, and for on-line and in-process applications (47–49).

A better understanding of FT-NIR spectral characteristics has also aided in this re-emergence of this spectroscopic technique. For example, the presence of harmonics in the overtones of the broad peaks in the FT-NIR spectra was solved by the introduction of appropriate mathematical equations. Figure 4 shows the FT-NIR absorption spectra of several commercially available triacylglycerols (TAG) containing three FA with one *cis* (triolein), one *trans* (trielaidin), two *cis* (trilinolein) and three *cis* (trilinolenin) double bonds per FA molecule. The first, second and third overtones are visible around 5800, 7100 and 8300 cm^{-1} wavenumbers, respectively. However, the broad nature of these peaks does not allow for qualitative analysis of the absorption peaks in the same way as in FTIR. The resolution of these broad peaks can now be studied using the Chemometric functions available in different software packages, including the OPUS software that we have used with the

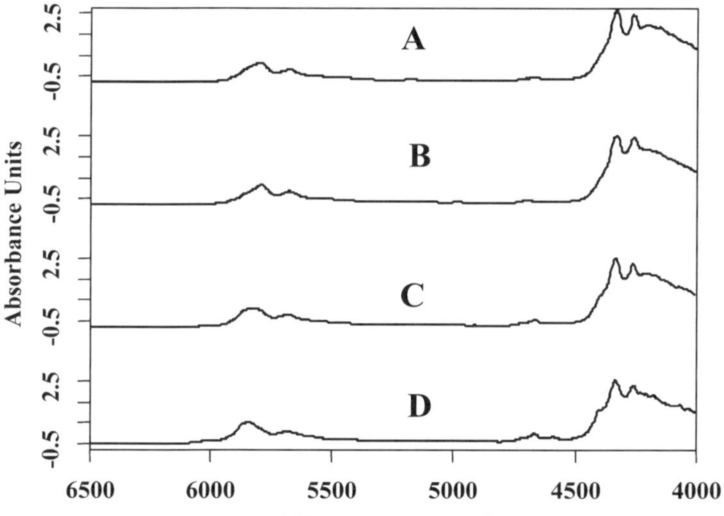

Fig. 4. The FT-NIR absorption spectra of triolein (A), trielaidin (B), trilinolein (C), and trilinolenin (D) in the spectral region from 4,000 to 10,000 cm^{-1}.

Bruker instruments (or PIROUETTE software from Infometrix, Inc., Woodinville, WA, and others). The second derivative spectra of the above TAGs show significant spectral differences between the TAG molecules in the 4500 to 4750 cm^{-1} and 5600 to 5900 cm^{-1} region (Fig. 5). The spectral differences associated with corresponding changes in the TAG molecule having 1, 2, and 3 double bonds in each FA of the TAG molecule are more clearly evident by overlaying the second derivative spectra of triolein, trilinolein, and trilinolenin (Fig. 6). The intensity of the second derivative peaks at 5830 cm^{-1} and 5870 cm^{-1} increased, while the intensity of the peaks at 5768 cm^{-1} and 5680 cm^{-1} decreased, as the number of double bonds in the TAG molecule increased from 3 to 9. These spectral features are vibrational characteristics of each material or mixture and do not change unless the chemical nature of the material or mixture is altered.

FT-NIR Classification of Fats and Oils

As outlined above under the section *Development of FT-NIR Classification Model*, there are two types of classification models, and we chose to use the factorization model to gain more detailed information of the different fats and oils. In the factorized analysis, each average spectrum was assessed with respect to different components (known as vectors) present within the sample. Once the average spectra of any given fat or oil was collected and stored (all edible oils and fats), Chemometric analysis was performed to generate the appropriate Classification Reference Model.

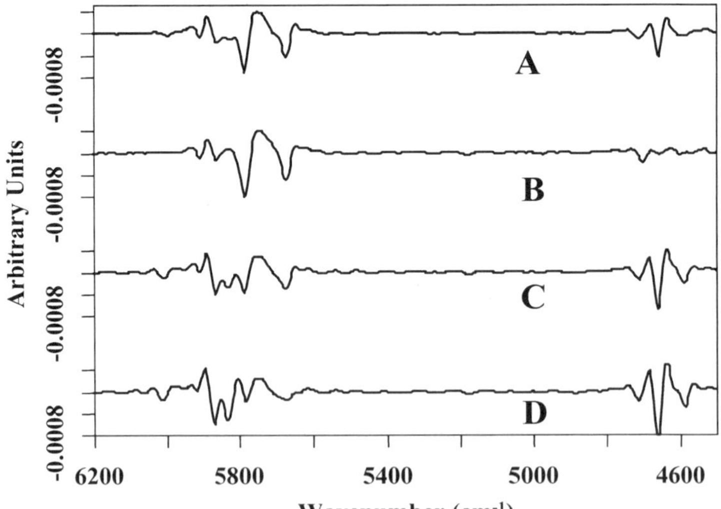

Fig. 5. Second derivative FT-NIR spectra of triolein (A), trielaidin (B), trilinolein (C), and trilinolenin (D) in the spectral region from 4,500 to 6,200 cm⁻

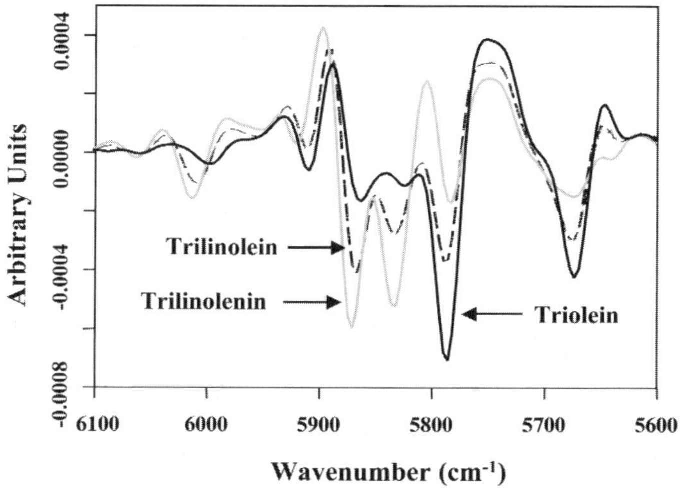

Fig. 6. Overlay of the second derivative FT-NIR spectra of triolein, trilinolein, and trilinolenin in the spectral region from 5,600 to 6,100 cm^{-1} (71). Reproduced with permission from AOCS Press.

In addition to studying chemical differences in various molecules, FT-NIR has the potential to assess some physical characteristics such as the shape and physical state of the sample under investigation.

Therefore, it is critical to exclude impurities, otherwise the classification or the quantification models would generate unreliable results. We recently experienced a significant shift in the second derivative FT-NIR spectrum of oil extracted from a salad dressing (Fig. 7). This shift was attributed to the residual solvents in the oil (i.e., chloroform and methanol), since removal of the solvent by warming gave a FT-NIR second derivative spectrum very similar to soybean oil (Fig. 7) which agreed with the GC results (70).

Once a classification model has been established, certain correlations between the individual samples, depending on their components in each sample, are generated in the Chemometric analysis. These correlations can be shown (for illustration purposes) as plots of vectors. Each classification model consists of many different vectors depending on the complexity of the model. Vectors are mathematical expressions used to quantify changes and differences between data sets and should not be confused with specific FA. Each vector may represent several relationships, and therefore, the plots of vectors should only be used for illustrative purposes, and classification and quantification should be based on the actual models. Figure 8 shows the plot of vectors 2 and 4 in the analysis of different fats, oils and their mixtures. Vectors 2 and 4 were selected for Figure 8 to illustrate the large differences between triolein, trielaidin, trilinolein, and trilinolenin and how oils consisting of mixtures of these FAs relate to TAGs consisting of single FAs.

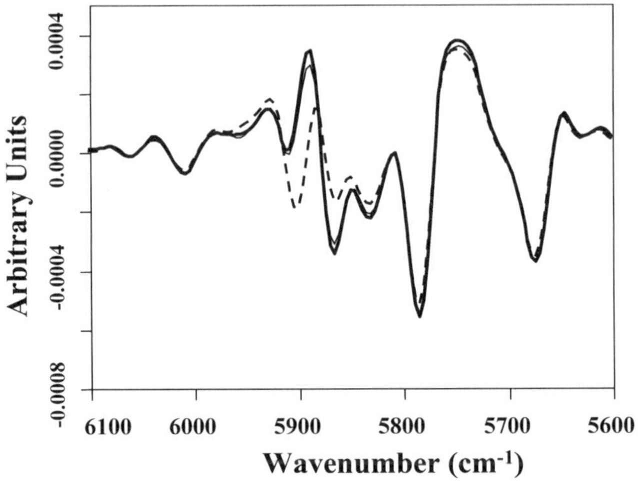

Fig. 7. Second derivative FT-NIR spectrum of the oil extracted from a commercial salad dressing that contained residual amounts of extraction solvent (chloroform and methanol) (dotted lines). The other two FT-NIR spectra are those of the oil after removal of the solvents and soybean oil (70). (Reproduced with permission from Lipid Technology).

Certain trends are very evident in Figure 8. For example, flax, walnut and canola oils showed characteristic similarity to trilinolenin, trilinolein, and triolein, respectively, because these oils contained significant amounts of the corresponding FAs. Mixtures of walnut or soybean oil with commercial shortening (a partially hydrogenated soybean oil product) altered the cluster from one predominant in unsaturated FAs toward increased *trans* and saturated FAs (Vector 2). Cluster "F" showed limited variation, since it represented two different samples of soybean oil with only slight differences in their FA composition, and scanned using two different FT-NIR spectrometers manufactured by Bruker Optics (Matrix-F and Vector 22/N). Increased partial hydrogenation resulted in the spectral convergence of different oils such as soybean oil and canola oil, as the fractions increased in total TFA and saturated FAs (cluster G and H).

Furthermore, the value of Vector 2 response obtained from the factorized analysis of the FT-NIR spectra was used successfully to determine the relative proportion of any two fats (or oils), as demonstrated in Figure 9. In this case, a number of mixtures were gravimetrically prepared using commercial lard and commercial shortening and analyzed by FT-NIR. Plotting the value of Vector 2 against the relative proportion of these two fats showed a highly significant relationship (Fig. 9). Although the initial plot was obtained by analyzing gravimetrically prepared fat mixtures by FT-NIR, a subsequent determination of any unknown mixture of these two fats can be readily assessed by FT-NIR and the developed mathematical rela-

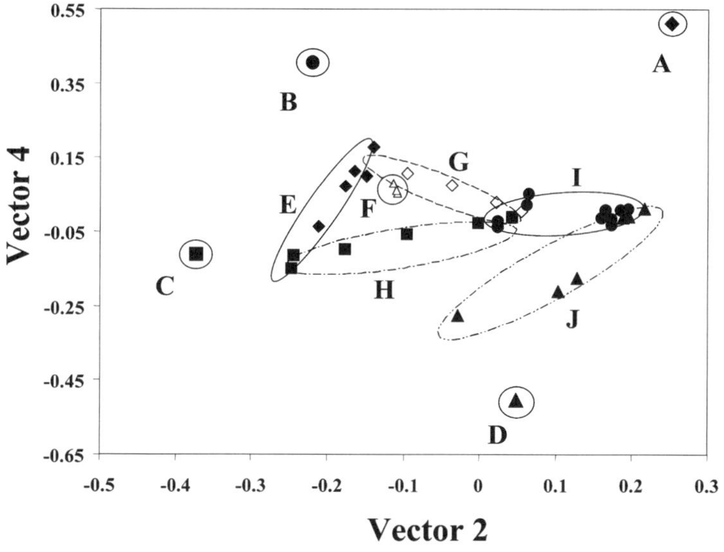

Fig. 8. Factorized analyses (vectors 2 and 4) of triolein, trielaidin, trilinolein, trilinolenin and several vegetable oils, fats and fat/oil mixtures. A, trielaidin; B, trilinolein; C, trilinolenin; D, triolein; E, flax oil and walnut oil mixtures; F, different soybean oils; G, walnut oil and commercial shortening mixtures; H, flax oil and commercial shortening mixtures; I, partially hydrogenated soybean oil fractions (Supplier A); and J, partially hydrogenated canola oil fractions (Supplier B) (71). Reproduced with permission from AOCS Press.

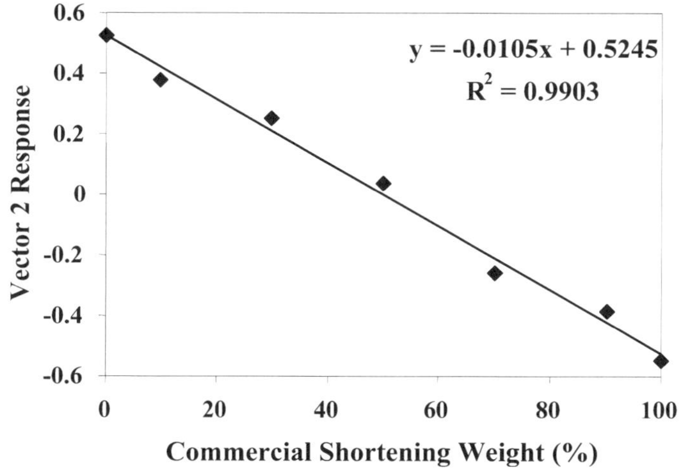

Fig. 9. FT-NIR response to changes in the concentrations of a simple mixture of 2 commercial fats, lard and shortening.

tionship. This type of FT-NIR assessment of fat and oil mixtures could be a valuable tool for quality control and assurance measurements. A number of examples come to mind, such as the adulteration of olive oil with other vegetable oils, vegetable ghee with pork fat, cocoa butter used in chocolate preparations with other fats, and milk fats with vegetable oils. The limit of detection will need to be established in each mixture using appropriate fats, oils and standards.

Typical Classification Report

Scanning and classification of unknowns by FT-NIR is rapid and easily achieved once the Classification Reference Model was established. Table 1 shows typical

TABLE 1
Classification of Various Oils and Fats

Sample name[a]	Hit Quality	Threshold	Total *trans* Content by GC (%)
Oils high in oleic acid (9*c*-18:1)			
Triolein	0.000001	0.015418	0
Canola oil (High oleic)	0.130415	0.012786	2.92
Olive oil	0.206076	0.031364	0.52
Sunflower oil (High oleic)	0.270968	0.009483	0.67
Canola oil	0.273320	0.009296	1.88
Oils high in linoleic acid (9*c*12*c*-18:2)			
Sunflower oil	0.000001	0.010479	0.74
Soybean oil A	0.145410	0.013488	2.08
Walnut oil A	0.163329	0.018309	1.99
Soybean oil B	0.174058	0.012568	0.41
Soybean oil C	0.186473	0.010179	2.08
Flax-walnut A	0.210679	0.009303	1.41[b]
Flax-walnut B	0.272688	0.011585	1.37[b]
Soybean oil PH A	0.280025	0.012384	18.96
Trilinolein	0.287943	0.012516	0
Soybean oil PH B	0.293452	0.015885	26.69
Sunflower oil (High oleic)	0.355010	0.009483	0.67
Fats high in *trans* FA			
Soybean oil PH D	0.000000	0.013628	50.25
Canola oil PH C	0.036650	0.022235	53.06
Soybean oil PH C	0.117174	0.010300	43.60
Canola oil PH D	0.133215	0.014515	59.49
Soybean oil PH B	0.281868	0.015885	26.69
Canola oil PH B	0.291071	0.019192	32.60
Soybean oil PH A	0.323771	0.012384	18.96
Canola oil PH A	0.355013	0.022775	23.89
Trielaidin	0.443282	0.022775	100

[a]All oils and fats were locally purchased products; partially hydrogenated (PH) soybean and canola oils were provided by two separate suppliers.
[b]Total *trans* FA content determined only by FT-NIR.

classification reports for oils high in oleic acid, linoleic acid, and partially hydrogenated fractions high in TFA (up to 60% total *trans*). Each of the classification reports includes the hit quality (a measure of the test samples' similarity to each reference sample), the threshold value (a value based on the standard deviation with a 99% confidence interval), and total *trans* content of selected fats and oil. In each case, the test sample was first compared to itself which accounts for a hit quality (or a Euclidean distance $D_{\text{test sample}}$) of, or near zero.

In the first set, oils high in oleic acid were examined. In this particular report triolein was compared to all samples within our Classification Reference Library. The closest match was to itself showing a hit quality of 0.000001; all other fats and oils within the reference library showed significantly higher hit quality values compared to triolein. The next closest match to triolein was a high oleic canola oil (canola oil C), followed by olive oil and then regular canola oil, and a sunflower oil, all of which had a high content of oleic acid (Table 1). The hit quality of common canola oil was larger than that of the high oleic acid sunflower variety. The relative order of these oils is in agreement with the GC data showing decreased levels of oleic acid.

The second example in the classification report is that of regular sunflower oil selected as oil high in linoleic acid. As expected, the reference library search identified regular sunflower oil in the reference library with very low hit quality value, but all remaining oils showed much higher hit quality values. Among the oils showing the next closest matches were soybean and walnut oils, and oil mixtures containing substantial amounts of these two oils. Trilinolein showed a poorer fit than the oils high in linoleic acid which is not surprising considering that sunflower oil contains only about 70% linoleic acid, not 100% as in trilinolein. It is worth noting that the sunflower oil high in oleic acid was even farther down the list of matching oils with a hit quality of 0.355 (Table 1), demonstrating the larger difference in FA composition between these two sunflower oils.

The third report shows the classification of fats having a high TFA content. Partially hydrogenated (PH) soybean oil with a TFA content of 50% was selected and the best fits were obtained from the reference library. A number of PH soybean and canola oil fractions are shown in Table 1 with increasing hit quality values. It is evident from these results that the deviation of the FA composition of these fats (specifically the TFA content) to that of the selected fat under consideration plays a greater role in determining the hit quality value than the source of the oil. Again the TAG consisting of all TFA, trielaidin, showed an even greater hit quality value than all the other TFA containing fats (Table 1).

All three reports clearly show that a positive identification of a material can be made when its FT-NIR spectrum (spectroscopic fingerprint) is compared to those in the Classification Reference Library. It is this type of information that can easily be used in quality control and quality assurance applications when the same material from the same source or different sources needs to be assessed. Substantial savings could be realized in terms of man power and waiting periods for the completion of wet chemical analyses. The FT-NIR analysis is carried out on neat samples without any prior preparation and subsequent lengthy analysis.

FT-NIR Quantification of FAs

FT-NIR is a non-destructive method that has been used to a limited extent for the quantitative analysis of selective FAs, or groups of FAs, such as *trans* and unsaturated FAs in edible oils (52–55). NIR measurements using dispersive type instruments are limited to predicting only the major FAs in vegetable oils (63–66). The FT-NIR method was not extended to the complete FA profile of fats and oils because of the large variations and low correlation coefficients observed for minor FAs. The majority of the above studies were carried out using dispersive instruments with lower resolutions which could be a major factor in the proper classification of individual FAs. Another factor is the temperature at which the fats and oils are recorded since temperature will influence the FT-NIR spectral intensities giving rise to this variation (71). There was a further concern that the sample matrix may affect quantitation, since in the FT-NIR method the whole sample is scanned in the neat form without prior isolation of each TAG component. This matrix dependence was recently demonstrated when traces of residual solvent affected the outcome; see Figure 7 (70).

Reason for Using GC as Primary Method and FT-NIR as Secondary Method

FAs generally occur in the form of mixed TAGs in fats and oils. To prepare response factors for every FA in a mixed TAG was for all purposes considered impossible because of the limited availability of pure TAGs. Several single FA TAGs such as triolein, trielaidin, trilinolein and trilinolenin are available from Nu Chek Prep Inc., but pure mixed TAGs are less common. Many more pure and mixed TAGs would be necessary in fairly large quantities (at least 2 mL/sample) to prepare all the references and validation samples required to cover the complete range of FAs present in all the fats and oils, in order to make FT-NIR the primary method of analysis and still establish a relatively universal calibration model. Therefore, it was decided to rely on the GC results as the primary reference of FA composition, and FT-NIR as the secondary method of analysis. Furthermore, by including a variety of fingerprints from fats/oils that varied in saturation, *cis* (up to 9 double bonds/molecule) and *trans* unsaturation (low, medium, high) content, a relatively universal calibration model could be developed applicable to most of the common fats and oils. Using this approach the accuracy of the GC results was very critical, since any errors would be directly passed onto the FT-NIR models.

A number of precautions were taken to insure the reliability of the GC data. This included purifying all FAME preparations by TLC prior to GC analysis, performing all GC separations using 100 m fused silica capillary GC columns, conducting generally more than one GC separation using different temperature programs and sample loads, and routinely checking for quantitation of GC analyses. Additional prior separations of the FAMEs were performed by Ag^+-TLC or Ag^+-HPLC; see details of the GC method in the Experimental Procedures section above.

Development of the FT-NIR Quantification Model

FT-NIR models were developed taking into consideration the characteristics of fats and oils, and the TFA content being either low, medium or high. Numerous edible oils and fats and their mixtures were used to form a pool of almost 60 different samples to develop the FT-NIR models. Each of the fats and oils and a number of mixtures thereof were analyzed by GC and FT-NIR. To enhance the accuracy of the model for the analysis of *trans* containing samples, three separate models were developed for low (<2%), medium (2 to 20%) and high (20 to 60%) *trans* content. Table 2 shows cross-validation error predictions for each of the low, medium and high *trans* models, respectively. The root mean square error of cross-validation

TABLE 2
Error Predictions Obtained from Cross-Validation for Low (<2%), Medium (<20%), and High (<60%) *trans* Content

Fatty Acid	Rank	R^{2+}	RMSECV	Range
Low *trans*				
16:0	4	0.85	2.05	3.58–24.21
18:0	4	0.85	1.25	1.97–14.64
9*c*-18:1	8	0.99	2.28	15.74–69.74
9*c*12*c*-18:2	7	0.98	2.61	10.68–71.15
9*c*12*c*15*c*-18:3	7	0.96	2.28	0.15–51.83
Medium *trans*				
16:0	6	0.88	1.75	3.58–24.21
18:0	5	0.84[a]	1.23	1.97–14.64
10*t*-18:1	7	0.93	0.35	0.02–4.95
11*t*-18:1	5	0.87	0.41	0.01–3.79
9*c*-18:1	7	0.99	2.21	15.75–69.74
9*c*12*c*-18:2	8	0.98	2.73	10.68–71.15
9*c*12*c*15*c*-18:3	8	0.96	2.12	0.15–51.83
High *trans*				
16:0	7	0.67	1.76	4.04–16.01
18:0	5	0.93	1.05	3.18–16.25
10*t*-18:1	7	0.91	0.91	0.46–10.26
11*t*-18:1	7	0.74	1.13	0.37–8.35
9*c*-18:1	7	0.98	2.1	6.26–51.93
9*c*12*c*-18:2	5	0.99	0.88	0.02–50.08
9*c*12*c*15*c*-18:3	8	0.97	0.37	0–8.17

[a]Values below 0.85 should be used with caution. We are currently investigating potential improvements to calibration curves with $R^2 \leq 0.85$. (Reproduced with permission from AOCS Press).

(RMSECV) values depended on the relative concentration of the FA in the sample, being lower at low concentrations. For example, the RMSECV for 10*t*-18:1 was 0.91 when present in the range of 0.46 to 10.26%, and 0.35 in the range of 0.02 to 4.95%.

All three models proved to be highly successful in predicting the content of the major FAs in these fats and oils as demonstrated by comparing the FT-NIR obtained by PLS1 analysis and the actual GC values. Figure 10 shows the correlation for GC and FT-NIR data for 18:0 (Fig. 10A), 9*c*-18:1 (Fig. 10B), 9*t*-18:1 (Fig. 10C), 10*t*-18:1 (Fig. 10D), 9*c*,12*t*-18:2 (Fig. 10E), and 11*t*,15*c*-18:2 (Fig. 10F) that ranged from trace levels of *trans* isomers to 71% for oleic acid (Fig. 10B). The comparisons between the actual (GC) and observed (FT-NIR) values for the selected FAs were highly significant with R^2 values of 0.67 to 0.98 (Table 2). A similar PLS1 analysis for all the other major FAs in the fats and oils provided equally significant R^2 values (values not included). The RMSECV for several of FAs are shown in Table 2 using the low, medium, and high *trans* calibration models. Further improvements to the calibration model should be possible, especially those with lower R^2 values, by studying all FAs individually and including samples having a wider range in each of the FAs.

The most demanding part of the development of the classification models was the acquisition of quantitative GC results for each of the samples, incorporating the GC results into the FT-NIR spectral database using the OPUS software system, and validating each of the samples with calibration curves. The quantitative FA composition of the fats or oils obtained by GC provided the FA ranges in the reference library. The need for accurate GC data for the development of the FT-NIR models should be stressed. Poor correlations between GC and FT-NIR results were often attributed to misidentified or coeluting GC peaks. It should be noted that FT-NIR spectral data are strictly based on the vibrational characteristics of the FAs, while GC results may be subject to errors of coelution or misidentification. The FT-NIR results are very much dependent on the model used, and if the model is not properly constructed and validated there is potential for inaccurate results. The predicted results of FAs from unknowns that are not within the range present in the reference library generally leads to underestimation of the FA content. This can be easily rectified by expanding the range of the FAs in the reference library using appropriate fats and oils and their GC quantitative analysis.

Quantitative analysis of unknown fats and oils including their FA composition is accomplished by scanning the sample and determining the FA composition based on previously developed calibration models, provided the FA ranges of the unknown are in the existing reference library. The results of five different fats and oils including soybean, canola, palm and flaxseed oils and shortening are presented in Table 3. The data showed a good correlation between the GC and FT-NIR results based on the low *trans* model. The FT-NIR technique was able to predict the FA composition of these oils, including minor *trans* isomers of 18:1, *c/t* isomers of 18:2, *c/c/t* isomers of 18:3, and the CLA isomers, some of which were as low as 0.2% of total FAME.

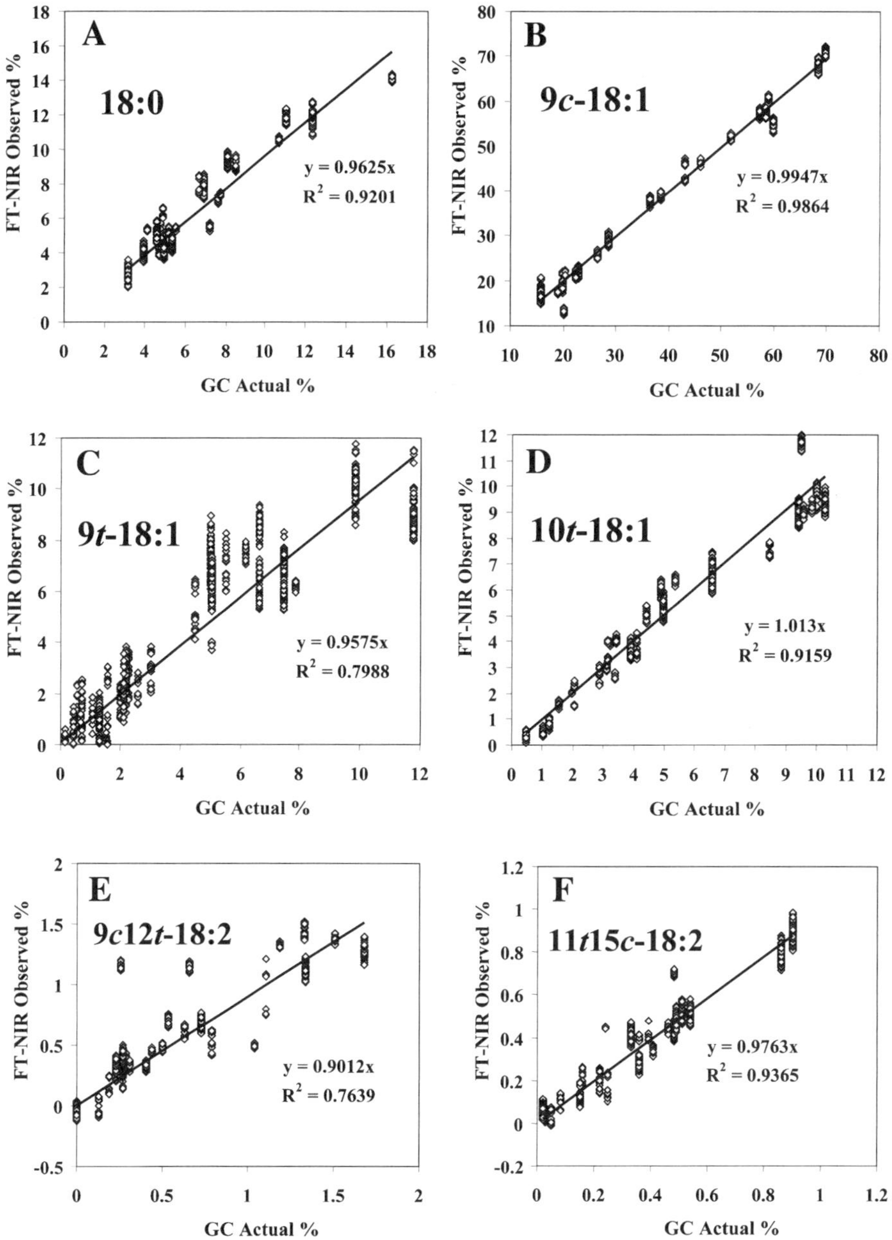

Fig. 10. Comparison of actual GC values and observed FT-NIR values for (A) 18:0 (low *trans* model), (B) 9*c*-18:1 (low *trans* model), (C) 9*t*-18:1 (high *trans* model), (D) 10*t*-18:1 (high *trans* model), (E) 9*c*12*t*-18:2 (high *trans* model), and (F) 11*t*15*c*-18:2 (high *trans* model).

TABLE 3
Comparison of FT-NIR and GC Results for Different Oils

	Soybean oil		Canola oil		Palm oil		Flaxseed oil		Shortening	
Fatty acid	FT-NIR	GC	FT-NIR	GC	FT-NIR	GC	FT-NIR	GC	FT-NIR	GC
14:0	0.35	0.43	0.80	0.24	1.07	1.06	0.40	0.43	0.57	0.21
16:0	11.08	10.61	2.70	4.22	44.23	44.27	4.84	4.93	16.00	16.01
9*c*-16:1	0.05	0.09	0.50	0.23	0.21	0.17	–0.05	0.05	0.57	0.10
17:0	0.04	0.12	0.05	0.07	0.11	0.11	0.06	0.06	0.19	0.11
9*c*-17:1	0.02	0.06	0.07	0.09	0.03	0.03	0.04	0.04	0.11	0.06
18:0	3.31	5.59	2.65	2.17	5.60	6.25	4.16	4.31	9.88	11.04
6*t*-8*t*-18:1	0.05	0.01	0.02	0.02	0.30	0.41	0.02	0.01	1.59	1.60
9*t*-18:1	0.10	0.03	0.12	0.06	0.36	0.46	0.05	0.01	2.16	2.11
10*t*-18:1	0.08	0.04	0.05	0.04	0.20	0.28	0.03	0.02	3.79	4.08
11*t*-18:1	0.05	0.01	0.01	0.01	0.14	0.23	0.02	0.01	2.98	3.79
12*t*-18:1	0.04	0.01	0.02	0.00	0.10	0.16	0.02	0.01	1.42	1.43
13*t*/14*t*-18:1	0.04	0.01	0.01	0.01	0.12	0.18	0.00	0.01	1.73	1.86
15*t*-18:1	0.00	0.00	-0.01	0.00	0.07	0.09	0.00	0.00	0.47	0.65
9*c*-18:1	20.89	22.35	58.93	58.52	36.64	35.25	20.64	20.20	22.74	24.46
11*c*-18:1	0.99	1.38	2.30	3.32	0.89	0.72	1.15	0.66	2.03	1.69
12*c*-18:1	0.06	0.01	0.02	0.01	0.06	0.08	0.02	0.01	3.32	4.11
13*c*-18:1	0.01	0.05	0.01	0.02	0.04	0.05	0.02	0.01	0.27	0.23
14*c*/16*t*-18:1	0.00	0.00	0.00	0.00	0.03	0.03	0.00	0.00	0.21	0.14
19:0	0.01	0.00	0.00	0.00	0.00	0.00	0.00	0.00	0.03	0.00
9*t*12*t*-18:2	0.00	0.00	0.00	0.00	0.01	0.01	0.00	0.00	0.10	0.11
9*t*13*c*/8*t*12*c*-18:2	0.03	0.00	0.01	0.00	0.01	0.01	0.01	0.00	1.04	0.92
9*c*12*t*-18:2	0.14	0.36	0.05	0.16	0.10	0.12	0.02	0.06	0.88	1.33
9*t*12*c*-18:2	0.12	0.32	0.09	0.12	0.08	0.08	0.05	0.01	0.81	1.05
11*t*15*c*-18:2	0.00	0.00	0.00	0.00	0.01	0.00	0.00	0.00	0.38	0.22
9*c*12*c*-18:2	50.04	48.75	16.94	18.77	8.65	8.72	15.76	15.40	19.66	19.68
20:0	0.34	0.47	0.53	0.71	0.33	0.29	0.25	0.17	0.33	0.40
cct-18:3	0.05	0.57	0.05	0.56	0.01	0.01	0.22	0.23	0.18	0.23
tcc-18:3	0.03	0.52	0.02	0.51	0.01	0.01	0.04	0.04	0.17	0.20
11*c*-20:1	0.34	0.21	0.65	1.34	0.18	0.11	0.47	0.18	0.28	0.20
9*c*12*c*15*c*-18:3	9.59	6.33	8.79	6.98	0.37	0.18	51.56	51.83	1.13	0.97
ΣCLA	0.02	0.07	0.02	0.05	0.00	0.00	0.00	0.02	0.07	0.17
11*c*14*c*-20:2	0.03	0.04	0.02	0.08	0.00	0.00	0.05	0.04	0.07	0.00
22:0	0.18	0.45	0.26	0.38	0.07	0.03	0.13	0.14	0.18	0.31
24:0	0.11	0.13	0.13	0.17	0.03	0.01	0.11	0.09	0.06	0.10
SFA	15.99	17.87	8.69	8.04	46.24	46.02	10.53	10.20	28.31	28.27
cis MUFA	22.02	24.14	62.40	63.54	38.04	36.40	21.91	21.16	30.17	31.00
cis PUFA	58.84	55.11	25.55	25.83	9.02	8.89	66.72	67.27	21.83	20.65
trans 18:1	0.38	0.11	0.22	0.14	1.27	1.82	0.14	0.05	14.75	15.71
c/*t* 18:2	0.31	0.74	0.14	0.32	0.21	0.22	0.08	0.09	3.28	3.79
c/*c*/*t* 18:3	0.08	1.21	0.09	1.18	0.02	0.02	0.27	0.29	0.58	0.42
Total TFAs	0.82	2.08	0.53	1.65	1.50	2.05	0.57	0.44	18.67	20.09

Abbreviations: CLA, conjugated fatty acids; DUFA, diunsaturated fatty acids; MUFA, monounsaturated fatty acids; PUFA, polyunsaturated fatty acids; TUFA, triunsaturated fatty acids. The calibration curve for some very minor FA was lower than zero, with the intercept giving negative values. Several minor FAs were not included.

The complete analysis including scanning, cleaning of the probe and subsequent processing of the data is generally achieved in a few minutes. The analysis of hard fats requires prior melting before FT-NIR spectra are recorded which provides an additional challenge since FT-NIR signals are affected by temperature (unpublished data). Furthermore, the present method is limited to neat fats and oils and does not apply to fat/water mixtures such as found in butter or margarine. To analyze such samples will require the development of a new FT-NIR model, or a prior extraction of the fat portion from the product.

Typical GC Separation of FAME and CLA Isomers

A typical GC separation of the FAMEs prepared by base-catalyzed methylation of partially hydrogenated soybean oil is shown in Figure 11. The GC chromatogram is separated in three sections showing the regions from 14:0 to 18:0 (Fig. 11A), 18:0 to 20:0 that includes all the *cis* and *trans* 18:1 and *c/t* 18:2 isomers (Fig. 11B), and all the FAMEs that elute after 20:0 which includes the *c/c/t* and *c/t/t* 18:3 and CLA isomers and the longer FAMEs (Fig. 11C). An expansion of the CLA region is shown in Figure 12 that will be discussed below in greater detail.

The TFA isomers distribution and the relative abundance of the *trans* containing 18:1, 18:2 and 18:3 isomers in fat and oil products is not always the same. Figure 13 shows a comparison of three fat/oil products showing the different types of TFA profiles encountered in the marketplace. The partially hydrogenated soybean oil sample (also shown in Fig. 11) had a total TFA content of 19.6% that consisted mainly of *trans*-18:1(13.9%) with lesser amounts of *c/t*-18:2 (5.2%) and *c/c/t*-18:3 isomers (0.5%). Contrary to the generally accepted view, in this case the relative abundance of the 10*t*- and 11*t*-18:1 isomers was more than the 9*t*- and 6*t*-8*t*-18:1 isomers (Fig. 13A). This was in contrast to a canola oil that contained 4.42% total TFAs mainly as *trans* 18:1 isomers (3.93%) with smaller amounts of *c/t*-18:2 (0.32%) and *c/c/t*-18:3 isomers (0.17%), and the *trans* 18:1 isomer distribution showed the more common profile where the relative concentration of 9*t*- and 6*t*-8*t*-18:1 was higher than 10*t*- and 11*t*-18:1 (Fig. 13B). Figure 13C shows a different canola oil, also purchased locally, that had a very different *trans* 18:1, 18:2 and 18:3 isomers distribution of 0.12, 0.29 and 1.22%, respectively. The high content of *trans*-containing 18:3 isomers in this oil sample clearly indicates that there are very different TFA-containing oils in the marketplace today. Such variation in TFA composition appears to indicate different refining processes or partial hydrogenation practices used by the industry to produce fat and oil products in compliance with the new TFA regulation.

The CLA isomers content in the vegetable oils ranged from 0.02 to 0.2% of total FAMEs. In refined vegetable oils and fats CLA isomers are mainly produced during the deodorization step in the refining process. Therefore, it was not surprising to find that the CLA isomers consisted of a random distribution of all positional and geometric isomers from 8,10- to 12,14-CLA (Fig. 12), that are unlike those present in ruminant fats where the 9*c*11*t*-CLA isomer (rumenic acid) predominates.

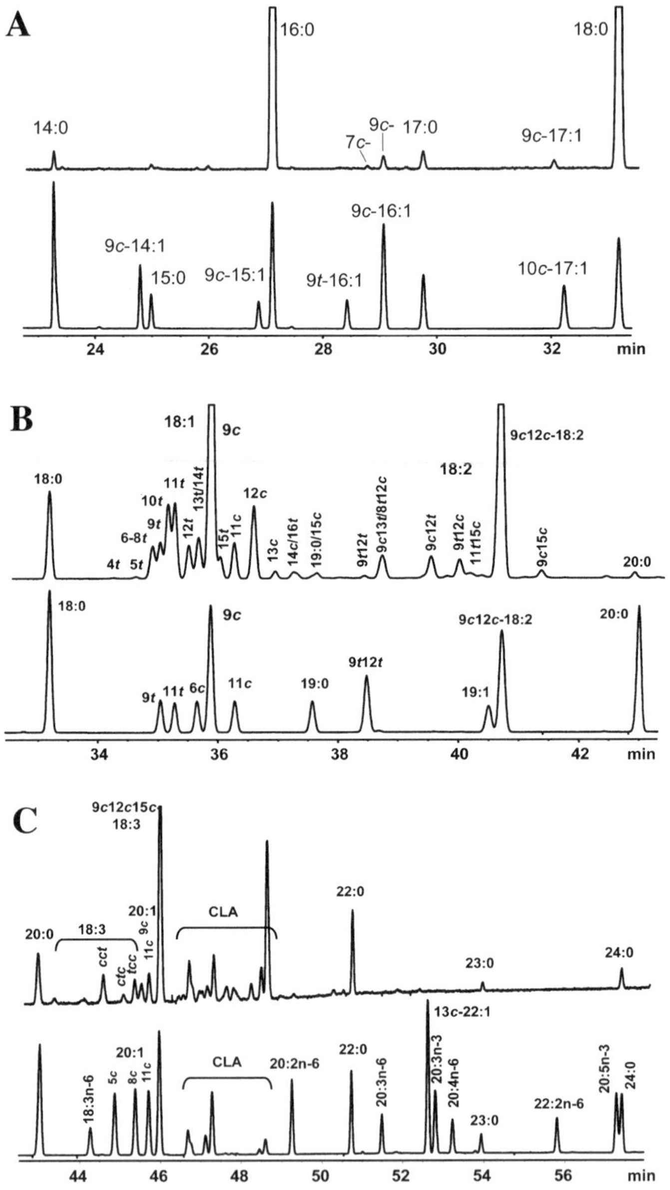

Fig. 11. GC chromatogram of a partially hydrogenated soybean oil compared to a GC standard mixture containing #463, CLA #UC-59M, 21:0, 23:0 and 26:0 from Nu Chek Prep. A 100 m CP Sil 88 capillary column was operated using a temperature program described in Materials and Methods. The GC chromatogram is presented in three sections from the methyl esters of (A) 14:0 to 18:0, (B) 18:0 to 20:0, and (C) 20:0 to 24:0.

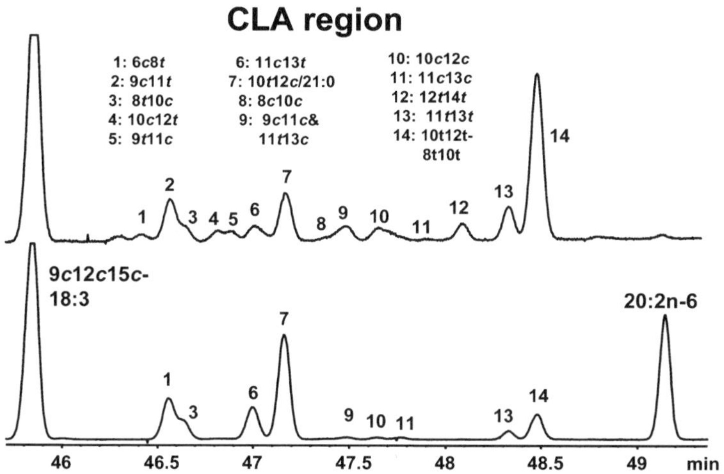

Fig. 12. Partial GC chromatogram of the CLA region from the same sample presented in Figure 11.

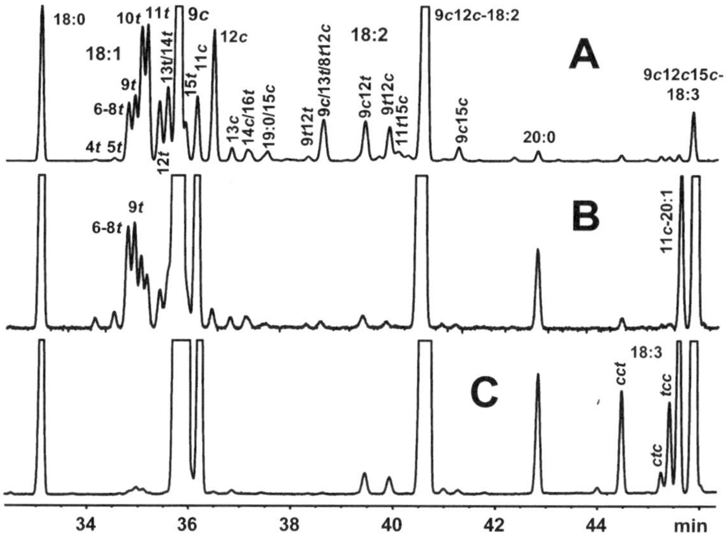

Fig. 13. Partial GC chromatograms from 18:0 to 9c12c15c-18:3 of partially hydrogenated soybean oil, and two canola oils that differed in their TFA distribution.

The CLA isomers formed during catalytic or heating processes are randomly formed by a combination of free-radical chain reactions and sigmatropic rearrangements (12), while the CLA isomers in ruminants are enzymatically produced by a combination of isomerization and reduction step (24). The FT-NIR models we reported did not include a wide range in the concentration of the CLA isomers (71). Based on the present findings the potential exists for developing CLA models using the FT-NIR technique.

It would not be prudent to exclude the CLA content produced during partial hydrogenation of vegetable oils from total TFA as presently permitted in the TFA labeling acts (1–5). The CLA composition in refined and partially hydrogenated products contains many CLA isomers with unknown physiological and biological effects. Fortunately, the CLA content in most industrial fats or vegetable oils is generally less than 0.2%.

Present Assessment of the FT-NIR Method for FA Determination and Future Developments Required

The FT-NIR models developed were based on Chemometric analyses of FT-NIR absorptions and accurate FA compositions obtained by GC. Three separate models were established to analyze fats and oils containing low, medium and high levels of TFA. These calibration models are capable of providing rapid and accurate analysis of the total TFA content in fats and oils for regulatory labeling purposes, as well as the composition of the individual TFA isomers. The method compares well to the current certified methods of *trans* fat analysis by ATR-FTIR that provides only the total *trans* content of a fat or oil, and not the individual isomer composition. The FT-NIR method described in this report is a secondary method relying on accurate GC data to serve as the primary method. The GC method remains essential to establish reliable FA compositional data to develop the FT-NIR models, but subsequent labor- and time-consuming GC analysis for routine fat and oil samples will not be necessary. While GC analyses need to be performed on every sample including the conversion to FAMEs, separation of the FAMEs on highly polar capillary columns and interpretation of the data, the FT-NIR model needs to be developed only once, and subsequent measurements require only minutes for complete analysis of neat fat and oil samples. The time and cost saving for routine FA determination and quality control are obvious. Based on the results presented here and published elsewhere (70–73), it would appear that the FT-NIR method has a great potential to compliment the GC technique for FA determination when routine analysis of many samples are required.

In the present TFA regulations only the total TFA content, excluding the CLA content has to be reported. However, that data may not provide the best information necessary to evaluate the risk of CHD, since the TFA isomer(s) responsible have not been identified. The FT-NIR method described here is a method capable of providing the detailed TFA isomer composition so far available only by GC, but is now possible for routine analysis at a fraction of the time. The marked differences in the relative abundance of the *trans* containing 18:1, 18:2 and 18:3 isomers and the *trans*-18:1 isomer distribution in commercial vegetable oils and fats, makes a more detailed analysis an even greater necessity to better understand what kind of TFA are actually being consumed. In most studies reported to date (6–10) only the total TFA content was reported, and not the relative abundance of *trans* containing 18:1, 18:2 and 18:3, and the *trans*-18:1 isomer distribution. In general it was assumed that TFA are mainly derived from 18:1 and the major isomer was 9*t*-18:1. The partial GC chromatograms shown in Figure 13 above would suggest that the TFA composition is much more complex.

A number of factors remain to be refined to collaboratively validate and establish FT-NIR as an official method for complete FA determination of fats and oils. The accuracy, reproducibility and limit of detection of minor FAs will need to be rigorously examined, as well as the effects, if any, of the position of FAs on the TAG molecule. Furthermore, the effect of low levels of non-TAG components such as plant sterols (or cholesterol) and tocopherols, requires investigation. Temperature and other parameters in the FT-NIR method will need to be strictly controlled, because their influence appears to have been the cause of the limited application of the FT-NIR method (unpublished data). It should be emphasized that the FT-NIR method described here is applicable for all fats and oils provided the FAs are in the range of the FAs in the model used. A model can easily be modified to include additional FAs or expand the FAs ranges. For higher accuracy it is recommended that product-specific quantitative FT-NIR models be developed. For this reason, the present models are not applicable for the direct determination of products such as salad dressing and margarine that contain substantial amounts of water. Specific FT-NIR models will need to be established to analyze these products directly.

Conclusion

A rapid method was developed for classification and quantification of the FA composition of edible oils and fats using FT-NIR. The FT-NIR spectra showed unique fingerprints for saturated FAs, *cis* and *trans* monounsaturated FAs, and all n-6 and n-3 PUFAs within TAG to permit qualitative and quantitative comparisons of fats and oils. The quantitative models were based on incorporating accurate GC data of the different fats and oils and FT-NIR spectral information into the calibration models using Chemometric analysis. FT-NIR classification models were developed based on Chemometric analyses of over 60 fats, oils and fat/oil mixtures that were used in the identification of similar materials. This database was used to prepare three calibration models—one suitable for the analysis of common fats and oils with low levels of TFAs, the other two for fats and oils with intermediate and high levels of TFAs. The FT-NIR method showed great potential to provide the complete FA composition of unknown fats and oils in minutes. Compared to the official GC method, the FT-NIR method was used to analyze fats and oils directly in their neat form that required no derivatization of the fats to volatile FAME or time-consuming GC separations and analyses. The FT-NIR method also compared well to the official ATR-FTIR method that unfortunately provides only the total TFA content, while FT-NIR also provides the complete FA profile. The FT-NIR method has the potential to be used for rapidly screening and monitoring fat products, TFA determinations for regulatory labeling purposes, and detection of contaminants.

Acknowledgments

The Industrial Research Assistance Program (IRAP) of National Research Council (NRC) of Canada provided partial funding during the initial stages of this investigation. Contribution number S233 from Food Research Program, Agriculture and Agri-Food Canada. The advice by Mr. David Hawkes, Industrial Technology Advisor from IRAP and Dr. Magdi M. Mossoba (US-FDA, Maryland, USA) is gratefully acknowledged as well as the technical assistance by A.R. Kamalian, C. Winsborough, S.L. Winsborough and M. Hernandez.

References

1. Regulations Amending the Food and Drug Regulations (Nutritional Labelling, Nutrient Content Claims and Health Claims), Department of Health, *Canada Gazette, Part II (January 1, 2003), Ottawa, Canada, Vol. 137*:154–409 (2003).
2. Ratnayake, W.M.N. and C. Zehaluk, *Trans* Fatty Acids in Foods and Their Labeling Regulations, in *Healthful Lipids* (Akoh, C.C. and O.-M. Lai, Eds.), pp. 1–32, AOCS Press, Champaign, IL (2005).
3. Department of Health and Human Services, Food and Drug Administration, Food Labeling; *Trans* Fatty Acids in Nutrition Labeling; Nutrient Content Claims, and Health Claims; Final Rule, July 11, 2003, *Federal Register 68 (No. 133)*:41434–41506 (2003).
4. ANVISA, National Agency of Sanitary Monitoring Brazil, Mercusul Regulamento Técnico Sobre Rotulagem Nutricional De Alimentos Embalados, RDC No 360, December 23, 2003.
5. Stender, S. and J. Dyerberg, Influence of *Trans* Fatty Acids on Health, *Ann. Nutr. Metabol. 48*:61–66 (2004).
6. Mensink, R.P. and M.B. Katan, (1990) Effects of Dietary *Trans* Fatty Acids on High-Density and Low-Density Lipoprotein Cholesterol Levels in Healthy Subjects, *N. Engl. J. Med. 323*:439–445.
7. Ascherio, A., M.B. Katan, P.L. Zock, M.J. Stampfer, and W.C. Willett, *Trans* Fatty Acids and Coronary Heart Disease, *N. Engl. J. Med. 340*:1994–1998 (1999).
8. Mensink, R.P., P.L. Zock, A.D.M. Kester, and M.B. Katan, Effects of Dietary Fatty Acids and Carbohydrates on the Ratio of Serum Total to HDL Cholesterol and on Serum Lipids and Apolipoproteins: A Meta-Analysis of 60 Controlled Trials, *Am. J. Clin. Nutr. 77*:1146–1155 (2003).
9. Mozaffarian, D., E.B. Rimm, I.B. King, R.L. Lawler, G.B. McDonald, and W.C. Levy, *Trans* Fatty Acids and Systemic Inflammation in Heart Failure, *Am. J. Clin. Nutr. 80*:1521–1525 (2004).
10. Lopez-Garcia, E., M.B. Schulze, J.B. Meigs, J.E. Manson, N. Rifai, M.J. Stampfer, W.C. Willett, and F.B. Hu, Consumption of *Trans* Fatty Acids is Related to Plasma Biomarkers of Inflammation and Endothelial Dysfunction, *J. Nutr. 135*:562–566 (2005).
11. Sébédio, J.L., M. Catte, M.A. Boudier, J. Prevost, and A. Grandgirard, Formation of Fatty Acid Geometric Isomers and Cyclic Fatty Acid Monomers During the Finish Frying of Frozen Prefried Potatoes, *Food Res. Internat. 29*:109–116 (1996).
12. Destaillats, F. and P. Angers, Thermally Induced Formation of Conjugated Isomers of Linoleic Acid, *Eur. J. Lipid Sci. Technol. 107*, 167–172 (2005).
13. Ohnishi, M. and G.A. Thompson, Jr., Biosynthesis of the Unique *Trans* -Δ^3-Hexadecenoic Acid Component of Chloroplast Phosphatidylglycerol: Evidence Concerning its Site and Mechanism of Formation, *Arch. Biochem. Biophys. 288*:591–599 (1991).

14. Precht, D. and J. Molkentin, *Trans*-Geometrical and Positional Isomers of Linoleic Acid Including Conjugated Linoleic Acid (CLA) in German Milk and Vegetable Fats, *Fett/Lipid 99*:319–326 (1997).
15. Ratnayake, W.M.N., G. Pelletier, R. Hollywood, S. Bacler, and D. Leyte, *Trans* Fatty Acids in Canadian Margarines: Recent Trends, *J. Am. Oil Chem. Soc. 75*:1587–1594 (1998).
16. Precht, D. and J. Molkentin, Recent Trends in the Fatty Acid Composition of German Sunflower Margarines, Shortenings and Cooking Fats with Emphasis on Individual C16:1, C18:1, C18:2 C18:3 and 20:1 *Trans* Isomers, *Nahrung 44*:222–228 (2000).
17. Satchithanandam, S., C.J. Oles, C.J. Spease, M.M. Brandt, M.P. Yurawecz, and J.I. Rader, *Trans*, Saturated, and Unsaturated Fat in Foods in the United States Prior to Mandatory *Trans*-Fat Labeling, *Lipids, 39*:11–18 (2004).
18. Kraft, J., M. Collomb, P. Möckel, R. Sieber, and G. Jahreis, Differences in CLA Isomer Distribution of Cow's Milk Lipids. *Lipids 38*:657–664 (2003).
19. Parodi, P.W., Milk Fat in Human Nutrition, *Austr. J. Dairy Technol. 59*:3–59 (2004).
20. Weggemans, R.M., M. Rudrum, and E.A. Trautwein, Intake of Ruminant *Versus* Industrial *Trans* Fatty Acids and Risk of Coronary Heart Disease—What is the Evidence? *Eur. J. Lipid Sci. Technol. 106*:390–397 (2004).
21. Aro, A., S. Männistö, I. Salminen, M.-L. Ovaskainen, V. Kataja, and M. Uusitupa, Inverse Association Between Dietary and Serum Conjugated Linoleic Acid and Risk of Breast Cancer in Postmenopausal Women, *Nutr. Cancer 38*:151–157 (2000).
22. Corl, B.A., D.M. Barbano, D.E. Bauman, and C. Ip, *Cis*-9, *Trans*-11 CLA Derived Endogenously from *Tra* (2003)*ns*-11 18:1 Reduces Cancer Risk in Rats, *J. Nutr. 133*:2893–2900.
23. Kramer, J.K.G., C. Cruz-Hernandez, M. Or-Rashid, and M.E.R. Dugan, The Use of Total *Trans*-11 Containing FA, Rather than Total n-7″ FA, is Recommended to Assess the Content of FA with a Positive Health Image in Ruminant Fats, *Lipids 39*:693–695 (2004).
24. Bauman, D. E. and J. M. Griinari, Nutritional Regulation of Milk Fat Synthesis, *Ann. Rev. Nutr. 23*:203–227 (2003).
25. Mossoba, M.M., J.K.G. Kramer, P. Delmonte, M.P. Yurawecz, and J.I. Rader, *Official Methods for the Determination of* Trans *Fat*, AOCS Press, Champaign, IL (2003).
26. Ratnayake, W.M.N., Overview of Methods for the Determination of *Trans* Fatty Acids by Gas Chromatography, Silver-Ion Thin-Layer Chromatography, Silver-Ion Liquid Chromatography, and Gas Chromatography/Mass Spectrometry, *J. AOAC International 87*:523–539 (2004).
27. AOAC Official Method 994.15, Total *Cis*- and *Trans*-Octadecenoic Isomers and General Fatty Acid Composition in Hydrogenated Vegetable Oils and Animal Fats (41.1.35A), in *Official Methods of Analysis*, 17th Ed., Chapter 41:32–36, AOAC International, Gaithersburg, MD (2000).
28. AOCS Official Method Ce 1h-05, Determination of *Cis*-, *Trans*-, Saturated, Monounsaturated and Polyunsaturated Fatty Acids in Vegetable and Non-ruminant Animal Oils and Fats by Capillary GLC, in *Official Methods and Recommended Practices of the AOCS*, 5th Edition, 2nd Printing, AOCS, Champaign, IL (2005).
29. Precht, D. and J. Molkentin, C18:1, C18:2 and C18:3 *Trans* and *Cis* Fatty Acid Isomers Including Conjugated *Cis* Δ9, *Trans* Δ11 Linoleic Acid (CLA) as Well as Total Fat Composition of German Human Milk Lipids, *Nahrung 43*:233–244 (1999).
30. Cruz-Hernandez, C., Z. Deng, J. Zhou, A.R. Hill, M.P. Yurawecz, P. Delmonte, M.M. Mossoba, M.E.R. Dugan, and J.K.G. Kramer, Methods for Analysis of Conjugated

Linoleic Acids and *Trans*-18:1 Isomers in Dairy Fats Using a Combination of Gas Chromatography, Silver-Ion Thin-Layer Chromatography/Gas Chromatography, and Silver-Ion Liquid Chromatography, *J. AOAC Internat. 87*:545–562 (2004).
31. Kramer, J.K.G., V. Fellner, M.E.R. Dugan, F.D. Sauer, M.M. Mossoba, and M.P. Yurawecz, Evaluating Acid and Base Catalysts in the Methylation of Milk and Rumen Fatty Acids with Special Emphasis on Conjugated Dienes and Total *Trans* Fatty Acids, *Lipids 32*:1219–1228 (1997).
32. Kramer, J.K.G. and J. Zhou, Conjugated Linoleic Acids and Octadecenoic Acids: Extraction and Isolation of Lipids, *Eur. J. Lipid Sci. Technol. 103*:594–600 (2001).
33. Kramer, J.K.G., C. Cruz-Hernandez, and J. Zhou, Conjugated Linoleic Acids and Octadecenoic Acids: Analysis by GC, *Eur. J. Lipid Sci. Technol. 103*:600–609 (2001).
34. Kramer, J.K.G., C.B. Blackadar, and J. Zhou, Evaluation of Two GC Columns (60-m SUPELCOWAX 10 and 100-m CP Sil 88) for Analysis of Milkfat with Emphasis on CLA, 18:1, 18:2 and 18:3 Isomers, and Short- and Long-Chain FA, *Lipids. 37*:823–835 (2002).
35. Sehat, N., M.P. Yurawecz, J.A.G. Roach, M.M. Mossoba, J.K.G. Kramer, and Y. Ku, Silver-Ion High-Performance Liquid Chromatographic Separation and Identification of Conjugated Linoleic Acid Isomers, *Lipids 33*:217–221 (1998).
36. Sehat, N., R. Rickert, M.M. Mossoba, J.K.G. Kramer, M.P. Yurawecz, J.A.G. Roach, R.O. Adlof, K.M. Morehouse, J. Fritsche, K.D. Eulitz, H. Steinhart, and Y. Ku, Improved Separation of Conjugated Linoleic Acid Methyl Esters by Silver-Ion High-Performance Liquid Chromatography, *Lipids 34*:407–413 (1999).
37. Yurawecz, M.P., J.A.G. Roach, N. Sehat, M.M. Mossoba, J.K.G. Kramer, J. Fritsche, H. Steinhart, and Y. Ku, A New Conjugated Linoleic Acid Isomer, 7 *trans*, 9 *cis*-Octadecadienoic Acid, in Cow Milk, Cheese, Beef and Human Milk and Adipose Tissue, *Lipids 33*:803–809 (1998).
38. Piperova, L.S., B.B. Teter, I. Bruckental, J. Sampugna, S.E. Mills, M.P. Yurawecz, J. Fritsche, K. Ku, and R.A. Erdman, Mammary Lipogenic Enzyme Activity, *Trans* Fatty Acids and Conjugated Linoleic Acids Are Altered in Lactating Dairy Cows Fed a Milk Fat-Depressing Diet, *J. Nutr. 130*:2568–2574 (2000).
39. Adam, M., M. Chew, S. Wasserman, A. McCollum, R.E. McDonald, and M.M. Mossoba, Determination of *Trans* Fatty Acids in Hydrogenated Vegetable Oils by Attenuated Total Reflection Infrared Spectroscopy: Two Limited Collaborative Studies, *J. Am. Oil Chem. Soc. 75*:353–358 (1998).
40. Mossoba, M.M., M.P. Yurawecz, P. Delmonte, and J.K.G. Kramer, Overview of Infrared Methodologies for *Trans* Fat Determination, *J. AOAC International 87*:540–544 (2004).
41. AOCS Official Method Cd 14d.99, Rapid Determination of Isolated *Trans* Geometric Isomers in Fats and Oils by Attenuated Total Reflection Infrared Spectroscopic, in *Official Methods and Recommended Practices of the AOCS*, 5th Edition, 2nd Printing, AOCS, Champaign, IL (1999).
42. AOAC Official Method 2000.10, Determination of Isolated *Trans* Unsaturated Fatty Acids in Fats and Oils, in *Official Methods of Analysis*, 17th Ed., 41.1.37A, AOAC International, Gaithersburg, MD (2002).
43. Mossoba, M.M., R.E. McDonald, D.J. Armstrong, and S.W. Page, Identification of minor C18 Triene and Conjugated Diene Isomers in Hydrogenated Soybean Oil and Margarine by GC-MI-FT-IR Spectroscopy, *J. Chromatogr. Sci. 29*:324–330 (1991).
44. Mossoba, M.M., J.K.G. Kramer, J. Fritsche, M.P. Yurawecz, K. Eulitz, Y. Ku, and J.I. Rader, Application of Standard Addition to Eliminate Conjugated Linoleic Acid and

Other Interferences in the Determination of Total *Trans* Fatty Acids in Selected Food Products by Infrared Spectroscopy, *J. Am. Oil Chem. Soc. 78*:631–634 (2001).

45. Christy, A.A., P.K. Egeberg, and E.T. Østensen, Simultaneous Quantitative Determination of Isolated *Trans* Fatty Acids and Conjugated Linoleic Acids in Oils and Fats by Chemometric Analysis of the Infrared Profiles, *Vibrational Spectroscopy 33*:37–48 (2003).
46. Milosevic, M., V. Milosevic, J.K.G. Kramer, H. Azizian, and M.M. Mossoba,Determining Low Levels of *Trans* Fatty Acids in Foods Using an Improved ATR-FTIR Procedure, *Lipid Technology 16*:252–255 (2004) .
47. Near-Infrared Spectroscopy, Principles, Instruments Applications, Siesler, H.W., Y. Ozaki, S. Kawata, and H.M. Heise, eds., Wiley-VCH Verlag GmbH, Weinheim, Germany (2002).
48. Ciurczak, E. and J. Drennen, *Near-Infrared Spectroscopy in Pharmaceutical and Medical Applications*, Marcel Dekker Inc., New York (2002).
49. Near-Infrared Spectroscopy in Agriculture, Roberts, C.A., Workman, J. and Reeves, J.B., eds., Number 44 in the series of *Agronomy, Am. Soc. Agronomy, Crop Sci. Soc. Am., and Soil Sci. Soc. Am.*, Madison, WI (2004).
50. van de Voort, F.R., J. Sedman, and T. Russin, Lipid Analysis by Vibrational Spectroscopy, *Eur. J. Lipid Sci. Technol. 103*:815–826 (2001).
51. Pasquini, C., Near Infrared Spectroscopy: Fundamentals, Practical Aspects and Analytical Applications, *J. Braz. Chem. Soc., 14*:198–219 (2003).
52. Li, H., F.R. van de Voort, J. Sedman, and A.A. Ismail, Rapid Determination of *Cis* and *Trans* Content, Iodine Value, and Saponification Number of Edible Oils by Fourier Transform Near-Infrared Spectroscopy, *J. Am. Oil Chem. Soc. 76*:491–497 (1999).
53. Li, H., F.R. van de Voort, A.A. Ismail, J. Sedman, R. Cox, C. Simard, and H. Buijs, Discrimination of Edible Oil Products and Quantitative Determination of Their Iodine Value by Fourier Transform Near-Infrared Spectroscopy, *J. Am. Oil Chem. Soc. 77*:29–36 (2000).
54. Li, H., F.R. van de Voort, A.A. Ismail, J. Sedman, and R. Cox, *Trans* Determination of Edible Oils by Fourier Transform Near-Infrared Spectroscopy, *J. Am. Oil Chem. Soc. 77*:1061–1067 (2000).
55. Cox, R., J. Lebrasseur, E. Michiels, H. Buijs, H. Li, F.R. van de Voort, A.A. Ismail, and J. Sedman, Determination of Iodine Value with a Fourier Transform Near-Infrared Based Global Calibration Using Disposable Vials: An International Collaborative Study, *J. Am. Oil Chem. Soc. 77*:1229–1234 (2000).
56. AOCS Official Method Cd 1e-01 Determination of Iodine Value by Pre-calibrated FT-NIR with Disposable Vials, in *Official Methods and Recommended Practices of the AOCS*, 5th Edition, 2nd Printing, AOCS, Champaign, IL (2001).
57. Moh, M.H., Y.B. Che Man, F.R. van de Voort, and W.J.W. Abdullah, Determination of Peroxide Value in Thermally Oxidized Crude Palm Oil by Near-Infrared Spectroscopy, *J. Am. Oil Chem. Soc. 76*:19–23 (1999).
58. Li, H., F.R. van de Voort, A.A. Ismail, and R. Cox, Determination of Peroxide Value by Fourier Transform Near-Infrared Spectroscopy. *J. Am. Oil Chem. Soc. 77*, 137–142 (2000).
59. Yildiz, G., R.L. Wehling, and S.L. Cuppett, Method for Determining Oxidation of Vegetable Oils by Near-Infrared Spectroscopy, *J. Am. Oil Chem. Soc. 78*:495–502 (2001).
60. Jensen, P.N., J. Christiansen, and S.B. Engelsen, Oxidative Changes in Pork Scratchings, Peanuts, Oatmeal and Muesli Viewed by Fluorescence, Near-Infrared and Infrared Spectroscopy, *Eur. Food Res. Technol. 219*:294–304 (2004).
61. Wesley, I.J., R.J. Barnes, and A.E.J. McGill, Measurement of Adulteration of Olive Oils by Near-Infrared Spectroscopy, *J. Am. Oil Chem. Soc. 72*:289–292 (1995).

62. Wesley, I.J., F. Pacheco, and A.E.J. McGill, Identification of Adulterants in Olive Oils, *J. Am. Oil Chem. Soc. 73*:515–518 (1996).
63. Sato, T., Y. Takahata, T. Noda, T. Yanagisawa, T. Morishita, and S. Sakai, Nondestructive Determination of Fatty Acid Composition of Husked Sunflower (*Helianthus annua* L.) Seeds by Near-Infrared Spectroscopy, *J. Am. Oil Chem. Soc. 72*:1177–1183 (1995).
64. Pérez-Vich, B., L. Velasco, and J.M. Fernández-Martínez, Determination of Seed Oil Content and Fatty Acid Composition in Sunflower Through the Analysis of Intact Seeds, Husked Seeds, Meal and Oil by Near-Infrared Reflectance Spectroscopy, *J. Am. Oil Chem. Soc. 75*:547–555 (1998).
65. Velasco, L., F.D. Goffman, and H.C. Becker, Development of Calibration Equations to Predict Oil Content and Fatty Acid Composition in Brassicaceae Germplasm by Near-Infrared Reflectance Spectroscopy, *J. Am. Oil Chem. Soc. 76*:25–30 (1999).
66. Mailer, R.J., Rapid Evaluation of Olive Oil Quality by NIR Reflectance Spectroscopy, *J. Am. Oil Chem. Soc. 81*:823–827 (2004).
67. Molette, C., P. Berzaghi, A. Dalle Zotte, H. Remignon, and R. Babile, The Use of Near-Infrared Reflectance Spectroscopy in the Prediction of the Chemical Composition of Goose Fatty Liver, *Poultry Sci. 80*:1625–1629 (2001).
68. Gomzález-Martín, I., C. Gomzález-Pérez, J. Hernández-Méndez, N. Álvarez-García, and S.M. Lázaro, Determination of Fatty Acids in the Subcutaneous Fat of Iberian Breed Swine by Near Infrared Spectroscopy. A Comparative Study of the Methods for Obtaining Total Lipids: Solvents and Melting with Microwaves, *J. Near-Infrared Spectroscopy 10*:257–268 (2002).
69. Garrido-Varo, A., J. García-Olmo, and M.D. Pérez-Martín, Applications in Fats and Oils, (Roberts, C.A., J. Workman, and J.B. Reeves, eds.), pp. 487–558, Number 44 in the series of Agronomy, *Am. Soc. Agronomy, Crop Sci. Soc. Am.,* and *Soil Sci. Soc. Am.*, Madison, WI (2004).
70. Azizian, H., J.K.G. Kramer, A.R. Kamalian, M. Hernandez, M.M. Mossoba, and S.L. Winsborough, Quantification of *Trans* Fatty Acids in Food Products by GC, ATR-FTIR and FT-NIR Methods, *Lipid Technol. 16*:229–231 (2004).
71. Azizian, H. and J.K.G. Kramer, A Rapid Method for the Quantification of Fatty Acids in Fats and Oils with Emphasis on *Trans* Fatty Acids Using Fourier Transform Near Infrared Spectroscopy (FT-NIR), *Lipids 40*:855–867 (2005).
72. Azizian, H. and J.K.G. Kramer, A Non-invasive Analytical Tool for Many Applications, *inform 16*:656–658 (2005).
73. Azizian, H., *United States Patent Application No. 10/840277* (Filed on May 7, 2004).
74. Wolff, R.L., Analysis of Alpha-Linolenic Acid Geometrical Isomers in Deodorized Oils by Capillary Gas-Liquid Chromatography on Cyanoalkyl Polysiloxane Stationary Phases: A Note of Caution, *J. Am. Oil Chem. Soc. 71*:907–909 (1994).
75. Grob, K. and M. Biedermann, The Two Options for Sample Evaporation in Hot GC Injectors: Thermospray and Band Formation. Optimization of Conditions and Injection Design, *Anal. Chem. 74*:10–16 (2002).
76. Precht, D. and J. Molkentin, Rapid Analysis of the Isomers of *Trans*-Octadecenoic Acid in Milk Fat, *Int. Dairy J. 6*:791–809 (1996).
77. Roach, J.A.G., M.M. Mossoba, M.P. Yurawecz, and J.K.G. Kramer, Chromatographic Separation and Identification of Conjugated Linoleic Acid Isomers, *Analytica Chimica Acta 465*:207–226 (2002).

15 Infrared Spectroscopy and Partial Least Square Calibration in the Simultaneous Quantification of Isolated *trans* and Conjugated Linoleic Acids

Alfred A. Christy

Department of Chemistry, Faculty of Mathematics and Natural Sciences, Agder University College, Serviceboks 422, N-4604 Kristiansand, Norway

Introduction

Oils and Fats, trans *and Conjugated Linoleic Acids (CLAs)*

The physical states of the glycerides that make up oils and fats vary from liquid to solid depending on the nature of the fatty acids incorporated into the glycerides and temperature. Solid fat contains a high proportion of long-chain saturated fatty acids. Liquids such as edible oils contain a high proportion of unsaturated fatty acids. Most of the unsaturated fatty acids incorporated in the triglycerides are found in the *cis* form. The most common fatty acids found in the edible oils are palmitic acid (C_{14}), myristic acid (C_{16}), stearic acid (18:0), oleic acid (9c, 18:1), and methylene-interrupted unsaturated fatty acids such as linoleic acid (9*c*12*c* 18:2) and linolenic acid (9*c*12*c*5*c* 18:3). In addition to these, some oils may contain small amounts of other fatty acids, non-methylene-interrupted conjugated linoleic acids (CLAs).

Hydrogenation is an industrial process to produce margarine or shortenings. During this process, the degree of unsaturation in edible oils is reduced at the same time that conversion of *cis* to *trans* isomerization is taking place. The molecules containing *trans* isomers can pack themselves in an orderly manner; hence they are high melting compared with their corresponding *cis* counterparts. This *cis-trans* isomerization during hydrogenation contributes to the net hardening of the edible oil. The margarines that are produced by hydrogenation are sold as butter substitutes. These butter substitutes contain up to 40% *trans* fatty acids and were found to have cholesterol-raising potential (1–4). The health authorities in several countries require the industries to specify the amount of *trans* fatty acids in their products. This prompted scientists to explore techniques and ways to determine the *trans* content not only of edible oils but also other food items and meat products.

CLAs are a mixture of geometrical and positional isomers of linoleic acid with conjugated double bonds. CLAs have gained considerable attention in recent years because of their health benefits (5–8). The conjugated isomers in milk and fats are generally found in *trans/cis*, *cis/trans* and *trans/trans* forms. There are several CLAs present in milk and fat from ruminants. Of these 9*c*11*t* and 10*t*12*c* dominate, with 9*c*11*t* comprising ~80–90% of all of the CLAs. The isomer 9*c*11*t* is also thought to be the most biologically active (5,6). These isomers can also be found in partially hydrogenated soybean and sunflower oils, which are rich in linoleic acids. The conjugated fatty acids are formed only from the methylene-interrupted linoleic acids, and again the 9*c*11*t* and 10*t*12*c* isomers dominate in the partially hydrogenated oils. The isomers 9*c*11*t* and 10*t*12*c* are in almost equal proportion in the partially hydrogenated oils. The concentrations of the other isomers are negligible. Quantitative determination of CLAs has been carried out by gas chromatography (GC) (9) and GC-mass spectrometry (MS) (10). The analyses involve long wet chemical methods and chromatographic analysis.

Infrared (IR) Spectrometry in the Determination of *trans* Fatty Acids and CLAs

The use of IR spectrometry and GC to determine the *trans* content of edible oils and fats was investigated extensively; the methods are standardized by AOCS (11). Two of the methods employ GC and the third employs conventional IR spectrometry. The quantitative determination of *trans* fatty acids by IR spectrometry was based on the weak absorption at ~969 cm^{-1} arising from the C-H out-of-plane deformation vibration. However, this band overlaps with the broad absorption band of triglycerides and C-H out-of-plane deformation bands arising from the CLAs in the sample and provides a spectral profile with a down sloping base line. The *trans-trans* isomers of the CLAs give a single absorption at ~988 cm^{-1} at the same time that *cis-trans* and *trans-cis* isomers give rise to two absorption bands at 981 and 947 cm^{-1}. In the presence of isolated *trans* fatty acids in the glycerides, the band at 981 cm^{-1} shifts to a slightly higher wave number.

There are two different AOCS methods for the quantification of *trans* fatty acids depending on the concentration. For samples that have a *trans* content >15%, the quantification is carried out by comparing the peak height of the *trans* band of a 2% sample in carbon disulfide solution against the peak height of the *trans* band of 2% elaidic acid in CS_2. For samples that have *trans* content <15%, the quantification is carried out after wet chemical processing. The samples are first saponified and esterified (as methyl esters) and the quantification is then carried out by comparing the peak height of the sample at 969 cm^{-1} against the peak height of the *trans* band of methyl elaidate at the same wave number. This procedure requires sample preparation in a solvent that is considered unpleasant and poisonous.

There are several drawbacks to the AOCS method. The results obtained in the two procedures mentioned above suffer accuracy problems. The method used for the samples with higher *trans* concentrations yields higher values. The deviation is higher when the *trans* concentration is <15%. The method used for the samples with *trans* content <15% results in 1.5–3% lower *trans* values (12).

However, there has been progress made in the ways the *trans* band has been utilized in the quantification of *trans* fatty acids in edible oils and fats. Mossoba et al. (1)

summarized the developments in the sample handling and computational techniques in quantifying the absorption at 669 cm^{-1}. Developments in IR instrumentation, sampling techniques, and multivariate data handling software are responsible for the refinements in the determination of *trans* fatty acids by IR spectroscopy. Attenuated total internal reflectance (ATR) is one of the techniques that took advantage of the high throughput of the modern FTIR spectrometers. Infrared spectra of neat edible oil samples were measured using the reflectance technique and the *trans* fatty acid contents (13,14) were determined. Lancer and Emkem (15) developed a quantitative method for the determination of isolated *trans* fatty acids based on the area of the peak representing the C-H out-of-plane deformation absorption in fatty acid methyl esters. They showed that the results were in good agreement with the results obtained using GC. Ulberth and Haider (16) used a transmission technique and standards prepared with methyl elaidate in methyl esters of fatty acids from a *trans*-free soybean oil and determined the *trans* content of fats after converting the fatty acids into their methyl esters. The authors used the base line technique in the quantification and also investigated the use of partial least squares (PLS) calibration in determining the *trans* content in fats. Mossoba et al. (1) used trielaidin in triolein to establish calibration plots to determine the *trans* content of partially hydrogenated soybean oils.

There have been attempts to improve accuracy in the determination using FTIR spectroscopy. The problems encountered in the determination of *trans* content in fats and oils by IR spectrometry using absorption at 969 cm^{-1} arise from the fact that it is difficult to find an ideal reference material that could remove the sloping of the *trans* band at 969 cm^{-1}. This is because the triglycerides of fats and oils contain a variety of combinations of fatty acids in the molecules, and the absorptions in the IR spectrum arising from the fatty acid groups are affected by the other fatty acids present in glyceride molecules. Researchers have been trying to find ways to improve the quantification of the absorption at 969 cm^{-1} by experimenting with modifications both in the sample preparation and in the methods of quantification.

The use of IR spectrometry to determine CLAs has not been fully explored. Apart from the peak assignments for the CH deformation vibrations in CLAs, no attempt has been made to use the peaks for the quantification. This may be due to the fact that the absorptions arising from the CH deformation bands of the CLAs overlap with the *trans* CH out-of-plane deformation band. Furthermore, the concentrations of CLA are relatively low compared with the *trans* fatty acid concentrations in natural samples. In addition, sloping of the IR profiles in the region from 1000 to 650 cm^{-1} complicates and distorts the shapes of the peaks.

Chemometrics and Its Application in the Determination of *trans* Fatty Acids

Chemometrics, a subject dealing with multivariate data analysis, has contributed and aided in solving problems in qualitative and quantitative analysis. The application of multivariate data analysis to near-IR spectral profiles in industrial applications is a success story. Several wet chemical analytical determinations in industry have been replaced by multivariate calibration models.

When spectral profiles are measured on natural samples, the profile reflects the total effect of the vibrational modes of the molecules in the system. The system is a complex mixture of several different components. It is often very difficult to find absorption at a particular wave number for the quantification of a particular component in the system. This is because the particular absorption is affected by the matrix of the system. When standards are made in a particular matrix that is different from the matrix found in the natural system, the quantitative analysis using one variable calibration approach is likely to suffer accuracy problems. Determination of isolated *trans* fatty acids in oils and fats falls into this category. When multivariate calibrations are applied in a particular area of the spectral profiles spanning the absorption of a component one is interested in quantifying, the changes in the profiles are modeled together with the real concentration of the component. This fact was used effectively by Van de Voort et al. (17) in the determination of *cis* and *trans* content of fats and oils. They used an automated transmission cell with calibration standards prepared from pure triglycerides of fatty acids to establish calibration with the PLS technique to predict *cis* and *trans* unsaturated double bonds in edible oils and fats. This technique illustrates the use of the chemometric approach in modeling features of the spectra arising from the different functional groups in the mixtures of the triglyceride standards.

The discussion in the preceding paragraphs clearly reveals that the oils and fats may contain both isolated *trans* and conjugated fatty acid moieties in their triglycerides. There have been attempts to quantify only the *trans* fatty acids using univariate and multivariate calibrations. This may be due to the fact that the concentrations of CLA in oils and fats are lower than the *trans* fatty acids content of the oils and fats. A survey of the literature revealed that there has been no study dealing with the simultaneous determination of both isolated *trans* and CLAs.

This chapter illustrates the simultaneous quantitative determination of *trans* fatty acids and CLAs by multivariate data calibration of IR spectral profiles of oils and fats. Because the PLS calibration technique is the main data-handling technique in this application, a short description of the technique is given below.

Calibrations: A Short Theoretical Description of the PLS Technique

Univariate and Multivariate Calibrations

When spectroscopic measurements are made (for example, in the IR) of a solution containing one component, the concentration of the component can be correlated to the absorption of the radiation at a selected wave number. If the system follows Beer-Lambert's law, then the concentration of the constituent can be determined from the absorbances (at the particular wave number) and concentrations of standard solutions spanning the unknown concentration. The procedure is called univariate calibration.

When a mixture of components is spectroscopically characterized, the absorption at a particular wave number depends on the absorptivities and concentrations of the components in the mixture. Then the determination of the concentrations of the different species in the mixture becomes a multidimensional problem. The problem can be solved by measuring absorbance of the mixture at selected wavelengths that together span the number of components in the mixture. If the mixture contains n components, these will provide n linearly independent equations and the concentration can be determined unambiguously if the absorptivities are known. This type of calibration problem is very sensitive to measurement noise and interference from unknown species. Measurement noise can be reduced and a more accurate solution can be found by measuring absorbance at more than n wave numbers. (On the other hand, if the measurements are made at less than n wave numbers there exists more than one solution to the calibration problem.) The calibration model describing the relation between the spectra and concentrations can be calculated by regression techniques, e.g., least squares techniques. The calibration problem discussed above is a direct-type calibration because the relation between the absorbance and absorptivities is known. In many areas of chemistry, this type of relation is not known. Natural samples such as food, coal, or mineral ores belong to this category. Interferences affecting the spectrum of natural samples cannot be isolated and described analytically. Therefore, an indirect calibration is necessary to estimate the model parameters. Spectral data from natural systems have, in general, a very low sample-to-variable ratio. For example, in the mid-IR, samples are measured over the range 4000–400 cm^{-1} with a 1 cm^{-1} interval. One thus obtains 3601 frequency elements while the number of samples is on the order 10 or 100. This implies strong collinearity between the variables. In this situation, standard regression techniques fail because of the problems involved with inversion of matrices of less than full rank. A solution may be found in calibration techniques that reduce the dimensionality of the variables to a few components. Many procedures were developed for this purpose. These techniques are collectively termed latent variable regression methods (18). Principal component (PC) and PLS regression (PLS) are two of the regression techniques most widely used in chemistry and related fields. Martens (19) compared a number of calibration methods for near-IR data of food samples. He concluded that PLS regression often gives the optimal results. Furthermore, the method has good predictive ability and can detect outliers. The PLS regression technique is called PLSl or PLS2 depending on the number of dependent variables used in the calibration. Thus PLSl is used for calibration with one dependent variable and PLS2 for two or more dependent variables. The validity of the PLS method was proved recently by Manne (20).

Multivariate Calibration by PLS1 Regression

In PLS1, the relation between spectral data and the dependent variable is modeled by data compression and calibration. In the data compression stage, the spectral data are modeled in terms of a set of common latent regression factors (t_1, t_2, $\ldots t_A$)

(scores). These are I dimensional orthogonal column vectors for a data set with I samples and M spectral variables. These factors describe the major variations in the spectral data and at the same time are relevant for predicting the dependent variable. The compression and calibration steps can be written as follows:

$$\mathbf{X} = \mathbf{1}\bar{\mathbf{x}}' + \sum_{1}^{A} \mathbf{t}_a \mathbf{p}_a' + \mathbf{E} \qquad [1]$$

$$\mathbf{y} = \mathbf{1}\bar{\mathrm{y}} + \sum_{1}^{A} \mathbf{t}_a \mathbf{q}_a + \mathbf{f} \qquad [2]$$

where $\bar{\mathbf{x}}$ is a column vector containing the mean of spectral variables and $\bar{\mathrm{y}}$ is the mean of dependent variable; $\boldsymbol{p}_a$ are spectral loadings, which are column vectors with M elements, and $\boldsymbol{q}_a$ is a column vector with A elements of calibration coefficients; **E** and **f** are residual matrices. For mean centered spectral and chemical data, these can be simply written

$$\mathbf{X} = \mathbf{T}\mathbf{P}' + \mathbf{E} \qquad [3]$$

$$\mathbf{y} = \mathbf{T}\mathbf{q} + \mathbf{f} \qquad [4]$$

In the PLS algorithm the vectors **t**, **p** and **q** are calculated starting with an equation relating the spectral data and dependent variable.

$$\mathbf{x} = \mathbf{y}\,\mathbf{w}_1' + \mathbf{E}' \qquad [5]$$

$\mathbf{w}_1$ is the column vector (weightings) and is estimated as follows:

$$\mathbf{w}_1' = \mathbf{y}'\mathbf{X}/||\mathbf{y}'\mathbf{X}|| \qquad [6]$$

$\mathbf{w}_1$ is then normalized to unity. The estimate $\hat{\mathbf{w}}_1$ is then used to calculate the $\mathbf{t}_1$, $\mathbf{p}_1$, and $\mathbf{q}_1$ (20). These estimates are again used to calculate the residuals of spectral and dependent variables using Eqs. 1 and 2. The process is repeated until A factors that minimize the prediction error are found.

The prediction of an unknown sample from the spectral variables x_i proceeds as follows. First, the score $\mathbf{t}_1$ is found by using the estimate for $\mathbf{w}_1$ and Eq. 7.

$$\mathbf{x}_i - \bar{\mathbf{x}} = \mathbf{t}_1\mathbf{w}_1 + \mathbf{e}_1 \qquad [7]$$

Then, the next score $\mathbf{t}_2$ is calculated from the solution of the above equation and the estimate for $\mathbf{w}_2$ using Eq. 8.

$$(\mathbf{x}_i - \bar{\mathbf{x}}) - \mathbf{t}_1\mathbf{p}_1' = \mathbf{t}_2\mathbf{w}_2 + \mathbf{e}_2 \qquad [8]$$

The process is repeated until the Ath factor. The dependent variable is predicted by

$$\hat{\mathbf{y}} = \bar{\mathrm{y}} + \sum_{1}^{A} \mathbf{t}_a\mathbf{q}_a' + \mathbf{f} \qquad [9]$$

Alternatively the same prediction can be written as

$$\mathbf{y} = \bar{\mathbf{y}} + \mathbf{x}'\mathbf{b} \qquad [10]$$

where the calibration coefficients vector **b** is determined by the equation

$$\mathbf{b} = \hat{\mathbf{w}}'(\hat{\mathbf{p}}\ \hat{\mathbf{w}})^{-1}\ \mathbf{q} \qquad [11]$$

Validation of the Calibration Model. To evaluate the prediction ability of the calibration model, the cross-validation technique is used (19). Prediction ability is then tested by predicting samples with a known dependent variable. Samples in the calibration set are grouped into smaller sets and the dependent variables of the samples in these sets are predicted from the calibration models of the remaining samples. Cross-validation of a calibration model makes it possible to select the optimal number of factors, i.e., the number giving the minimum prediction error for the calibration set.

Error of Prediction. The standard error of prediction (SEP) of a cross-validated calibration model is calculated by the following equation (19):

$$\mathrm{SEP} = [(1/\mathrm{I_p}) \sum_{\mathrm{m}=1}^{\mathrm{n_v}} \sum_{\mathrm{i}=1}^{\mathrm{Iv_m}} \mathrm{f}_i^2]^{1/2} \qquad [12]$$

Here, Iv_m is the number of predicted samples in the cross validation group m, n_v is the number of groups used in the cross-validation, and I_p is the total number of prediction samples; f_i^2 is the residual variance of the sample i after A factors.

Outlier Detection. To obtain a good calibration model, one has to remove outlying samples, those that are extreme compared with the others in the calibration set. The outlying property may be due to interferences in the spectral data or to measurement error in the dependent variable. Interferences in spectral data extract additional factors, thus increasing the complexity and reducing the predictive ability of the calibration model. Such samples are outliers in the spectral data. On the other hand, the same samples can be well described by the calibration model but the predicted data are far away from the experimental values. These samples are outliers in the dependent variable.

Data Transfer. Calibration of multivariate data using multivariate data programs requires the data to be presented in a format suitable for digital computation. The first requirement is for communication between the instrument measuring the data and the computer running the application software. In the application described in this paper, data used for the multivariate calibration were transferred from the PE spectrum one in ASCII format for further transformation and multivariate data analysis. The data acquired was imported to an Excel table column and then to a data matrix formed in the SIRIUS software.

Variable Reduction. In multivariate data analysis, the data handling is done by use of computers. Multivariate calibration problems require handling and inverting matrices with many entries, which requires a great deal of computer memory and time. By means of variable reduction, the number of spectral variables in the calibration sample set may be reduced by a factor of 10 without losing much of the information contained in the data. This provides smaller matrices for calibration modeling. However the reduction that can be performed without a significant loss of information is data dependent and the reduction has to be done in a trial-and-error manner. A procedure called maximum entropy (21,22) can be used for this purpose.

Target Rotation. PLS calibration of multivariate data provides latent variables that are orthogonal to each other. None of the PLS components correlate optimally with a given dependent variable used in the calibration. However, an optimally correlated component is defined by the regression vector for each dependent variable (target). By normalizing the regression vector for each target and subsequently projecting the samples onto these normalized regression vectors, score vectors are obtained that are proportional to the predicted target vectors. The score vectors can be used to obtain the covariance graphs between the modeled data and the targets. The procedure is called PLS target rotation (18,23) and provides components that are directly related to the dependent variable. Target rotation extracts correlation information and displays it in a single (two-dimensional) plot called the target projection plot. These plots are easy to interpret compared with single factors from multivariate calibrations. The target rotation procedure is as follows.

The spectral data of the calibration sample sets are decomposed as

$$\mathbf{X} = \hat{\mathbf{X}} + \mathbf{E} \qquad [13]$$

$\hat{X}$ is the modeled spectral data obtained using the PLS regression. The covariance between spectral data and the dependent variable is obtained as

$$(1/n)\sum_i (x_{ip} - \bar{x}_p)(y_i - \bar{y}) = \mathrm{Cov}(x_p, y) \qquad [14]$$

Covariance between modeled spectral data and the dependent variable is given by

$$(1/n)\sum_i(\hat{x}_{ip} - \bar{x}_p)(y_i - \bar{y}) = \mathrm{Cov}(\hat{x}_p, y) \qquad [15]$$

Experimental

Standards, Calibration Samples, and Test Samples

Standards of trielaidin (9*t*-98%) and trienolaidin (9*t*12*t*-98%) were purchased from Sigma. The triglyceride standards of 9*c*11*t*; 10*t*12*c* CLAs and, 1,2-palmitate 3-elaidate were purchased from Larodan Chemicals. The *trans* content of pure trielaidin was taken as 100%, making the *trans* content of trienolaidin 200% and that of 1,2-palmitate 3-elaidate 33.3%. The CLA content of the triglyceride of the 9*c*11*t* isomer was taken as 100%. The CLA content of the 10*t*12*c* isomer is also therefore 100%. Virgin olive oil was purchased from an authorized dealer. Other edible oils were purchased from grocery shops. Fat from beef was extracted by heating portions of meat containing fat in a microwave oven. The melted fat was decanted and put in a small plastic cup and frozen until the spectral measurement was made.

Two sets of calibration mixtures containing trielaidin (9*t*), trienolaidin (9*t*12*t*), and triglycerides of the 9*c*11*t* and 10*t*12*c* CLAs and 1,2-palmitate 3-elaidate were prepared in virgin olive oil by weighing. The first set of the mixtures contained 27 samples in which the concentration of the *trans* content and/or the CLAs was allowed to vary from 0 to 2.5%. The second set contained 30 samples. Here, the concentrations were allowed to vary in the range 0–30%. Trilinolaidin and 1,2-palmitate 3-elaidate were used to introduce variations and model the absorptions arising from different *trans* isomers, and the triglyceride of 10*t*12*c* CLA was used to introduce variations and model the absorptions arising from different CLAs.

Test samples for the prediction of *trans* and CLA were prepared in four different batches. The first batch of samples was prepared by adding 1–2.5% of trielaidin and/or triglyceride of 9*c*11*t* CLA in virgin olive oil to test the model in the range 1–2.5%. The second batch of samples was prepared in the same manner but with the addition of 0–30% trielaidin and/or triglyceride of 9*c*11*t* CLA for a prediction in the high concentration range. The third batch contained one sample extracted from beef fat. The fourth batch of samples was prepared from the beef fat from the third batch. Portions of olive oil containing trielaidin and/or triglyceride of the 9*c*11*t* CLA were added to aliquots of the beef fat from the third batch. The predicted *trans* and CLA contents of the beef fat in the third batch were used to calculate the final *trans* and/or CLA contents of the synthetic samples. The *trans* and CLA contents in all of the samples were calculated as percentages of *trans* fatty acids (9*t*, 18:1) and as percentages of CLA (18:2), respectively.

Spectral Measurements and Data Handling

The IR spectra of the calibration mixtures, test mixtures, beef fat, and the mixtures prepared from the beef fat were measured using a Perkin Elmer spectrum one FTIR spectrometer equipped with a DTGS detector and a SPECAC, zinc selenide, single reflectance attenuated total reflectance (ATR) accessory. The accessory contains a plate with a microconical-shaped trough with a single reflectance crystal attached to the bottom. The crystal has a diameter of ~2.5 mm; a sample volume of 20 μL is sufficient to cover the crystal to a depth of 1 mm. This layer of the sample is more than enough to measure a representative IR spectrum.

A few microliters of each sample mixture were dropped onto the single reflectance crystal using a blunt thin glass rod. A total of 32 scans were collected in the range 1000–850 cm^{-1} at a resolution of 4 cm^{-1}. The reflectance spectrum was then expressed as a ratio vs. the previously scanned background spectrum of the reflectance crystal. The spectra of the test samples were measured in the same way. The test samples that contained high concentrations of *trans* isomers were placed in an oven set at 50°C before the measurements to avoid solidification on the reflectance crystal. The IR spectra of the samples were then measured in the same manner. The crystal was cleaned in between the measurements using dichloromethane and acetone.

The IR spectra of the calibration mixtures, test samples, and beef fat were imported into SIRIUS (24), a multivariate data analysis program, to form a table containing rows of different oils and fats (objects) and columns containing wave numbers (variables). The table was then expanded by two more columns to contain the amounts (percentage by weight) of *trans* and CLA fatty acid contents. This table represents a matrix containing spectral intensities as independent variables and *trans* and CLA fatty acid contents as dependent variables. The spectral part of the matrix then underwent double derivation. Cross-validated PLS calibrations were then performed on data matrix with raw data as independent variables and *trans* and CLA contents as dependent variables. The same was done with the profiles of the double derivated data as independent variables and *trans* and CLA fatty acids contents as depended variables. The calibrations were performed using PLS1 (PLS calibration for one dependent variable) and PLS2 (PLS calibration for two or more dependent variables). A cross-validation procedure (24) was used for validating the calibration models. The test samples were then predicted for their *trans* and CLA contents. The *trans* and CLA contents of the beef fat were then reconfirmed by calculation.

Results and Discussion

The reflectance spectra of triglycerides of pure *trans* fatty acids and pure CLA are shown in Figure 1. The spectra clearly reveal how the IR profiles of these two classes of fatty acids overlap in the region shown in the figure. The raw spectral profile and the second derivative of the reflectance spectrum of one of the mixtures containing triglycerides of *trans* and CLAs in virgin olive oil are shown in Figures 2a and b, respectively. The raw spectral profiles of the mixtures containing *trans* fatty acids and CLAs do not show the features of the CH bending absorption characteristics at low concentrations (<0.3%). The features become visible at relatively higher concentrations. In contrast, the features become evident in the second derivative of the spectral profiles even at lower concentrations. The absorption at 967 cm^{-1} arising from the *trans* CH out-of-plane deformation vibration and the absorptions at 946 and 982 cm^{-1} arising from the CH out-of-plane deformation vibrations in the CLA are clear in Figure 2b compared with Figure 2a.

The PLS1 and PLS2 algorithms were explored in establishing calibration models. The PLS1 calibrates the spectral data with one dependent variable at a time and PLS2 calibrates the spectral data with two or more dependent variables. In this application, we found that PLS1 gave lower prediction errors than PLS2 for *trans* and CLA fatty acids. In the case of a low concentration range, the cross-validated PLS calibration model for the *trans* content explained 45% of the total variance in the second derivative profiles (independent variables) and 98% variance in the *trans* (dependent variable) content. The variances explained are ~47% in the second derivative profiles and 96% in CLA content. This clearly indicates that only the spectral parts that are relevant for the explanation of the variance in the dependent variables are extracted by the PLS calibrations. The profiles arising from the presence of CLA are irrelevant for the prediction of *trans* content; the profiles arising from the presence of *trans* fatty acids are irrelevant in the prediction of CLA content. Furthermore, this indicates that the spectral profile of one component is not affected by the other component when the concentrations are low. The variances explained in the independent and dependent variables are high in the high concentration ranges. It appears that when the concentrations of these components are high in the mixture, the spectral profile of one component is affected by the spectral profile of the other component.

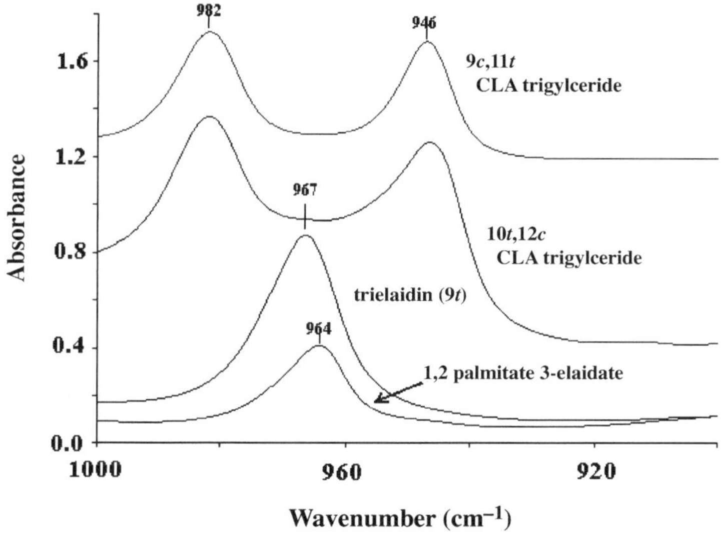

Fig. 1. Infrared spectra of triglycerides of certain *trans* fatty acids in the region 1000–850 cm^{-1} showing their CH out-of-plane deformation absorptions.

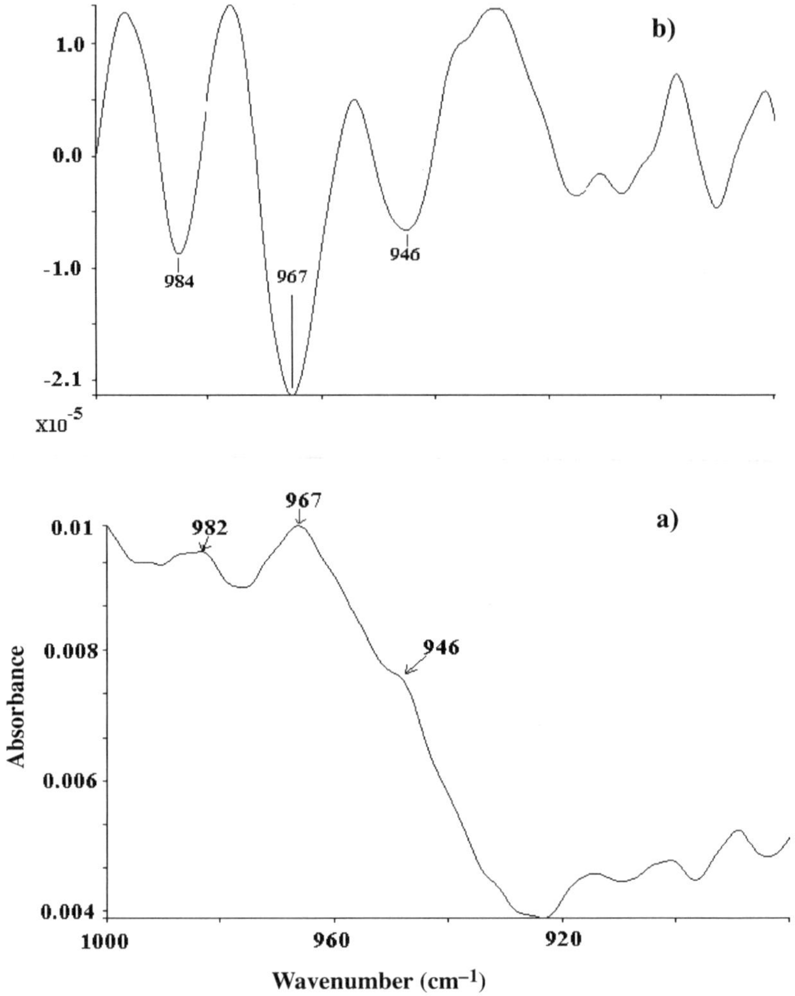

Fig. 2. (A) Raw and (B) second derivative profiles of an infrared spectrum of a mixture of triglyceride of the 9*c*11*t* conjugated linoleic acid isomer (acid concentration is 0.3% by weight) and triglyceride of the 9*t* 18:1 fatty acid (acid concentration is 2.0% by weight) in the region 1000–850 cm^{-1}.

The correlations between measured and predicted values of the dependent variables modeled by the second derivative profiles for the low concentration range are shown in Figures 3a and b. The cross-validated calibration models predict the concentrations of *trans* fatty acids and CLA within an error limit of 0.1%. Similar plots are shown for the high concentration range in Figures 4a and b. The errors of prediction are 0.9% for *trans* content and 0.5% for CLA content.

Infrared measurements are generally affected by baseline variations. Irrespective of the sampling technique, the baseline shifts and variations in background absorptions in the spectral profiles affect quantitative determinations. The picture changes in most of the cases in which the spectral profiles undergo double derivation. The process not only removes the baseline effect from the spectral profiles but also identifies some hidden absorptions in the broad IR spectral profiles (25). The double derivation of the spectral profiles may be necessary when the concentrations are low. A similar comparison was made with the spectral profiles obtained for the *trans* and CLA content in the high concentration range. The predictions and prediction errors were similar for calibrations with raw spectral profiles and spectral profiles after double derivation. However, we preferred to establish the calibration models using spectral profiles subjected to double derivation to avoid day-to-day variations in the spectral baselines.

The loading plots for the calibration models are shown in Figures 5 and 6. Figure 5 shows the loading plots for the calibration models established in the low concentration range of the dependent variables. In the case of *trans* content determination, the variable at 967 cm^{-1} has a negative loading. The second derivative profiles normally give a valley at the peak position in the raw spectral profiles. In the case of CLA determination, the variables at 946 and 981 cm^{-1} have negative loadings. These are in agreement with the spectral profiles that correlate with the variations in *trans* and CLA contents. The loading plots for the calibrations with raw spectral profiles in the high concentration range of *trans* and CLA are shown in Figures 6a and b. The *trans* content model has a positive loading at 967 cm^{-1}, and the CLA content model has positive loadings at 946 and 981 cm^{-1}, which are exactly in the same regions as the absorptions of the CH deformation vibrations of the CLA.

The added and predicted *trans* and CLA content in olive oil and beef fat are shown in Tables 1 and 2. The samples prepared in the 4th batch clearly show that the *trans* content exceeds 2.5%, which is not covered by the low concentration calibration model. At the same time, CLA contents in the samples are in the low concentration range. The *trans* contents of the samples were predicted by the high concentration calibration model. The CLA contents of the samples were predicted by the low concentration calibration model. The accuracy in the analytical results determined confirms that the *trans* and CLA concentrations predicted for beef in the 3rd batch was correct. The prediction values again show that the models can predict the concentrations of *trans* and CLA content with reasonable accuracy. The *trans* and CLA contents of beef fat are in the range reported for this fat in the literature (26–28).

The *trans* and CLA contents of natural samples are relatively low and the concentration range was reported in the literature. When applying calibration models, it is always rewarding to prepare calibration samples with concentrations spanning the concentration range. Furthermore, preparing samples with close variations in *trans* and CLA contents can further reduce prediction errors. Oils that are free from *trans* and CLA can be included to model variations of the spectral profiles in the region 1000–850 cm^{-1}.

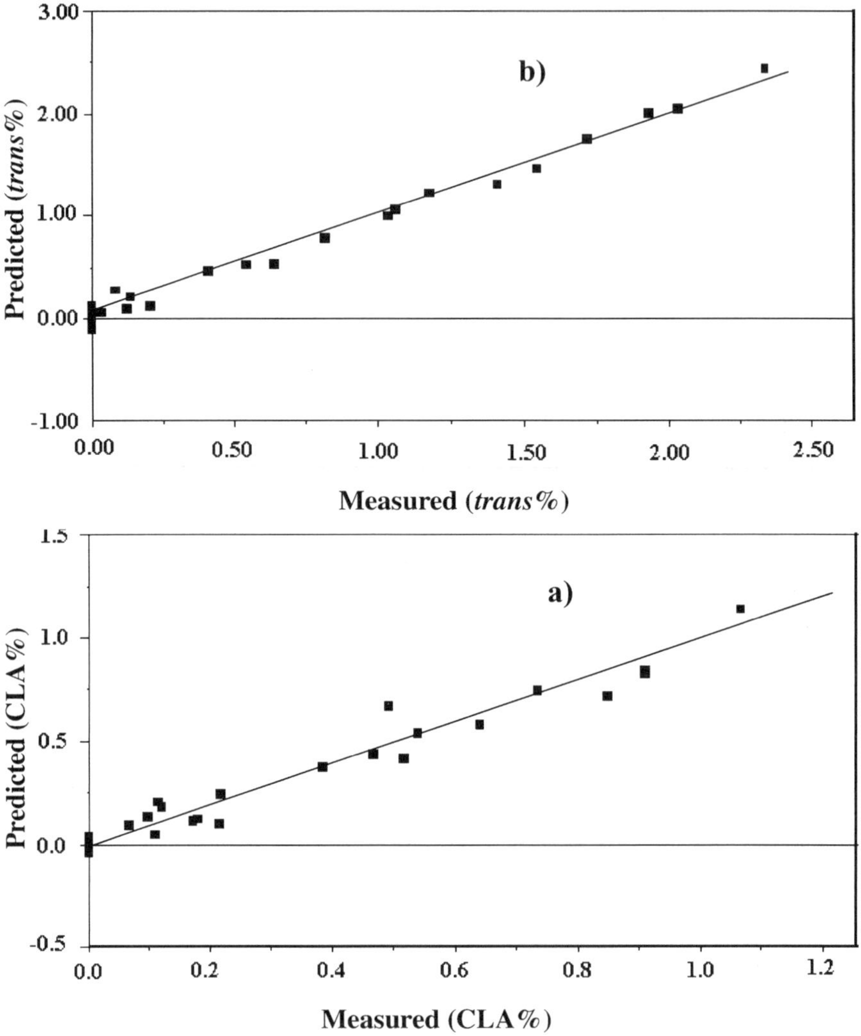

Fig. 3. Calibration plots showing the correlations between (A) measured and predicted conjugated linoleic acid content and (B) measured and predicted *trans* fatty acid content for the mixtures in the low concentration series.

One of the requirements in using spectroscopic data in calibration models is that the measurements should be reproducible. Difficulty in reproducing the spectral profile of a sample affects the prediction of a quantity in the sample using PLS calibration. Even in the internal reflectance technique with one reflection, the spectral profiles can vary with time and small movements of the sampling accessory. This problem can be avoided by measuring the background spectrum before the spectral

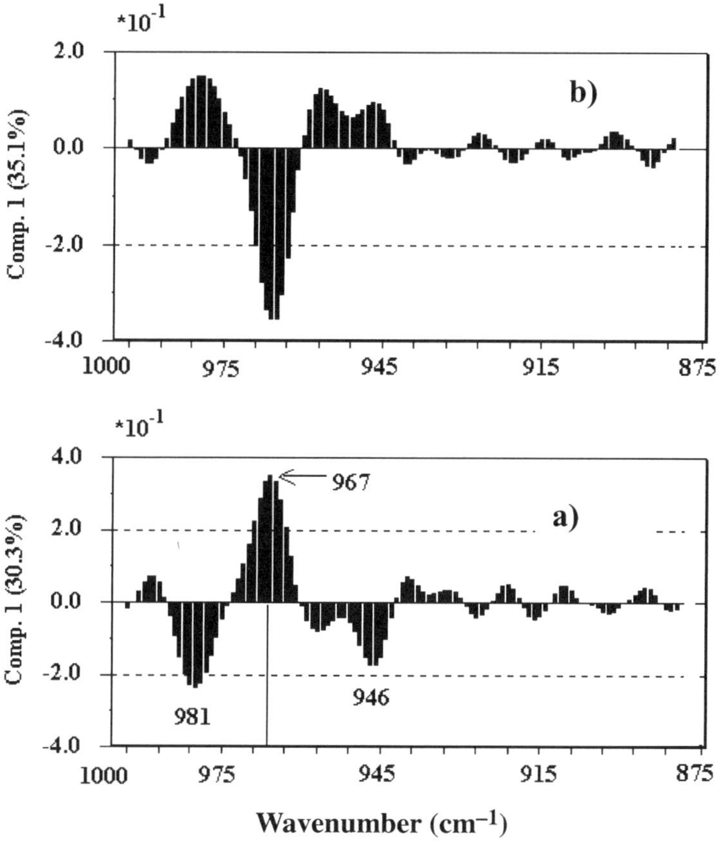

Fig. 4. Calibration plots showing the correlations between (A) measured and predicted conjugated linoleic acid content and (B) measured and predicted *trans* fatty acid content for the mixtures in the high concentration series.

measurement of each sample and by using data with a double derivative instead of raw spectral profiles.

The ATR accessory used in the experiments provides a simple way to measure IR spectra of the samples. As mentioned earlier, a few microliters of a sample mixture are sufficient to acquire the IR spectrum. Other ATR accessories using several reflections require quite large quantities of the standards of the *trans* and CLA fatty acids for IR measurements in the range 0–30%. However, an IR spectrum measured using several reflections techniques provides relatively noise-free spectral profiles compared with an IR spectrum measured using the single reflection technique. However, the volume of sample needed for IR measurement is several hundred times larger compared with the volume required for IR measurements with one reflection accessory. The noise in the spectrum can also be reduced using a sensitive

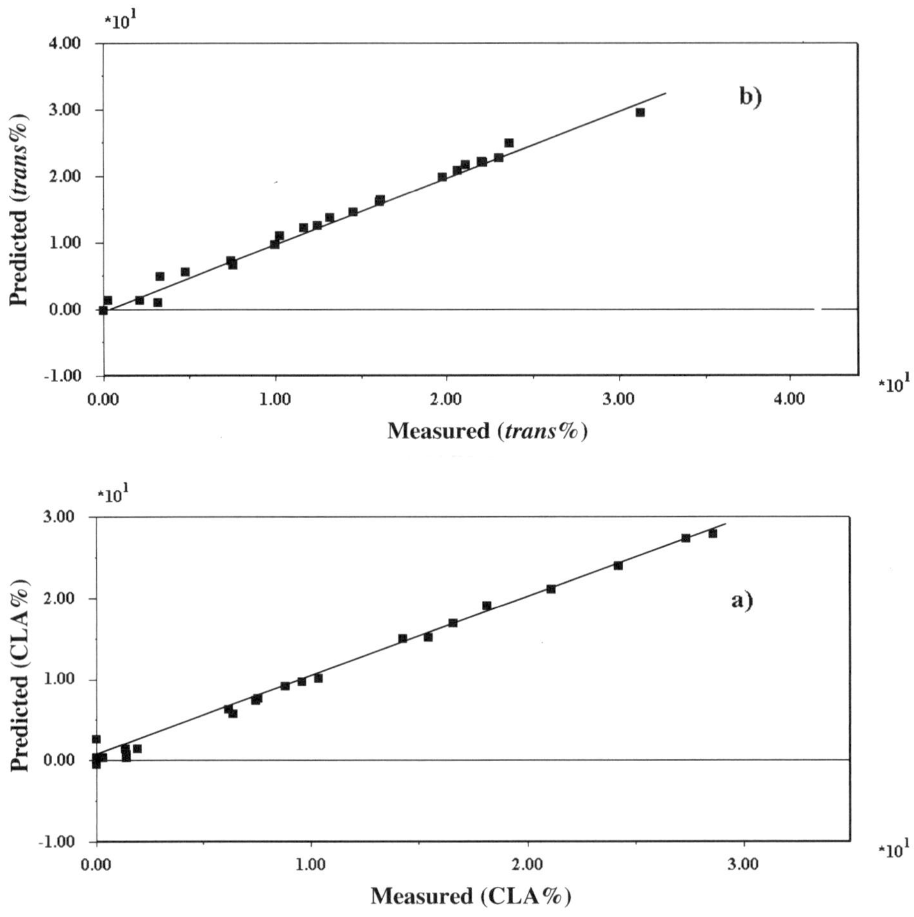

Fig. 5. Loading plots for the calibration models established for the mixtures in the low concentration series with second derivative profiles. (A) Loading plot for the calibration model with conjugated linoleic acid concentration as the dependent variable. (B) Loading plot for the calibration model with trans fatty acid concentration as the dependent variable.

HgCdTe detector compared with a deuterated triglycine sulfate detector. The noise in the measurement data affects the accuracy in prediction.

Conclusion

We showed in this chapter that the total *trans* and CLA contents of oils and fats can be determined simultaneously by PLS calibration. The approach eliminates the steps

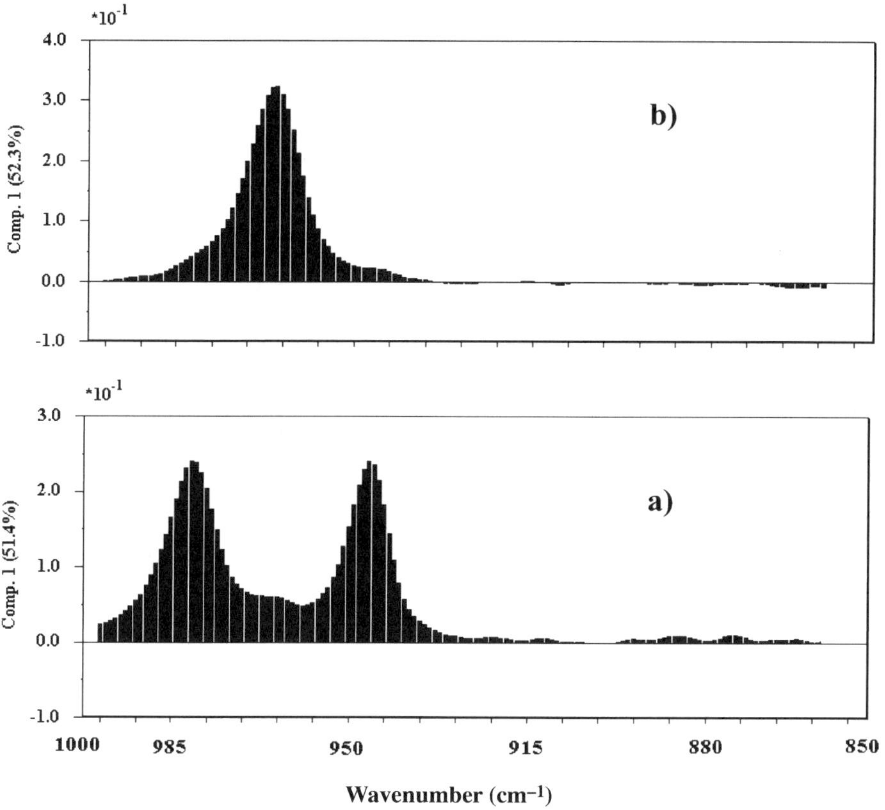

Fig. 6. Loading plots for the calibration models established for the mixtures in the high concentration series with raw profiles. (A) Loading plot for the calibration model with conjugated linoleic acid concentration as the dependent variable. (B) Loading plot for the calibration model with *trans* fatty acid concentration as the dependent variable.

needed in wet chemical methods and provides quantitative values to the isolated *trans* and CLA contents of edible oils and fats.

The method provides the total *trans* and CLA content in a sample. The breakdown of *trans* content into its isomer components or CLA content into its isomer components is not possible using this approach. A wet chemical method in combination with chromatography is still the method of choice. The variations allowed for the *trans* and CLA content in this work can cover the concentrations of these fatty acids in natural and laboratory prepared samples.

The results predicted for the samples in the 4th batch showed that the *trans* and CLA contents of the fat sample were predicted with reasonable accuracy (see Table 2). This increases confidence in the predicting abilities of the models.

TABLE 1
Added and Predicted Fatty Acids in Olive Oil, and Predicted *trans* and Conjugated Linoleic Acid (CLA) Content in Beef Fat

Sample	Added *trans*	Predicted *trans* (%)	Added CLA	Predicted CLA
1	1.7	1.7	2.0	1.9
2	1.5	1.5	1.6	1.7
3	2.8	2.6	2.6	2.4
4	2.3	2.2	1.7	1.5
5	18.3	18.5	13.8	13.1
6	9.6	9.0	12.9	13.0
7	12.8	12.3	20.7	20.1
8	6.7	6.3	3.5	3.3
Animal fats				
Beef		4.3		0.2

TABLE 2
Calculated and Predicted *trans* and Conjugated Linoleic Acid (CLA) Content in Animal Fats (4th Batch of Samples)

Sample Beef	Calculate *trans* %	Predicted *trans* % 4.3 (from batch 3)	Calculated CLA%	Predicted CLA% 0.2 (from batch 3)
A	2.1	2.2	2.0	2.1
B	3.1	3.4	1.3	1.2
C	2.7	2.7	1.0	1.0
D	3.1	3.0	1.2	1.0
E	2.7	3.2	1.4	1.6

The results presented here indicate that the determination of the concentrations of *trans* and CLA contents by multivariate calibration and prediction is a better way of quantification than the method based on the evaluation of the peak at 967 cm^{-1}.

Acknowledgments

Elsevier Science B.V., Amsterdam, the Netherlands is thanked for their kind permission to reproduce text and figures from *Vibrational Spectroscopy, 33*, 2003, 37–48.

References

1. Mossoba, M.M., M.P. Yurawecz, and R.E McDonald, Rapid Determination of Total *trans* Content of Neat Hydrogenated Oils by Attenuated Total Reflection Spectroscopy, *J. Assoc. Off. Anal. Chem. 73*:1003–1009 (1996).
2. Mensink, R.P., and M.B. Katan, Effect of Dietary *trans* Fatty Acids on High-Density Lipoprotein Cholesterol Levels in Healthy Subjects, *N. Engl. J. Med. 323*:439–445 (1990).

3. Willet, W.C., M.J. Meir, J.E. Manson, G.A. Golditz, F.E. Speizer, B.A. Rosner, L.A. Sampson, and C.H. Hennekens, Intakes of *trans* Fatty Acids and Risk of Coronary Heart Disease Among Women, *Lancet 341*:581–585 (1993).
4. Judd, J.T., B.A. Clevidence, R.E. Museing, J. Witts, M.E. Sunkin, and J.J. Podczasy, Dietary *trans* Fatty Acids: Effects of Plasma Lipids and Lipoproteins of Healthy Men and Women, *Am. J. Clin. Nutr. 59*:861–865 (1994).
5. Ip, C., S.F. Chin, J.A. Scimeca, and M.W. Pariza, Mammary Cancer Prevention by Conjugated Dienoic Derivatives of Linoleic Acid, *Cancer Res. 51*:6118–6124 (1991).
6. Ip, C., S.F. Chin, J.A. Scimeca, and H.J. Thompson, CLA, A Powerful Anticarcinogen from Animal Fat Sources, *Cancer 74* (3 Suppl.):1050–1054 (1994).
7. Nicolosi, R.J., K.V. Courtemanche, L. Laitinen, J.A. Scimeca, and P.J. Huth, Effect of Feeding Diets Enriched in CLA on Lipoproteins and Aortic Atherogenesis in Hamsters, Circulation 88 (Suppl.):2458 (1993).
8. Shanta, N.C., E.A. Decker, and Z. Ustunol, CLAs Concentration in Processed Cheese, *J. Am. Oil Chem. Soc. 69*:425–428 (1992).
9. Lavillonniere, F., J.C. Martin, P. Bougnoux, and J.L. Sébédio, Analysis of CLA Isomers and Content in French Cheeses, *J. Assoc. Off. Anal. Chem. 75*:343–352 (1998).
10. McGuire, M.K., Y. Park, R.A. Behre, L.Y. Harrison, T.D. Shultz, and M.A. McGuire, CLA Concentrations of Human Milk and Infant Formula, *Nutr. Res. 17*:1277–1283 (1997).
11. *Official Methods and Recommended Practices of the American Oil Chemists' Society*, 4th ed., American Oil Chemists' Society, Campaign, IL, 1989.
12. Firestone, D., and P. LaBouliere, Determination of Isolated *trans* Isomers by Infrared Spectrophotometry, *J. Assoc. Off. Anal. Chem. 48*:437–443 (1965).
13. Belton, P.S., R.H. Wilson, H. Sadehgi-Jorabegi, and K.E. Peers, A Rapid Method for the Estimation of Isolated *trans* Double Bonds in Oils and Fats Using FTIR Combined with ATR, *Lebensm.-Wiss. Technol. 21*:153–157 (1988).
14. Dutten, H.J., Analysis and Monitoring of *trans*-Isomerization by IR ATR Spectrometry, *J. Am. Oil. Chem. Soc. 51*:406–409 (1974).
15. Lancer, A.C., and E.A. Emkem, Comparison of FTIR and Capillary GC Methods for Quantitation of *trans* Unsaturation in Fatty Acids Methyl Esters, *J. Am. Oil Chem. Soc. 65*:1483–1487 (1988).
16. Ulberth, F., and H-J. Haider, Determination of Low Level *trans* Unsaturation in Fats by Fourier Transform Infrared Spectroscopy, *J. Food Sci. 57*:1444–1447 (1992).
17. Van de Voort, F.R., A.A. Ismail, and J. Sedman, A Rapid, Automated Method for the Determination of *cis* and *trans* Content of Fats and Oils by Fourier Transform Infrared Spectroscopy, *J. Am. Oil. Chem. Soc. 72*:873–880 (1995).
18. Kvalheim, O.M., and T.V. Karstang, Interpretation of Latent-Variable Regression Models, *Chemom. Intell. Lab. Syst. 7*: 39–51 (1989).
19. Martens, H.A., Multivariate Calibration, Ph.D. Thesis, Technical University of Norway, Trondheim, 1985.
20. Manne, R., Analysis of Two Partial-Least-Squares Algorithm for Multivariate Calibration, *Chemom. Intell. Lab. Sys. 2*:187–197 (1987).
21. Full, W.E., T. Ehrlich, and S.K. Kennedy, Optimal Configuration and Information Content of Sets of Frequency Distribution, *J. Sedim. Petr. 54*:117–126 (1984).
22. Karstang, T.V., and R.J. Eastgate, Multivariate Calibration of X-Ray Diffractometer by Partial-Least-Squares Regression, *Chemom. Intell. Lab. Syst. 2*:209–219 (1987).

23. Kvalheim, O.M, Interpretation of Multivariate Data: Examples from Petroleum Geochemistry. Ph.D. Thesis, University of Bergen, Norway, 1987.
24. Kvalheim, O.M, A Partial Least Square Approach to Interpretative Analysis of Multivariate Data, *Chemom. Intell. Lab. Syst. 3*:189–197 (1988).
25. Christy, A.A., I.A. Darout, and N. Skaug, Quantitative Analysis in Diffuse Reflectance Infrared Spectrometry: Thiocyanate Levels in Miswak Aqueous Extracts, *Trends Appl. Spectrosc. 3*:25–33 (2001).
26. Wolff, R.L., Content and Distribution of *trans*-18:1 Acids in Ruminant Milk and Meat Fats. Their Importance in European Diets and Their Effect on Human Milk, *J. Am. Oil Chem. Soc. 72*:259–272 (1995).
27. Gunstone, F., CLA (CLA), Guest Contributions, Britannia Food Ingredients, Ltd. http://www.britanniafood.com/english/index.htm.
28. Chin, S.F., W. Liu, J.M. Storkson, Y.L. Ha, and M.W. Pariza, Dietary Sources of Conjugated Dienoic Isomers of Linoleic Acid, a Newly Recognized Class of Anticarcinogens, *J. Food Compos. Anal. 5*:185–197 (1992).

16 Investigation of Protein-Lipid Interactions by Vibrational Spectroscopy

Guangtao Meng[a], Nazlin K. Howell[b], and Eunice C.Y. Li-Chan[a]

[a]Food, Nutrition and Health Program, The University of British Columbia, Food, Nutrition and Health Building, Vancouver, BC, Canada V6T 1Z4 and [b]School of Biomedical and Molecular Sciences, University of Surrey, Guildford, Surrey, UK, GU2 7XH

Introduction

Proteins and lipids are two major components in cells that are of both plant and animal origin; the interactions between them play an important role in the organization of a large number of biological structures in living cells and tissues. Naturally occurring protein-lipid complexes were found to have various functions in cellular membranes and in the transport and metabolism of lipids (1). Some examples of protein-lipid complexes of animal origin are found in the cell membranes and lipoproteins of plasma (2,3), milk fat globule membrane (4), and egg yolk (5–7). Examples of plant origin include wheat grains (8,9), peanuts (10), and oil bodies of oil seeds such as soybean (11,12), sunflower seed, and rapeseed, as well as some fruits and vegetables (13,14).

Protein-lipid interactions also occur in processed and stored food products, causing the formation of so-called "induced protein-lipid complexes." An example is the interaction between fatty acid peroxides in lipids with the muscle proteins of some foods such as fish, meat, and meat products during processing and storage (15,16). The interaction of fish protein (e.g., carp myofibrils) with lipids (fish oil) during freeze-drying and storage was investigated by Kunimoto et al. (17), and the interaction of myoglobin in meat products with oxidized lipids during storage was reported by Kanner and Karel (18). Protein-lipid complexes also occur in peanut products, particularly during storage of peanut seeds and peanut flour when lipid oxidation produces various reactive groups that interact with proteins (10).

Protein-lipid complexes may be formed during food processing operations such as homogenization and mixing of milk and milk products. In the processing of ice cream, lipids interact with caseins to form protein-lipid complexes, whereas in butter-making, fat globules concentrate at an air-water interface and interact with proteins to form complexes at the interface (19,20). Several studies on dough and the bread gluten network confirmed the presence of induced protein-lipid complexes, in which the polar ends of lipids are bound to gliadin by hydrophilic interactions, whereas the nonpolar ends are bound to glutenin by hydrophobic interactions; the presence of these protein-lipid associations cements the gluten network and contributes to the structure of the gas-retaining complexes that are essential for good gas retention, adequate loaf volume, and satisfactory bread structure (9,21). Soy films and soymilk are examples of complexes of lipids with the major soybean globular proteins, the 7S and 11S globulins. Microscopic examination of soy films revealed a structure consisting of a continuous protein matrix in which lipid droplets are dispersed (11,12).

Although their structural organization may differ, the physicochemical characteristics of protein-lipid complexes that form as a result of processes such as heating, mixing, shearing and storage, are quite similar to those of protein-lipid complexes that occur in living systems (9). The types of bonding in protein-lipid interactions include covalent bonds as well as non-covalent bonds involving hydrogen bonding, electrostatic, hydrophobic, and van der Waals forces. The interactions may vary depending on pH, ionic strength, and temperature, and usually more than one type of bonding is involved, such that the complexes cannot be readily dissociated by simple manipulations of pH, ionic strength, or ultracentrifugal force fields (22).

Various methods have been applied to investigate the nature of protein-lipid interactions. These methods include electron microscopy (19), nuclear magnetic resonance (NMR) (23–25), fluorescence probe (26, 27), and electron spin resonance (ESR) (25,28,29) spectroscopy. Because of experimental difficulties in characterizing the spatial structure and mode of membrane binding, molecular modeling techniques and computer simulation methods with which to study these designed models are becoming essential complementary tools. Several models were proposed to represent the structural organization of protein-lipid complexes in biological systems (30).

Vibrational spectroscopy, which consists of two physically different yet conceptually complementary methods, infrared (IR) and Raman spectroscopies, is also being applied increasingly to the study of protein-lipid interactions. Most molecules have vibrational modes with frequencies that lie in the mid-IR spectral range between 3300 and 650 cm^{-1}. Variations in the positions, widths, and strengths of these modes with composition and structure allow molecular species to be identified uniquely, including the functional groups of interest in food and biological systems.

Compared with other spectroscopic techniques, vibrational spectroscopy offers certain advantages, which have gained it a firm place within the spectroscopic arsenal used to investigate protein-lipid interactions in foods and biological materials. First, the experimental accessibility to a large number of IR and Raman active transitions that originate from specific functional groups/moieties of proteins or lipids can provide information on their interactions. Second, vibrational spectroscopic data are obtained in a noninvasive manner from intrinsic "molecular probes"; that is, neither IR nor Raman spectroscopy requires extrinsic probe molecules such as spin labels or fluorescent probes. Third, these techniques are not limited by molecular size, allowing the study of high-molecular-weight complexes that are common in foods and other biomaterials. Fourth, the molecular events that are monitored by the vibrational spectroscopic experiment are atomic motions on a picosecond time scale, and thus provide an instantaneous "snapshot" of all molecular conformations

(31). Finally, the technique is versatile with respect to the state of the sample; it is applicable, for example, to solid, liquid, or emulsion systems.

The objective of this chapter is to provide an overview of the analytical technique based on vibrational spectroscopy to the reader who is interested in its potential for investigating protein-lipid interactions. The principles, history, and development of IR and Raman spectroscopy will be described briefly. The application of these two branches of vibrational spectroscopy will be illustrated through selected examples that are relevant to the investigation of protein-lipid interactions in food systems.

Infrared Spectroscopy

Principles

IR spectroscopy is based on the absorption of radiation corresponding to the IR frequency range of the electromagnetic spectrum. The intensity of IR absorption, which is due to vibrations of molecules and atoms within molecular segments, is governed by the Beer-Lambert law:

$$I = I_0 e^{-\varepsilon c d}$$

where I_0 and I denote the intensities of the incident and transmitted beams, respectively; ε is the molecular absorptivity coefficient, and c and d are the concentration of the sample and the cell path length, respectively. IR spectra are customarily presented as the percentage transmission (T) or the absorbance (A) plotted vs. the frequency in wavenumber (cm^{-1}).

History and Development of IR Spectroscopy

IR spectroscopy has roots dating to 200 years ago, when William Herschel discovered the infrared region of the electromagnetic spectrum in 1800 (31). It is one of the earliest experimental methods recognized to be potentially useful in providing information on structural features of chemical and biological molecules and the complexes induced by interactions between them. For example, in 1950, it was demonstrated that a strong correlation existed between the position of certain bands in IR spectra of homopolypeptides and their secondary structure (32). However, before Fourier transform infrared (FTIR) spectroscopy, IR spectra were measured with a dispersive instrument utilizing a moving slit in front of the detector; alternatively, moving prisms or gratings were applied to a stationary slit to disperse radiation from the source into spectrometer elements. Because the slits excluded ~95% of the initial radiation, both the sensitivity and resolution of the early techniques based on dispersive IR were poor.

In recent years, the Fourier transform method was applied to IR spectroscopy in the mid-IR region. The principle of interferometry used in FTIR involves splitting radiation from the source into two optical paths. The radiation from each path is reflected by mirrors and returned along the same path to recombine constructively or destructively, depending on the phase difference of the two optical paths or the distance from the beam splitter. FTIR offers a fast and powerful tool for improving vibrational spectroscopy analysis. Aside from extensive computing and data manipulation capabilities, FTIR also has a number of other advantages, i.e., improved signal-to-noise ratio, multiplexing capabilities, a significant reduction in scan times, higher energy throughput, and superior wavelength accuracy compared with dispersive-based instrumentation (33).

Various sampling methods are applicable to IR spectroscopy. Among the IR window materials available for transmission experiments in aqueous samples, calcium fluoride (CaF_2) is most commonly used. Barium fluoride (BaF_2) has a lower spectral cut-off point (at ~800 cm^{-1}) compared with CaF_2, thus enabling additional IR spectral bands to be observed, but it is significantly more soluble in an aqueous solution. Materials insoluble in water (such as ZnSe, AgCl, KRS-5 or Irtran™) are available, but are characterized by a high refractive index, which results in major reflection losses and persistent interference fringes in the spectra. However, antireflection-coated windows are available that partially solve this problem (34).

Another widely used sampling technique is based on attenuated total reflection (ATR) as an alternative to transmission. For ATR measurements, the sample is prepared on the surface of a trapezoidal-shaped IR-transparent crystal. The IR beam is guided through the crystal in such a way that some reflections take place at the surface. Because the IR beam provides an evanescent wave entering the medium of lower refractive index (i.e., the region containing water, buffer and other molecules), the deposition of IR-absorbing matter on the crystal surface causes the IR light to be partially absorbed. The penetration depth of the IR radiation in this arrangement is strictly dependent upon the wavelength and may be up to a few micrometers. The IR spectrum thus measured contains only information on a very thin layer of the sample that is in close proximity to the surface of the crystal. This allows the spectrum of a substance in a water solution to be obtained relatively easily, without much interference from IR absorption of the bulk water.

The sensitivity of FTIR spectroscopy allows a microscope to be coupled to the interferometer. The obligatory mirror optics of the microscope, which allows access to the full IR spectral range, are typically coupled to a second optical system with visible light. Objects can thus be visualized as with a conventional microscope, and spectra can be taken of the selected spots. Light throughput is consequently small compared with the normal FT-IR instruments. Nevertheless, the reduced detector size (a consequence of the small sample spot) allows high-quality spectra to be recorded.

Applications of IR Spectroscopy to Study Protein-Lipid Interactions

IR spectroscopy has long been recognized as a useful method for providing information on structural features of peptides and proteins. Detailed analyses of the structure-sensitive amide bands were performed to establish a correlation between the frequencies of these bands and the type of secondary structure, such as purely α-helical or β-sheet structures (35,36). Nine such IR bands exist; they are termed amide A, amide B, and amides I–VII, in order of decreasing frequency (Table 1) (36,37). Of all the amide bands, the amide I band occurring in the region of 1600–1700 cm^{-1}, which represents primarily the C=O stretching vibration of the amide groups, was found to be the most intense and most useful for the analysis of the secondary structure of proteins (Table 2) (36,38,39).

An early application of IR spectroscopy in studying lipids was the determination of fat and moisture in dairy products. The fat content is usually estimated from the strong absorption bands of the carbonyl group at 1745 cm^{-1} (νC=O) and of the CH_2 group at 2870 cm^{-1} (νC-H) (40). IR spectroscopy was also used for quantitation of the degree of unsaturation of fatty acids (41), determination of fat oxidation

TABLE 1
Characteristic Infrared Bands of the Peptide Bond[a]

Wavenumber (cm^{-1})	Vibration	Nomenclature (amide)
~3300	N-H stretching in resonance with 1st amide II overtone	A
~3100	N-H stretching in resonance with 1st amide II overtone	B
1610–1695	C=O stretching	I
1480–1575	N-H bending and C-N stretching	II
1220–1320	C-N stretching and N-H bending	III
625–765	O-C-N bending, mixed with other modes	IV
640–800	Out-of-plane N-H bending	V
535–605	Out-of-plane C=O bending	VI
~200	Skeletal torsion	VII

[a]*Source:* Refs. 36,37.

TABLE 2
Characteristic Infrared Frequencies of Typical Secondary Structure of Proteins in Amide I Region in H_2O and D_2O Environments[a]

Wavenumber (cm^{-1})		
H_2O	D_2O	Secondary structure
1695–1675	1680–1670	β-Sheet
1690–1650	1690–1650	γ-Turns
1685–1655	1675–1640	β-Turns
1670–1660	1670–1660	3_{10}-Helix
1666–1658	1658–1652	α_{II}-Helix
1660–1652	1648–1640	Irregular structures
1658–1650	1655–1646	α_I-Helix
1638–1632	1636–1630	β-Sheet
1625–1615	1625–1615	Intermolecular β-sheet

[a]*Source:* Refs. 36,38,39.

and fat crystallinity, or the identification of hydroxyl groups (42,43). Using FTIR spectroscopy with ATR cells, Safar et al. (44) studied the character of edible oils, butters, and margarines, using standard fatty acids to assign the wavenumber positions for typical functional groups in the lipids (Table 3).

Protein-lipid interactions were studied by IR spectroscopy based on the distinct vibrational spectral characteristics of protein vs. lipid molecules. For example, FTIR was used for investigating the conformation of proteins adsorbed to the oil-water interface in β-lactoglobulin-stabilized emulsions (45). The emulsifying properties of β-lactoglobulin depend greatly on the temperature and pH, probably because of changes in its structure under such circumstances. A study using differential scanning calorimetry (DSC) of β-lactoglobulin adsorbed to the oil-water interface in an emulsion (46) showed no denaturation transition for the adsorbed protein, indicating the loss of the secondary structure of β-lactoglobulin during or after adsorption. Other investigations demonstrated that whey proteins adsorbed to an oil-water interface are much more easily digested by proteinases than the native protein in solution, suggesting that a conformational change occurs in the protein due to protein-lipid interaction (47,48). This protein denaturation was also confirmed by the slow change of surface viscosity and the time-dependent polymerization of the adsorbed β-lactoglobulin on the oil-water interface (49,50).

In research reported by Fang and Dalgleish (45), FTIR spectroscopy was used to describe the structural changes of β-lactoglobulin during heat treatment at different pH values both in solution and at oil-water interfaces. This study used the second-derivative spectra of the FTIR absorption in the amide I region, which has the advantage of revealing the component bands in considerable detail (51,52). The peak intensity of the second derivative is proportional to the original peak intensity,

TABLE 3
Characteristic Infrared Bands of Edible Oils, Butter, and Margarines[a]

Wavenumber (cm^{-1})	Vibration	Origin
3450	O-H stretching	Intermolecular bonded (water)
3005	C-H symmetrical stretching	(*cis* olefin) -CH=CH-
2953	C-H asymmetric stretching	Aliphatic ($-CH_3$)
2922	C-H asymmetric stretching	Aliphatic ($-CH_2$)
2853	C-H symmetric stretching	Aliphatic ($-CH_2$)
1743	C=O stretching	(C=O) ester
1640	O-H deformation	(O-H) water
1462	C-H scissoring	Aliphatic ($-CH_2$)
1377	C-H symmetric deformation	Aliphatic ($-CH_3$)
1238	C-H out-of-plane bending	Aliphatic ($-CH_2$)
1162	C-O stretching	(C-O) ester
1025	C-O-C stretching	(C-O-C) ester
966	C-H out-of-plane bending	*trans* (-CH=CH-)
722	C-H rocking	Aliphatic ($-CH_2$)

[a]*Source:* Ref. 44.

which makes it suitable for semiquantitative analysis. The results showed that denaturation of β-lactoglobulin during heat treatment and upon adsorption to an oil-water interface appeared to occur via similar intermediate structures, which began with the loss of β-sheet structure. However, although heat denaturation generated more intermolecular β-sheet and a small amount of unordered structure, adsorption of the protein to the oil-water interface induced a larger amount of unordered structure and a small amount of intermolecular β-sheet. The denaturation of β-lactoglobulin on the interface was a much slower process than heat denaturation; even though some changes were detectable shortly after the adsorption of the protein, more extensive denaturation occurred during storage of the emulsions for 72 h.

The effect of temperature on soy lecithin–stabilized emulsions was studied using FTIR (53). Oil-in-water (o/w) 4% (wt/vol) soy lecithin emulsions were prepared with 6% (vol/vol) medium-chain triglycerides and 94% (vol/vol) water using a 2-stage homogenizer set at a pressure of 3000 psig. Two granular deoiled products from soybean lecithin, Lecigran and Lecimulthin, were used as emulsifiers. Emulsions made with Lecigran and Lecimulthin were compared with a control emulsion, with no emulsifier added. After preparation, the emulsions were cooled to 4°C, held at this temperature for 1 h, and the spectra were then collected. The emulsions and reference water were then raised to 22, 37, 52, 67 and 82°C, held at each temperature for 1 h, and the spectra were collected. The 4 regions used for more detailed investigation in the spectra of the emulsions were those contributing to -OH vibration (3600–3200 cm^{-1}), $-CH_2$ stretching (3000–2800 cm^{-1}), H-O-H bending vibrations (3000–4000 cm^{-1}), and P=O, C-O-C, and P-O-C vibrations (1200–1050 cm^{-1}). The control emulsion was greatly affected by temperatures other than room temperature. This was due to the lack of an emulsifier, resulting in a destabilization of the emulsion at high temperatures. The control emulsion spectrum had the highest peak intensities for the -OH region because of reduced interaction at the o/w interface. Both of the emulsions with Lecimulthin and Lecigran remained stable throughout the temperature range. Peak intensities for the emulsion containing Lecimulthin were lower than those for the emulsion made with Lecigran due to greater water bonding. This study demonstrated that FTIR is a potentially powerful tool for measuring emulsifier-water interactions that could be used in the rapid determination of emulsion stability in food systems.

IR spectroscopy was used to study the interactions between lipid and protein to verify the structural basis of antioxidative activity of zein (maize prolamins). One unique property of cereal proteins, especially prolamins, is their antioxidative activity in the powder system, that is, the ability to inhibit lipid oxidation. The reason for this functionality of zein was believed to be the physical shielding of lipid molecules from oxygen by the zein matrix. The physical interaction of lipids and zein was demonstrated using FTIR spectroscopy in the powder system (54). The mixing of lipid (linolenic acid ethyl ester, LAE) with zein powder caused decreases in the α-helix and intermolecular hydrogen bonded β-sheet of zein when stored in the "humid" state, suggesting the strong interaction of LAE and zein molecules. On the other hand, zein could not inhibit the oxidation of lipids in the dry state, which might be due to prevention of the interaction between zein and LAE under these conditions. To confirm this, the effect of heating in the temperature range from 25 to 160°C on the interaction between lipid and zein in a dry powder system was investigated using FTIR spectroscopy (55). The heat treatment of the powdered zein with and without LAE caused increases in the α-helix, β-turn, and β-sheet, concomitant with decreases in the intermolecular hydrogen-bonded β-sheet and random coil. Such changes in the secondary structure were more drastic for the powder with LAE. The heating of the zein-LAE mixed powder also caused decreases in the peaks originating from LAE in the FTIR spectra. These results suggest that the heat treatment induced structural changes resulting from the interaction of the zein and LAE in the powder system, which may also have affected the antioxidative activity of dry powder zein as measured by the peroxide value. When a zein-LAE mixed powder was heated before storage, the oxidation of LAE was inhibited for 7 d, whereas LAE was oxidized within 1 d in the absence of heat treatment.

In recent years, the use of FTIR to monitor biochemical changes in living cells has gained considerable importance (56,57). The changes in the cells and tissues, which are subtle and often not obvious from histopathological studies, were shown to be well-resolved using FTIR-microspectroscopy (FTIR-MSP) and FTIR spectroscopy (58,59). FTIR-MSP can be used to study changes in levels of various metabolites during processes such as cell maturation and tissue differentiation (60,61). Researchers also showed that ATR-FTIR spectroscopy is a reliable method for studying membrane proteins and lipids (62).

Raman Spectroscopy

Principles

Raman spectroscopy is a branch of vibrational spectroscopy that is based on the shifts in the wavelength or frequency of an exciting incident beam of radiation that result from inelastic scattering upon interaction between the photons and the sample molecules.

Differences Between IR and Raman Spectroscopy

Raman spectroscopy is a technique that complements IR spectroscopy. The discrete vibrational transitions that occur in the ground electronic state of molecules, which correspond to various stretching and bending deformation modes of individual chemical bonds, contribute to both Raman and IR spectroscopy. Although both techniques are based on the interactions of electromagnetic radiation with molecules, an IR spectrum measures the absorption (or transmission) of energy from an incident beam in the IR region, whereas the Raman spectrum measures the inelastic scattering or "shift" of incident radiation in the ultraviolet visible (UV-vis) or near-IR (NIR) region. A Raman spectrum is obtained by plotting the intensity of scattered radiation as a function of the Raman shift, whereas an IR spectrum is obtained as the absorption or transmission of energy as a function of the frequency.

IR absorption requires a change in the intrinsic dipole moment with molecular vibration, whereas Raman scattering depends on changes in the polarizability of functional groups. In this case, polar groups such as C=O, N-H, and O-H have strong IR stretching vibrations, whereas nonpolar groups such as C=C, C-C and S-S have intense Raman bands. Hence, because water is a polar molecule, water solutions always show strong absorption in the IR spectrum. Thus, IR spectroscopy is most commonly used for the analysis of dry or nonaqueous samples including oils. For aqueous samples, very short path length sample cells or an ATR cell could be used for IR spectroscopy, but careful subtraction of the water baseline spectrum is required for accurate measurement. In contrast, water has weak Raman scattering properties and produces less interference in Raman spectroscopy. As a result, Raman spectroscopy is usually more suitable for the in vivo or in situ study of biological systems, including foods, which are primarily aqueous in nature.

History and Development of Raman Spectroscopy

In terms of history, Raman spectroscopy is younger than IR spectroscopy; it is based on a discovery by Sir Chandrasekhra Venkata Raman in 1928. It is amazing that Sir Raman used very crude instrumentation to make the remarkable discovery of the scattering phenomenon bearing his name. He used sunlight as the source, a telescope as the collector, and his eyes as the detector. This discovery won him the Nobel Prize in 1930 (63).

Developments in the Raman technique were attributed to advances made in instrumentation including new laser sources and detector systems. Lasers were introduced into the Raman method in 1962, which prompted the manufacturers to produce commercial spectrometers with laser sources. The early models used weak but stable helium-neon lasers with one excitation frequency in the red. These were soon replaced by more powerful krypton and argon ion lasers, which provided a range of stable frequencies throughout the visible region, and could be extended into the UV region by frequency-doubling devices. Both organic chemists and biochemists were quick to apply this new technique to study a range of biological molecules. It was also in the early 1960s that a double monochromator was demonstrated to remove stray light more efficiently than a single monochromator. Later, a triple monochromator was introduced, which was even more efficient in removing stray light. Holographic gratings appeared in 1968, which added to the efficiency of the collection of Raman scattering in commercial Raman instruments (63). Together with the introduction of multichannel detectors, including charge-coupled devices (CCD), these developments have brought commercial dispersive Raman instruments to the present state of the art of Raman measurements.

In addition, Raman spectra can now also be obtained by FT-Raman spectroscopy, pioneered in the 1980s by Hirschfeld and Chase (64). FT-Raman instruments are commercially available either as units interfaced to FTIR spectrometers or as dedicated FT-Raman instruments. Before the introduction of FT-Raman instruments, the use of conventional Raman spectroscopy was greatly limited by fluorescence problems that plagued classical visible laser techniques. The FT-Raman technique has largely solved this problem through the use of NIR lasers such as the Nd:YAG laser with excitation at 1064 nm. In the NIR region, the photon energy is usually not sufficient to cause transitions between electronic states that give rise to fluorescence. Besides reducing the interference from fluorescence, the FT-Raman technique can also achieve better frequency precision and higher resolution compared with the conventional Raman technique, facilitating further analysis such as spectral subtraction. However, the use of a Nd:YAG laser causes a 16-fold reduction in signal compared with a visible laser at 514.5 nm because the cross section of Raman scattering is inversely related to the energy of the incident radiation, through a ν^4 relation (65). Hence, higher sample concentrations and higher laser power are usually necessary with NIR lasers to achieve spectra with acceptable signal-to-noise ratios, and great care must be taken to avoid overheating or "cooking" the samples. Alternative approaches to the fluorescence problem and low sensitivity were used therefore for various applications, including resonance Raman spectroscopy or surface-enhanced Raman spectroscopy.

Raman microscopy was developed in the 1970s. This technique provides the capability of obtaining analytical quality Raman spectra with 1-μm spatial resolution using samples in the picogram range (65). The emergence of confocal Raman MSP in particular has made it possible to acquire Raman spectral maps or images of samples by taking advantage of this spatial resolution. For the confocal technique, the excitation laser is focused on a small area (spot size 2 μm) of the sample through a microscope objective. The detected Raman scattering signal is limited by a pinhole to the area surrounding the focus. The smaller the pinhole, the better is (especially) the depth resolution, but at the expense of signal intensity (66,67). Confocal Raman spectroscopy was demonstrated as a potential technique with which to study the microstructure of complex biomaterials including foods (68).

Compared with IR sampling, the Raman sampling techniques are intricate and versatile. First, for colorless samples, because glass does not absorb Raman-scattered light in the visible region, conventional glass or quartz tubes (cylindrical or capillary) are widely used for Raman sampling. Raman samples could be aqueous solutions (because as described above, Raman scattering of water is very weak), other liquids including oils or emulsions, solids, or even a gas. Solid samples can also be made into pellets by mixing with KBr, similar to the procedure used in IR spectroscopy. For colored samples, because of the absorption of the energy of the laser beam, decomposition by local heating may occur. In such cases, the laser power should be reduced. In addition, the following procedures could be taken: changing the wavelength of the exciting laser source, defocusing the laser beam on the sample, diluting the sample concentration in a pellet or in solution to avoid energy absorption, applying a cooling system, rotating the sample, and rotating or oscillating the laser beam on a fixed sample. Some special sampling cells are also available for Raman spectroscopy, including thermostated cells, high-temperature cells, low-temperature cells, UV resonance cells, and fiber optics (65). The development of optical fibers, which serve both to transmit the incident exciting beam and to collect the Raman scattering signal, provides the opportunity for applications of Raman spectroscopy involving noncontact or remote sampling (69).

Applications of Raman Spectroscopy to Study Protein-Lipid Interactions

The pioneering work applying Raman spectroscopy to food science was reviewed in 1996 (70). Assignments of Raman bands in spectra of protein and edible oil are summarized in Tables 4 and 5 (71–74). However, although Raman spectroscopy has been widely used to study proteins and lipids individually, research on studying protein-lipid interactions in food systems using this technique is in its infancy, and applications in this field are only beginning to appear in the literature. Figure 1 shows a typical Raman spectrum of bovine serum albumin (BSA), corn oil, and protein-lipid interactions at their interface (75).

In the 1970s, Larsson and colleagues (76,77) applied Raman spectroscopy to study model systems containing both lipid and protein, and the changes in the environment of hydrocarbon chains of different lipids and phosphate esters. These early studies focused on the changes in the lipid components, and assumed that the contribution of protein components to the spectrum was insignificant. They emphasized that the region of 2800–3000 cm^{-1}, which is attributed to the C-H stretching vibrations, is very sensitive to structural changes in lipid-water and lipid-protein-water phase systems. The ratio of the intensity of the peaks at 2850 and 2885 cm^{-1}, corresponding to symmetric and asymmetric stretching vibrations of the CH_2 groups, respectively, was found to be related to the disorder of the hydrocarbon chains. Thus, the peak at 2850 cm^{-1} was reported to be dominant in the liquid state of the hydrocarbon chains, whereas the peak at 2885–2890 cm^{-1} dominated for the hydrocarbon chains in crystalline form.

In another early study, the change in intensity of the 2930 cm^{-1} band of ribonuclease relative to that of the 2850 cm^{-1} CH_2 stretching band of egg lecithin was used to monitor the changes in the environment of the aliphatic amino-acid residues upon interaction with the phospholipid (78). This study showed that marked amplification near 2930 cm^{-1} may reflect exposure of apolar protein side-chains to an aqueous milieu. Use of the 2850 cm^{-1} band of lecithin as an internal standard was claimed to be justified due to its thermal stability over the range of temperature used to monitor ribonuclease unfolding; however, possible changes in this band due to ribonuclease-lecithin interaction upon heating were not considered.

Although relatively few publications exist on the study of protein-lipid interactions in food systems by Raman spectroscopy, it has been much more widely used in the study of biological systems such as the cell membrane or phospholipid bilayers.

TABLE 4
Assignments of Major Raman Bands in Food Proteins[a]

Wavenumber (cm^{-1})	Vibration	Origin
510	S-S stretch *gauche-gauche-gauche* conformation	Cystine
525	S-S stretch *gauche-gauche-trans* conformation	Cystine
545	S-S stretch *trans-gauche-trans* conformation	Cyst(e)ine and
630–670	C-S stretch *gauche* conformation	Methionine
700–745	C-S stretch *trans* conformation	Cysteine
2550–2580	S-H stretch	
850/830	Fermi resonance between ring fundamental and overtone	Tyrosine
760,880,1360	Indole ring	Tryptophan
1006	Ring breathe	Phenylalanine
1400–1430	C=O stretch of COO^-	Aspartic and
1700–1750	C=O stretch of COOH or COOR	glutamic acid
1450,1465	C-H bending	Aliphatic
2800–3000	C-H stretching	residues
1655 ± 5	Amide C=O stretch, N-H wag (α-Helix)	Amide I
1670 ± 3	Amide C=O stretch, N-H wag (Anti-parallel β-sheet)	
1665 ± 3	Amide C=O stretch, N-H wag (Disordered structure, solvated)	
1685	Amide C=O stretch, N-H wag (Disordered structure, non-hydrogen bonded)	
>1275	N-H in-plane bend, C-N stretch (α-Helix)	Amide III
1235 ± 5	N-H in-plane bend, C-N stretch (Anti-parallel β-sheet)	
1245 ± 4	N-H in-plane bend, C-N stretch (Disordered structure, solvated)	
1235	N-H in-plane bend, C-N stretch (Disordered structure, non-hydrogen bonded)	

[a]*Source:* Refs. 71–73.

The study of interactions between ferricytochrome c and dimyristoylphosphatidic acid (DMPA) liposomes and its cationic binary and ternary lipid-protein complex (79) is a good example. Ferricytochrome c is a water-soluble heme protein involved with mitochondrial electron transport, which can form complexes with cardiolipin and phosphatidic acid, two of the anionic lipids found in the inner mitochondrial membrane. Previous studies suggested that even weak electrostatic interaction of ferricytochrome c with a zwitterionic lipid, dipalmitoylphosphatidylcholine (DPPC), can lead to demonstrable acyl chain perturbations within the bilayer (80). Raman spectroscopy was shown to be a sensitive, noninvasive technique for studying the acyl chain perturbation. Spectral peak height intensity ratios and line widths were used to determine various order/disorder parameters, which are characteristic of bilayer assemblies (81–83). For example, the intensity ratio, I_{2850}/I_{2880}, was shown to

TABLE 5
Assignments of the Major Raman Bands of Edible Oils[a]

Wavenumber (cm^{-1})	Vibration	Origin
3015	=C-H asymmetric stretching	RCH=CHR
2970	C-H asymmetric stretching	$-CH_3$
2940	C-H asymmetric stretching	$-CH_2$
2900	C-H symmetric stretching	$-CH_3$
2860	C-H symmetric stretching	$-CH_2$
1750	C=O stretching	RC=OOR
1670	C=C stretching	*trans* RCH=CHR
1660	C=C stretching	*cis* RCH=CHR
1445	C-H deformation	$-CH_2$
1310	C-H twisting or =C-H deformation	$-CH_2$ or *trans* RCH=CHR
1275	=C-H deformation	*cis* RCH=CHR
1100–1000	C-C stretching	$-(CH_2)_{n-}$
900–800	C-C stretching	$-(CH_2)_{n-}$

[a]*Source:* Refs. 70, 74.

reflect lateral chain-chain interactions within the bilayer, where 2850 and 2880 cm^{-1} were assigned to the methylene CH_2 symmetric and asymmetric vibrations, respectively. Furthermore, by using deuterated dimyristoylphosphatidylcholine ($DMPC-d_{54}$) and non-deuterated DMPA, the effects of ferricytochrome c at different ionic

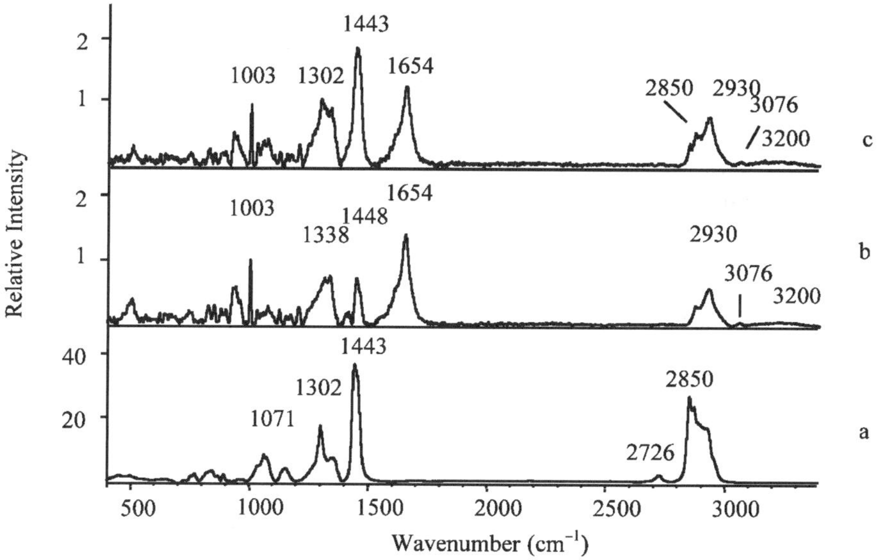

Fig. 1. Raman spectra of BSA, mineral oil and their interface. (a) Mineral oil; (b) 25% BSA aqueous solution; (c) BSA/mineral oil interface (75).

environments, pH, and temperature conditions on the individual lipid species could be monitored because the acyl C-D and C-H vibrations were observed in different spectral regions.

More recently, FT-Raman spectroscopy was used to study protein-lipid interactions in a lysozyme-corn oil system (84). Emulsions were made by homogenization of lysozyme (25% in D_2O) and corn oil. Raman spectra of various layers from the emulsions, as well as the lysozyme solution and oil, were measured. The difference spectra, obtained by subtracting the individual lysozyme or corn oil spectra from the spectrum of the emulsion, were used to ascertain the groups involved in the protein-lipid interaction. The hypothesis was that if there were no changes in the protein or oil molecules due to specific interactions, the difference spectrum would be expected to resemble the original spectrum of oil or lysozyme on its own. On the other hand, if some differences were observed between the resultant difference spectrum and the individual spectra, then some interaction between protein and lipid molecules may have occurred. The results indicated hydrophobic interactions involving both protein and lipid components in the emulsion. The interactions with corn oil in the emulsion caused changes in the protein spectrum compared with the original spectrum of aqueous lysozyme solution, including reduced intensity of tryptophan bands at 760, 1013, 1340, and 1360 cm^{-1}, a reduced intensity ratio of the tyrosine doublet at 850 and 830 cm^{-1}, and increased intensity of the C-H bending band at 1455 cm^{-1}. The corn oil structure was also affected, as shown by decreased intensity at 2855 cm^{-1} (lipid CH_2 symmetric stretch) and 3011 cm^{-1} (unsaturated fatty acid hydrocarbon chain =C-H stretch) and a higher intensity ratio of the C-H stretching band at 2900 cm^{-1} to bands at 2885 cm^{-1} and 2933 cm^{-1}. Raman spectral analysis suggested that the presence of lipids can alter the molecular structure of proteins and result in changes in exposure of hydrophobic groups, secondary structures, and conformation of disulfide groups. The protein-lipid interaction and the restructuring of water molecules surrounding the proteins may play important roles in protein denaturation. This research also showed that Raman spectroscopy accompanied by detailed spectral analysis can provide a valuable tool for the study of protein-lipid interactions in food emulsions and mixtures as well as in the biological membranes and tissues.

Another study using FT-Raman spectroscopy to investigate protein-lipid interaction dealt with the effect of lipids (fish oil) on frozen stored cod collagen (85). For this purpose, collagen prepared from fresh cod fillets was mixed with 500 ppm cod oil. The prepared model was incubated overnight (18 h) at 4°C. Raman spectra of collagen in the presence or absence of fish oil were measured on a FT-Raman spectrometer with excitation from a Nd:YAG laser at 1064 nm. A decrease in the intensity of the amide I region at 1660 cm^{-1} indicated an alteration in the protein secondary structure in the presence of fish oil. The tyrosine doublet ratio (I_{850}/I_{830} cm^{-1}) decreased from 1.8 for fresh collagen to 1.3 in the presence of fish oil, indicating greater "buriedness" of the tyrosine residues resulting from interaction with fish oil. A decrease in the intensity of the peak at 1340 cm^{-1} indicated exposure of buried tryptophan residues in proteins in the presence of fish oil. Further changes of the hydrophobic groups were observed in the CH stretching bands at 2940, 2888, and 2976 cm^{-1}, CH_3-C skeletal stretch at 937 cm^{-1}, phenylalanine ring band at 1034 cm^{-1}, isopropyl antisymmetric stretch and CN stretch (backbone) at 1128 cm^{-1}, and CH_3 antisymmetric rock (aliphatic) at 1160 cm^{-1} as well as tryptophan bands at 1554 and 1451 cm^{-1}. The effect of oil on frozen stored collagen was also investigated by DSC in the same study. Results showed that in the presence of fish oil, the enthalpy increased by 8.7% and an extra peak was observed at T_m of 44.6°C, in addition to the one at 28.2°C found in fresh collagen in the absence of fish oil. The DSC result suggested a protein-lipid interaction, which is consistent with the Raman spectra analysis. Based on the above description, the authors suggested that collagen significantly affects the texture and rheological properties of fish muscle. This is caused by insolubility of proteins, which results from changes in the secondary structure and formation of intramolecular cross-linkages during frozen storage. The effect of lipids and lipid oxidation products in gadoid and fatty fish is of great importance in this process.

In addition to collagen, the major changes in fish muscle protein, particularly myosin, during frozen storage were also attributed to the effect of oxidizing lipids in addition to ice crystal damage. To ascertain the role of lipid oxidation products and ice crystals in stored frozen cod, Badii and Howell (86) treated cod fillets with either antioxidants (vitamin C, vitamin E, or a mixture of vitamins C + E + citrate) or antioxidants with cryoprotectants (sucrose + sorbitol). Untreated and treated cod samples were stored at –1 0°C; cod fillets stored at –3 0°C were used as a control. Stored frozen samples were analyzed at intervals for changes in protein solubility, thermodynamic parameters, rheological analysis, and structure by FT-Raman spectroscopy. Results indicated that protein denaturation and texture changes were minimized in the presence of antioxidants alone and in combination with cryoprotectants.

A comparison of the Raman spectra (500–1800 cm^{-1}) indicated a decrease in α-helix and an increase in β-sheet secondary structure in cod muscle proteins, when stored frozen in the absence of antioxidants or cryoprotectants (86). The intensity of tryptophan bands at 1340 and 1554 cm^{-1} also decreased at –1 0°C in the absence of antioxidants and cryoprotectants (86), indicating an exposure of hydrophobic groups due to protein denaturation and muscle cell damage (87). In contrast, in the presence of antioxidants, bands in the CH stretching region centered at 2937 or 3067 cm^{-1} (aromatic) and CH bend at 1450 cm^{-1} (aliphatic) had similar intensity at –1 0°C, as those stored at –3 0°C (control) (86). In addition, the tyrosine doublet ratio I_{850}/I_{830} was higher for samples stored at –1 0°C in the presence of vitamin C, indicating exposed tyrosine residues, which can interact with water molecules. Thus, both antioxidants on their own and mixtures of antioxidants and cryoprotectants are effective treatments for minimizing protein denaturation and toughening in fish. The protection of proteins in stored frozen lean fish from damage by oxidizing long-chain polyunsaturated fatty acids (PUFA) was reported for the first time by Badii and Howell (86) and for fatty fish (Atlantic mackerel) by Saeed and Howell (88).

The effect of oxidized lipids including methyl linoleate or fish oils on amino acids and proteins in frozen stored fish muscle may be explained by results obtained in extensive studies by Howell and Saeed (89,90) using ESR and NMR spectroscopy as well as FT-Raman spectroscopy. The interaction of proteins with lipids and lipid oxidation products may result in the loss of specific amino acids such as cysteine, lysine, histidine, and methionine. Lipid oxidation products including primary free radicals can cross-link with proteins, resulting in polymerization. Saeed et al. (29) provided direct evidence of the transfer of free radicals from oxidizing lipids (methyl linoleate of fish oil) to amino acids (arginine and lysine) and proteins (lysozyme, ovalbumin and myosin) in emulsions, using ESR spectroscopy. Interestingly, there was a strong radical signal on the carbon, which increased for up to ~7 d, after which it decreased due to interaction with other radicals and formation of secondary products. The disappearance of the radical signal coincided with an increase in fluorescence intensity due to cross-linking of amino acids and proteins either with themselves or with lipid oxidation products; this would form the basis of aggregation and texture changes.

Recently, using ESR spectroscopy with spin-trapping, it was found that the aromatic amino acids in methyl linoleate-β-lactoglobulin emulsions are particularly sensitive to oxidation and the amino acid most likely to be damaged is tyrosine, resulting in dimerization through the carbon atom on the aromatic ring to form di-tyrosine (25,90). The formation of dityrosine was confirmed by 2-dimensional NMR and high-performance liquid chromatography. FT-Raman spectra of the protein-oxidized lipid mixtures, compared with protein-fresh methyl linoleate samples, exhibited changes in the amide region, showing a reduction in α-helix and an increase in β-sheet, indicating protein denaturation. Unfolding of the molecule exposed tryptophan residues. The tyrosine doublet ratio I_{850}/I_{830} was lower in the oxidized sample than in the unoxidized protein, suggesting that the phenolic hydroxyl oxygen was ionized or strongly hydrogen bonded to a negatively charged molecule. FT-Raman spectra thus confirmed changes in the amide and aromatic groups of β-lactoglobulin in line with changes found by ESR and NMR spectroscopy.

Raman MSP was used recently to study the protein-lipid interactions at the interface between a biphasic system of mineral or corn oil and an aqueous BSA solution (75). With the help of the microscope, the incident laser beam was focused on various positions within the interface to collect Raman spectra of at least three different positions for each interface as well as the individual oil and aqueous phases. By comparing the spectra, a gradient of distribution of protein and oil at different positions within the interface was observed. The protein-lipid interaction in the interface was studied by subtracting the spectrum of BSA or oil from that of the interface. The difference spectrum obtained by subtracting the spectrum of mineral or corn oil from that of the BSA/oil interface indicated interactions involving different functional groups of the BSA and the oil molecules. Against mineral oil, the BSA spectrum showed reduced intensity of the tryptophan band at 750 cm^{-1} and reduced intensity ratio of the tyrosine doublet at 850 to 830 cm^{-1}, indicating changes in the microenvironment of these hydrophobic residues. A negative band at

2850 cm^{-1} indicated the involvement of the CH groups in the mineral oil. However, the amide regions, normally assigned to protein secondary structure, were not affected. When the spectrum of BSA was subtracted from the BSA/mineral oil interface spectrum, the resultant difference spectrum showed changes of symmetric and antisymmetric CCC stretches at 980 and 1071 cm^{-1}, respectively. In contrast, the difference spectrum of BSA/corn oil interface – BSA showed a decrease of CH_2 symmetric stretching at 2850 cm^{-1} and a decrease of unsaturated fatty acid hydrocarbon chain stretch at 3010 cm^{-1}. Raman MSP is a useful tool with which to study the nature of protein-lipid interactions at an oil-water interface.

Conclusions

Both IR and Raman spectroscopies can provide information on the vibrational transitions within molecular species. Although IR spectroscopy is a better-established technique with a longer history of use, resulting from the earlier availability and lower cost of commercial FTIR instrumentation, recent advances in Raman instrumentation, particularly in the area of Raman microscopy and FT-Raman spectroscopy, have led to growing acceptance of the complementary nature of these analytical tools in vibrational spectroscopy for the study of food and biological systems, including the investigation of protein-lipid interactions.

Acknowledgments

The authors gratefully acknowledge the financial support of the Natural Sciences and Engineering Research Council (NSERC) of Canada and The Royal Society, UK.

References

1. Shimizu, M., A. Ametan, S. Kaminogawa, and K. Yamauchi, The Topology of α-Casein Adsorbed to an Oil/Water Interface: An Analytical Approach Using Proteolysis, *Biochim. Biophys. Acta 869*:259–264 (1986).
2. Christie, W.W., *Lipid Analysis Isolation, Separation, Identification and Structural Analysis of Lipids*, 2nd ed., Pergamon Press, Oxford, 1982, pp. 177–197.
3. Carlson, J.W., A. Jonas, and S.G. Stephen, Imaging and Manipulation of High-Density Lipoproteins, *Biophys. J. 73*: 1184–1189 (1997).
4. McWilliams, M., *Foods Experimental Perspectives*, 3rd ed., Merrill/Prentice Hall, Upper Saddle River, NJ, 1997, pp. 332–335.
5. Parkinson, T.L., The Effect of Pasteurization and Freezing on the Low Density Lipoproteins of Egg, *J. Sci. Food Agric. 28*:806–810 (1977).
6. Horn, T.D., Egg Protein, in *Applied Protein Chemistry*, edited by R.A. Grant, Applied Science, London, 1980, pp. 69–86.
7. Gunstone, F.D., J.L. Harwood, and F.B. Padley, *The Lipid Handbook*, 2nd ed., Chapman and Hall, London, 1994, pp. 436–464.
8. Hoseney, R.C., K.F. Finney, and Y. Pomeranz, Functional (Breadmaking) and Biochemical Properties of Wheat Flour Components. VI. Gliadin-Lipid-Glutenin Interaction in Wheat Gluten, *Cereal Chem. 47*:135–140 (1970).

9. Chobert, J.M., and T. Haertle, Protein-Lipid and Protein-Flavour Interactions, in *Food Proteins and Their Applications*, edited by S. Damodaran and A. Paraf, Marcel Dekker, New York, 1997, pp. 143–170.
10. St Angelo, A.J., and E.E. Graves, Studies of Lipid-Protein Interaction in Stored Raw Peanuts and Peanut Flours, *J. Agric. Food Chem. 3 4*:643–646 (1986).
11. Farnum, C., D.W. Stanley, and J.I. Gray, Protein-Lipid Interactions in Soy Films, *J. Can. Inst. Food Sci. Technol. 9*:201–206 (1976).
12. Beckwith, A.C., Interaction of Phosphatidylcholine Vesicles with Soybean 7S and 11S Globulin Proteins, *J. Agric. Food Chem. 3 2*:1397–1402 (1984).
13. Murphy, D.J., and I. Cummins, Seed Oil-Bodies Isolation, Composition and Role of Oil-Body Apolipoproteins,*Phytochemistry 2 8*:2063–2069 (1989).
14. Tzen, J.T.C., Y.Z. Cao, P. Laurent, C. Ratnayake, and A.H. Hung, Lipids, Proteins and Structure of Seed Oil Bodies from Diverse Species, *Plant Physiol. 1 0 1*:267–276 (1993).
15. Gardner, H.W., Lipid Hydroperoxide Reactivity with Proteins and Amino Acids: A Review, *J. Agric. Food Chem. 2 7*:220–229 (1979).
16. Pokorny, J., Z. Reblova, L. Kourimska, F. Pudil, and A. Kwiatkowska, Effect of Interactions with Oxidized Lipids on Structure Change and Properties of Food Proteins, in *Food Proteins: Structure and Functionality*, edited by K.D. Schwenke and R. Mothes, VCH Publishers, Weinheim, 1993, pp. 232–235.
17. Kunimoto, M., K. Matsumoto, and K. Zama, The Interaction of Fish Proteins and Lipids During Freeze-Drying and Storage. I. The Interaction of Carp Myofibrils and Fish Oil, *Bull. Fac. Fisheries (Hokkaido Univ.) 36*:50–56 (1985).
18. Kanner, J., and M. Karel, Changes in Lysozyme Due to Reactions with Peroxidizing Methyl Linoleate in a Dehydrated Model System, *J. Agric. Food Chem. 24*:468–472 (1976).
19. Berger, K.G., and G.W. White, An Electron Microscopic Investigation of Fat Destabilization in Ice Cream, *J. Food Technol. 6*:285–294 (1971).
20. Evans, E.W., Interactions of Milk Components in Food Systems, in *Interactions of Food Components*, edited by G.G. Birch and M.G. Lindley, Elsevier Applied Science Publishers, London, 1986, pp. 241–275.
21. Finney, K.F., Fractionating and Reconstituting Techniques to Relate Functional (Bread Making) to Biochemical Properties of Wheat Flour Components, *Cereal Sci. Today 1 6*:342–356 (1971).
22. Kamat, V.B., E.G. Graham, and M.F. Davis, Vegetable Protein: Lipid Interactions, *Cereal Chem. 5 5*:295–307 (1978).
23. Pointer-Keenan, C.D., D.K. Lee, K. Hallok, A. Tan, R. Zand, and A. Ramamoorthy, Investigation of Interaction of Myelin Basic Protein with Phospholipid Bilayers Using Solid-State NMR Spectroscopy, *Chem. Phys. Lipids 1 3 2*:47–54 (2004).
24. Strandberg, E., S. Morein, D.T.S. Rijkers, R.M.J. Liskamp, P.C.A. van der Wel, and J.A. Killian, Lipid Dependence of Membrane Properties and Snorkeling Behavior of Aromatic and Charged Residues in Trans-Membrane Peptides, *Biochemistry 41*:7190–7198 (2002).
25. Saeed, S., D. Gillies, G. Wagner, and N.K. Howell, ESR and NMR Spectroscopy Studies on Protein Oxidation and Formation of Dityrosine in Emulsions Containing Oxidised Methyl Linoleate, *Food Chem. Toxicol.* in press.
26. Sklar, L.A., B.S. Hudson, and R.D. Simoni, Conjugated Polyene Fatty Acids as Fluorescent Probes: Binding to Bovine Serum Albumin, *Biochemistry 16*:5100–5108 (1977).

27. Kato, A., and S. Nakai, Hydrophobicity Determined by a Fluorescence Probe Method and Its Correlation with Surface Properties of Proteins. *Biochim. Biophys. Acta 624*:13–20 (1983).
28. Marsh, D., L.I. Horvath, M.J. Swamy, S. Mantripragada, and J.H. Kleinschmidt, Interaction of Membrane-Spanning Proteins with Peripheral and Lipid-Anchored Membrane Proteins: Perspectives from Protein-Lipid Interactions. (Review) *Molecular Membrane Biology 1 9*:247–255 (2002).
29. Saeed, S., S.A. Fawthrop, and N.K. Howell, Electron Spin Resonance (ESR) Study on Free-Radical Transfer in Food Lipid-Protein Interaction, *J. Sci. Food Agric. 79*:1809–1816 (1999).
30. Efremov, R.G., D.E. Nolde, A.G. Konshina, N.P. Syrtcev, and A.S. Arseniev, Peptides and Proteins in Membranes: What Can We Learn via Computer Simulations? *Curr. Med. Chem. 1 1*:2421–2442 (2004).
31. Mantsch, H.H., Historical Survey of Infrared and Raman Spectroscopy of Biological Materials, in *Infrared and Raman Spectroscopy of Biological Materials*, edited by H.-U. Gremlich, and B. Yan, Marcel Dekker, Inc., New York, 2001, pp. 1–14.
32. Elliot, A., and E.J. Ambrose, Structure of Synthetic Polypeptides, *Nature 165*:921–922 (1950).
33. Iñón, F.A., S. Garrigues, and M. de la Guardia, Nutritional Parameters of Commercially Available Milk Samples by FTIR and Chemometric Techniques, *Anal. Chim. Acta 513*:401–412 (2004).
34. Fabian, H., and W. Mäntele, Infrared Spectroscopy of Proteins. In *Handbook of Vibrational Spectroscopy, Vol. 5*, edited by J.M. Chalmers and P.R. Griffiths, John Wiley & Sons Ltd, Chichester, UK, 2002, pp. 3399–3425.
35. Miyazawa, T., Infrared Spectra and Helical Conformations, in *Poly-α-Amino Acids*, edited by G.D. Fasman, Marcel Dekker, New York, 1967, pp. 69–103.
36. Krimm, S., and J. Bandekar, Vibrational Spectroscopy and Conformation of Peptides, Polypeptides and Proteins, *Adv. Protein Chem. 38*: 181–346 (1986).
37. Susi, H., Infrared Spectroscopy Conformation, *Methods Enzymol. 2 2*:455–472 (1972).
38. Arrondo, J.L.R., A. Muga, J. Castresana, and F.M. Goñi, Quantitative Studies of the Structure of Proteins in Solution by Fourier-Transform Infrared Spectroscopy, *Prog. Biophys. Mol. Bio. 59*: 23–56 (1993).
39. Arrondo, J.L.R., and F.M. Goni, Structure and Dynamics of Membrane Proteins as Studied by Infrared Spectroscopy, *Prog. Biophys. Mol. Biol. 72*:367–405 (1999).
40. Biggs, D.A., G. Johnsson, and L.O. Sjaunja, Analysis of Fat, Protein, Lactose and Total Solids by Infrared Absorption,*Bull. Int. Dairy Fed. 208*:21–29 (1987).
41. Lanser, A.C., and E.A. Emken, Comparison of FTIR and Capillary Gas Chromatographic Methods for Quantitation of *trans* Unsaturation in Fatty Acid Methyl Esters, *J. Am. Oil Chem. Soc. 65*:1483–1487 (1988).
42. Mostafa, A.N., and J.M. Deman, Application of Infrared Spectroscopy in the Study of Polymorphism of Hydrogenated Canola Oil, *J. Am. Oil Chem. Soc. 26*:1481–1483 (1985).
43. Kochhar, S.P., and J.B. Rossell, Analysis of Oils and Fats Using Infrared Spectroscopy, *Int. Analyst 5*:23–26 (1987).
44. Safar, M., D. Bertrand, P. Robert, M.F. Devaux, and C. Genot, Characterization of Edible Oils, Butters and Margarines by Fourier Transform Infrared Spectroscopy with Attenuated Total Reflectance, *J. Am. Oil Chem. Soc. 71*:371–377 (1994).

45. Fang, Y., and D.G. Dalgleish, Conformation of βlg Studied by FTIR: Effect of pH, Temperature and Adsorption to the Oil-Water Interface, *J. Colloid Interface Sci. 196*:292–298 (1997).
46. Corredig, M., and D.G. Dalgleish, A Differential Microcalorimetric Study of Whey Proteins and their Behaviour in Oil-in-Water Emulsions, *Colloids Surf. B: Biointerfaces 4*:411–422 (1995).
47. Shimizu, M., T. Kamiya, and K. Yamauchi, The Adsorption of Whey Proteins on the Surface of Emulsified Fat, *Agric. Biol. Chem. 45*:2491–2496 (1981).
48. Agbooa, S.O., and D.G. Dalgleish, Enzymatic Hydrolysis of Milk Proteins Used for Emulsion Formation. 1. Kinetics of Protein Breakdown and Storage Stability of the Emulsions, *J. Agric. Food Chem. 44*:3631–3636 (1996).
49. Dickinson, E., S.E. Rolfe, and D.G. Dalgleish, Surface Shear Viscometry as a Probe of Protein-Protein Interactions in Mixed Milk Protein Films Adsorbed at the Oil-Water Interface, *Int. J. Biol. Macromol. 12*:189–194 (1990).
50. Dickinson, E., and Y. Matsumura, Time-Dependent Polymerization of β-Lactoglobulin through Disulphide Bonds at the Oil-Water Interface in Emulsions, *Int. J. Biol. Macromol. 13*:26–30 (1991).
51. Susi, H., and D.M. Byler, Protein Structure by Fourier Transform Infrared Spectroscopy: Second Derivative Spectra, *Biochem. Biophys. Res. Commun. 115*:391–397 (1983).
52. Dong, A.C., P. Huang, and W. Caugher, Protein Secondary Structures in Water from Second-Derivative Amide I Infrared Spectra, *Biochemistry 29*:3303–3308 (1990).
53. Whittinghill, J.M., J. Norton, and A. Proctor, A Fourier Transform Infrared Spectroscopy Study of the Effect of Temperature on Soy Lecithin-Stabilized Emulsions. *J. Am. Oil Chem. Soc. 76*:1393–1398 (1999).
54. Mizutani, Y., Y. Matsumura, K. Imamura, K. Nakanishi, and T. Mori, Effects of Water Activity and Lipid Addition on Secondary Structure of Zein in Powder Systems, *J. Agric. Food Chem. 51*:229–235 (2003).
55. Mizutani, Y., Y. Matsumura, H. Murakami, and T. Mori, Effects of Heating on the Interaction of Lipid and Zein in a Dry Powder System, *J. Agric. Food Chem. 52*:3570–3576 (2004).
56. Diem, M., L. Chiriboga, P. Lasch, and A. Pacifico, IR Spectra and IR Spectral Maps of Individual Normal and Cancerous Cells, *Biopolymers 67*:349–353 (2002).
57. Holman, H.N., M.C. Martin, E.A. Blakely, K. Bjornstad, and W.R. McKinney, IR Spectroscopic Characteristics of Cell Cycle and Cell Death Probed by Synchrotron Radiation Based Fourier Transform IR Spectromicroscopy, *Biopolymers 57*:329–335 (2000).
58. Cohenfor, M.A., T.A. Godwin, F. Cahn, P. Bhandare, T.A. Caputo, and B. Rigas, Infrared Spectroscopy of Normal and Abnormal Cervical Smears: Evaluation by Principal Component Analysis, *Gynecol. Oncol. 66*:59–65 (1997).
59. Romeo, M., M. Quinn, F. Burden, and D. McNaughton, Influence of Benign Cellular Changes in Diagnosis of Cervical Cancer Using IR Microspectroscopy, *Biopolymers 67*:362–366 (2002).
60. Chiriboga, L., P. Xie, H. Yee, D. Zarou, D. Zakim, and M. Diem, Infrared Spectroscopy of Human Tissue. IV. Detection of Dysplastic and Neoplastic Changes of Human Cervical Tissue via Infrared Microscopy, *Cell Mol. Biol. 44*:219–229 (1998).
61. Chiriboga, L., P. Xie, H. Yee, D. Zarou, D. Zakim, and M. Diem, Infrared Spectroscopy of Human Tissue. I. Differentiation and Maturation of Epithelial Cells in the Human Cervix, *Biospectroscopy 4*:47–53 (1998).

62. Tatulian, S.A., Attenuated Total Reflection Fourier Transform Infrared Spectroscopy: A Method of Choice for Studying Membrane Proteins and Lipids, *Biochemistry 42*:11898–11907 (2003).
63. Ferraro, J.R., and K. Nakamoto, Chapter 1. Basic Theory, in *Introductory Raman Spectroscopy*, Academic Press Inc., New York, 1994, pp. 1–94.
64. Hirschfeld, T., and B. Chase, FT-Raman Spectroscopy: Development of Justification, *Appl. Spectrosc. 40*:133–137 (1986).
65. Ferraro, J.R., and K. Nakamoto, Chapter 2. Instrumentation and Experimental Techniques, in *Introductory Raman Spectroscopy*, Academic Press Inc., New York, 1994, pp. 95–134.
66. Everall, N.J., Modeling and Measuring the Effect of Refraction on the Depth Resolution of Confocal Raman Microscopy, *Appl. Spectrosc. 54*:773–782 (2000).
67. Barbillat, J., E. Silva, B. Lenain, M. Manfait, S. Sharonov, and P. Valisa, Raman Confocal Spectral Imaging, in *ICORS Hong Kong 94*, edited by N.T. Yu and X.Y. Li, Wiley, Singapore, 1994, pp. 1102–1106.
68. Pudney, P.D.A., T.M. Hancewicz, and D.G. Cunningham, The Use of Confocal Raman Spectroscopy to Characterize the Microstructure of Complex Biomaterials: Foods, *Spectroscopy 16*:217–225 (2002).
69. Coates, J., A Review of Sampling Methods for Infrared Spectroscopy, *Applied Spectrosc. 52*:49–91 (1998).
70. Li-Chan, E.C.Y., The Applications of Raman Spectroscopy in Food Science, *Trends Food Sci. Technol. 7*:361–370 (1996).
71. Li-Chan, E., S. Nakai, and M. Hirotsuka, Raman Spectroscopy as a Probe of Protein Structure in Food Systems, in *Protein Structure-Function Relationships in Foods*, edited by R.Y. Yada, R.L. Jackman, and J. L. Smith, Blackie Academic and Professional, Chapman & Hall, Glasgow, 1994, pp. 163–197.
72. Peticolas, W.L., Raman Spectroscopy of DNA and Proteins, *Methods Enzymol. 246*:389–416 (1995).
73. Tu, A.T., Peptide Backbone Conformation and Microenvironment of Protein Side Chains, in *Spectroscopy of Biological Systems,* edited by R.J.H. Clark and R.E. Hester, John Wiley and Sons, New York, 1986, pp. 47–112.
74. Baeten, V., P. Hourant, M. Morales, and R. Aparicio, Oil and Fat Classification by FT-Raman Spectroscopy, *J. Agric. Food Chem. 46*: 2638–2646 (1998).
75. Meng, G.T., J.C.K. Chan, D. Rousseau, and E.C.Y. Li-Chan, Study of Protein-Lipid Interactions at the Bovine Serum Albumin-Oil Interface by Raman Micro-Spectroscopy, *J. Agric. Food Chem. 53*:845–852 (2005).
76. Larsson, K., and R.P. Rand, Detection of Changes in the Environment of Hydrocarbon Chains by Raman Spectroscopy and Its Application to Lipid-Protein Systems, *Biochim. Biophys. Acta 326*:245–255 (1973).
77. Larsson, K., Raman Spectroscopy for Studies of Interface Structure in Aqueous Dispersions, in *Lipids, Vol. 2: Technology*, edited by R. Paoletti, G. Jacini and R. Porcellati, Raven Press, New York, 1976, pp. 355–360.
78. Verma, S.P., and D.F.H. Wallach, Raman Spectra of Some Saturated, Unsaturated and Deuterated C_{18} Fatty Acids in the HCH-deformation and CH-stretching Regions, *Biochim. Biophys. Acta 486*:217–227 (1977).
79. Vincent, J.S., and I.W. Levin, Raman Spectroscopic Studies of Dimyristoylphosphatidic Acid and Its Interactions with Ferricytochrome c in Cationic Binary and Ternary Lipid-Protein Complexes, *Biophys J. 59*:1007–1021 (1991).

80. Vincent, J.S., and I.W. Levin, Interaction of Ferricytochrome c with Zwitterionic Phospholipid Bilayers: A Raman Spectroscopic Study, *Biochemistry 27*:3438–3446 (1988).
81. Levin, I.W., Vibrational Spectroscopy of Membrane Assemblies, in *Advances in Infrared and Raman Spectroscopy*, edited by R.J.H. Clark and R.E. Hester, Heyden and Son Ltd., London. 1984 (11), pp. 1–48.
82. Levin, I.W., Structural and Dynamical Properties of Model and Intact Membrane Assemblies by Vibrational Spectroscopy, in *Chemical Biological and Industrial Applications of Infrared Spectroscopy,* edited by J.R. Durig, John Wiley and Sons, New York, 1985, pp. 173–197.
83. Gaber, B.P., and W.L. Peticolas, On the Quantitative Interpretation of Biomembrane Structure by Raman Spectroscopy, *Biochim. Biophys. Acta 4 6 5*:260–274 (1977).
84. Howell, N.K., H. Herman, and E.C.Y. Li-Chan, Elucidation of Protein-Lipid Interactions in a Lysozyme-Corn Oil System by Fourier Transform Raman Spectroscopy, *J. Agric. Food Chem. 4 9*:1529–1533 (2001).
85. Badii, F., and N.K. Howell, Elucidation of the Effect of Formaldehyde and Lipids on Frozen Stored Cod Collagen by FT-Raman Spectroscopy and Differential Scanning Calorimetry, *J. Agric. Food Chem. 5 1*:1440–1446 (2003).
86. Badii, F., and N.K. Howell, Effect of Antioxidants, Citrate and Cryoprotectants on Protein Solubility and Texture of Frozen Cod (*Gadus morhua*), *J. Agric. Food Chem. 5 0*:2053–2061 (2002).
87. Badii, F., P. Zhdan, and N.K. Howell, Elucidation of Protein Aggregation in Frozen Cod and Haddock by Transmission Electron Microscopy/Immunocytochemistry, Light Microscopy and Atomic Force Microscopy, *J. Sci. Food Agric. 84*:1919–1928 (2004).
88. Saeed, S., and N.K. Howell, Rheological and Differential Scanning Calorimetry Studies on Structural and Textural Changes in Frozen Atlantic Mackerel (*Scomber scombrus*), *J. Sci. Food Agric. 84*:1216–1222 (2004).
89. Howell, N.K., and S. Saeed, The Effect of Antioxidants on the Production of Lipid Oxidation Products and Transfer of Free Radicals in Oxidised Lipid-Protein Systems. In: *Antioxidants in Human Health and Disease*, edited by T.K. Basu, N.J. Temple and M.L. Garg, CAB International, Oxford, UK, 1999, pp. 43–54.
90. Howell, N.K, S. Saeed, and D. Gillies, ESR and NMR Studies on Mixtures of β-Lactoglobulin and Methyl Linoleate in Emulsions. In: *Applications of Magnetic Resonance in Food Science*, edited by P.S. Belton, The Royal Society of Chemistry, Cambridge, UK, 2005.

Part V. Applications

17 Fat Replacers: An Overview

William E. Artz[a], Symon M. Mahungu[b], and Steve L. Hansen[c]

[a]Department of Food Science and Human Nutrition, University of Illinois, Urbana, IL, USA; [b]Department of Food Technology, Egerton University, Njoro, Kenya; [c]Cargill Analytical Services, Minnetonka, MN, USA

Introduction

In 1998 the food industry achieved and then exceeded the Healthy People 2000 goal of offering more than 5000 reduced-fat processed food products (1). More than 1000 reduced- or low-fat products were introduced each year during the 1990s (2,3). The three most popular reduced-fat product categories included (i) fat-free or low-fat milk, (ii) salad dressing, sauces, or mayonnaise, and (iii) cheese/dairy products (4). A survey by the 1998 Calorie Control Council National (4) indicated that these product categories are consumed by ~50% of those who consumed low-fat products. In addition, consumption of fat-reduced or fat-free margarine/spreads, chip/snack foods, meat products, and ice cream/frozen desserts was reported by more than a third of the consumers who use reduced fat products. Because ~88% of the adult population consume low- or reduced-fat foods and beverages, the large majority of the population has demonstrated substantial interest in foods that are promoted as low in fat (1,4). Results from a survey by the Calorie Control Council indicated that ~67% of adults believe a need exists for food ingredients that can replace the fat in food products, and >50% find "reduced in both fat and calories" an appealing descriptor (4).

The two terms, fat replacer and fat substitute, were differentiated in a paper by Miraglio (5). The author used the term fat replacer to denote an ingredient that replaces some or all the functions of fat and that may or may not provide nutritional value, whereas a fat substitute replaces all of the functions of a fat with essentially no energy contribution. In this chapter, these terms will be used interchangeably because they both serve the same primary purpose, i.e., a reduction in the fat content of the diet.

Health Issues

In spite of the successful development and commercialization of low-fat food products and widespread acceptance and consumption of these products by consumers, as of 2001, obese and overweight adults comprised 58% of the population in the United States (6). The obesity problem is not limited to the United States. In 2000 the World Health Organization reported that >1 billion adults are overweight and that >300 million of these were clinically obese. Throughout the entire world, the problem of obesity is increasing, not only in industrialized nations, but also in urban areas of developing nations (7). A reduction in our caloric intake and a simultaneous increase in our energy expenditure via regular exercise is the primary solution recommended by knowledgeable nutritionists and medical experts. However, application of the appropriate food technologies designed to reduce fat content and caloric density could help ameliorate the severity of the problem.

The macronutrient composition of a diet can influence hunger, satiety, food intake, body weight, and body composition (8). Fat, rather than carbohydrates, has been the macronutrient most associated with overeating and obesity. Fat is often consumed in excess because its palatability and caloric density are high. Low-fat foods in combination with the appropriate fat substitute can potentially reduce caloric intake by making less palatable low-fat foods more desirable.

Fats contribute to the appearance, taste, mouth-feel, lubricity, texture, and flavor of many food products; they provide essential fatty acids and are carriers of fat-soluble vitamins (9–11). The amount and type of fats present in foods determine the characteristics of that food and can affect consumer acceptance. An ideal fat replacer should be completely safe and physiologically inert; it should achieve a substantial fat and caloric reduction while maintaining the desired functional and sensory properties of a conventional high-fat product (12). Historically, dietary fats and oils have been considered a primary source of energy without regard to the health effects of their specific complement of fatty acids and sterols (13). In 1995, fats accounted for ~38% of the total calories in the diet of Western populations, particularly in the United States (9). However, dietary fat intakes > 11% of the total caloric intake developed after the domestication of mammals and the subsequent selective breeding of genetically fatter animals (14), indicating that a high-fat diet has become the norm only relatively recently in the history of *Homo sapiens*.

Although there are many nutrition recommendations that remain controversial, there is a consensus among health and nutrition professionals that most Americans should reduce their dietary fat intake and alter the composition of the fat consumed (15). The relation of dietary fat and cholesterol to coronary heart disease is supported by extensive and consistent clinical, epidemiologic, metabolic, and animal evidence. Studies strongly indicate that the formation of atherosclerotic lesions in coronary arteries is increased in proportion to the levels of total and low density lipoprotein (LDL) cholesterol in blood, which in turn, are increased by diets high in total fat (13). A reduction in the relative amounts of high-fat food products in the diet can be an effective means of reducing caloric intake and is consistent with public health goals to reduce the risk of chronic diseases (16–18). Thus, dietary fat is one of the major nutrition concerns of Americans. In response to the rising consumer demand for reduced-fat foods, the food industry has developed a multitude of nonfat, low-fat, and reduced-fat versions of regular food products (4). To generate reduced-fat or fat-free products that have the same organoleptic characteristics as those of the regular fat version, food manufacturers frequently employ fat substitutes in the formulation of these foods (19). These fat substitutes are made from carbohydrates, protein, or fat, or a combination of these components. Many of the carbohydrate- and protein-based fat substitutes have received GRAS (Generally Recognized As Safe) status from the Food and Drug Administration (FDA) (11). In January 1996, the U.S. FDA approved olestra (currently termed Olean) for use in savory snacks (20–22). This fat substitute is chemically referred to as sucrose polyester or sucrose fatty acid polyester; its Code of Federal Regulation (CFR) reference number is CFR 172.867 (23). There are other fat substitutes under development or in the market (4) that have the potential to partially replace some, but not all calories from fat.

Fat substitutes could replace a significant proportion of dietary fat and as such become macronutrient substitutes (17). Hence, the safety of these materials must be established via extensive safety testing before FDA approval and introduction into the food supply (11). Appropriate methods of safety evaluation must be used. Traditional methods for the safety evaluation of macronutrient substitutes are inappropriate because an evaluation of concentrations high enough to provide a 100-fold safety factor are not feasible (17).

Labeling

The labels on fat-modified products must conform to the Nutrition Labeling and Education Act criteria for the use of fat- and calorie-related terms (24). The FDA requires that the labels on foods containing fat substitutes, e.g., salatrim or olestra, list the analytical fat amount on the nutrition fats label with a footnote indicating the amount that is bioavailable.

Food products that are labeled *fat free* and *low fat* must contain $\leq$0.5 and $\leq$3 g of fat per serving, respectively. *Reduced* or *less fat* may be used on the labels of products that contain 25% less fat than regular (full-fat) products. Although the fat labeling claims do not provide any indication of the caloric content of the food item, products containing one third fewer calories or one half the fat of the reference food may be labeled as *light*. If 50% of calories in a food are derived from fat, the fat content of the *reduced-fat* version must be reduced by 50%. The terms *calorie free* and *low calorie* can be used only on products with <5 and 40 calories per serving, respectively, and *reduced* or *fewer calories* can be used only on products that have 25% of the calories in the regular product (1,4).

Chemistry

This section addresses the synthesis and/or preparation and analysis of some of the major fat substitutes in use or under development that have potential as fat substitutes or fat replacers. Although many of the fat substitutes are included in this chapter, the absence of a fat substitute from this presentation implies nothing about the utility, safety, or potential of that substitute. Some of the fat substitutes that will be discussed are those that contain fatty acids attached to a molecule other than glycerol, such as olestra, or those in which attachment has been modified to reduce the susceptibility of the compound to fatty acid release via lipase; the esterified propoxylated glycerols are a good example of the latter. The other categories of fat substitutes include carbohydrate-based biopolymers, modified proteins, and structured triacylglycerols.

Esterified Propoxylated Glycerols (EPGs). Fatty acid EPGs (ARCO Chemical Company, Newtown Square, PA) were developed for use as fat substitutes. Glycerol is propoxylated with propylene oxide to form a polyether polyol. Fatty acids are then esterified to the polyol (25–35). The resulting triacylglycerol is similar to natural fats in structure and functionality. Fatty acid EPG is a low-to-noncaloric oil, which is heat stable, only very slightly digestible, and nontoxic. Preparation of propoxylated glycerides for use as fat substitutes involves transesterifying propoxylated glycerol with fatty acid esters in a solvent-free system (ARCO Chemical Technology Limited Partnership, Wilmington, DE) (27) to avoid reagents unacceptable in food systems.

Fatty Acid Partially Esterified Polysaccharide (PEP). The ARCO Chemical Technology Limited Partnership, (Wilmington, DE) has patented a fatty acid PEP that is partially esterified with fatty acids (36). It is nonabsorbable, indigestible, and nontoxic. Suitable oligo/polysaccharide materials include xanthan gum, guar gum, gum arabic, alginates, cellulose hydrolysis products, hydroxypropyl cellulose, starch hydrolysis products ($n < 50$), karaya gum, and pectin. The preferred level of esterification involves one or more hydroxyl groups per saccharide unit with one or more fatty acids.

Carbohydrate Fatty Acid Esters. The carbohydrate-fat combination fat substitutes include those derived from polydextrose, sugar alcohols, altered sugars, starch derivatives, cellulose, and gums. They can also be made from components from rice, wheat, corn, oats, tapioca, or potato and can replace from 50 to 100% of the fat in foods (13).

Reports on the synthesis and analysis of carbohydrate fatty acid esters have been published by several groups (37–40). Carbohydrate fatty acid polyesters with a degree of substitution (DS: number of hydroxyl groups esterified with long-chain fatty-acids) of 4–14 are lipophilic, nondigestible, nonabsorbable, fat-like molecules with the physical and chemical properties of conventional fats and oils and are referred to as low-calorie fat substitutes (9,41,42). Swanson published work on carbohydrate fatty acid esters synthesized from a variety of carbohydrate sources under a variety of catalytic conditions (9,41–43), including glucose, sucrose, raffinose, stachyose, and verbascose fatty acid esters (37). The synthesis of trehalose octaoleate and sorbitol hexaoleate (38) was reported.

Oleic acid esters of erythritol, pentaerythritol, adonitol, and sorbitol were prepared by transesterification with an excess of methyl oleate to form complete esters (44). The esters formed were erythritol tetraoleate, pentaerythritol tetraoleate, adonitol pentaoleate, and sorbitol hexaoleate. These esters were not susceptible to in vivo lipolysis by lipolytic enzymes of rat pancreatic juice, suggesting potential application as low-calorie oils (37,44). Chung et al. (45) also reported the preparation of a sugar alcohol fatty acid ester made with sorbitol.

Enzymatic methods for the synthesis of carbohydrate fatty acid esters were discussed in detail by Riva (46). One of the most promising enzymes tested, particularly for fatty acid esterification of the alkylated glycosides, was a lipase from the yeast *Candida antartica*, which had been immobilized on macroporous resin beads. Mutua and Akoh (47) reported the synthesis of glucose and alkyl glycoside fatty acid esters in organic solvents using *C. antartica* as a catalyst.

Sucrose Polyester (SPE)-Olestra (Olean). Probably the most extensively studied and publicized of the low-to-noncaloric fat substitutes are the sucrose fatty acid esters. Typically, these esters are prepared from the reaction of sucrose with long-chain fatty acid methyl esters (48). Depending upon the reaction conditions, from 1 to 8 fatty acids can be attached. Olestra is the generic name for the mixture of hexa-, hepta-, and octaesters of sucrose formed with long-chain fatty acids. Because digestive enzymes do not release the fatty acids, olestra is noncaloric (15).

Olestra was approved by the FDA for use in savory snacks in 1996 (20). A review by Akoh (9) also described various methods that are used to prepare sucrose polyesters. SPE have been prepared (80–90% yields) by reacting sucrose octaacetate (SOAC) and methyl palmitate. This solvent-free method was improved upon by Volpenhein (49).

Yamamoto and Kinami (50) reported on a method to produce sugar fatty acid esters. The methods of Volpenhein (49) and Yamamoto and Kinami (50) required molecular distillation to remove unreacted fatty acid methyl esters. Akoh and Swanson (43) reported a SPE synthesis procedure that produced yields between 96.6 and 99.8% of the purified SPE.

Alkyl Glycosides Fatty Acid Esters. Alkyl glycoside fatty acid esters are nonionic, nontoxic, odorless, biodegradable compounds with emulsification properties. Direct esterification of reducing sugars such as glucose and galactose often results in excessive sugar degradation and charring. Therefore, alkylation is necessary to convert reducing sugars with reactive C-1 anomeric centers to nonreducing, less reactive, anomeric C-1 centers (37,38).

Alkyl glycoside fatty acid esters could be used to replace fat in food products such as frying oils and Italian salad dressings (Curtice-Burns, Rochester, NY) (51). Alkyl glycosides can be formed by reacting a reducing saccharide with a monohydric alcohol. Soybean, safflower, corn, peanut, and cottonseed oils are preferred because they contain C_{16}–C_{18} fatty acids that do not volatilize at the temperatures used for interesterification.

Albano-Garcia et al. (52) reported a solventless synthesis of methyl glucoside esters of coconut fatty acids. Akoh and Swanson (37,38) synthesized novel alkyl glycoside polyesters such as methyl glucoside polyesters, methyl galactoside polyesters, and octyl glucoside polyesters by a solvent-free interesterification. To achieve high yields, the alkyl glycosides free hydroxyl groups were first protected by acetylation before interesterification.

Glucosides containing from 1 to 50 alkoxy groups can be used as fat substitutes at substitution ranges of 10–100% in low-calorie salad oils, plastic shortenings, cake mixes, icing mixes, mayonnaise, salad dressings, and margarines (Procter & Gamble, Cincinnati, OH) (53).

Starch-Based Fat Replacers. There are several starch-derived fat replacers available that are essentially maltodextrins. They are produced upon partial enzymatic or acid-catalyzed hydrolysis of starch and are fully digestible. The low DE (dextrose equivalent) maltodextrins have fat binding functional properties, unlike the high DE products. Starches and sugars contain 4 kcal/g, but because they are generally used at concentrations of much less than 100%, the actual caloric content is typically 0.5–2 kcal/g. They include, for example, Lycadex-100 and Lycadex-200 from Roquette Freres; Crestar SF from Euro Centre Food; N-Lite, Instant N-Oil and N-Oil II from National Starch Company; Maltrin M040 from Grain Processing Corporation; Paselli SA-2 from Avebe America, Incorporated; Tapiocaline from Tripak; Star Dri, Stellar and Sta-Slim from A.E. Staley; Amalean I from American Maize Products, and Rice-trin from Zumbro. Typically, a smooth viscous solution or a soft gel is formed when the product is hydrated.

Lycadex is an enzymatically hydrolyzed cornstarch-based maltodextrin. The typical concentration used is 20% and it is nongelling. Typical applications include sauces and salad dressings.

Maltrin M040 is a maltodextrin (DE = 5) made from cornstarch (54–56); it is soluble in hot water. A solution containing 30–50% solids produces a thermoreversible gel with a bland flavor, smooth mouth feel, and texture similar to hydrogenated oils for butterfat. According to the manufacturer, it can be used for dairy products, frozen foods, sauces, salad dressings, confectionery products, and dry mixes.

Paselli SA2 is a maltodextrin derived from potato starch (54,57). It forms a shiny, white, thermoreversible gel with a smooth, fat-like texture and neutral taste that is also suited for acid formulations. The gel is stable with fat- and oil-containing products. A gel is formed at concentrations >20% and the gel strength increases with an increase in concentration.

N-Oil is a tapioca-based maltodextrin (54,57). An instantized version, N-Oil II, can be used with foods that either require no heat processing, or very modest thermal processing, such as high-temperature, short-time thermal processing. N-Oil is heat stable, and it can withstand high temperatures, high shear, and acidic conditions. Upon cooling, a solution of hot water–soluble N-Oil develops a texture similar to that of hydrogenated shortening.

Amalean I is a modified high-amylose cornstarch used at a relatively low concentration (8%) compared with the maltodextrins. A gel for fat replacement can be prepared upon heating a paste or slurry to 88–90°C for 3 min. An Amalean II product is available for enhancing batter aeration.

StaSlim is a modified starch designed to provide a creamy texture for a wide range of food products, but it is particularly well-suited for salad dressings. In contrast to other products, it is usually processed as a warm liquid in the formulation. It is prepared as a slurry at a concentration of 3–20% and heated with agitation to 65–71°C to completely solubilize the starch. It is recommended for use in salad dressings, cheese products, and soups.

Stellar is a microparticulated cornstarch gel used in pastries, snack foods, frostings, and fillings, cheese products, margarines, meat products, salad dressings, and other selected products (54,57). The water is bound sufficiently such that water migration from the cake to the filling is slowed and staling is retarded. The small, intact, starch granules duplicate the mouth feel of fat. The particle aggregates are 3–5 μm in diameter, slightly larger than the protein-based fat replacer products, but approximately the same size as the fat crystals they are designed to replace. The microparticulate character is required to maintain the smooth, creamy texture. If it is heated sufficiently to completely gelatinize and disperse the starch (105°C), much of the fat substitute or mimetic functionality is lost permanently. Applications include low-fat margarines, salad dressings, soups, confectionery products, baked goods, frostings, fillings, and selected dairy and meat products (58).

The use of an unmodified, pregelatinized, waxy starch (waxy starches contain nearly 100% amylopectin) to produce a blueberry muffin with only 3% fat, yet with the textural characteristics comparable to a muffin containing 15% fat, was reported (59). The waxy starch assists in providing a uniform cell size, the appropriate crumb structure, and a strong tendency to retain moisture, which extends the shelf-life and produces a desirable moist mouthfeel. The use of this fat substitute increases the moisture content; thus, the potential problem of mold growth during storage should be considered (e.g., with the addition of 0.1% potassium sorbate). This problem must be examined carefully when using any fat substitute that increases the relative amount of moisture in the food product.

Polydextrose is a low-calorie polymer that can be used to replace some of the fat in a food product. It is formed by the random polymerization of glucose, sorbitol, and citric acid. Sorbitol provides an upper molecular weight limit and reduces the formation of water-insoluble material. Most of the linkages are glycosidic and the 1–6 bond, in particular, predominates. Litesse, which is a polydextrose-type product and Veri-Lo, which consists of fat-coated carbohydrate-based gel particles, are both from Pfizer.

Biopolymer/fiber-based products include gums, celluloses, hemicelluloses, pectins, β-glucans, and lignins. These are isolated from a wide variety of sources, including cereals, fruits, legumes, nuts, and vegetables. Gums are typically not used as fat substitutes directly; rather, they are used at low concentrations (0.1–0.5%) to form gels that increase product viscosity. Agar, alginate, gum arabic, carrageenan, konjac, guar gum, high and low methoxy pectin, xanthan gum, and cellulose derivatives could all be used.

Fiber-based products available include Nutrio-P-fiber from the Danish Sugar Factory; Nutricol derived from Konjac flour; P-150 C and P-285 F derived from pea fiber from Grindsted Products; Avicel, a microcrystalline cellulose, and carrageenan are both from FMC. Slendid is a pectin that forms gels with fat-like melt ability from Hercules. Quaker oatrim is a product of Rhone-Poulenc Food Ingredients derived from oats.

Cellulose derivatives used include α-cellulose, carboxymethyl cellulose, hydroxypropyl cellulose, microcrystalline cellulose, and methyl cellulose (54). The gel produced from the cellulose derivatives has several desirable fat-like functional properties, which include creaminess, fat-like mouthfeel, stability, texture modification, increased viscosity, and the glossy appearance of high-fat emulsions. FMC, for example, has cellulose-based gels (60) as well as gums, which increase emulsion viscosity and, hence product stability. Because the gel mimics some of the rheological properties associated with high-fat emulsions, it allows reduction of the oil content.

Oatrim is a cold water dispersible amylodextrin rich in β-glucans (61), produced upon partial enzymatic hydrolysis of the starch. It is derived from whole oat or debranned whole oat flour. An oatrim-based gel (25% solids) can function as a fat replacer. In addition, the β-glucan components in the oatrim have a demonstrated hypocholesterolemic effect (62).

Protein-Based Fat Substitutes. The protein-based fat substitute that has received the most attention has been Simplesse, which consists of microparticulated milk and/or egg white proteins, sugar, pectin, and citric acid (15). The protein is particulated during a combined pasteurization and homogenization process that produces microparticles of uniform size and spherical shape, ~1 μm in diameter (63). The nutritive quality of the protein is essentially unaltered during the preparative processing. Most of the protein-based fat substitutes have a similar basis, that is, the protein is particulated into spherical particles with some combination of heat and shear, with or without added components such as gums, stabilizers, and emulsifiers. The simulated fat produced from an aqueous suspension of microparticulated protein particles of the appropriate size and characteristics have shear thinning characteristics and a smoothness and creaminess similar to fat. Particles with significantly smaller or larger dimensions than 0.5–2 μm do not seem to have the desired sensory characteristics, unless the particles are very soft, hydrated, and compressible, which can extend the size range considerably. An aqueous microparticulate protein suspension contains fewer than the normal 4 kcal/g for protein, generally 1–2 kcal/g. The protein-based substitutes can be used to formulate a variety of products, including cheesecake, puddings, sauces, pie fillings, sour cream, ice cream, cream cheese, mayonnaise, dips, and spreads. Simplesse was affirmed as GRAS by the FDA for use in several products. The main limitation, which is the limitation for most of the currently approved fat substitutes, is that it cannot be used for heated product applications, particularly frying.

Another version is Simplesse 100 (54), which is a thixotrophic fluid derived from whey protein concentrate. It can be used in a wide variety of dairy and bakery products. It is also available as a readily hydratable powder. Simplesse 300 is a mixture of egg white and milk proteins that can be used in food products that require heating. Of course, it cannot be used as a frying medium.

Once investigators realized that microparticulation could be used to produce a homogenous suspension of small protein particles that had fat-like properties, other protein-based fat replacer microparticulation processes and systems were examined. Research efforts on this problem were reviewed by Cheftel and Dumay (64). Membrane processing (ultrafiltration, followed by diafiltration) has been used to concentrate casein micelles four- to ninefold to produce a product that can be used to partially or completely replace fat in ice cream, chocolate mousse, dairy product–based spreads, and sauces (65). The micelles are reported to function as microparticles with a diameter of ~0.1–0.4 μm.

Protein precipitation under carefully controlled conditions can produce a microparticulated product that can also be used as a fat substitute (66). A dilute solution of water-soluble protein (1–5%) is precipitated with heat and/or a change in pH to the isoelectric point of the protein. Starches, gums, or emulsifying agents, for example, can be used to enhance product characteristics and prevent extensive aggregation.

A solution of alcohol-soluble (70–80% aqueous ethanol) proteins (prolamines from corn, wheat, or rice, for example) can be precipitated by dilution with water to produce a microparticulated spherical protein precipitate (67). Gums should be used to prevent extensive aggregation. To produce a concentrated protein suspension, ultrafiltration, followed by diafiltration can be used. Freeze-drying can be used to produce a powdered precipitate.

One of the first thermomechanical coagulation processes for the microparticulation of protein was described by Singer et al. (68). A whey protein concentrate is first dispersed in water, acidified, and heated under high-shear conditions. The resultant dispersion contains small spherical particles (0.1–2.5 μm) with a creamy and smooth texture. Protein sources other than whey protein can also be used (64).

In addition to the dual heating/high-shear particulation processes used to make Simplesse-type products, extrusion can be used to prepare a microparticulated protein-based fat substitute (64). The extrusion conditions are very mild compared with typical extrusion conditions. For example, a whey protein concentrate (20%) within a pH range of 3.5–3.9 (extruder conditions: barrel temperature of 85–100°C and a screw speed of 100–200 rpm) can be extruded to produce a semisolid spread with a smooth and creamy texture. No die is used; thus, there is no pressure buildup and subsequent expansion. The particles are rather large (12 μm) on average. If the extrusion is done at a higher pH (4.5–6.8), the protein solubility is reduced, the percentage of large particles (>20 μm) is much greater, and the resultant texture is coarse and grainy. The nitrogen solubility index of the protein remains relatively high, 43–47% for the acid product and 69–70% for the neutral product. The addition of 0.5% xanthan gum improved the creamy texture and reduced the particle size.

Reduced-Calorie Fat-Based Fat Replacers. The objective for these products is similar to that for the protein and carbohydrate-based fat substitutes, a substantial reduction in calories, rather than a complete elimination of fat in the product. Examples of fat-based replacers include caprenin, captrin, and salatrim.

Caprenin is a reduced-calorie triacylglycerol (Procter & Gamble) formed by the esterification with the fatty acids, caprylic, capric, and behenic. Because behenic acid is absorbed only partially, the caprenin contains ~5, rather than the normal 9 kcal/g. Caprenin has functional properties similar to those of cocoa butter and is intended to replace some of the cocoa butter in selected confectionery products. It is digested, absorbed, and metabolized by the same pathways as other triacylglycerols. Captrin from Stepan Food Ingredients (69) is a randomized triacylglycerol made from linear saturated fatty acids primarily C_8–C_{10} in length. Some of the proposed uses include baked goods, confections, dairy product analogs, snack foods, and soft candy.

Another fabricated triacylglycerol, similar to caprenin, is Salatrim, which is a triacylglycerol comprised of a mixture of long-chain (primarily stearic acid) and short-chain (acetic, propionic and butyric) fatty acids randomly esterified to glycerol (70). It contains ~5 kcal/g, rather than the 9 kcal/g contained in regular fats and oils. An entire symposium on Salatrim was published in the February, 1994 issue of the *Journal of Agriculture and Food Chemistry*. The research reported included structural characterization(s) of the oil, an analysis of the oil in food products, and an extensive series of papers on the metabolism and toxicology of the oil in various animal and human model systems. Salatrim's utility is similar to that of caprenin as a fat replacer in reduced fat systems; it could be used as a cocoa butter substitute in confectionery products and in baked products and filled dairy products. Caprenin and Salatrim will be of little use for deep fat frying applications, however, because release of the smaller-molecular-weight fatty acids will likely cause some undesirable flavor effects.

Salatrim has been prepared by interesterification of saturated long-chain fatty acid (LCFA) triacylglycerols, and short-chain fatty acid (SCFA) triacylglycerols (71). The SCFA (triacetin, tripropionic, tributyrin) were mixed with LCFA (hydrogenated canola oil, cottonseed oil, and soybean oil) and the mixture combined with a catalyst.

An alternative to regular cooking oils that was developed and marketed in Japan has essentially the same calories as regular fats and oils, although the oil is metabolized much differently. Diacylglycerol cooking oils have a 1,3 configuration. The taste and texture are similar to those of triacylglycerol cooking oils. However, the 1,3-diacylglycerols are not hydrolyzed to 2-monoacylglycerols in the gut; therefore the absorption and metabolism of 1,3-diacylglycerols differs from that of triacylglycerols. The physiological differences include lower postprandial lipemia; compared with normal triacylglycerol cooking oils, an increased percentage of the fatty acids is oxidized rather than stored as body fat. Preliminary studies suggest that these differences in energy partitioning between diacylglycerols and triacylglycerols can be exploited to reduce the amount of body fat stored from the consumption of cooking oil and other food items with added fats and oils. It has been widely sold in Japan since its introduction in early 1999, and the product is being test-marketed in the United States. Increased use of DAG oil may provide an additional means of reducing obesity, while concurrently achieving desirable food product characteristics and maintaining good food product quality (72,73).

Food Applications

There are numerous factors that must be considered when selecting a fat substitute, in addition to the obvious and critical sensory quality questions. Is any thermal processing applied to the product? How severe is the thermal processing (pasteurization vs. sterilization)? How pH sensitive is the fat substitute? How long will the product be stored, i.e., are there undesirable textural or flavor changes that occur during long-term storage or during some excessively turbulent shipping? Will it be refrigerated? Must it be refrigerated? What are the home preparation steps involved? Because water activity changes generally occur, is the product microbiologically stable? Are there "opportunities" for abuse in the home, i.e., if opened and left on the counter overnight, is food poisoning a possibility?

A comparison of yogurts produced with one of seven selected fat substitutes was made with a control yogurt sample containing milk fat. The overall acceptability and organoleptic quality were comparable to the control for five of the seven fat substitutes, Litesse, N-Oil II, Lycadex-100, Lycadex-200, and Paselli SA2. Interestingly, the overall acceptability, including the flavor and aroma, generally improved after 20 d of storage (56,75,75), in contrast to the control. Yogurt made with Simplesse was very similar to the control yogurt made with anhydrous milk fat (76), except that the flavor intensity of the Simplesse-containing product was less sour and there was more serum separation than in the control.

Frozen dairy dessert samples made with fat mimetics (Simplesse or N-Lite D) and 2.1% milk fat were compared with control ice milk samples containing 4.8% milk fat (77). Overall, the protein- or Simplesse-containing product was more similar to the control sample than the sample containing the carbohydrate-based fat mimetic. The Simplesse-based ice milk samples had rheological properties that were generally the same as those of the control. However, the maximum overrun was >40% greater for the Simplesse-containing sample than the control.

Milk fat content can be reduced substantially in frozen dairy dessert products with the addition of gums or maltodextrin gels (54,78). Low-DE maltodextrins, such as Paselli SA2, can be used with frozen dairy desserts if the maltodextrins are thermostable, cold water soluble, and able to form stable mixtures with other components. Other thermostable, cold water–soluble carbohydrate-based fat substitutes include some of the gums, such as carrageenan, guar gum, and carboxymethylcellulose.

Mono- and diglyceride emulsifiers can be used to replace as much as 50% of the fat in baked goods because they can be used as a 50/50 emulsifier/water mixture (54,79). Additional emulsifiers used similarly include sodium stearoyl lactylate, sorbitan monostearate, and polysorbate 60 at high hydration levels. A hydrated blend of emulsifiers was developed specifically for fat replacement that included stearyl monoglyceridyl citrate, glycol monostearate, and lactylated monoglycerides. Some of the products that can be used to make acceptable frozen dessert products include oatrim, Maltrin 40, and N-Oil. Simplesse, protein hydrolysates, Sta-Slim 143, and sucrose fatty acid esters have been used as fat substitutes in yogurt, sour cream, cream cheese, cheese spread, and frozen dairy desserts. Additional applications were discussed by Frye and Setser (54), and Sharp (80) discussed some of the technical limitations regarding baked goods and reduced-fat formulations.

Low-fat shortbread cookies were prepared using carbohydrate-based fat substitutes (81). A combination of carbohydrate-based fat substitutes (Litesse, N-Flate, Rice™Trin, Stellar, or Trim-choice) and emulsifiers (diacetyl-tartaric esters of momnoglycerides, glycerol monostearate, sodium stearoyl-2-lactylate) were used. The principal effects of fat substitutes on shortbread cookie attributes were an increase in moisture content, greater toughness, and reduced specific volume. Zoulias et al. (82) examined four different types of fat mimetics using cookies. A fat reduction of 35–50% was achieved with acceptable cookies.

Olestra can be used as a partial fat substitute in shortening, margarines, and frying oils (83–85). Cheddar cheese has been prepared with milk fat sucrose polyesters (86). There were significant differences in moisture, pH, or whey titratable acidity between the control cheese and cheeses containing milk fat sucrose fatty acid polyester. The latter contained fat globules that were smaller and more uniform in size than a control cheese with no added sucrose polyester. Bachmann (87) reviewed various types of cheese analogs that were developed using fat substitutes.

A patent was awarded for the incorporation of the alkyl glycoside fatty acid polyester into food products (Curtice-Burns, Rochester, NY) (88). The fat substitute could replace fats in such products as shortening, margarine, butter, salad and cooking oils, mayonnaise, salad dressing, and confectionery coatings or "invisible fats" in such foods as oilseeds, nuts, dairy, and animal products. Substitution at 10–100% is possible; however, <100% is preferred, preferably in the range of 33–75%.

Simplesse and selected maltodextrins can be used in soft margarine applications for reduced-calorie spreads. The product has the mouthfeel, flavor, and spreading properties of a soft margarine (54). Simplesse has also been used in low-fat milk (89). The addition of 3% Simplesse to skim milk gave milk an appearance that was similar to that of 0.5% fat milk, and a mouthfeel similar to that of milk containing 1% fat.

Gums are used extensively in salad dressings to provide the viscosity associated with high-fat products (90). Low pH stability by each gum is also required. Maltodextrins can be used, providing better product characteristics than gums (54). An aqueous 25% maltodextrin solution can replace 30–50% of the fat in salad dressings. Simplesse, as well as the fat-based substitutes, can be used in salad dressings and mayonnaise.

Meat applications for fat substitutes include breakfast sausages, hamburger, hot dogs, gravies, and soups. Some of the products used include Paselli SA2, N-Oil, carrageenan, hydrolyzed vegetable protein, cellulose gum, alginate, and maltodextrins (54,91). Carrageenan was used to produce a ground beef patty similar in product characteristics to a high-fat product (92). Spices and a mixture of hydrolyzed vegetable protein and salt (1:2) were added (0.375%) to enhance beef flavor intensity, and 0.5% carrageenan and 10% water were added to enhance other product organoleptic attributes; this was the formulation adapted by McDonald's as their "McLean Deluxe." Others reported that both hydrolyzed soy protein (93) and oat bran (94) can be effective as the basis for a fat reduction system for ground meat. Fat substitutes can even have an effect on fat particle packing structure, affecting the shape and size of the particle (95).

One rather creative use of fat substitutes was the suggestion that sucrose fatty acid esters could be used to inhibit lipoxygenase and thereby improve food quality (96). There was an increase in the binding strength of the sucrose fatty acid monoester and soybean lipoxygenase-1 (L-1) as the fatty acid carbon chain length increased from 8 to 12. Thermodynamic analysis of the binding constants indicated that the binding was hydrophobic in character. Sucrose fatty acid esters can also suppress lipase activity and even have an antibacterial effect in some cases.

Toxicology

As with any new food ingredient, fat substitutes must be tested on animals before they are tested on humans. Ordinarily in animal toxicological tests, food additives are fed at dietary levels severalfold in excess of the concentrations that will occur in foods destined for human consumption (15). This is done to provoke potential toxic responses and to establish safety factors. Because the amount of a fat substitute that could occur in the human diet is very large relative to other food additives such as added colors or flavors, feeding the fat substitutes at very high levels could result in spurious results because this would require reducing a large part of the nutrients in the diet. Munro (97) pointed out that responses that "at first glance may be considered to be of toxicological significance may on further investigation be the result of dietary nutrient imbalance or physiological perturbation induced by the test material when fed at excessive exposure levels." Diets with a large component percentage of fat substitutes could become unpalatable, leading to a poor consumption rate and concomitant poor growth, which might be interpreted as a sign of toxicity.

Measurements of growth, of blood and urine chemistry, plus gross and histologic examination of tissues are often made. In addition, when appropriate, carcinogenicity, genotoxicity, and reproductive and developmental toxicity testing may also be performed. Even if animal testing proves negative, the FDA recognizes that confirmatory studies in humans are an important part of establishing the safety of macronutrient substitutes (15).

In toxicological studies, the potential effects of fat substitutes that may not be evident in standard toxicological tests also have to be considered based on physiologic effects that may be specific to the chemical or physical properties or the mechanism or site of action of the substitute. There is also a need for confirmatory human studies in normal as well as at-risk populations, such as people with diabetes or compromised gastrointestinal (GI) tracts, or abnormalities that could possibly be caused by the fat substitute under consideration. For nonabsorbable fat substitutes, effects on GI epithelium, colonic microflora ecology, bile acid physiology, pancreatic function, and laxation effects should be considered (13,15,97,98). For fat substitutes that are absorbed, absorption, distribution, metabolism, and elimination of the substitute should be considered.

The most exhaustively studied fat substitute has been olestra. Olestra is neither hydrolyzed nor absorbed (44,99). Olestra is not toxic, carcinogenic, mutagenic, or teratogenic; when fed to animals at doses up to 10% of the total diet, there were no noted toxic effects on weight gain, hematology, urinalysis, or tissue pathology (100). Because it is not absorbed, the only organ that olestra contacts is the GI tract. It has no significant effect on gastric emptying, total transit time, fecal water, or pH of pancreas, fecal bile acids, or interohepatic circulation of bile acids. It was reported that for specific GI symptoms (gas, diarrhea, abdominal cramping), there were no significant differences between humans who consumed chips with olestra or triacylglycerols (21,101). The gut microflora do not metabolize olestra under anaerobic conditions, but during waste treatment, it is degraded aerobically in sludge-amended soils (102,103).

References

1. Wylie-Rosett, J., Fat Substitutes and Health—An Advisory from the Nutrition Committee of the American Heart Association, *Circulation 105*:2800–2804 (2002).
2. American Dietetic Association, Position of American Dietetic Association: Fat Replacers, *J. Amer. Diet Assoc. 98*:463–468 (1998).
3. U.S. Department of Health and Human Services, Healthy People 2000: National Health Promotion and Disease Prevention Objectives (Nutrition 2.15 Availability of reduced fat processed food), Washington, DC: U.S. Public Health Service, U.S. Department of Health and Human Services, 1990.
4. Calorie Control Council, Fat Replacers: Food Ingredients for Healthy Eating, available at: http://www.caloriecontrol.org/fatreprint.html, accessed December 11, 2004.
5. Miraglio, A.M., Nutrient Substitutes and Their Energy Values in Fat Substitutes and Replacers, *Am. J. Clin. Nutr.* 62:1175S–1179S (1995).
6. Mokdad, A.H., E.S. Ford, B.A. Bowman, W.H. Dietz, F. Vinicor, S.B. Virginia, and J.S. Marks, Prevalence of Obesity, Diabetes, and Obesity-Related Health Risk Factors, *J. Am. Med. Assoc. 389*:76–79 (2003).
7. World Health Organization, Controlling the Global Obesity Epidemic, WHO official website: www.who.int/nut/obs.htm, (2003) accessed December 11, 2004.
8. Rolls, B.J., Carbohydrates, Fat and Satiety, *Am. J. Clin. Nutr. 61*:960S–967S (1995).
9. Akoh, C.C., Lipid Based Fat Substitutes, *Crit. Rev. Food Sci. Nutr. 35*:405–430 (1995).

10. Artz, W.E., and S.L. Hansen, Current Development in Fat Replacers, in *Food lipids and Health*, edited by R.E. McDonald, R.E. and D.B. Min, Marcel Dekker, Inc., New York, 1996.
11. Artz, W.E. and S.L. Hansen, The Chemistry and Nutrition of Nonnutritive Fats, in *Deep Frying, Chemistry, Nutrition and Practical Applications*, edited by E.G. Perkins and M.D. Erickson, AOCS Press, Champaign, IL, 1996.
12. Grossklaus, R., Fat Replacers—Requirements from a Nutritional Physiological Point of View, *Eur. J. Med. Res. 1*:59–64 (1995).
13. Glueck, C.J., P.A. Streicher, E.K. Illig, and K.D. Weber, Dietary Fat Substitutes, *Nutr. Res. 14*:1605–1619 (1994).
14. Garn, S.M., From the Miocene to Olestra—A Historical Perspective on Fat Consumption, *J. Am. Diet. Assoc. 97*:S54–S57 (1997).
15. Gershoff, S.N., Nutrition Evaluation of Dietary Fat Substitutes, *Nutr. Rev. 53*:305–307 (1995).
16. Borzelleca, J.F., Macronutrient Substitutes: Safety Evaluation, *Regul. Toxicol. Pharmacol. 16*:253–264 (1992).
17. Borzelleca, J.F., A Proposed Model for Safety Assessment of Macronutrient Substitutes, *Regul. Toxicol. Pharmacol. 23*:S15–S18 (1996).
18. Degraaf, C., T. Hulshof, J.A. Weststrate, and J. Hautvast, Nonabsorbable Fat (Sucrose Polyester) and the Regulation of Energy Intake and Body Weight, *Am. J. Physiol. 39*:R1386–R1393 (1996).
19. Miller, G.D., and S.M. Groziak, Impact of Fat Substitutes on Fat Intake, *Lipids 31*:S293–S296 (1996).
20. Akoh, C.C., New Developments in Low Calorie Fats and Oils Substitutes, *J. Food Lipids 3*:223 (1996).
21. Freston, J.W., D.J. Ahnen, S.J. Czinn, D.L. Earnest, M.J. Farthing, S.L. Gorbach, R.H. Hunt, R.S. Sandler, and M.M. Schuster, Review and Analysis of the Effects of Olestra, a Dietary Fat Substitute, on Gastrointestinal Function and Symptoms, *Regul. Toxicol. Pharmacol. 26*:210 (1997).
22. Zorich, N.L., D. Biedermann, K.A. Riccardi, L.J. Bishop, and T.G. Filloon, Randomized, Double-Blind, Placebo-Controlled, Consumer Rechallenge Test of Olean Salted Snacks, *Regul. Toxicol. Pharmacol. 26*:209–218 (1997).
23. Code of Federal Regulations, Foods and Drugs, Title 21, Parts 170–199, Office of the Federal Register, National Archives and Records Administration, U.S. Government Printing Office, Washington, DC, 1997.
24. U.S. Food and Drug Administration, A Food Labeling Guide—Appendix A, October, 2004 update. Available at: http://www.cfsan.fda.gov/~dms/flg-6a.html, accessed December 11, 2004.
25. Anonymous, Fat Substitute Update, *Food Technol. 44*:92, 94, 97 (1990).
26. Arciszewski, H., Fat Functionality, Reduction in Baked Foods, *INFORM 2*:392, 394, 396, 398–399 (1991).
27. Cooper, C.F., Preparation of Propoxylated Glycerides as Dietary Fat Substitutes, European Patent 353,928 (1990).
28. Duxbury, D.D., and N.M. Meinhold, Dietary Fats and Oils, *Food Process. 52*:58 (1991).
29. Dziezak, J.D., Fats, Oils, and Fat Substitutes, *Food Technol. 43*:66–74 (1989).
30. Gillis, A., Fat Substitutes Create New Issues, *J. Am. Oil Chem. Soc. 65*:1708–1712 (1988).

31. Hassel, C.A., Nutritional Implications of Fat Substitutes, *Cereal Foods World 38*:142–144 (1993).
32. White, J.F., and M.R. Pollard, Esterified Epoxide-Extended Polyols as Nondigestible Fat Substitutes of Low-Caloric Value, Eur. Patent 254,547 (1988).
33. White, J.F., and M.R. Pollard, Non-Digestible Fat Substitutes of Low-Calorie Value, U.S. Patent 4,861,613 (1989).
34. White, J.F., and M.R. Pollard, Non-Digestible Fat Substitutes of Low-Calorie Value, Eur. Patent 325,010 (1989).
35. White, J.F., and M.R. Pollard, Low-Calorific and Non-Digestive Substitute of Fat/Oil, China Patent 1,034,572 (1989).
36. White, J.F., Partially Esterified Polysaccharides (PEP) Fat Substitutes, U.S. Patent 4,959,466 (1990).
37. Akoh, C.C., and B.G. Swanson, Synthesis and Properties of Alkyl Glycoside and Stachyose Fatty Acid Polyesters, *J. Am. Oil Chem. Soc. 66*:1295–1301 (1989).
38. Akoh, C.C., and B.G. Swanson, Preparation of Trehalose and Sorbitol Fatty Acid Polyesters by Interesterification, *J. Am. Oil Chem. Soc. 66*:1581–1587 (1989).
39. Drake, M.A., C.W. Nagel, and B.G. Swanson, Sucrose Polyester Content in Foods by a Colorimetric Method, *J. Food Sci. 59*:655 (1994).
40. Rios, J.J., M.C. Perezcamino, G. Marquezrui, and M.C. Dobarganes, Isolation and Characterization of Sucrose Polyesters, *J. Am. Oil Chem. Soc. 71*:385–390 (1994).
41. Akoh, C.C., Synthesis of Carbohydrate Fatty Acid Polyesters, in *Carbohydrate Polyesters as Fat Substitutes*, edited by C.C. Akoh and B.G. Swanson, Marcel Dekker, Inc., New York, 1994.
42. Akoh, C.C., Oxidative Stability of Fat Substitutes and Vegetable Oils by the Oxidative Stability Index Method, *J. Am. Oil Chem. Soc. 71*:211–216 (1994).
43. Akoh, C.C., and B.G. Swanson, Optimized Synthesis of Sucrose Polyesters: Comparison of Physical Properties of Sucrose Polyesters, Raffinose Polyesters and Salad Oils, *J. Food Sci. 55*:236–243 (1990).
44. Mattson, F.H., and R.A. Volpenhein, Hydrolysis of Fully Esterified Alcohols Containing from One to Eight Hydroxyl Groups by the Lipolytic Enzymes of Rat Pancreatic Juice, *J. Lipid Res. 13*:325–328 (1972).
45. Chung, H-Y., J. Park, J.-H. Kim, and U.Y. Kong, Preparation of Sorbitol Fatty Acid Polyesters, Potential Fat Substitutes: Optimization of Reaction Conditions by Response Surface Methodology, *J. Am. Oil Chem. Soc. 73*:637–643 (1996).
46. Riva, S, Enzymatic Synthesis of Carbohydrate Esters, in *Carbohydrate Polyesters as Fat Substitutes*, edited by C.C. Akoh, and B.G. Swanson, Marcel Dekker, Inc., New York, 1994.
47. Mutua, L.N., and C.C. Akoh, Synthesis of Alkyl Glycoside Fatty Acid Esters in Nonaqueous Media by *Candida sp*. Lipase, *J. Am. Oil Chem. Soc. 70*:43–46 (1993).
48. Gardner, D.R., and R.A. Sanders, Isolation and Characterization of Polymers in Heated Olestra and an Olestra/triglyceride Blend, *J. Am. Oil Chem. Soc. 67*:788–792 (1990).
49. Volpenhein, R.A., Synthesis of Higher Polyol Fatty Acid Polyesters Using Carbonate Catalysts, U.S. Patent, 4,517,360 (1985).
50. Yamamoto, T., and K. Kinami, Production of Sucrose Fatty Acid Polyester, U.S. Patent, 4,611,055 (1986).
51. Meyer, R.S., J.M. Root, M.L. Campbell, and D.B. Winter, Low Caloric Alkyl Glycoside Fatty Acid Polyester Fat Substitutes, U.S. Patent 4,840,815 (1989).

52. Albano-Garcia, E., R.G. Lorica, M. Pama, and L. de Leon, Solventless Synthetic Methods for Methyl Glucoside and Sorbitol Esters of Coconut Fatty Acids, *Phillipp. J. Coconut Stud. 5*:51–57 (1980).
53. Ennis, J.L., P.W. Kopf, S.E. Rudolf, and M.F. van Buren, Esterified Alkoxylated Alkyl Glycosides Useful in Low Calorie Fat-Containing Food Compositions, Eur. Patent 415,636 (1991).
54. Frye, A.M., and C.S. Setser, Bulking Agents and Fat Substitutes, in *Low-Calorie Foods Handbook,* edited by A.M. Altschul, Marcel Dekker, Inc., New York, NY, 1993.
55. Tamime, A.Y., M.N.I. Barclay, G. Davies, and E. Barrantes, Production of Low-calorie Yogurt Using Skim Milk Powder and Fat-substitute. 1. A Review, *Milchwissenschaft 49*:85–88 (1994).
56. Tamime, A.Y., E. Barrantes, and A.M. Sword, The Effect of Starch Based Fat Substitutes on the Microstructure of Set-style Yogurt Made from Reconstituted Skimmed Milk Powder, *J. Soc. Dairy Technol. 49*:1–10 (1996).
57. Anonymous, Fats, Oils and Fat Substitutes, *Food Technol. 43*:66–74 (1989).
58. Anonymous, Carbohydrate-Based Ingredient Performs like Fat for Use in a Variety of Food Applications, *Food Technol. 45*:262, 263, 276 (1991).
59. Hippleheuser, A.L., L.A. Landberg, and F.L. Turnak, A System Approach to Formulating a Low-Fat Muffin, *Food Technol. 49*:92–96 (1995).
60. Penichter, K.A., and E.J. McGinley, Cellulose Gel for Fat-Free Food Applications, *Food Technol. 45*:105 (1991).
61. Anonymous, USDA's Oatrim Replaces Fat in Many Food Products, *Food Technol. 44*:100 (1990).
62. Inglett, G.E., and S.B. Grisamore, Maltodextrin Fat Substitute Lowers Cholesterol, *Food Technol. 45*:104 (1991).
63. Singer, N.S., and R.H. Moser, Microparticulated Proteins as Fat Substitutes, in *Low-Calorie Foods Handbook,* edited by A.M. Altschul, Marcel Dekker, Inc., New York, NY, 1993.
64. Cheftel, J.C., and E. Dumay, Microcoagulation of Proteins for Development of Creaminess, *Food Rev. Int. 9*:473–502 (1993).
65. Habib, M., and J.S. Podolski, Concentrated, Substantially Non-Aggregated Casein Micelles as a Fat/Cream Substitute, European Patent Application 0,334,226 (1989).
66. Mai, J., D. Breitbar, and C.D. Fischer, Proteinaceous Material, European Patent Application 0,400,714 (1990).
67. Stark, L.E., and A.T. Gross, Hydrophobic Protein Microparticles and Preparation Thereof, World Patent Application WO 90/03123 (1990).
68. Singer, N.S., S. Yamamoto, and J. Latella, Protein Product Base, U.S. Patent 4,734,287 and European Patent Application 0,250,623 (1988).
69. Anonymous, Stepan Seeks GRAS Status for Captrin, *INFORM 5*:1167 (1994).
70. Smith, R.E., J.W. Finely, and G.A. Leveille, Overview of Salatrim, a Family of Low-Calorie Fats, *J. Agric. Food Chem. 42*:432–434 (1994).
71. Klemann, L.P., K. Aji, M.M. Chrysam, R.P. D'Amelia, J.M. Henderson, A.S. Huang, M.S. Otterburn, and R.G. Yarger, Random Nature of Triacylglycerols Produced by the Catalyzed Interesterification of Short- and Long-Chain Fatty Acid Triglycerides, *J. Agric. Food Chem. 42*:442 (1994).
72. Flickinger, B.D. and N. Matsuo, Nutritional Characteristics of DAG Oil, *Lipids* 38:129–132 (2003).

73. Katsuragei, Y., T. Yasukawa, N. Matsuo, B.D. Flickinger, I. Tokimitsu, and M.G. Matlock, *Diglycerol Oil,* AOCS Press, Champaign, IL, 2004.
74. Barrantes, E., A.Y. Tamime, G. Davies, and M.N.I. Barclay, Production of Low-Calorie Yogurt Using Skim Milk Powder and Fat-Substitute. 2. Compositional Quality, *Milchwissenschaft 49*:135 (1994).
75. Barrantes, E., A.Y. Tamime, and A.M. Sword, Production of Low-Calorie Yogurt Using Skim Milk Powder and Fat-Substitute. 3. Microbiological and Organoleptic Qualities, *Milchwissenschaft 49*:85 (1994).
76. Barrantes, E., A.Y. Tamime, D.D. Muir, D.D., and A.M. Sword, The Effect of Substitution of Fat by Microparticulate Whey Protein on the Quality of Set-Type, Natural Yogurt, *J. Soc. Dairy Technol. 47*:61 (1994).
77. Schmidt, K., A. Lundy, J. Reynolds, J., and L.N. Lee, Carbohydrate or Protein Based Fat Mimicker Effects on Ice Milk Properties, *J. Food Sci. 58*:761 (1993).
78. White, C.H., Low-Fat Dairy Products, in *Low-Calorie Foods Handbook,* edited by A.M. Altschul, Marcel Dekker, Inc., New York, 1993.
79. Vetter, J.L. Low-Calorie Bakery Foods, in *Low-Calorie Foods Handbook,* edited by A.M. Altschul, Marcel Dekker, Inc., New York, 1993.
80. Sharp, T., Technical Constraints in the Development of Reduced-Fat Bakery Products, *Proc. Nutr. Soc. 60*:489–496 (2001).
81. Sanchez, C., C.W. Klopfenstein, and C.E. Walker, Use of Carbohydrate-Based Fat Substitutes and Emulsifying Agents in Reduced-Fat Shortbread Cookies, *Cereal Chem. 72*:25–29 (1995).
82. Zoulias, El., V. Oreopoulou and E. Kounalaki, Effect of Fat and Sugar Replacement on Cookie Properties, *J. Sci. Food Agric. 82*:1637–1644 (2002).
83. Hollenbach, E.J. and N.B. Howard, Emulsion Concentrate for Palatable Polyester Beverage, U.S. Patent 4,368,213 (1983).
84. Robbins, M.B., and S.S. Rodriguez, Low Calorie Baked Products, U.S. Patent 4,461,782 (1984).
85. Roberts, B.A., Oleaginous Compositions, U.S. Patent 4,446,165 (1984).
86. Crites, S.G., M.S. Drake, and B.G. Swanson, Microstructure of Low-Fat Cheddar Cheese Containing Varying Concentrations of Sucrose Polyesters, *Lebensm.-Wiss. Technol. 30*:762–766 (1997).
87. Bachmann, H.P., Cheese Analogues: A Review, *Int. Dairy J. 11* (4–7 Special Issue SI):505–515 (2001).
88. Winter, D.B., R.S. Meyer, J.M. Root, and M.L. Campbell, Process for Producing Low Calorie Foods from Alkyl Glycoside Fatty Acid Polyesters, U.S. Patent 4,942,054 (1990).
89. Phillips, L.G., and D.M. Barbano, The Influence of Fat Substitutes Based on Protein and Titanium Dioxide on the Sensory Properties of Lowfat Milks, *J. Dairy Sci. 80*:2726–2731 (1997).
90. Quesada, L.A., and W.L. Clark, Low-Calorie Foods: General Category, in *Low-Calorie Foods Handbook,* edited by A.M. Altschul, Marcel Dekker, Inc., New York (1993).
91. Keeton, J.T., Low-Fat Meat Products—Technological Problems with Processing, *Meat Sci. 36*:261–276 (1994).
92. Egbert, W.R., D.L. Huffman, and C.-M. Chen, Development of Low-Fat Ground Beef: Low-Fat Hamburgers Introduced by McDonald's Corp., *Food Technol. 45*:64–73 (1991).

93. McMindes, M.K., Applications of Isolated Soy Protein in Low-Fat Meat Products, *Food Technol. 45*:61–64 (1991).
94. Pszczola, D.E., Oat-Bran-Based Ingredient Blend Replaces Fat in Ground Beef and Pork Sausage, *Food Technol. 45*:60–66 (1991).
95. Xu, W., A. Nikolov and D.T. Wasan, Shear-induced Fat Particle Structure Variation and the Stability of Food Emulsions: II. Effects of Surfactants, Protein and Fat Substitutes, *J. Food Eng. 66*:107–116 (2005).
96. Nishiyama, J., Y. Shizu, and T. Kuninori, Inhibition of Soybean Lipoxygenase-1 by Sucrose Esters of Fatty Acids, *Biosci. Biotechnol. Biochem. 57*:557–560 (1993).
97. Munro, I.C., Issues to be Considered in the Safety Evaluation of Fat Substitutes, *Food Chem. Toxicol.* 28:751–784 (1990).
98. Vanderveen, J.E., and W.H. Glinsmann, Fat Substitute: A Regulatory Perspective, *Annu. Rev. Nutr. 12*:473–481 (1992).
99. Miller, K.W., K.D. Lawson, and D.H. Tallmage, Disposition of Ingested Olestra in the Fischer 344 Rat, *Fund. Appl. Toxicol. 24*:229–237 (1995).
100. Bergholtz, C.M., Safety Evaluation of Olestra, a Non-Absorbed Fatlike Fat Replacement, *Crit. Rev. Food Sci. Nutr. 32*:141–163 (1992).
101. Cheskin, L.J., R. Miday, N. Zorich, and T. Filloon, Gastrointestinal Symptoms Following Consumption of Olestra or Regular Triglyceride Potato Chips—A Controlled Comparison, *J. Am. Med. Assoc. 279*:150–152 (1998).
102. Haighbaird, S.D., J. Bus, C. Engelen, and R.N. Hill, Biodegradation of Noncaloric Fat Substitutes Sucrose Polyesters in Sewage Sludge Amended Soil, *Chemosphere 35*:413–425 (1997).
103. Thomson, A.B.R., R.H. Hunt, and N.L. Zorich, Review Article: Olestra and Its Gastrointestinal Safety, *Aliment. Pharmacol. Ther. 12*:1185–1200 (1998).

18 Phospholipids: Structures and Physicochemical Activities

Marilyn C. Erickson

Center for Food Safety and Department of Food Science and Technology, University of Georgia, Griffin, GA

Introduction

Phospholipid is a generic term that refers to lipids containing a phosphoric acid residue. This chapter will focus on these constituents by presenting an overview of their chemical and physical structures. Phospholipids are major constituents of bilayers and membranes, and their presence affects a variety of barrier properties including fluidity, phase transitions, heat capacity, miscibility/solubility, permeability, surface tension, and dipole potential. Because changes in these properties are observed upon physical and chemical modification of phospholipids, a description of the processes (hydrogenation, hydroxylation, hydration, complexation, oxidation, and conjugation) that lead to phospholipid alterations is included.

Phospholipid Chemical Structures

Two major structural groups comprise phospholipids: phosphoglycerides and sphingomyelin. Like triacylglycerols, the backbone of phosphoglycerides is glycerol (a three-carbon alcohol), but only the primary and secondary alcohol residues of glycerol are esterified to fatty acids (long-chain carboxylic acids). The third site is esterified to a phosphate group, which in turn is linked to choline, ethanolamine, serine, inositol, or glycerol (Fig. 1). Sphingomyelin, on the other hand, contains as its backbone, the long-chain base, sphingosine (*trans*-D-erythro-1,3-dihydroxy-2-amino-4-octadecene), an 18-carbon moiety (Fig. 2). Normally, the amino function is amidated with a long-chain fatty acyl chain, the phosphoric acid group is attached to the alcohol group, and a choline group is esterified to the phosphoric acid moiety. Although sphingomyelin may represent a major lipid in certain membranes of animals, it is of minor importance in plants and probably absent from bacteria.

Variation in the structure of phosphoglycerides contributes to the complexity of phospholipids. Although the diacyl forms of the phosphoglycerides predominate, a vinyl ether (O-CH=CH-R) residue can also replace the carboxyl acid ester on carbon 1 of glycerol to form a specialized form of phospholipid, a plasmalogen (1). In the human body, 20% of the phospholipids are plasmalogens with choline and

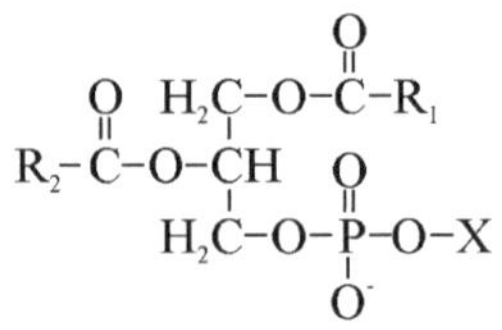

R_1 and R_2 are fatty acids

X is one of the following substituents

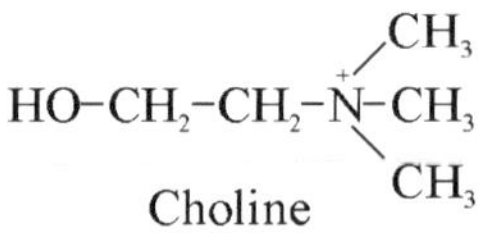

Choline

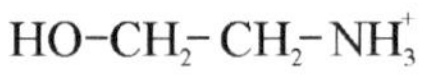

Ethanolamine

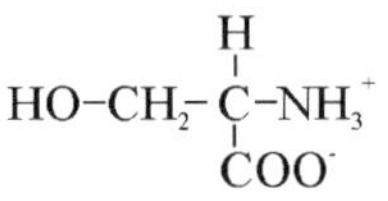

Serine

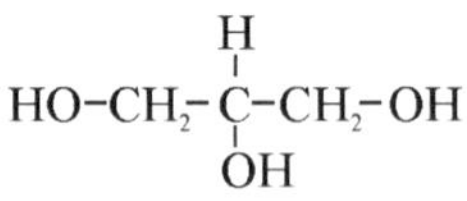

Glycerol

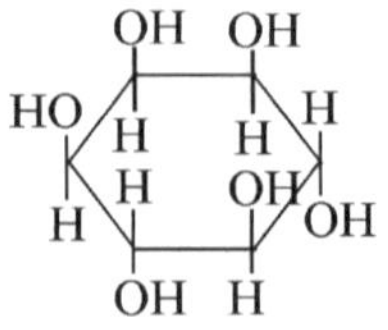

Inositol

Fig. 1. Basic structure of glycerol-based phospholipids.

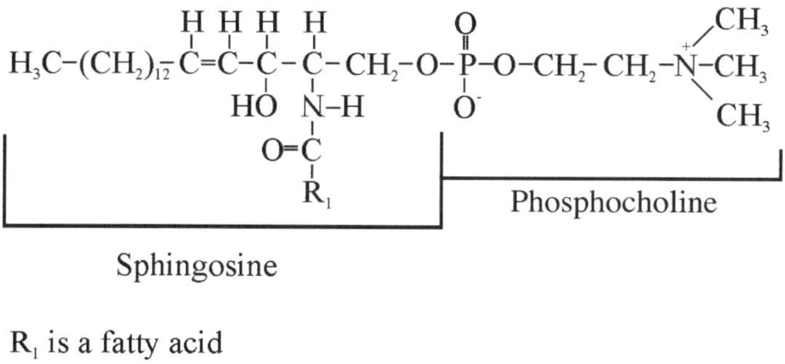

Fig. 2. Structure of sphingomyelin.

ethanolamine plasmalogens as the most common forms. Phosphonolipids, major constituents of three phyla and synthesized by phytoplankton, contain a covalent bond between the phosphorus atom and the carbon of the nitrogenous base and comprise another phosphoglyceride variant (2). Given these variations along with the wide variability in substituents at the C-1 and C-2 positions and the number of different substituents associated with phosphoric acid, the number of permutations in structure can be as great as 1000. Despite this potential complexity, some common denominators exist. Specifically, positional specificity predominates in the phosphoglycerides, with the C-1 position containing primarily (>95%) saturated residues and the C-2 position having >95% unsaturated residues.

Phospholipid Physical Structures

Phospholipids are characterized by the presence of a polar or hydrophilic (water-loving) head group and a nonpolar or hydrophobic (water-hating) fatty acid region. Consequently, with the exception of single-chain lysophospholipids and some charged lipids, such as phosphatidylglycerol, few phospholipids dissolve in water and then only up to a critical concentration (10^{-5}–10^{-10} mol/L). Above this critical concentration, the amphiphilic nature of phospholipids forces the molecules to self-aggregate in aqueous environments to form a variety of macromolecular structures. This ability to adopt different structures is referred to as lipid polymorphism; however, the common feature is that the apolar portions of the amphiphiles form a hydrophobic core, whereas the polar groups build an aggregate surface and interact with the aqueous medium. The most common form of aggregate is the fluid lamellar phase, also known as the bilayer; it consists of two opposing hydrocarbon monolayers and an enclosed aqueous compartment. Inverted hexagonal and cubic phases, on the other hand, are examples of nonlamellar phases in which a water core is not present (3). The particular phase adopted by the phospholipid component depends on temperature, pressure, lipid molecular structure, and concentration, and the composition of the aqueous dispersing medium. Further information on other phospholipid structures may be found in the review of Seddon and Cevc (4).

Phospholipid Bilayer/Membrane Properties

Fluidity/Diffusion/Transbilayer Movement

Lipid bilayers play important roles in cells as barriers for maintaining concentrations and as matrices to support membrane proteins. One of the primary integral properties of a bilayer is its fluidity, and this property in turn is governed by a complex pattern of the components' mobilities. Motions of three different types may be exhibited by individual lipid molecules: lateral, rotational, and transversal diffusion (5). Lateral diffusion refers to the two-dimensional translocation of a lipid molecule in the plane of the bilayer; rotational diffusion is the movement of a lipid molecule around an axis, and transversal diffusion is the redistribution of a lipid molecule between the two leaflets of the bilayer. Although the last-mentioned process is extremely slow in phospholipid bilayers due to the unfavorable passage of a hydrophilic headgroup across the hydrophobic membrane core, the presence of docosahexaenoic acid (22:6) in the phospholipid supports faster flip-flop (6). Transversal phospholipid diffusion may also be mediated in membranes by the presence of flippase proteins (7,8).

As lipid molecules interact, both static and dynamic parameters that affect bilayer/membrane fluidity are encountered. The "order parameter" is a static component and is a measure of the angular range of rotational motion, whereas "microviscosity," a dynamic component, measures the rate of rotational motion.

Numerous physical and chemical factors regulate the fluidity processes of membranes including temperature, pressure, membrane potential, fatty acid composition, protein and/or cholesterol incorporation, and Ca^{2+} concentration. For example, the addition of cholesterol to dimyristoylphosphatidylcholine bilayers changes the lipid packing in the bilayers and reduces the lateral diffusion of the phospholipids to one-fourth that found in the absence of cholesterol (9). In the case of Ca^{2+}, on the other hand, decreases in membrane fluidity are observed through the nonspecific interaction of the electrolyte with acidic phospholipids, causing structural rearrangements (10).

Phase Transitions

As temperatures decrease, phospholipids change from a fluid liquid crystalline state to a frozen gel state. This phase transition may be monitored by a variety of techniques, including nuclear magnetic resonance (NMR), electron spin resonance, fluorescence and differential scanning calorimetry (DSC). In particular, DSC has the capability of determining both the enthalpy (energy required to melt the acyl chains) and cooperativity (number of molecules undergoing the transition simultaneously). Unfortunately, phase transitions are more difficult to determine in membranes due to the entropy/enthalpy compensations encountered because enthalpy lost by lipids is absorbed by membrane proteins partitioning into the more fluid phase of the bilayer (11).

Different phospholipids have different phase transitions; hence, in bilayers containing mixtures of lipids, separation of the component into crystalline domains (lateral phase separation) can occur when the temperature is below the phase transition of the phospholipid with the highest melting temperature. Local gel states, in turn, could subsequently affect protein function by restricting protein mobility in the bilayer matrix. Such localized phase transitions were observed in certain plant food tissues exposed to refrigerator temperatures with an ensuing reduction in product quality (12).

Transition temperatures of phospholipids in membranes are influenced by several compositional factors. The longer the chain length of the fatty acid within a phospholipid class, the higher the transition temperature. The presence of *cis* double bonds in the fatty acid, however, inhibits hydrocarbon packing in the gel state and causes the phase transition to occur at a lower temperature. Differential effects on bilayer phase transition are observed in the presence of free fatty acids. Oleic acid had negligible effects on the bilayer phase transition, whereas palmitic acid increased the bilayer phase transition temperatures (13). Differential effects on bilayer properties also occur with cholesterol incorporation into the bilayer (14). In small amounts ($\leq$3 mol%), a softening of the bilayers in the transition region occurred. Above this cholesterol concentration, a rigidification of the bilayer, characterized as a liquid-ordered phase, occurred. This phase is liquid in the sense that the molecules diffuse laterally as in a fluid, but at the same time, the lipid-acyl chains have a high degree of conformational order.

Heat Capacity

Most phospholipids that can be found in cell membranes have chain-melting temperatures close to a temperature regime of biological relevance. At this melting temperature, the heat capacity reaches a maximum, and therefore heat capacity can be considered another characteristic property of the phospholipid bilayer system. To measure heat capacities in bilayers, calorimetric techniques were used. Using such methods, it was found that relaxation times of lipid systems close to the chain melting transition are proportional to the excess heat capacity. For example, relaxation times in the range of up to 45 s were found in multilamellar systems, which are extremely cooperative and therefore display a very pronounced and narrow heat capacity maximum. Extruded vesicles, on the other hand, display a much broader melting profile, a lower maximum heat capacity, and have relaxation times much faster, i.e., in the range of 3 s (15).

In biomembranes, calorimetric experiments have currently been unable to measure heat capacity maxima. The lack of these anomalies does not imply that there are no heat capacity events, but rather that they are difficult to distinguish from the baseline. One complication is that it is difficult or even impossible to distinguish heat capacity events originating from lipids and from proteins. In addition, the diverse phospholipid composition within biomembranes gives rise to a continuous process of melting events. Therefore, it is conjectured that fluctuations do not occur on a global level but rather locally at domain interfaces or in the lipid interface of proteins. These local events may modify protein activity as was noted with melting and phospholipase A_2 (PLA_2) activity (16).

Miscibility/Solubility

Although it is tempting to consider bilayers and membranes as bulk solutions, they must be viewed as an anisotropic fluid. Insight into the interactions among different types of lipid components was gained from studying temperature-composition phase diagrams. An index of lipid phase diagrams complete with bibliographic references to the source material is found in the review by Koynova and Caffrey (3) and may also be found in the LIPIDAG database (http://www.lipidat.chemistry.ohio-state.edu). Although much of the index is devoted to simple binary or pseudo-binary systems, a sizable fraction refers to ternary, quaternary, and higher-order systems. Many of these more complex systems include nonlipid materials such as amino acids, peptides, proteins, anesthetics, and other additives. Less conventional "phase diagrams" are also included in which lipid phase behavior is reported as a function of pH, osmotic strength, ionic strength, and of temperature or pressure.

Recently, an immense amount of research has focused on immiscibility in bilayers and membranes as it contributes to the establishment of domains. When two lipid components differ in their fluid-phase/ordered-phase transition temperatures by several degrees, fluid-phase and ordered-phase domains can emerge in the temperature range between the two transition temperatures (17). For example, raft-like assembly in mixed bilayers was favored by the presence of long, saturated sphingo- or phospholipids with high phase transition temperature as well as physiological amounts of cholesterol. Moreover, it was demonstrated that in model membranes made of ternary mixtures of a saturated sphingolipid, an unsaturated phospholipid, and cholesterol, the sphingolipid-enriched domains are in the liquid-ordered phase, whereas the phosphoglyceride-rich domains correspond to the liquid-disordered phase (18). Even when saturated phosphoglycerides are included in mixtures, sphingomyelin is preferred as a partner for cholesterol in lipid rafts (19). On the other hand, oleoyl-sphingomyelin, unlike saturated forms of sphingomyelin, does not form segregated domains with cholesterol because of its greater miscibility with phosphatidylcholine (20). Another anisotropic lipid domain pattern evident in phospholipid bilayers is the ripple, a corrugated structure with defined periodicity ranging from 100 to 300 Å, depending on the lipid (21). Ripples appear in a temperature range below the main phase transition and above a low enthalpy transition called the pretransition; they were found to exist in the fluid-phase/ordered-phase coexistence temperature range of a binary phospholipid mixture. In particular, the fluid-phase domains elongate parallel to the ripples (22). The implication of such observations is yet to be fully explored, but it is conjectured that the domain structures that emerge from phase separation in lipid bilayer membranes provide insights into the general organization and functional properties of biological membranes. For example, the possibility exists that some membrane proteins preferentially associate with the ordered lipid chains in complexes (23).

Permeability

One of the most important functions exhibited by phospholipid bilayers and membranes is their ability to act as a barrier to other constituents, in which case, bilayer permeabilities are of relevance. To measure permeabilities of phospholipid bilayers to water, light-scattering techniques are employed whereby the swelling of the bilayers is monitored when osmotic gradients are applied. Under such conditions, water permeability coefficients of liquid-crystalline lipid bilayers range from 10^{-2} to 10^{-4} cm/s (24). Moreover, as the degree of fatty acid unsaturation increases in a phospholipid bilayer or as free fatty acids are added to the bilayer (25,26), the water permeability increases. In contrast, cholesterol reduces water permeability by increasing the order in the hydrocarbon region. Because bilayer structure is also affected by the absence of a carbonyl group in the phospholipid molecule (ether-linked phospholipid bilayer) or by the presence of branched fatty acids, reduced rates of water permeation were recorded under these conditions (27,28). In addition, the influence of the physical state of the bilayer on water permeation is demonstrated by lower permeabilities occurring in the gel-phase than in the liquid-crystalline bilayer (29). Increased complexity to this issue of water permeability is illustrated by the observance of phase changes in phospholipid systems that are triggered by an increase in the osmotic pressure and water permeation. Upon the induction of these structural phase transformations, nonlinear water permeabilities occur.

Oxygen is another constituent whose bilayer permeability is of interest because oxygen transport across biological membranes is fundamental to all aerobic organisms. To assess oxygen permeability, oxygen concentrations in phospholipid bilayers were estimated from the paramagnetic enhancements in relaxation of lipid-soluble spin labels that are induced by interaction with the electron spin of molecular oxygen (30). As an example, in fluid lipid bilayers of dimyristoylphosphatidylcholine, oxygen permeability coefficients of 210 cm/s were measured and appeared to be controlled by the penetration of water via the transmembrane polarity profile (31).

Compared with water, permeability coefficients for nonelectrolytes (uncharged polar solutes) in phospholipid bilayers are at least two orders of magnitude smaller. Properties of the lipid matrix, however, affect the permeation of nonelectrolytes in the same manner as they affect water. That is, decreased unsaturation of phospholipids or increased cholesterol content results in lower permeability coefficients. Because the nonelectrolyte permeability increases as the solubility in a hydrocarbon environment increases, the rate-limiting step in diffusion was ascribed to the initial partitioning of the molecule into the lipid bilayer (32).

Permeability coefficients for electrolytes through phospholipid bilayers are extremely small, and rates of $<10^{-10}$ cm/s are commonly observed. For permeation to proceed, a counterflow of ions of equivalent charge must occur; otherwise, a membrane potential will occur that is equal and opposite to the chemical potential of the diffusing species (32). Transient existence of hydrogen-bonded wires in phospholipid bilayers was hypothesized to account for the greater permeability of H^+ or OH^- (~10^{-4} cm/s) compared with Na^+ and K^+ (10^{-14} cm/s), whereby proton flux, as opposed to monovalent ion flux, occurs as a consequence of H–O–H•••O–H bond rearrangements (33,34).

Ionic permeation of lipid bilayers can be described by two alternative mechanisms: the solubility-diffusion mechanism and the pore mechanism (35). In the former, ions such as halide ions, partition and diffuse across the hydrophobic phase, whereas in the latter, ions traverse the bilayer through transient hydrophilic defects caused by thermal fluctuations. Permeation of monovalent cations, such as potassium, on the other hand, was ascribed to a combination of both mechanisms. In any event, ion permeability appears to be related to the order in the hydrocarbon region, and increased order leads to decreased permeability, as well as to the phospholipid polar headgroup charge. Depending on the surface charge of the latter group, anions and cations could be attracted or repelled to the lipid-water interface.

Curvature/Stress/Surface Tension

The spontaneous curvature of individual phospholipid molecules in a monolayer and the resultant curvature stress within bilayer membranes is thought to affect both membrane protein activity and the energetics of topological changes in membranes such as those required for membrane fusion. Consequently, to determine curvature elasticity of lipid bilayer membranes, several experimental methods were explored, e.g., by amplitude and frequency of thermal fluctuations in the membrane contour (36), by tether formation of large vesicular membranes (37), and by X-ray diffraction of osmotically stressed systems (38). Through the measurement of monolayer spontaneous curvatures and bending modulus, some indication of phase preference was achieved. For example, phosphatidylcholines with zero curvature form flat lamellar L_α phases, whereas phosphatidylethanolamines with negative curvature induce or form the reverse hexagonal H_{II} phase, and lysolipids with large polar groups and positive curvature form either micelles or the hexagonal H_I phase (39). Through insertion of phospholipid molecules having an intrinsic curvature different from the phospholipids comprising a bilayer, membrane tensions or stresses are introduced into the bilayer. These stresses often, but not always, lead to vesicle shape change and ultimately could affect the structure and function of intrinsic membrane proteins (40).

Dipole Potential

Another intrinsic property of phospholipid bilayers is the dipole potential. The dipole potential is considered to be the potential formed between the hydrated headgroups at the membrane surface and the low polar interior of the membrane. It arises primarily from the aligned dipolar residues of phospholipid molecules with the participation of water molecules at the level of the phospholipid's carbonyl and phosphate groups (41). Not to be discounted, though, is the contribution of the acyl chains to membrane dipole potential, arising from their function as a determinant of lipid packing density (42). Ultimately, the derived potential produces a virtual positive charge in the bilayer and is thought to be responsible for the substantial difference in the translocation rates between positively and negatively charged hydrophobic ions, the insertion of excreted peptides, and the activity of certain membrane proteins (41).

Phospholipid Modification

Physical

Modification of phospholipids by external forces such as temperature and pressure leads to volume changes and phase transition shifts. In the case of volume changes, the thermal expansivity of the lipid is composed of contributions from the headgroups, the glycerol backbone, and the acyl chains, weighted with their partial volumes. In contrast, application of static magnetic fields leads to reorientation of phospholipid molecules through the diamagnetic anisotropic properties of the phospholipids. Phospholipid reorientation, in turn, results in deformation of imbedded ion channels. Patch-clamp studies of calcium channels and to a lesser extent sodium channels have provided support for this hypothesis (43).

Chemical

Several chemical processes may be applied to phospholipids and in so doing, the properties of the phospholipids are altered. The chemical processes that will be discussed in more detail below include the following: hydrogenation, hydroxylation, hydration, complexation, oxidation, and conjugation.

Hydrogenation. Addition of hydrogen to double bonds in the fatty acid chains of phospholipids is a process known as hydrogenation. Although more commonly applied to triacylglycerols, hydrogenation of phospholipids improves phospholipid properties as exemplified by hydrogenated lecithin (phosphatidylcholine) functioning well as an emulsifier and as an inhibitor of fat bloom in chocolate (44). The process of hydrogenation involves the mixing of lipid with a catalyst such as nickel, heating, and then exposing the mixture to hydrogen at high pressures with agitation. Unfortunately, phospholipids are not as easily hydrogenated as triacylglycerols; consequently, they require higher temperatures and pressures. For example, hydrogenation of lecithin is carried out at 75–80°C in at least 70 atm pressure (45). Lower temperatures and pressures may be employed, however, when chlorinated solvents or mixtures of these solvents with alcohol are used in the presence of a palladium catalyst (46). Improved hydrogenation of phospholipids is also encountered when fine-grained nickel catalysts are used instead of moderate-grained catalysts (47).

Hydroxylation. Addition of hydroxyl groups to the double bonds of unsaturated phospholipids is referred to as hydroxylation. Such a process applied to lecithin employs hydrogen peroxide in glacial acetic acid and sulfuric acid and improves the stability of lecithin and its dispersability in water and aqueous media (48). Such products were advocated as useful in candy manufacture in which sharp moldings can be obtained when the hydroxylated product is used with starch molds.

Hydrolysis. Both chemical and enzymatic hydrolysis of phospholipids occurs. With chemical acid hydrolysis, the severity of the acid conditions determines the outcome. Under mild conditions, the alk-1-enyl bonds of plasmalogens are cleaved to form long-chain aldehydes. With increasing strength of acid and heating, both phosphatidylinositol and diphosphatidylglycerol are hydrolyzed to form the diglyceride and the phosphorylated head group (inositol phosphate and glyceroldiphosphate). Total hydrolysis into each of the component parts of phospholipids is accomplished with strong acid (49).

Formation of lysophospholipids from unsaturated phosphatidylcholines by hypochlorous acid was also demonstrated using matrix-assisted laser desorption ionization-time of flight (MALDI-TOF) mass spectrometry and ^{31}P NMR spectroscopy. In contrast, saturated phosphatidylcholines were not affected by hypochlorous acid. It has been conjectured that the formation of chlorohydrin groups on the unsaturated fatty acids renders the ester group more accessible to hydrolyzing agents. Based upon the levels of hypochlorous acid generated with acute inflammatory responses, formation of lysophospholipids by this mechanism could be relevant in vitro (50).

Fatty acids and glycerolphosphates are the primary products of mild alkaline hydrolysis (37°C, 0.025–0.1 M NaOH in methanolic or ethanolic solutions) of ester bonds in phospholipids with susceptibility following the order: diacyl > alk-1-enyl, acyl > alkyl, acyl. Spingomyelin, on the other hand, is not affected unless subjected to stronger alkaline conditions. Stronger conditions, however, also lead to further hydrolysis of glycerolphosphates. Irrespective of the conditions, phospholipids in membrane vesicles are more susceptible to hydrolysis than phospholipids present as monomers or Triton X-100 micelles. Hydrolysis of phosphatidylethanolamine is also three times faster than hydrolysis of phosphatidylcholine (51).

Enzymatic hydrolysis of phospholipids is achieved through a series of enzymes known as phospholipases and classified according to the bond that is cleaved; PLA_1 and PLA_2 preferentially hydrolyze the acyl groups at the *sn*-1 and *sn*-2 position, whereas phospholipase C targets the ester linkage between the glycerol backbone and the phosphoryl group, and phospholipase D cleaves the ester linkage on the other side of the phosphoryl group. Although benefits such as improved emulsifying properties were achieved with in vitro application of phospholipases to a phosphatidylcholine mixture, greater interest lies with the in vivo activities of phospholipases. In particular, hydrolysis products of phospholipase serve as secondary messengers in cell signaling pathways in both plants and animals (52,53). With such a critical role, conditions conducive to phospholipase activity including interfacial activation were studied intensively. Both enzyme and substrate hypotheses were advanced for PLA_2 activation. In the enzyme hypothesis, conformational changes in PLA_2 are required, but although data provide evidence that conformational changes may be occurring, the nature of those conformational changes is not well understood. Greater understanding, however, is available regarding the substrate hypothesis, in which the physical properties of the membrane, including membrane fluidity, curvature, and surface charge are considered the major determinants of activation. For example, the highly cationic enzyme PLA_2 (pI > 10.5) has a marked preference for anionic phospholipid interfaces and increasing proportions of the anionic phospholipid, phosphatidic acid, in the membrane enhances interfacial binding and membrane hydrolysis (54,55). The presence of calcium also facilitates binding of PLA_2 by as much as 10-fold (56). Increases in activity were also ascribed to the accumulation of hydrolysis products, free fatty acids, and lysolipids. During these accumulations, small depressions, indicative of phase separation, are noted in the membrane, and it is at these depressions that increases in activity are noted (57,58). Enhanced binding of phospholipases through a similar mechanism was advanced for induction of PLA_2 activity by peroxidation of membrane phospholipids (59). Localized changes in the interfacial water activity at these binding sites were suggested as the mechanism of control of PLA_2 (60).

Hydration. The nature and behavior of water at the surface of bilayers and biological membranes has been of considerable interest for some time. The significance of phospholipid hydration is illustrated by its role in controlling many membrane processes such as membrane transport, ion conductance, and insertion of proteins and other molecules into membranes, and their translocation across the membrane. Dehydration of phospholipids is also suggested to play a role in membrane fusion events.

A number of different methods for measuring the amount of water absorbed by phospholipids have been utilized including gravimetric, X-ray diffraction, neutron diffraction, NMR, and differential scanning calorimetry (61). Inherent in any of these methods is the dependence on sample preparation with variations influencing the water content of phospholipid bilayers (62).

Penetration of water molecules into lipid bilayers is not homogeneous, and although lipid headgroups are charged, the organization of water molecules in their hydration shell was proposed to be similar to an idealized clathrate structure of water around apolar solutes (63). A number of factors determine the number of water molecules in the hydration shell: the type of lipid headgroup, the phase state of lipids, acyl chain composition, the presence of double bonds, and the presence of sterols (64–67). In general, the headgroup hydration increases as the distance between adjacent headgroups is increased. In the case of phosphatidylcholine, up to 34 water molecules were bound by the phospholipids when directly mixed with bulk water, whereas a maximum of 18 water molecules were bound by phosphatidylethanolamine (68–70). With increasing distance between headgroups, direct and indirect interactions among polar groups at the membrane interface become weaker and the number of interlipid links decreases. At the same time, carbonyl and, to a lesser extent, phosphate oxygen atoms become more accessible to water, and phospholipid hydration increases. Detailed analyses also revealed that as the distance between phospholipid headgroups increases, intramolecular water bridges become more common than intermolecular water bridges.

Hydration of a phospholipid appears to be cooperative. Incorporation of the first three to four water molecules on each phospholipid occurs on the phosphate of the lipid headgroup and is exothermic, whereas the remaining water molecules are incorporated endothermically (71). Although differences in electrical conductivity also exist for these two distinct types of bound water, the behavior of the bound water appears to be independent of the headgroup composition (72).

Through neutron diffraction experiments on phospholipid multilayers, water was shown to penetrate into the bilayer headgroup region, but appreciable quantities of water do not reach the hydrocarbon core (73–75). For example, in dilauroylphosphatidylethanolamine bilayers, ~7 and 10 water molecules were found in the gel and liquid crystalline phases, respectively; however, about half of these water molecules were located between adjacent bilayers and the other half in the headgroup region (76,77).

Aggregation of phospholipids adversely affects hydration and occurs when the phospholipid concentration exceeds its critical micelle concentration (cmc), which is dependent on the free energy gained when an isolated amphiphile in solution enters an aggregate. For diacyl phospholipids in water, the cmc in general is quite low, but it depends on both the chain length and the headgroup. For a given chain length, the solubility of charged phospholipids is higher, whereas the cmc of a single-chain phospholipid is higher than that of a diacyl phospholipid with the same headgroup and the same chain length (78).

The presence of monovalent and/or divalent cations in the fluid phase changes the hydration properties of phospholipids. For example, the most extensively studied divalent cation, Ca^{2+}, binds to the phosphate group of phosphatidylserine (79) and liberates water from between bilayers and from the lipid polar groups (78). Monovalent cations, such as Na^+, K^+, or Cs^+, also decrease the fluid spaces between adjacent charged phosphatidylserine and phosphatidylglycerol bilayers as a result of screening of the charge (80,81). A comparison of the effect of different ions reveals a high degree of specificity of ion-lipid interactions, which affects hydration. In infrared (IR) spectroscopic studies of 1-palmitoyl-2-oleoyl-phosphatidylcholine, Ba^{2+}, Sr^{2+}, Na^+, and K^+ only weakly affected hydration, Mg^{2+} and Ca^{2+} caused partial dehydration, and Cu^{2+} and Zn^{2+} caused considerable dehydration of the phosphate and carbonyl groups (82). Such drastic dehydration of lipid headgroups, as occurs with complexation of these latter ions to phospholipids, is suggested to increase fusogenic potency of lipid membranes (83). Membranes of anionic lipids as opposed to membranes of zwitterionic lipids, however, would be more typically affected by these and other metal cations because of the stronger attractive Coulombic forces. For instance, a strong dehydration effect is observed upon cation binding to the acidic phospholipids, in which up to eight water molecules are expelled from the interface once cation-phospholipid association has taken place (84–86).

Complexation to Ions. Electrostatic interactions play a dominant role in the process of ion-membrane binding with phospholipid affinity for cations following the sequence: lanthanides > transition metals > alkaline earths > alkaline metals. Electrostatic forces also play a strong role in lipid-anion binding with affinity for anions by phosphatidylcholine following the sequence: $ClO_4^- > I^- \geq SCN^- > NO_3^- \geq Br^- > Cl^- > SO_4^{2-}$. With anions, however, size also plays an important role in the process of binding, partly as a result of the transfer of the local excess charges from the anion to the phospholipid headgroups and vice versa. An increasing density of net negative charge on the membrane, however, decreases the strength of anion binding to the phospholipid membrane (80). The decreasing density of net negative charge that occurs upon the lateral expansion of the lipid bilayer with a phase transition, on the other hand, decreases the apparent pK value of the anionic phospholipids (87).

Results of NMR, IR spectroscopy, and neutron diffraction studies strongly imply that inorganic cations interact predominantly with the phosphodiester groups of phospholipid head groups (85,86,88–90). Inorganic anions, on the other hand, interact specifically with the trimethylammonium residues of phosphatidylcholine headgroups (91,92). Various degrees of binding exist between phospholipids and ions. Complete displacement of the water molecules from the region between an ion and its binding site corresponds to an inner-sphere complex. Outer-sphere complex formation, on the other hand, exists when only one water molecule is shared between the ion and its ligand. Forces involved in the inner-sphere complex formation include ion-dipole, ion-induced dipole, induced dipole-induced dipole, ion-quadrupole forces, in addition to the Coulombic electrostatic forces. Hydrogen bonding can also participate in inner-sphere complex formation, whereas outer-sphere complexes may be stabilized by "through-water" hydrogen bonding.

Complexation to Cholesterol. The notion that transient microheterogeneities in mammalian cell membranes control many important cellular processes such as signal transduction, membrane fusion, and membrane trafficking is becoming widely accepted. Evidence is also beginning to emerge that suggests that microheterogeneities may also be responsible for producing the diseased states. Hence, the fundamental relations that exist between different lipids and their mixing behavior are of paramount interest.

To assess whether discrete sterol-phospholipid complexes are formed within bilayers, several studies were conducted. In mixtures of cholesterol and either dipalmitoylphosphatidylcholine or dilauroylphosphatidylcholine, complexes preferentially existed in the 2:1 and 1:1 stoichiometries, respectively. Differences in observed populations of complexes were attributed to differences in packing geometries and phospholipid conformations due to the differing tail lengths of the two phosphatidylcholine lipids (93). Selective recognition of phospholipids by sterols was also demonstrated by Sugahara et al. (94) in that a sterol favored C-18 over C-14 phospholipids as nearest neighbors especially when the sterol concentration in the bilayer was high. In the presence of sphingolipids, however, cholesterol interacts preferentially with this lipid compared with phospholipids possibly as a result of a stabilizing hydrogen bond that occurs between the amide group of sphingomyelin and the hydroxyl group of cholesterol (95).

Complexation to Protein. Considerable evidence exists in the literature for the active role of phospholipid molecules in intrinsic membrane proteins. This role includes both modulation of protein conformation and activity (96–98). Through high-resolution structural studies of membrane proteins, details of these lipid-protein interactions are emerging (99). For example, the hydrophobic surface of a membrane protein is covered by a shell of disordered lipids, referred to as boundary or annular lipids; it is equivalent to the solvent layer around a water-soluble protein. Strong evidence for the presence of these annular lipids is the close relation between the number of lipid molecules estimated to surround a membrane protein and the circumference of the protein. Lipid molecules in the annular shell rapidly exchange with the bulk lipids; a lipid molecule remaining in the annular shell typically exchanges at a rate of ~1–2 × 10^7/s at 30°C (100). Binding to the annulus shows relatively little structural specificity, although the presence of a charged or polar headgroup is required to provide good localization of the molecule at the lipid-water interface and to interact with charged residues on the protein flanking the transmembrane region. Distinct from these annular lipids, however, are tightly bound lipid molecules found in deep clefts of protein transmembrane α-helices. Although it has not yet been demonstrated experimentally, it is likely that binding at the nonannular sites will show much more specificity than binding at the annular sites. In any event, packing of transmembrane α-helices is likely to be affected by the structure of the surrounding lipids. In particular, the chain length of the lipids in a bilayer is an important variable determining the activities of membrane proteins. Low activities are seen in too thin and too thick bilayers, consistent with changes in the conformation of the membrane protein, necessary to achieve hydrophobic matching with the bilayer.

Oxidation. Through both enzymatically controlled and random autoxidation processes, phospholipids are susceptible to oxidation. In the case of autoxidation, the reaction is initiated by the interaction of an active oxygen species, such as the hydroxyl free radical or the protonated form of superoxide, with the unsaturated fatty acids of phospholipids. Through this interaction, hydrogen is abstracted from the unsaturated fatty acid of the phospholipid and a lipid free radical is formed. Enzymatic abstraction of hydrogen from an unsaturated fatty acid, on the other hand, occurs when Fe^{3+} at the active site of lipoxygenase, is reduced to Fe^{2+}. The lipid free radical from both enzymatic and nonenzymatic reactions, in turn, reacts with molecular oxygen to form a lipid peroxyl radical. In subsequent lipid-lipid propagation interactions, the lipid peroxyl radical abstracts a hydrogen atom from an adjacent molecule to form a lipid hydroperoxide and a new lipid free radical. Further magnification of oxidation may occur through branching reactions (also known as secondary initiation) in which Fe^{2+} interacts with a hydroperoxide to form a lipid alkoxyl radical and hydroxyl radical, which will then abstract hydrogens from unsaturated fatty acids.

The ramifications of phospholipid oxidation in biological and food systems are immense. On a molecular level, lipid peroxidation was manifested in a decreased hydrocarbon core width and molecular volume (101). In food systems, on the other hand, hydroperoxides, generated during phospholipid oxidation, decompose to aldehydes and ketones. Although these breakdown products are often described by the terms "rancid" and "warmed-over," specific oxidation products may be desirable flavor components (102,103), particularly when formed in more precise (i.e., less random) reactions by the action of lipoxygenase enzymes (104–106) and/or by the modifying influence of tocopherol on autoxidation reactions (107). In biological systems, oxidized phospholipids form the precursors of bioactive fatty acids (108,109) and in some cases, levels of these bioactive phospholipids were shown to be increased in atherosclerotic lesions (110). The complexity of the involvement of phospholipid oxidation products in atherogenesis is illustrated by the demonstration that different mechanisms exist for the activation of cells by different oxidized phospholipids. In addition, it appears that different concentrations of lipids are required for the activity of particular phospholipid oxidation products.

The major factor affecting the oxidative susceptibility of a phospholipid is its fatty acid composition. In addition to decreasing the carbon-hydrogen dissociation energy (111), increasing unsaturation also physically affects oxidation by generating smaller-sized vesicles (112). The larger curvature in the outer bilayer leaflet of these vesicles increases lipid-lipid spacing and hence facilitates penetration by oxidants. In other cases, increased lipid packing promotes oxidation of phospholipids. Stimulation of phospholipid oxidation by trivalent metal ions (Al^{3+}, Sc^{3+}, Ga^{3+}, In^{3+}, Be^{2+}, Y^{3+}, and La^{3+}) was attributed to the capacity of the ions to increase lipid packing and promote the formation of rigid clusters or displacement to the gel state, processes that bring phospholipid acyl chains closer together to favor propagation steps (113,114). Substitution of an enol ether bond for the ester bond at position 1 of the glycerol backbone in plasmalogen phospholipids, on the other hand, leads to inhibition of lipid oxidation. Apparently, plasmalogens do not readily propagate oxidation of polyunsaturated fatty acids because the enol ether double bond binds either to iron (115) or to initiating peroxyl radicals (116).

When present in bilayers or membranes, phospholipids were shown to oxidize more quickly than emulsified triacylglycerols (117) apparently because propagation is facilitated by the arrangement of phospholipid fatty acids in the membrane. The presence of phospholipids, however, does not preclude acceleration of lipid oxidation. When present as a minor component of oil systems, solubilized phospholipids have limited the oxidation of the triacylglycerols (118,119). The order of effectiveness of individual phospholipids was as follows: sphingomyelin = lysophosphatidylcholine = phosphatidylcholine = phosphatidylethanolamine > phosphatidylserine > phosphatidylinositol > phosphatidylglycerol (120) with both the amino and hydroxyl groups in the side chain participating in the antioxidant activity (121). Variation within the phospholipid classes toward oxidation was also ascribed to the iron-trapping ability of the polar headgroup (122). For example, phosphatidylserine was shown to inhibit lipid oxidation of phosphatidylcholine hydroperoxides induced by a ferrous-ascorbate system (123).

Conjugation. Covalent modification of phospholipids for purposes of increasing the therapeutic agent's in vivo efficacy is a relatively new avenue of research. For example, synthesis of "pseudophospholipids" containing either valproic acid or ibuprofen instead of a fatty acid at the *sn*-1 position, was achieved (124). The surface properties and aggregation behavior resembled those properties of natural phospholipids; however, the drug molecules were resistant to hydrolysis by phospholipase A_2.

References

1. Nagan, N., and R.A. Zoeller, Plasmalogens: Biosynthesis and Functions, *Prog. Lipid Res. 40*:199–229 (2001).
2. Moschidis, M.C., Phosphonolipids, *Prog. Lipid Res. 23*:223–246 (1985).
3. Koynova, R., and M. Caffrey, An Index of Lipid Phase Diagrams, *Chem. Phys. Lipids 115*:107–219 (2002).
4. Seddon, J.M., and G. Cevc, Lipid Polymorphism: Structure and Stability of Lyotrophic Mesophases of Phospholipids, in *Phospholipids Handbook*, edited by G. Sevc, Marcel Dekker, Inc., New York, 1993, pp. 403–454.
5. Smith, R.L., and E. Oldfield, Dynamic Structure of Membranes by Deuterium NMR, *Sci. 225*:280–288 (1984).
6. Armstrong, V.T., M.R. Brzustowicz, S.R. Wassall, L.J. Jenski, and W. Stillwell, Rapid Flip-Flop in Polyunsaturated (Docosahexaenoate) Phospholipid Membranes, *Arch. Biochem. Biophys. 414*:74–82 (2003).
7. Hermann, A., A. Zachowski, and P.F. Devaux, Protein-Mediated Phospholipid Translocation in the Endoplasmic Reticulum with a Low Lipid Specificity, *Biochemistry 29*:2023–2027 (1990).
8. Kubelt, J., A.K. Menon, P. Müller, and A. Herrmann, Transbilayer Movement of Fluorescent Phospholipid Analogs in the Cytoplasmic Membrane of *E. coli*, *Biochemistry 41*:5605–5612 (2002).
9. Orädd, G., G. Lindblom, and P.W. Westerman, Lateral Diffusion of Cholesterol and Dimyristoylphosphatidylcholine in a Lipid Bilayer Measured by Pulsed Field Gradient NMR Spectroscopy, *Biophys. J. 83*:2702–2704 (2002).
10. Shinitzky, M., Membrane Fluidity and Cellular Functions, in *Physiology of Membrane Fluidity*, edited by M. Shinitzky, CRC Press, Boca Raton, FL, 1984, Chap. 1.

11. Silvius, J.R., Thermotrophic Phase Transitions of Pure Lipids in Model Membranes and Their Modification by Membrane Proteins, in *Lipid-Protein Interactions, Vol. 2*, edited by P.C. Jost and O.H. Griffith, Wiley, New York, 1982, Chap. 7.
12. Stanley, D.W., Biological Membrane Deterioration and Associated Quality Losses in Food Tissues, *Crit. Rev. Food Sci. Nutr. 30*:487–553 (1991).
13. Inoue, T., S-I. Yanagihara, Y. Misono, and M. Suzuki, Effect of Fatty Acids on Phase Behavior of Hydrated Dipalmitoylphosphatidylcholine Bilayer: Saturated Versus Unsaturated Fatty Acids, *Chem. Phys. Lipids 109*:117–133 (2001).
14. Lemmich, J., K. Mortensen, J.H. Ipsen, T. Hønger, R. Bauer, and O.G. Mouritsen, The Effect of Cholesterol in Small Amounts on Lipid-Bilayer Softness in the Region of the Main Phase Transition, *Eur. Biophys. J. 25*:293–304 (1997).
15. Grabitz, P., V.P. Ivanova, and T. Heimburg, Relaxation Kinetics of Lipid Membranes and Its Relation to the Heat Capacity, *Biophys. J. 82*:299–309 (2002).
16. Burack, W., Q. Yuan, and R. Biltonen, Role of Lateral Phase Separation in the Modulation of Phospholipase A_2 Activity, *Biochemistry 32*:583–589 (1993).
17. Bagatolli, L.A., and E. Gratton, Two Photon Fluorescence Microscopy of Coexisting Lipid Domains in Giant Unilamellar Vesicles of Binary Phospholipid Mixtures, *Biophys. J. 78*:290–305 (2000).
18. Brown, D.A., and E. London, Structure and Function of Sphingolipid- and Cholesterol-Rich Membrane Rafts, *J. Biol. Chem. 275*:17221–17224 (2000).
19. Scherfeld, D., N. Kahya, and P. Schwille, Lipid Dynamics and Domain Formation in Model Membranes Composed of Ternary Mixtures of Unsaturated and Saturated Phosphatidylcholines and Cholesterol, *Biophys. J. 85*:3758–3768 (2003).
20. Epand, R.M., and R.F. Epand, Non-Raft Forming Sphingomyelin-Cholesterol Mixtures, *Chem. Phys. Lipids 132*:37–46 (2004).
21. Janiak, M.J., D.M. Small, and G.G. Shipley, Temperature and Compositional Dependence of the Structure of Hydrated Dimyristoyl Lecithin, *J. Biol. Chem. 254*:6068–6078 (1979).
22. Leidy, C., T. Kaasgaard, J.H. Crowe, O.G. Mouritsen, and K. Jørgensen, Ripples and the Formation of Anisotropic Lipid Domains: Imaging Two-Component Supported Double Bilayers by Atomic Force Microscopy, *Biophys. J. 83*:2625–2633 (2002).
23. Anderson, R.G.W., and K. Jacobson, Cell Biology: A Role for Lipid Shells in Targeting Proteins to Caveolae, Rafts, and Other Lipid Domains, *Science 296*:1821–1825 (2002).
24. Fettiplace, R., I.G.H. Gordon, S.B. Hladky, J. Requens, H.B. Zingshen, and D.A. Haydon, Techniques in the Formation and Examination of Black Lipid Bilayer Membranes, in *Methods in Membrane Biology, Vol. 4*, edited by E.D. Korn, Plenum Press, New York, 1974, pp. 1–75.
25. Blok, M.C., L.L.M. van Deenen, and J. De Gier, Effect of the Gel to Liquid Crystalline Phase Transition on the Osmotic Behavior of Phosphatidylcholine Liposomes, *Biochim. Biophys. Acta 433*:1–12 (1976).
26. Langner, M., and S.W. Hui, Effect of Free Fatty Acids on the Permeability of 1,2-Dimyristoyl-*sn*-glycero-3-phosphocholine Bilayer at the Main Phase Transition, *Biochim. Biophys. Acta 1463*:439–447 (2000).
27. Shinoda, K., W. Shinoda, T. Baba, and M. Mikami, Comparative Molecular Dynamics Study of Ether- and Ester-Linked Phospholipid Bilayers, *J. Chem. Phys. 121*:9648–9654 (2004).
28. Shinoda, W., M. Mikami, T. Baba, and M. Hato, Dynamics of a Highly Branched Lipid Bilayer: a Molecular Dynamics Study, *Chem. Phys. Lett. 390*:35–40 (2004).

29. Sparr, E., and H. Wennerström, Responding Phospholipid Membranes—Interplay Between Hydration and Permeability, *Biophys. J. 81*:1014–1028 (2001).
30. Hyde, J.S., and W.K. Subczynski, Simulation of Electron Spin Resonance Spectra of the Oxygen-Sensitive Spin Label Probe CTPO, *J. Magn. Reson. 56*:125–130 (1984).
31. Dzikovski, B.G., V.A. Livshits, and D. Marsh, Oxygen Permeation Profile in Lipid Membranes: Comparison with Transmembrane Polarity Profile, *Biophys. J. 85*:1005–1012 (2003).
32. Poznansky, M., S. Tang, P.C. White, J.M. Milgram, and M. Selenen, Non-Electrolyte Diffusion Across Lipid Bilayer Systems, *J. Gen. Physiol. 67*:45–66 (1976).
33. Deamer, D.W., Proton Permeability in Biological and Model Membranes, in *Intracellular pH: Its Measurement, Regulation and Utilization in Cellular Functions*, edited by R. Nuccitelli and D.W. Deamer, Liss, New York, 1982, pp. 173–187.
34. Krishnamoorthy, I., and G. Krishnamoorthy, Probing the Link Between Proton Transport and Water Content in Lipid Membranes, *J. Phys. Chem. B 105*:1484–1488 (2001).
35. Paula, S., A.G. Volkov, and D.W. Deamer, Permeation of Halide Anions Through Phospholipid Bilayers Occurs by the Solubility-Diffusion Mechanism, *Biophys. J. 74*:319–327 (1998).
36. Servuss, R.M., W. Harbich, and W. Helfrich, Measurement of the Curvature-Elastic Modulus of Egg Lecithin Bilayers, *Biochim. Biophys. Acta 436*:900–903 (1976).
37. Waugh, R.E., J. Song, S. Svetina, and B. Zeks, Local and Nonlocal Curvature Elasticity in Bilayer Membranes by Tether Formation from Lecithin Vesicles, *Biophys. J. 61*:974–982 (1992).
38. Fuller, N., and R.P. Rand, The Influence of Lysolipids on the Spontaneous Curvature and Bending Elasticity of Phospholipid Membranes, *Biophys. J. 81*:243–254 (2001).
39. Szule, J.A., N.L. Fuller, and R.P. Rand, The Effects of Acyl Chain Length and Saturation of Diacylglycerols and Phosphatidylcholines on Membrane Monolayer Curvature, *Biophys. J. 83*:977–984 (2002).
40. Traïkia, M., D.E. Warschawski, O. Lambert, J-L. Rigaud, and P.F. Devaux, Asymmetrical Membranes and Surface Tension, *Biophys. J. 83*:1443–1454 (2002).
41. Brockman, H., Dipole Potential of Lipid Membranes, *Chem. Phys. Lipids 73*:57–79 (1994).
42. Peterson, U., D.A. Mannock, R.N.A.H. Lewis, P. Pohl, R.N. McElhaney, and E.E. Pohl, Origin of Membrane Dipole Potential: Contribution of the Phospholipids Fatty Acid Chains, *Chem. Phys. Lipids 117*:19–27 (2002).
43. Rosen, A.D., Mechanism of Action of Moderate-Intensity Static Magnetic Fields on Biological Systems, *Cell Biochem. Biophys. 39*:163–174 (2003).
44. Julian, P.L., Treating Phosphatides, U.S. Patent 2,629,662 (1953).
45. Jacini, G., Hydrogenation of Phosphatides, U.S. Patent 2,870,179 (1959).
46. Cole, R.D., Hydrogenated lecithin, U.S. Patent 2,907,777 (1959).
47. Szukalska, E., Effect of Phospholipid Structure on Kinetics and Chemistry of Soybean Oil Hydrogenation with Nickel Catalysts, *Eur. J. Lipid Sci. Technol. 102*:739–745 (2000).
48. Wittcoff, H., Hydroxyphosphatides, U.S. Patent 2,445,948 (1948).
49. Hanahan, D.J., J. Ekholm, and C.M. Jackson, The Structure of Glyceryl Ethers and the Glyceryl Ether Phospholipids of Bovine Erythrocytes, *Biochemistry 2*:630–641 (1963).
50. Arnhold, J., A.N. Osipov, H. Spalteholz, O.M. Panasenko, and J. Schiller, Formation of Lysophospholipids from Unsaturated Phosphatidylcholines Under the Influence of Hypochlorous Acid, *Biochim. Biophys. Acta 1572*:91–100 (2002).

51. Kensil, C.R., and E.A. Dennis, Alkaline Hydrolysis in Model Membranes and the Dependence on Their State of Aggregation, *Biochem. 20*:6079–6085 (1981).
52. Hurley, J.H., Y. Tsujishita, and M.A. Pearson, Floundering About at Cell Membranes: a Structural View of Phospholipid Signaling, *Curr. Opin. Struct. Biol. 10*:737–743 (2000).
53. Ryu, S.B., Phospholipid-Derived Signaling Mediated by Phospholipase A in Plants, *Trends Plant Sci. 9*:229–235 (2004).
54. Kinkaid, A.R., R. Othman, J. Voysey, and D.C. Wilton, Phospholipase D and Phosphatidic Acid Enhance the Hydrolysis of Phospholipids in Vesicles and in Cell Membranes by Human Secreted Phospholipase A_2, *Biochim. Biophys. Acta 1390*:173–185 (1998).
55. Tatulian, S.A., Toward Understanding Interfacial Activation of Secretory Phospholipase A_2 (PLA_2): Membrane Surface Properties and Membrane-Induced Structural Changes in the Enzyme Contribute Synergistically to PLA_2 Activation, *Biophys. J. 80*:789–800 (2001).
56. Hixon, M.S., A. Ball, and M.H. Gelb, Calcium-Dependent and -Independent Interfacial Binding and Catalysis of Cytosolic Group IV Phospholipase A_2, *Biochemistry 37*:8516–8526 (1998).
57. Hyvönen, M.T., I. Öörni, P.T. Kovanen, and M. Ala-Korpela, Changes in a Phospholipid Bilayer Induced by the Hydrolysis of a Phospholipase A_2 Enzyme: A Molecular Dynamics Simulation Study, *Biophys. J. 80*:565–578 (2001).
58. Nielsen, L.K., K. Balashev, T.H. Callisen, and T. Bjørnholm, Influence of Product Phase Separation on Phospholipase A_2 Hydrolysis of Supported Phospholipid Bilayers Studied by Force Microscopy, *Biophys. J. 83*:2617–2624 (2002).
59. RashbaStep, J., A. Tatoyan, R. Duncan, D. Ann, T.R. RushpaRehka, and A. Sevanian, Phospholipid Peroxidation Induces Cytosolic Phospholipase A_2 Activity: Membrane Effects Versus Enzyme Phosphorylation, *Arch. Biochem. Biophys. 343*:44–54 (1997).
60. Rao, C.S., and S. Damodaran, Surface Pressure Dependence of Phospholipase A_2 Activity in Lipid Monolayers Is Linked to Interfacial Water Activity, *Colloids Surf. B: Biointerfaces 34*:197–204 (2004).
61. McIntosh, T.J., and A.D. Magid, Phospholipid Hydration, in *Phospholipids Handbook*, edited by G. Cevc, Marcel Dekker, New York, 1993, pp. 553–577.
62. Klose, G., B. Konig, H.W. Meyer, G. Schulze, and G. Degovics, Small-Angle X-Ray Scattering and Electron Microscopy of Crude Dispersions of Swelling Lipids and the Influence of the Morphology on the Repeat Distance, *Chem. Phys. Lipids 47*:225–234 (1988).
63. Alper, H.E., D. Bassolino-Klimas, and T.R. Stouch, The Limiting Behavior of Water Hydrating a Phospholipid Monolayer: A Computer Simulation Study, *J. Chem. Phys. 99*:5547–5559 (1993).
64. Jendrasiak, G.L., and J.H. Hasty, The Hydration of Phospholipids, *Biochim. Biophys. Acta 337*:79–91 (1974).
65. McIntosh, T.J., Hydration Properties of Lamellar and Non-Lamellar Phases of Phosphatidylcholine and Phosphatidylethanolamine, *Chem. Phys. Lipids 81*:117–131 (1996).
66. Murzyn, K., T. Róg, G. Jezierski, Y. Takaoka, and M. Pasenkiewicz-Gierula, Effects of Phospholipid Unsaturation on the Membrane/Water Interface: A Molecular Simulation Study, *Biophys. J. 81*:170–183 (2001).
67. Klymchenko, A.S., Y. Mély, A.P. Demchenko, and G. Duportail, Simultaneous Probing of Hydration and Polarity of Lipid Bilayers with 3-Hydroxyflavone Fluorescent Dyes, *Biochim. Biophys. Acta 1665*:6–19 (2004).
68. Small, D.M., Phase Equilibria and Structure of Dry and Hydrated Egg Lecithin, *J. Lipid Res. 8*:551–557 (1967).

69. LeNeveu, D.M., R.P. Rand, V.A. Parsegian, and D. Gingell, Measurement and Modification of Forces Between Lecithin Bilayers, *Biophys. J. 18*:209–230 (1977).
70. Gruner, S.M., M.W. Tate, G.L. Kirk, P.T.C. So, D.C. Turner, D.T. Keane, C.P.S. Tilcock, and P.R. Cullis, X-Ray Diffraction Study of the Polymorphic Behavior of *N*-Methylated Dioleoylphosphatidylethanolamine, *Biochemistry 27*:2853–2866 (1988).
71. Markova, N., E. Sparr, L. Wadso, and H. Wennerstrom, A Calorimetric Study of Phospholipid Hydration. Simultaneous Monitoring of Enthalpy and Free Energy, *J. Phys. Chem. B 104*:8053–8060 (2000).
72. Jendrasiak, G.L., and R.L. Smith, The Interaction of Water with the Phospholipid Head Group and Its Relationship to the Lipid Electrical Conductivity, *Chem. Phys. Lipids 131*:183–195 (2004).
73. Zaccai, G., J.K. Blasie, and B.P. Schoenborn, Neutron Diffraction Studies on the Location of Water in Lecithin Bilayer Model Membranes, *Proc. Natl. Acad. Sci. USA 72*:376–380 (1975).
74. Simon, S.A., and T.J. McIntosh, Depth of Water Penetration into Lipid Bilayers, *Methods Enzymol. 127*:511–521 (1986).
75. McDaniel, R.V., and T.J. McIntosh, Neutron and X-Ray Diffraction Structural Analysis of Phosphatidylinositol Bilayers, *Biochim. Biophys. Acta 983*:241–246 (1989).
76. McIntosh, T.J., and S.A. Simon, Area per Molecule and Distribution of Water in Fully Hydrated Dilauroylphosphatidylethanolamine Bilayers, *Biochemistry 25*:4948–4952 (1986).
77. McIntosh, T.J., and S.A. Simon, Area per Molecule and Distribution of Water in Fully Hydrated Dilauroylphosphatidylethanolamine Bilayers. Corrections, *Biochemistry 25*:8474 (1986).
78. Tanford, C, *The Hydrophobic Effect,* 2nd ed., John Wiley, New York, 1980.
79. Hauser, H., E.G. Finer, and A. Darke, Crystalline Anhydrous Ca-Phosphatidylserine Bilayers, *Biochem. Biophys. Res. Commun. 76*:267–274 (1977).
80. Loosley-Millman, M.E., R.P. Rand, and V.A. Parsegian, Effects of Monovalent Ion Binding and Screening on Measured Electrostatic Forces Between Charged Phospholipid Bilayers, *Biophys. J. 40*:221–232 (1982).
81. Hauser, H., and G.G. Shipley, Interaction of Divalent Cations with Phosphatidylserine Bilayer Membranes, *Biochemistry 23*:34–41 (1984).
82. Binder, H., and O. Zschörnig, The Effect of Metal Cations on the Phase Behavior and Hydration Characteristics of Phospholipid Membranes, *Chem. Phys. Lipids 115*:39–61 (2002).
83. Binder, H., K. Arnold, A.S. Ulrich, and O. Zschörnig, Interaction of Zn^{2+} with Phospholipid Membranes, *Biophys. Chem. 90*:57–74 (2001).
84. Dluhy, R.A., D.G. Cameron, H.H. Mantsch, and R. Mendelsohn, Fourier Transform Infrared Spectroscopic Studies of the Effect of Calcium Ions on Phosphatidylserine, *Biochemistry 22*:6318–6325 (1983).
85. Casal, H.L., H.H. Mantsch, and H. Hauser, Infrared Studies of Fully Hydrated Saturated Phosphatidylserine Bilayers. Effect of Li^+ and Ca^{2+}, *Biochemistry 26*:4408–4416 (1987).
86. Casal, H.L., A. Martin, H.H. Mantsch, F. Paltauf, and H. Hauser, Infrared Studies of Fully Hydrated Unsaturated Phosphatidylserine Bilayers. Effect of Li^+ and Ca^{2+}, *Biochemistry 26*:7395–7401 (1987).
87. Träuble, H., Membrane Electrostatics, in *Structure of Biological Membranes*, edited by S. Abrahamsson and I. Pascher, Plenum Press, New York, 1977, pp. 509–550.

88. Hauser, H., M.C. Phillips, B.A. Levine, and R.J.P. Williams, Ion-Binding to Phospholipids. Interaction of Calcium and Lanthanide Ions with Phosphatidylcholine, *Eur. J. Biochem. 58*:133–144 (1975).
89. Nolden, P.W., and T. Ackermann, A High-Resolution NMR Study (^{1}H, ^{13}C, ^{31}P) of the Interaction of Paramagnetic Ions with Phospholipids in Aqueous Dispersions, *Biophys. Chem. 4*:297–304 (1976).
90. Herbette, L., C. Napolitano, and R.V. McDaniel, Direct Determination of the Calcium Profile Structure for Dipalmitoyllecithin Multilayers Using Neutron Diffraction, *Biophys. J. 46*:677–685 (1984).
91. Jendrasiak, G.L., Halide Interaction with Phospholipids: Proton Magnetic Resonance Studies, *Chem. Phys. Lipids 9*:133–146 (1972).
92. MacDonald, P.J., and J. Seelig, Anion Binding to Neutral and Positively Charged Lipid Membranes, *Biochemistry 27*:6769–6775 (1988).
93. Pandit, S.A., D. Bostick, and M.L. Berkowitz, Complexation of Phosphatidylcholine Lipids with Cholesterol, *Biophys. J. 86*:1345–1356 (2004).
94. Sugahara, M., M. Uragami, and S.L. Regen, Selective Sterol-Phospholipid Associations in Fluid Bilayers, *J. Am. Chem. Soc. 124*:4253–4256 (2002).
95. Slotte, J.P., Sphingomyelin-Cholesterol Interactions in Biological and Model Membranes, *Chem. Phys. Lipids 102*:13–27 (1999).
96. Brown, M.F., Modulation of Rhodopsin Function by Properties of the Membrane Bilayer, *Chem. Phys. Lipids 73*:159–180 (1994).
97. Burack, W.R., and R.L. Biltonen, Lipid Bilayer Heterogeneities and Modulation of Phospholipase A_2 Activity, *Chem. Phys. Lipids 73*:209–222 (1994).
98. Sintes, T., and A. Baumgärtner, Protein Attraction in Membranes Induced by Lipid Fluctuations, *Biophys. J. 73*:2251–2259 (1997).
99. Lee, A.G., Lipid-Protein Interactions in Biological Membranes: A Structural Perspective, *Biochim. Biophys. Acta 1612*:1–40 (2003).
100. Marsh, D., and L.I. Horváth, Structure, Dynamics and Composition of the Lipid-Protein Interface. Perspectives from Spin-Labelling, *Biochim. Biophys. Acta 1376*:267–296 (1998).
101. Mason, R.P., M.F. Walter, and P.E. Mason, Effect of Oxidative Stress on Membrane Structure: Small-Angle X-Ray Diffraction Analysis, *Free Radic. Biol. Med. 23*:419–425 (1977).
102. Josephson, D.B., R.C. Lindsay, and D.A. Stuiber, Identification of Compounds Characterizing the Aroma of Fresh Whitefish (*Coregonus clupeaformis*), *J. Agric. Food Chem. 31*:326–330 (1983).
103. Lindsay, R.C., Fish Flavors, *Food Rev. Int. 6*:437–455 (1990).
104. Josephson, D.B., R.C. Lindsay, and D.A. Stuiber, Variations in the Occurrences of Enzymically Derived Volatile Aroma Compounds in Salt- and Freshwater Fish, *J. Agric. Food Chem. 32*:1344–1347 (1984).
105. Hsieh, R.J., and J.E. Kinsella, Lipoxygenase-Catalyzed Oxidation of N-6 and N-3 Polyunsaturated Fatty Acids: Relevance to and Activity in Fish Tissue, *J. Food Sci. 51*:940–945, 996 (1986).
106. Hsieh, R.J., and J.E. Kinsella, Lipoxygenase Generation of Specific Volatile Flavor Carbonyl Compounds in Fish Tissues, *J. Agric. Food Chem. 37*:279–286 (1989).
107. Karahadian, C., and R.C. Lindsay, Action of Tocopherol-Type Compounds in Directing Reactions Forming Flavor Compounds in Autoxidizing Fish Oils, *J. Am. Oil Chem. Soc. 66*:1302–1308 (1989).

108. Morrow, J.D., and L.J. Roberts, The Isoprostanes. Current Knowledge and Directions for Future Research, *Biochem. Pharmacol. 51*:1–9 (1996).
109. Salomon, R.G., W. Sha, C. Brame, K. Kaur, G. Subbanagounder, J. O'Neill, H.F. Hoff, and L.J. Roberts, Protein Adducts of Iso[4]levuglandin E2, a Product of the Isoprostane Pathway, in Oxidized Low Density Lipoprotein, *J. Biol. Chem. 274*:20271–20280 (1999).
110. Berliner, J.A., G. Subbanagounder, N. Leitinger, A.D. Watson, and D. Vora, Evidence For a Role of Phospholipid Oxidation Products in Atherogenesis, *Trends Cardiovasc. Med. 11*:142–147 (2001).
111. Adachi, S., T. Ishiguro, and R. Matsuno, Autoxidation Kinetics for Fatty Acids and Their Esters, *J. Am. Oil Chem. Soc. 72*:547–551 (1995).
112. Li, Q.T., M.H. Yee, and B.K. Tan, Lipid Peroxidation in Small and Large Phospholipid Unilamellar Vesicles Induced by Water-Soluble Free Radical Sources, *Biochem. Biophys. Res. Commun. 273*:72–76 (2000).
113. Verstraeten, S.V., L.V. Nogueira, S. Schreier, and P.I. Oteiza, Effect of Trivalent Metal Ions on Phase Separation and Membrane Lipid Packing: Role in Lipid Peroxidation, *Arch. Biochem. Biophys. 338*:121–127 (1997).
114. Verstraeten, S.V., and P.I. Oteiza, Effects of Al^{3+} and Related Metals on Membrane Phase State and Hydration: Correlation with Lipid Oxidation, *Arch. Biochem. Biophys. 375*:340–346 (2000).
115. Zommara, M., N. Tachibana, K. Mitsui, N. Nakatani, M. Sakono, I. Ikeda, and K. Imaizumi, Inhibitory Effect of Ethanolamine Plasmalogen on Iron- and Copper-Dependent Lipid Peroxidation, *Free Radic. Biol. Med. 18*:599–602 (1995).
116. Reiss, D., K. Beyer, and B. Engelmann, Delayed Oxidative Degradation of Polyunsaturated Diacyl Phospholipids in the Presence of Plasmalogen Phospholipids In Vitro, *Biochem. J. 323*:807–814 (1997).
117. Slabyj, B.M., and H.O. Hultin, Oxidation of a Lipid Emulsion by a Peroxidizing Microsomal Fraction from Herring Muscle, *J. Food Sci. 49*:1392–1393 (1984).
118. Hildebrand, D.H., J. Terao, and M. Kito, Phospholipids Plus Tocopherols Increase Soybean Oil Stability, *J. Am. Oil Chem. Soc. 61*:552–555 (1984).
119. King, M.C., L.C. Boyd, and B.W. Sheldon, Effect of Phospholipids on Lipid Oxidation of a Salmon Oil Model System, *J. Am. Oil Chem. Soc. 69*:237–242 (1992).
120. King, M.C., L.C. Boyd, and B.W. Sheldon, Antioxidant Properties of Individual Phospholipids in a Salmon Oil Model System, *J. Am. Oil Chem. Soc. 69*:545–551 (1992).
121. Saito, H., and K. Ishihara, Antioxidant Activity and Active Sites of Phospholipids as Antioxidants, *J. Am. Oil Chem. Soc. 74*:1531–1536 (1997).
122. Tadolini, B., P. Motta, and C.A. Rossi, Iron Binding to Liposomes of Different Phospholipid Composition, *Biochem. Mol. Biol. Int. 29*:299–305 (1993).
123. Yoshida, K., J. Terao, T. Suzuki, and K. Takama, Inhibitory Effect of Phosphatidylserine on Iron-Dependent Lipid Peroxidation, *Biochem. Biophys. Res. Commun. 179*:1077–1081 (1991).
124. Kurz, M., and G.K.E. Scriba, Drug-Phospholipid Conjugates as Potential Prodrugs: Synthesis, Characterization, and Degradation by Pancreatic Phospholipase A_2, *Chem. Phys. Lipids 107*:143–157 (2000).

19 Waxes and Sterols: Structures and Chemistry

Edward J. Parish[a] and Angela D. Bell[b]

Department of Chemistry, Auburn University, Auburn, AL 36849

Structure of Waxes

Waxes are esters formed from long-chain carboxylic acids and alcohols. *Beeswax*, for example, is composed of a 16-carbon carboxylic acid and a 30-carbon alcohol. It is the structural component of beehives. The word *wax* is derived from the Old English word *weax*, which means "material of the honeycomb." *Carnauba wax*, widely used in car waxes and floor polishes, has a 32-carbon carboxylic acid and a 34-carbon alcohol. It is a particularly hard wax because of its relatively high molecular weight. Waxes are esters.

In nature, plants and animals produce waxes to serve various purposes (Fig. 1). The surfaces of certain leaves and fruits are covered with wax to minimize excessive water loss by evaporation. It also acts as a protective layer against parasites. Some vertebrates secrete wax in order to keep their fur lubricated as well as water-repellent. Bird feathers are coated with wax to repel water. A waterproof, waxy layer is secreted by insects on the outside of their exoskeletons. *Spermaceti* is found in the head of the sperm whale which helps to regulate the animal's buoyancy for deep diving. It may also serve to amplify high-frequency sounds for locating prey. In contrast to these waxes, the paraffin wax, used to seal preserves and produce candles, is not a true wax. Instead, it is a mixture of high molecular weight alkanes.

For many years, natural waxes were used in manufacturing cosmetics, adhesives, varnishes, and waterproofing materials. For most of these uses, synthetic materials have now replaced natural waxes. Synthetic waxes encompass a very broad

$CH_3(CH_2)_{24}C(=O)O(CH_2)_{29}CH_3$	$CH_3(CH_2)_{30}C(=O)O(CH_2)_{33}CH_3$	$CH_3(CH_2)_{14}C(=O)O(CH_2)_{15}CH_3$
a major component of beeswax structural material of beehives	a major component of carnauba wax coating on the leaves of a Brazilian palm	a major component of spermaceti wax from the heads of sperm whales

Fig. 1. Examples of naturally occurring waxes.

spectrum of chemical type. These waxes range through polyethylene, polymers of ethylene oxide, derivatives of montan wax, alkyl esters of monocarboxylic acids, alkyl esters of hydroxyacids, polyhydric alcohol esters of hydroxyacids, Fischer-Tropsch waxes, hydrogenated waxes, and long-chain amides. Though 25–30 million pounds of Fischer-Tropsch waxes are consumed each year in the United States of America, separate production figures for most of these are not available (1).

Chemistry and Analysis of Waxes

The physical properties of waxes are more dependent on the molecular structure than the molecular size and chemical constitution. When compared to the more loosely packed waxes containing branched and cyclic molecules, paraffin wax essentially has a close-packed structure with *n*-alkanes. Unlike gums and resins, waxes have a more crystalline structure and may have well developed macrocrystals as in Chinese insect wax and spermaceti.

Ranging from hydrocarbons, esters, ketones, aldehydes, alcohols to acids, the chemical components of waxes are largely aliphatic long-chain molecules. In the United States, petroleum waxes are expended in the largest quantity of total wax consumption. Hydrocarbons present in petroleum waxes are mainly alkanes and some are unsaturated, branched chain compounds.

The predominant esters are those of saturated acids with twelve to twenty-eight carbon atoms combining with saturated alcohols of similar chainlength. Several marine waxes are unusual in having a significant degree of unsaturation. Jojoba wax is a predominantly unsaturated ester and is a liquid. Also, straight-chain alcohols and some wax esters, like wool wax, have a high proportion of cholesterol and lanosterol.

An even straight-chain of carbon atoms is typical of primary alcohols, acids, and esters. Ketones, secondary alcohols, and hydrocarbons mainly have an odd number of carbon atoms. Cycloparaffins are found in lignites and petroleum waxes. While most components in waxes are straight-chain, carnauba wax is an exception. It contains ~30% of a fraction comprising esters of hydroxy and methoxycinnamic acids (1).

Materials and Methods

Several common examples of chemical procedures used in the analysis and characterization of waxes are described below (1–3).

Acid value. The number of milligrams (mg) of potassium hydroxide required to neutralize one gram (1g) of the wax or waxy material is the acid value (Eq. 1). In a conical flask, four to five grams (4–5g) of the wax are melted and a mixture of equal parts of ethanol and toluene is added at its boiling point. The mixture is refluxed until all the wax has dissolved. A few drops of phenolphthalein are added after the mixture has been removed from the heat. With 0.5 M potassium hydroxide, the acid material is titrated.

$$\text{acid value} = \frac{(V_w)(56.104)}{w} \qquad [1]$$

where V_w = number of mL of potassium hydroxide used in the titration and w = mass of wax.

Saponification Number. The number of milligrams of potassium hydroxide required to hydrolyze 1 g of wax is the saponification number (Eq. 2). In a conical flask, 2 g of the wax and 10 mL of toluene are added. To avoid boiling the solvent, the flask is heated only to melt the wax. 25 mL of 0.5 M alcoholic potassium hydroxide are added and the solution is refluxed for 2 h. A few drops of phenolphthalein are added and the residual potassium hydroxide is titrated with 0.5 M hydrochloric acid. A blank titration is also performed on 25 mL of 0.5 M alcoholic potassium hydroxide plus toluene.

$$\text{saponification number} = \frac{(V_b - V_w)(56.104)}{w} \qquad [2]$$

where w = weight of sample taken, V_b = number of mL of hydrochloric acid used in the blank, and V_w = number of mL of hydrochloric acid used in the actual analyses.

Ester Value. The ester value is the difference between the saponification number and the acid value. The amount of alkali consumed in the saponification of the esters is indicated here. The ester values range from 70 for beeswax to 125 for spermaceti.

Iodine Number. The iodine number provides two implications. It expresses the amount of iodine that is absorbed by the wax and is a measure of degree of unsaturation. The amount of wax which will absorb 0.3–0.4 g of iodine is dissolved in 10 mL of carbon tetrachloride and 25 mL of Wijs solution added. The solution is allowed to stand for 60 minutes in the dark. Through the addition of potassium iodide solution and water, the excess iodine monochloride is reduced to free iodine. The liberated iodine is titrated with sodium thiosulfate. Iodine values range from 2 for Chinese insect wax to 30 for wool wax.

Acetyl Number. The acetyl number specifies the milligrams of potassium hydroxide essential for the saponification of the acetyl group assimilated by one gram of wax on acetylation. This technique is used for the estimation of hydroxylated esters, free alcohols, and free hydroxyl acids.

For 2 h, 10 g of the wax are heated with 40 g of acetic anhydride. The resulting mixture is poured into a beaker containing 50 mL of hot water. The solution is boiled for 30 min. Once the mixture separates into 2 layers, the oily layer is boiled with three successive portions of fresh water. All free acetic acid is removed. The acetylated wax is carefully separated from water and dried thoroughly. Standard alcoholic potassium hydroxide is used to saponify 2 g of acetylated wax. This solution is evaporated almost to dryness and the soap is dissolved in water. A portion of standard sulfuric acid, equivalent to the alkali used for saponification, is added and the solution is warmed gently until the fatty acids separate as a layer. The solution is filtered and washed with boiling water until the filtrate is no longer acid.

$$\text{acetyl number} = \frac{(\text{mL M/10 potassium hydroxide})(5.61)}{\text{mass of acetylated product taken}} \qquad [3]$$

To understand wax biosynthesis, manufacture, and application, a basic knowledge of the chemical analysis of natural waxes is essential. Natural waxes are much more complicated in chemical composition whereas synthetic waxes are constant and depend on the manufacturing process. Generally, natural waxes are isolated by chemical extraction, separated by chromatographic methods, and analyzed by means of mass spectrometry (MS). Both gas chromatography (GC) and high-pressure or -performance liquid chromatography (HPLC) techniques are employed. There are numerous textbooks detailing their general principles (4–7).

After isolation, the individual classes of waxes must be identified. Combined analytical approaches, for example, GC-MS, have been used to analyze individual wax classes due to the composition complexity of these materials. For the analysis of this class of compounds, mass spectrometry is a major analytical method. When utilizing the electron impact-MS (EI-MS), the wax molecules tend to give cleavage fragments rather than parent ions. Thus, soft (chemical) ionization (CI) and fast atom bombardment (FAB) have been frequently used to give additional information for wax analysis.

In electron impact-MS (EI-MS), positively charged ions are produced during the collision of sample molecules with a beam of energetic electrons in the ionization chamber. This process generally extracts an electron from the molecule, leading to a molecular ion with an odd number of electrons, a radical ion, which is denoted by a plus sign and a dot. If this positively charged radical ion has sufficient energy, it will fragment. Thus, the abundance of ions is directly dependent on the energy of the processes involved. An ionizing voltage of approximately 15 eV usually produces an intact ionized molecule, whereas higher electron energy gives more fragmentation.

Structure of Sterols

The elucidation of the basic structure of the steroids was based on the pioneering work of many investigators. In 1815, Chevreul, a French chemist, while studying the composition of fats, observed that a white crystalline compound could be obtained from the unsaponifiable material of certain animal fats. Some 45 years earlier, Poulletier de la Salle isolated the identical compound from gallstones. In recognition of the initial source material, it was named cholesterine (Greek: *chole* = bile and *stereos* = solid). This name was used in the continental literature although the more descriptive cholesterol is used in the English literature. Investigations into the occurrence of cholesterol and other sterols in the unsaponifiable portion of fats have led to their structural characterization.

Cholesterol-like compounds have been found in the lipids of a variety of plant and animal sources. *Sterols* is the collective name embraced for all crystalline unsaponifiable alcohols with properties resembling those of cholesterol. As a general rule, these compounds are 3-monohydroxysteroids having 27, 28, or 29 carbon atoms and nearly all have one or more double bonds. Most commonly, a double bond is found at position 5 and additional double bonds at C_{22} and C_{27} are also prevalent. 3β-Hydroxyl groups are characteristic of all naturally occurring sterols (Fig. 2).

Cholesterol is one of the most widely dispersed compounds in the animal kingdom. Vertebrates innately possess cholesterol. It is often present in invertebrates and has been discovered in some algae. Though present in all tissues of higher animals, the liver, brain, skin, and adrenals are particular sites of abundance.

Phytosterols are sterols of vegetative origin which occur in small amounts in all parts of plants. Seeds and pollen offer these compounds in a relatively great abundance. Sitosterol (*sito* = grain) is the most widely distributed sterol of plants. Calabar beans and soybeans have both sitosterol and stigmasterol. Stigmasterol has an ethyl group at C_{24} and an external double bond at C_{22}-C_{23} in addition to the double bond at C_5-C_6. Isolated from yeast, ergosterol is a long known and important sterol which is also accompanied by an array of other sterols. Initially, ergosterol was isolated from ergot; however, it also occurs in most fungi, including yeast, in lichens, algae, and some vegetable oils. Ergosterol differs from cholesterol in having an additional methyl group in the side chain at C_{24} and two additional double bonds at C_7-C_8 and C_{22}-C_{23}. Lanosterol was first isolated from wool fat. Lanosterol has double bonds that occur at C_8-C_9 and C_{24}-C_{25} and three additional methyl groups at C_4, C_4, and C_{14}. Thus, lanosterol is 4, 4, 14α-trimethyl-$\Delta^{8,24}$-cholestadien-3β-ol. Lanosterol is an intermediate in the biosynthetic pathway to cholesterol.

As a class of compounds, oxysterols can be defined as sterols bearing a second oxygen function, in addition to that at C_3, and having an iso-octyl or modified iso-octyl side chain (8). These compounds have demonstrated a variety of biological properties, including cytotoxicity, atherogenicity, carcinogenicity, mutagenicity, hypocholesterolemia, and effects on specific enzymes. Widely distributed in nature, they have been found in animal tissues and foodstuffs. They have also been isolated from drugs used in folk medicine for the treatment of cancer. Certain oxysterols have shown significant activity in the inhibition of DNA synthesis in cultured cells. In cholesterol biosynthesis, several oxygenated derivatives of cholesterol and sterol intermediates have been found to be potent inhibitors of sterol biosynthesis within

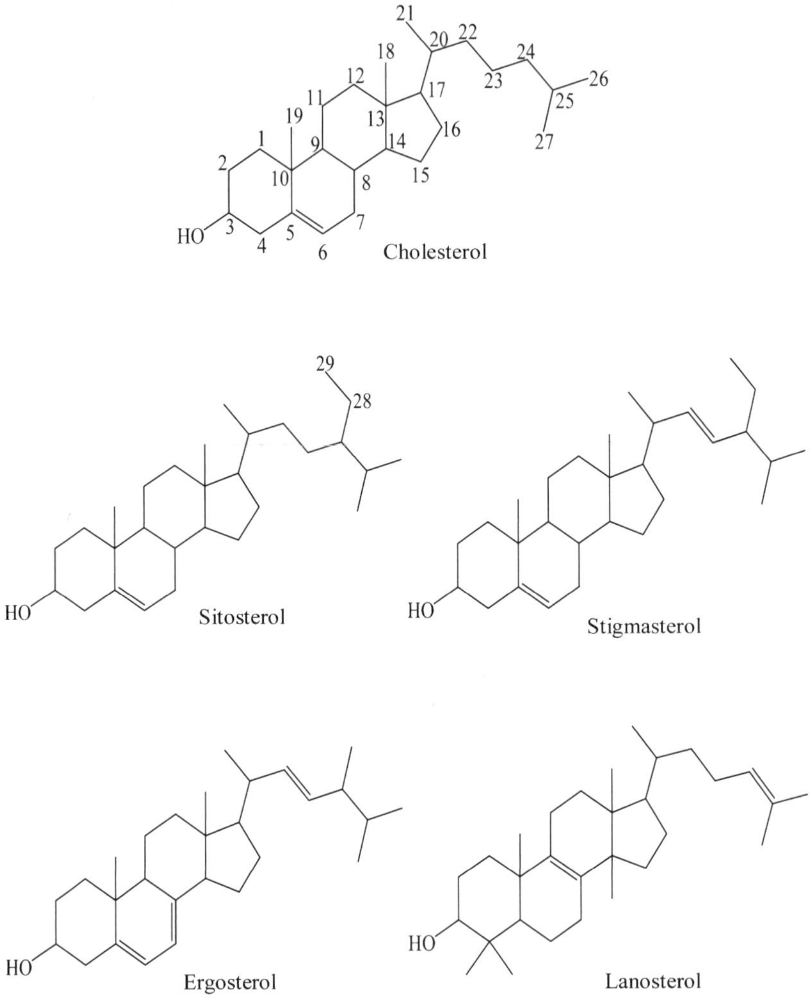

Fig. 2. Examples of naturally occurring sterols.

animal cells in culture. When oxygenated derivatives of cholesterol and lanosterol were used for the specific inhibition of cholesterol biosynthesis in mammalian cells, in many cases, the cellular levels of HMG-CoA reductase, a key regulatory enzyme in sterol biosynthesis, decreased (9). Structures of common oxysterols are shown (Fig. 3). While a large number of oxysterols were evaluated for their abilities to repress HMG-CoA reductase activity, evidence for the existence of a specific cytosolic receptor protein was discovered. After further evaluation of sterol activities, a good correlation was found between the actions of certain oxysterols on HMG-CoA reductase in L cells and their affinity for an oxysterol binding protein (9).

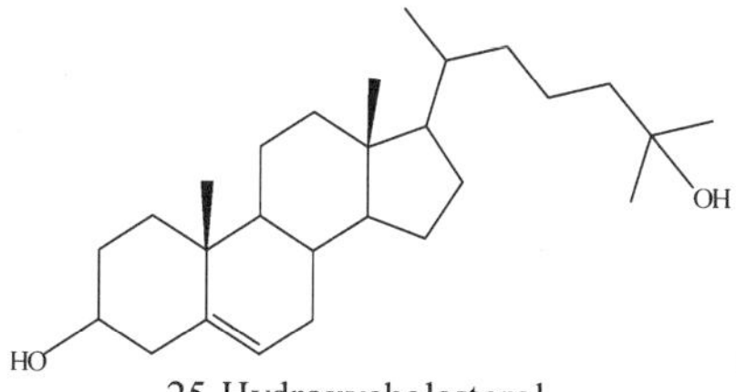

25-Hydroxycholesterol

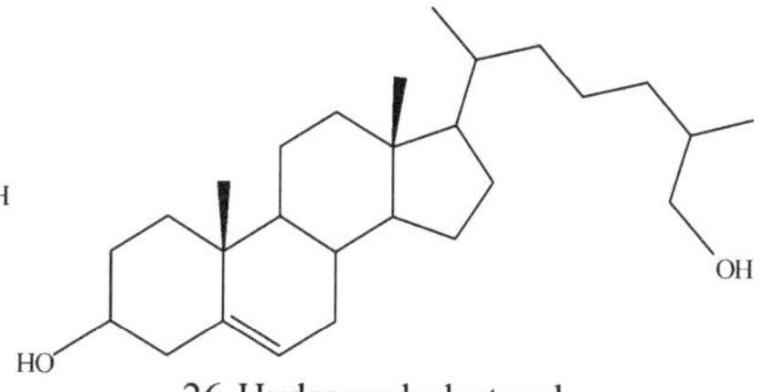

26-Hydroxycholesterol

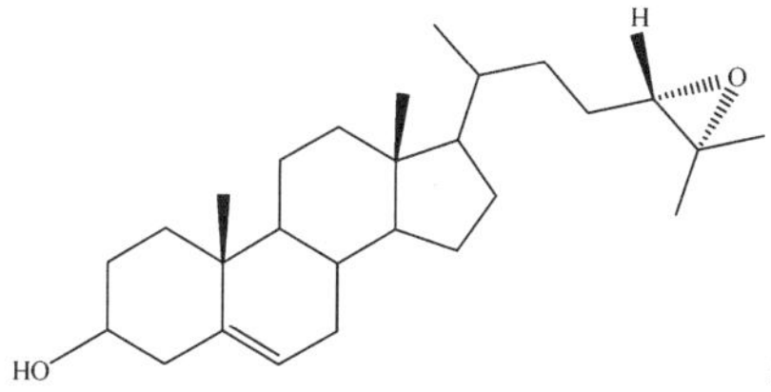

24(S), 25-Epoxycholesterol

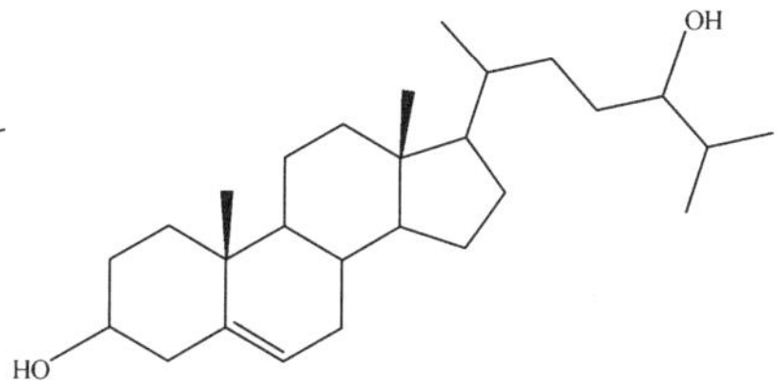

24-Hydroxycholesterol

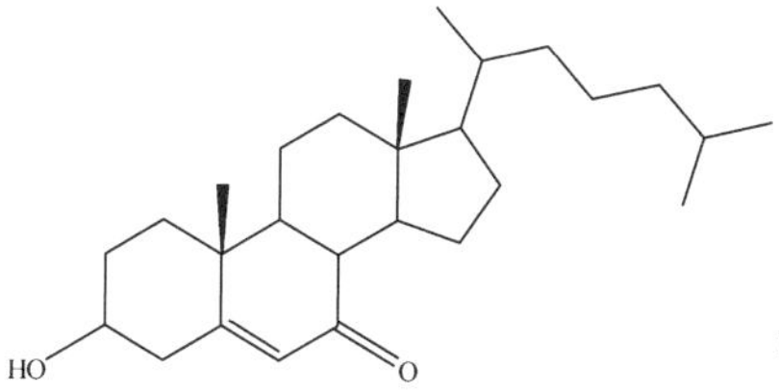

7-Ketocholesterol

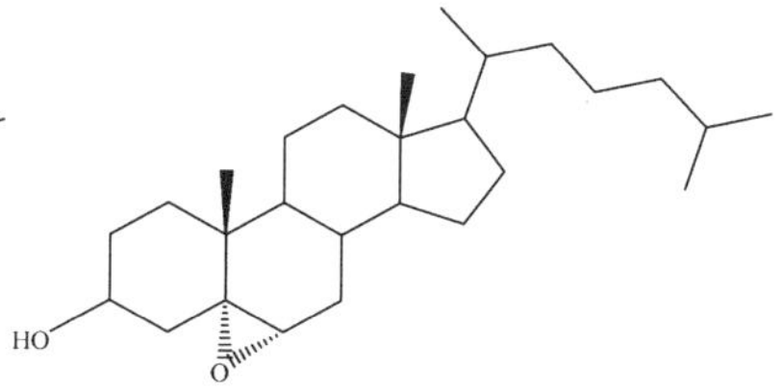

5α, 6α-Epoxycholesterol

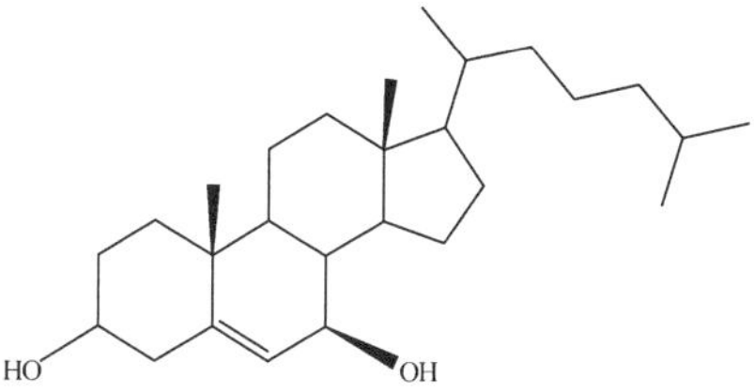

7β-Hydroxycholesterol

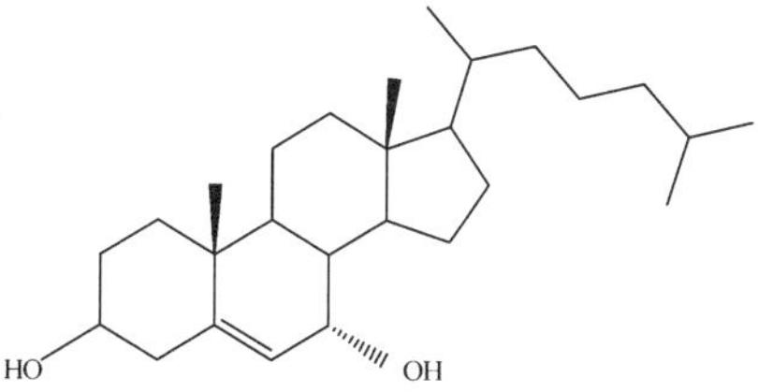

7α-Hydroxycholesterol

Fig. 3. Structures of common oxysterols.

Mammalian systems produce oxysterols. Derivatives of cholesterol, hydroxylated in the 7α-, 25-, or 26-positions, are produced in the liver during bile acid biosynthesis and in side-chain hydroxylation in the 20α- and 22R-positions in the initial step in the conversion of cholesterol to steroid hormones in endocrine organs. Also, all cells produce 32-hydroxylanosterol and 32-oxolanosterol during the conversion of lanosterol to cholesterol. Another mode of oxysterol biosynthesis has been described that uses the isopentenoid pathway to produce side-chain oxygenated derivatives of cholesterol and lanosterol. Such compounds are derived from squalene 2,3-epoxide by the introduction of a second oxygen function to form squalene 2,3:22, 23-dioxide prior to cyclization. Thus, this intermediate has been shown to form 24(S), 25-epoxylanosterol, 24(S),25-epoxycholesterol, and 25-hydroxycholesterol in mammalian systems. 24(S),25-Epoxycholesterol has been isolated from cultured mouse L cells, Chinese hamster lung fibroblasts, and human liver. These results add support to the hypothesis that oxysterols may be natural regulators of cholesterol biosynthesis in mammalian cells. Many oxygenated sterols are known to exist in and have been isolated from plants; these may be precursors to sterols required for growth and/or reproduction or may be secondary plant metabolites. Another major source of oxysterols is the autoxidation of cholesterol, lanosterol, and other related sterols (9,10).

Chemistry and Analysis of Sterols

The chemistry of sterols is a major topic which we cannot properly explore in one chapter. The reader is directed to major works in this area for a complete description and review of sterol chemistry (11–15). However, there are some basic topics which we will discuss.

Sterols containing a double bond at C_5-C_6 (e.g. cholesterol) often contain impurities which can be removed by converting to the C_5-C_6 dibromide. The dibromide is more crystalline and can be purified by recrystallization. The C_5-C_6 double bond can then be restored by treatment with zinc in acetic acid (Fig. 4).

One of the most basic chemical reactions that 3β-hydroxy-Δ^5-sterols may undergo is oxidation (Fig. 5). For example, chemical oxidation may give rise to 3-ketones or 3,6-diketones, depending on reaction conditions (11,12,16).

Materials and Methods

Early analysis of sterols relied on color tests. One of the most common is the Lieberman-Burchard color test which has been extensively used to detect the presence of sterols from biological samples (11,12). Another common method for the detection of 3β-hydroxy sterols is the digitonin precipitation. Cholesterol forms a very stable and almost insoluble complex with digitonin, a glycosidic saponin which is available commercially (11,12).

Thin layer chromatography (TLC) is primarily used as an analytical method to detect compounds. Silica gel G glass plates, commercially available, are commonly employed (16,17). The eluted products are easily visualized by sulfuric acid-alcohol and molybdic acid mixtures and heating. Also, small samples (<20 mg) can be separated by means of preparative TLC. For an analytical TLC plate, the thickness of the stationary phase, usually silica gel, is less than 0.25 mm; for preparative TLC, a much thicker stationary phase (>0.5 mm) is required. No more than 1 mg of substrate should be loaded on 1-cm analytical plates and 2 mg on preparative plates.

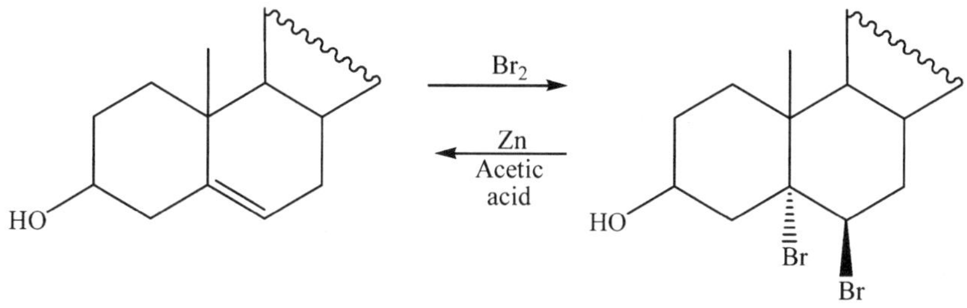

Fig. 4. The bromination of 3β-hydroxy-Δ^5-sterols.

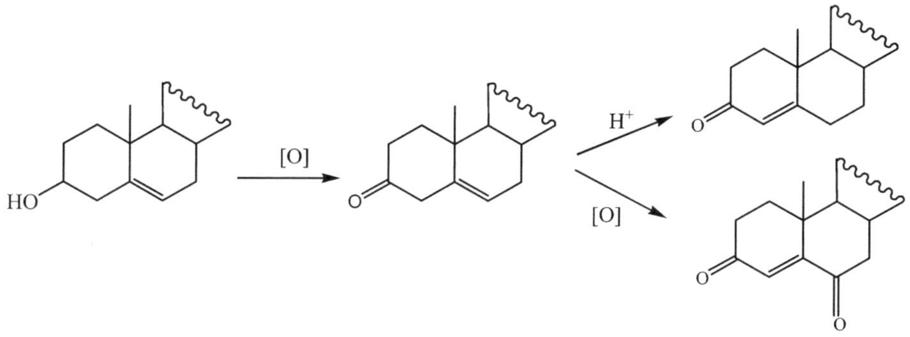

Fig. 5. Products obtained from the chemical oxidation of 3β-hydroxy-Δ^5-steroids.

Otherwise, overloading will prevent the achievement of a good separation. Nondestructive detection methods such as iodine and UV detection are usually used to locate individual compounds. The silica gel containing pure individual compounds is scraped from the plates and eluted with polar solvents.

Recently, major references have appeared that describe, in detail, the spectroscopic identification and quantitation of sterols (16,18). Most frequently, gas-liquid chromatography (GLC) is used to quantitate sterols in mixtures (18). Flame ionization detection (FID), electron capture detection (ECD), and mass detection (MD) are three different methods of detection that have been employed for quantitation (19–22). However, FID is by far the most commonly employed method. This is due to its relative insensitivity to temperature changes during analysis and to monitor structural differences in sterols. The use of capillary GC, particularly in combination with direct on-column injection, has increased the capability of GC as a useful and powerful tool in the analysis of sterols.

HPLC, adsorption or reversed phase, is becoming the most commonly used technique for the separation of individual sterols from mixtures. Quantification of sterols by HPLC is somewhat limited. GLC provided less selectivity than adsorption or reversed phase HPLC. Ultraviolet (UV) detectors are employed most frequently

(23,24). These detectors have limited sensitivity and cannot be considered universal sterol detectors. The UV absorption properties of most sterols differ significantly. Consequently, for complex sterol mixtures, it would not be possible to select a single wavelength for quantitation. However, HPLC coupled to variable-wavelength detectors or multidiode detectors can be used to quantitate specific sterols in a mixture.

Proton nuclear magnetic resonance (^1H NMR) spectroscopy is a key technique for the structural elucidation of sterols (18). A revolutionary change in NMR spectroscopy was initiated by the development of reliable high-field superconducting magnets together with the introduction of Fourier transform (FT) techniques in the early 1970s. The high-field superconducting magnets and FT instruments can now plot high-resolution ^1H NMR spectra which provide very much more information about the structure and stereochemistry of compounds. The ^1H NMR spectra of sterols are usually recorded in solution in deuteriochloroform ($CDCl_3$) with tetra-methylsilane (TMS), added as an internal standard. ^{13}C nuclear magnetic resonance (^{13}C NMR) spectra are considerably more useful than ^1H NMR spectra for structural analyses of complex molecules due to the great sensitivity of ^{13}C chemical shifts to structural changes (16,18). Each carbon atom in the molecule can usually be examined individually. The ^{13}C chemical shifts of many steroids have been compiled and reviewed.

MS, particularly the combination of MS with GC (GC/MS), has become an indispensable method for studying sterols and their biosynthesis. The most commonly used method of ionization for MS of sterols is EI. EI-MS has been discussed previously. Most analyses have been obtained with an ionizing voltage of 70 eV, which produces a high level of ions from fragmentation of sterols.

Infrared (IR) and UV spectroscopy have played vital roles in the elucidation of organic compound structures. In the past, these methods have provided much valuable data for the identification of sterols. The major information produced in the IR or UV spectrum, in reference to sterols, relates to the presence of hydroxyl and carbonyl groups and the location of olefinic bonds.

IR and UV spectroscopy now play a less important role in sterol structure elucidation with the advent of ^1H and ^{13}C NMR spectroscopy which provide all the information needed to identify and locate the hydroxyl and olefinic functionalities of a sterol. Thus, the IR and UV spectra of sterols now serve either as prior indicators of structural features to be solved by NMR or as confirmatory evidence of structures elucidated by the other more refined spectral methods available. However, during separation on HPLC, using a UV detector, UV spectroscopy is still important for the detection of sterols. Also, UV spectroscopy can provide a sensitive and accurate means for the quantification of certain sterols with a conjugated diene structure. The double bond(s) of a sterol is a UV absorbing chromophore; the wavelength(s) of maximum absorption and the molar absorption value depend upon the location of the bond(s) in the molecule and the nature of double-bond conjugation in the case of dienes or trienes. An isolated double bond produces "end absorption" in the range 190–220 nm. Ethanol is usually chosen as the solvent for UV spectroscopy of sterols.

The single-crystal X-ray diffraction of sterols provides a precise technique for the elucidation of such details and it has been used very successfully for the unambiguous assignment of structure to a number of sterols. The analysis can be achieved with a suitable crystal with dimensions in the range from about 0.1 to 0.5 mm. Therefore, it is a technique very well suited to investigations when the substance is available in only small amounts, provided that suitable crystals can be obtained for the compound or an appropriate halogenated derivative. In addition to providing proof of sterol structure, X-ray analysis also gives information about the conformation of the sterol molecule. This is proving valuable in considerations of sterol and steroid function, especially in relation to interactions with protein receptors and cell membrane biochemistry.

The X-ray structure of steroids plays a primary role in governing their interactions and activities. X-ray crystallographic studies provide the most reliable and precise data concerning molecular structure. By combining solid state data with physical chemical data on structures in solution and molecular energy calculations, a reasonable picture of dynamic properties of steroids can be constructed. This information, examined with biochemical, pharmacological, and physiological data, can give a better understanding of the molecular mechanisms of biosynthesis, metabolism, membrane transport, receptor binding, and nuclear interaction.

To obtain a detailed discussion of the identification of sterols with NMR (25, 26), mass (27), UV (28), infrared, and X-ray (29) techniques, the reader is referred to recent comprehensive reviews in this area.

References

1. R. J. Hamilton. Commercial Waxes: Their Composition and Application. In: *Waxes: Chemistry, Molecular Biology, and Functions* (R. J. Hamilton, ed.). The Oily Press, Ayr, Scotland, 1995, p. 257, 315.
2. A. H. Warth. *Chemistry and Technology of Waxes*. New York, Reinhold, 1956.
3. R. Sayers. *Wax—An Introduction*. European Wax Federation and Gentry Books, London, 1983.
4. *Kirk-Othmer Encyclopedia of Chemical Technology*, Vol. 24. Wiley, New York, 1984, p. 446–481.
5. P.E. Kolattukudy. *Chemistry and Biochemistry of Natural Waxes*. Elsevier-North Holland, Amsterdam, 1976.
6. J.C. Touchstone and M. F. Dobbins. *Practice of Thin Layer Chromatography*, 2nd ed. Wiley, New York, 1983.
7. W.W. Christie. *Gas Chromatography and Lipids, A Practical Guide*. The Oily Press, Ayr, Scotland, 1989.
8. E.J. Parish, V. B. B. Nanduri, H. H. Fohl, and F. R. Taylor. Oxysterols: Chemical Synthesis, Biosynthesis, and Biological Activities. *Lipids. 21:* 27–80 (1986).
9. E.J. Parish, S. C. Parish, and S. Li. Regulation of HMG-CoA Reductase by Side-chain Oxysterols and Their Derivatives. In: *Biochemistry and Function of Sterols* (E. J. Parish and W. D. Nes, eds.). CRC Press, Inc., Boca Raton, USA, 1997, p. 193.
10. F. R. Taylor. Oxysterol Regulation of Cholesterol Biosynthesis. In: *Regulation of Isopentenoid Metabolism* (W. D. Nes, E. J. Parish, and J. M. Trzaskos, eds.). Am. Chem. Soc., Washington, D. C., USA, 1992, p. 81.

11. L. F. Fieser and M. Fieser. *Natural Products Related to Phenanthrene*. Reinhold Publishing Corp., New York, 1949.
12. L. F. Fieser and M. Fieser. *Steroids*. Reinhold Publishing Corp., New York, 1959.
13. J. Fried and J. A. Edwards. *Organic Reaction in Steroid Chemistry*. Van Nostrand Reinhold, New York, 1972.
14. R. T. Blickenstaff, A. C. Ghosh, and G. C. Wolf. *Total Synthesis of Steroids*. Academic Press, New York, 1974.
15. J. B. Dence. *Steroids and Peptides*. Wiley, New York, 1976.
16. W. D. Nes and E. J. Parish, eds. *Analysis of Sterols and Other Biologically Significant Steroids*. Academic Press, San Diego, CA, 1989.
17. E. J. Parish, S. Chitrakorn, and S. Lowery. Selective Oxidation of Steroidal Allylic Alcohols Using Pyrazole and Pyridinium Chlorochromate. *Lipids 19:* 550–552 (1984).
18. L. J. Goad and T. Akihisa. *Analysis of Sterols*. Blackie Academic and Professional, New York, 1997.
19. F. Bernini, G. Sangiovanni, S. Pezzotta, G. Galli, R. Fumagalli, and P. Paoletti. Selected Ion Monitoring Technique for the Evaluation of Sterols in Cerebrospinal Fluid. *J. Neuro-Oncol. 4:* 31–40 (1986).
20. B. A. Knight. Quantitative Analysis of Plant Sterols. *Lipids 17:* 204–211 (1982).
21. G. W. Patterson. Chemical and Physical Methods in the Analysis of Plant Sterols. In: *Isopentenoids in Plants* (W. D. Nes, G. Fuller, and L. S. Tsai, eds.). Dekker, New York, 1984, p. 293–312.
22. W. D. Nes. A Comparison of Methods for the Identification of Sterols. In: *Methods in Enzymology*, Vol. III (J. H. Laio and H. C. Rilling, eds.). Academic Press, New York, 1985, p. 3–29.
23. M. P. Kautsky. *Steroid Analysis by HPLC*. Dekker, New York, 1981.
24. R. C. Heupel. Varietal Similarities and Differences in the Polycyclic Isopentenoid Composition of Sorghum. *Phytochemistry 24:* 2929–2936 (1985).
25. W. K. Wilson and G. J. Schroepfer Jr. ^{1}H and ^{13}C NMR Spectroscopy of sterols. In: *Molecular Structure and Biological Activity of Steroids* (M. Bohl and W. L. Duax, eds.). CRC Press, Boca Raton, FL, 1992, p. 33–42.
26. T. Akihisa. ^{13}C-NMR identification of sterols. In: *Analysis of Sterols and Other Biologically Significant Steroids* (W. D. Nes and E. J. Parish, eds.). Academic Press, San Diego, CA, 1989, p. 251–262.
27. A. Rahier and P. Beneveniste. Mass Spectral Identification of Phytosterols, in *Analysis of Sterols and Other Biologically Significant Steroids* (W. D. Nes and E. J. Parish, eds.). Academic Press, San Diego, CA, 1989, p. 223–236.
28. P. Acuna-Johnson and A. C. Oehlschlager. Identification of Sterols and Biologically Significant Steroids by UV and IR Spectroscopy. In: *Analysis of Sterols and Other Biologically Significant Steroids* (W. D. Nes and E. J. Parish, eds.). Academic Press, San Diego, CA, 1989, p. 267–278.
29. W. L. Duax, J. F. Griffin, and G. Cheer. Steroid Conformational Analysis Based on X-ray Crystal Structure Determination. In: *Analysis of Sterols and Other Biologically Significant Steroids* (W. D. Nes and E. J. Parish, eds.). Academic Press, San Diego, CA, 1989, p. 207–229.

Index

D

E

G

H

J

K

L

T